ENVIRONMENTAL SOURCES AND EMISSIONS HANDBOOK

ENVIRONMENTAL
SOURCES AND EMISSIONS
HANDBOOK

Marshall Sittig

NOYES DATA CORPORATION

Park Ridge, New Jersey   London, England

1975

363.738
S62e

Published in the United States of America by
Noyes Data Corporation
Noyes Building, Park Ridge, New Jersey 07656

TP

To
Dr. Richard S. Kochman
without whose help
this book could not have been written.

FOREWORD

Environmental pollution with its far-reaching effects is only now beginning to be understood. In this practical handbook the origins of both air and water pollution are surveyed.

Significant emphasis is placed on intermedia transfers which can result when an air pollutant is intentionally or accidentally transferred to an aqueous stream or vice versa. Sometimes pollutants react with each other or initiate deleterious chain reactions. Yet this same reactivity may be the key to efficient removal; by adding certain chemicals or flocculants or microorganisms, toxic substances may be carried off physically or converted not only to nontoxic substances, but also into useful products.

Operators or potential operators of processes which produce pollutants will find this volume quite useful. Besides discussing all sorts of pollution sources, it should help to define industry-wide emission practices and magnitudes. The detailed descriptions and tables are mostly based on studies conducted by industrial and engineering firms or university research teams under the auspices of various government agencies.

As is the case with some other handbooks in this series the entries are arranged in an alphabetical and encyclopedic fashion.

CONTENTS AND SUBJECT INDEX

Contents and Subject Index

INTRODUCTION

This volume will survey the origins of both air and water pollution. In addition, it will cover aspects of intermedia pollution which results, for example, when an air pollutant is removed and transferred to an aqueous stream.

This volume will explain the existence of particular pollutants from whatever source on one hand and the process sources of various pollutants on the other hand. This volume should be of use to the federal, state or municipal official who is faced with the presence of a particular pollutant or with emissions from a specific industry. Whether such an official has a metal plating plant or a racetrack in his or her neighborhood, this volume should enable that official to know what is issuing from that facility and how much.

This volume should be of use to the manufacturer of pollution control equipment or of chemicals for pollution control. It should help to set forth the markets, their characteristics and their dimensions. This volume should be useful to the operators or potential operators of processes which produce pollutants. It should help to define industry-wide emission practices and magnitudes.

This volume has attempted to concentrate on quantities and compositions of raw waste effluents. In some cases data have been included on effluents following in-plant control (the only data available perhaps). While parenthetical notes may be included on control techniques or their effectiveness, space limitations do not permit a thorough discussion of control techniques or their effects on emissions in this volume. The reader is referred to the companion *Pollutant Removal Handbook* for such information.

Taken together with the earlier *Pollutant Detection and Monitoring Handbook* and the *Pollutant Removal Handbook,* these three volumes permit one to determine

—the nature of emissions
—an estimate of their quantities
—the quantities of pollutants in effluent streams as well as
    ambient bodies of air and water
—and, finally, to choose and apply effective pollutant removal
    processes.

Table 1 summarizes some of the major air and water pollutants and the media in which they occur.

TABLE 1: MAJOR POLLUTANTS, SOURCES AND PRIMARY MEDIA

SOURCE	ORGANICS	SUSP. SOLIDS	SULFATES, PHOS., NITRATES	HEAVY METALS	ACIDITY-ALKALINITY	PATHOGENS	THERMAL	RADIOACTIVITY	PESTICIDES	NITROGEN OXIDES	SULFUR OXIDES	CARBON MONOXIDE	HYDROCARBONS	PARTICULATES	RADIOACTIVITY
	WATER									AIR					
BASIC METALS	●	●	●	●						●	●			●	
FOOD PROC.	●	●	●											●	
CHEM. & ALLIED	●	●	●	●	●				●		●		●	●	
PULP & PAPER	●	●	●		●						●	●		●	
PETROLEUM					●					●	●	●	●	●	
COAL, OIL COMB.										●	●	●	●	●	
NUCLEAR POWER							●	●							●
DOM. SEWAGE	●	●	●		●	●									
MOBILE SOURCES										●		●	●	●	
AGRICULTURE									●						

Source: Report EPA 600/5-73-003

For further details on the pollution problems presented by specific industries, the reader is referred to the following volumes:

> *Mercury Pollution Control* by H.R. Jones, Park Ridge, N.J., Noyes Data Corp. (1971)
>
> *Sulfuric Acid Manufacture and Effluent Control* by Marshall Sittig, Park Ridge, N.J., Noyes Data Corp. (1971)
>
> *Environmental Control—Organic and Petrochemical Industries* by H.R. Jones, Park Ridge, N.J., Noyes Data Corp. (1971)
>
> *Environmental Control in the Inorganic Chemical Industry* by H.R. Jones, Park Ridge, N.J., Noyes Data Corp. (1972)
>
> *Detergents and Pollution—Problems and Technological Solutions* by H.R. Jones, Park Ridge, N.J., Noyes Data Corp. (1972)
>
> *Fine Dust and Particulate Removal* by H.R. Jones, Park Ridge, N.J., Noyes Data Corp. (1972)
>
> *Pollution Control in the Nonferrous Metals Industry* by H.R. Jones, Park Ridge, N.J., Noyes Data Corp. (1972)
>
> *Waste Disposal Control in the Fruit and Vegetable Industry* by H.R. Jones, Park Ridge, N.J., Noyes Data Corp. (1973)
>
> *Pollution Control in the Textile Industry* by H.R. Jones, Park Ridge, N.J., Noyes Data Corp. (1973)
>
> *Pollution Control and Chemical Recovery in the Pulp and Paper Industry* by H.R. Jones, Park Ridge, N.J., Noyes Data Corp. (1973)
>
> *Pollution Control in the Petroleum Industry* by H.R. Jones, Park Ridge, N.J., Noyes Data Corp. (1973)
>
> *Pollution Control in Metal Finishing* by M.R. Watson, Park Ridge, N.J., Noyes Data Corp. (1973)

Pollution Control in Meat, Poultry and Seafood Processing by H.R.
Jones, Park Ridge, N.J., Noyes Data Corp. (1974)
Pollution Control in the Dairy Industry by H.R. Jones, Park Ridge,
N.J., Noyes Data Corp. (1974)
Pollution Control in the Organic Chemical Industry by Marshall
Sittig, Park Ridge, N.J., Noyes Data Corp. (1974)
Oil Spill Prevention and Removal Handbook by Marshall Sittig, Park
Ridge, N.J., Noyes Data Corp. (1974)

The reader is also referred to the brief reference or references at the end of each section
of this volume. In order to hold this volume to a manageable size, those references are
in many cases review references which in turn will lead the reader to another more detailed bibliography.

AIR AND WATER POLLUTION

Tables 2 and 3 summarize the national sources of air and water pollution by Standard
Industrial Classification (SIC) sectors. The significant SIC sectors for the mathematical
model are shown. Where possible, the data were standardized for the base year, 1971.
Sectors are presented for which quantitative information was found. The tables were prepared by multiplying pollution coefficients per unit of industrial output by the physical
output information.

TABLE 2: NATIONAL SOURCES OF AIR POLLUTION

		Pollutants (Millions of Pounds/Year)					
SIC	Name	Sulfur Oxides	Nitrogen Oxides	Particulates	Carbon Monoxide	Hydro-carbons	Radio-activity
01	Agriculture Crops			- - - [a]		- - -	
02	Agriculture – Livestock						
08	Forestry			549	2,100	1,777	
10	Metal Mining			25,500			
11	Anthracite Mining	- - -					
12	Bituminous Coal and Lignite Mining	- - -					
13	Oil and Gas Extraction						
14	Non-Metal Mining and Quarrying			15,504			
201, 202	Meat and Dairy Products			- - -		- - -	
203	Processed Fruits and Vegetables						
204	Grain Mill Products			3,334			
Misc. 20	Misc. Food Products			2			
22	Textile Mill Products			202			
24	Lumber and Wood Products			34(P)[b]		13(P)	
26	Paper and Allied Products	220		6,650	3,078		
281	Industrial Inorganic Chemicals	1,171		45	2,744	1,235	
282	Plastic Materials and Synthetics					32	
283	Drugs						
2873	Nitrogenous Fertilizers						
2874	Phosphate Fertilizers						
2879	Agriculture Chemicals nec[c]						
2895	Carbon Black			1,875	5,540	985	

(continued)

TABLE 2: (continued)

SIC	Name	Sulfur Oxides	Nitrogen Oxides	Particulates	Carbon Monoxide	Hydro-carbons	Radio-activity
				Pollutants (Millions of Pounds/Year)			
Misc. 28	Misc. Chemicals						
291	Petroleum Refining	4,800	602	970	30,773	3,025	
295	Paving and Roofing Material			1,068	1	1	
31	Leather and Leather Products						
324	Cement Hydraulic			14,624			
325	Structural Clay Products			431			
327	Concrete Gypsum and Plaster		33	7,897			
Misc. 32	Misc. Fiber Glass			- - -			
331	Blast Furnace and Basic Steel Production	1,000		21,669	1,177	2,382	
332	Iron and Steel Foundries		8	1,511	2,007		
333	Primary Non-Ferrous Metals	7,432		366			
334	Secondary Non-Ferrous Metals			377			
336	Non-Ferrous Foundries – Castings			66.			
34	Fabricated Metal Products					10	
36	Electric and Electronic Equip.					32	
37	Transportation Equip.						
40	Railroad Transportation	215	294	98	275	196	
42	Warehousing and Trucking	640	9,344	589	7,936	1,510	
44	Water Transportation						
45	Air Transportation	23	66	46	3,812	628	
491	Electric – Power Generation Services	40,000		50,500	198	127	- - -
492	Gas Production and Distribution		- - -			- - -	
4952	Sanitary Systems Sewers						
4953	Refuse Disposal Systems	200	800	2,800	14,400	4,000	
54	Food Stores						
5541	Gasoline Sales[d]	5,540	14,984	1,015	173,576	33,422	
72	Dry Cleaning					- - -	
	Totals	61,241	26,131	157,722	247,617	49,375	
	1970 EPA Data	68,000	46,080	50,000	294,000	70,000	

[a] Data gap. This sector contributes to this pollutant, but quantitative information is incomplete.

[b] Partial data. This sector contributes more of this pollutant than shown here but quantified information is incomplete.

[c] Not elsewhere classified.

[d] Includes all pollution generated by private automobile use.

Source: Report EPA 600/5-73-003

TABLE 3: NATIONAL SOURCES OF WATER POLLUTION

Pollutants (Millions of Pounds/Year)

SIC	Name	Sulfur Compounds (as S)	Nitrogen Compounds (as N)	Heavy Metals	Organics BOD_5	Suspended Solids	Acidity Alkalinity	Phosphorous Compounds (as P)	Radio-activity
01	Agriculture – Crops		8,250					- - -	
02	Agriculture – Livestock		1,000						
08	Forestry								
10	Metal Mining								
11	Anthracite Mining	2,200				7,000[d]			
12	Bituminous Coal and Lignite Mining								
13	Oil and Gas Extraction			- - -					
14	Non-Metal Mining and Quarrying								
201, 202	Meat and Dairy Products				2,298	1,187			
203	Processed Fruits and Vegetables				988	467			
204	Grain Mill Products								
Misc. 20	Misc. Food Products				48	13 (P)[b]			
22	Textile Mill Products				330	459	- - -		
24	Lumber and Wood Products				- - -	- - -			
26	Paper and Allied Products				7,898	3,286	- - -		
281	Industrial Inorganic Chemicals				471	4,450	- - -	11	
282	Plastic Materials and Synthetics				579	331	- - -		
283	Drugs				25	46			
2873	Nitrogenous Fertilizers	1	4		6	26			
2874	Phosphatic Fertilizers				- - -	47		41	
2879	Agriculture – Chemicals nec[c]				58	6			
2895	Carbon Black								
Misc. 28	Misc. Chemicals				99	184			
291	Petroleum Refining	39	70		268	164	- - -	6	
295	Paving and Roofing Material								
31	Leather and Leather Products				187	530	- - -		
324	Cement, Hydraulic								
325	Structural Clay Products								
327	Concrete, Gypsum and Plaster								
Misc. 32	Misc. Fiber Glass					- - -	- - -		
331	Blast Furnace and Basic Steel Production			18	238	1,870	- - -		
332	Iron and Steel Foundries				14	47	- - -		
333	Primary Non-ferrous Metals				5	256			
334	Secondary Non-ferrous Metals					- - -			
335	Non-ferrous Drawing, Rolling and Extruding				7	49			
336	Non-Ferrous Foundries – Cast					- - -			
34	Fabricated Metal Products			- - -			- - -		
36	Electric and Electronic Equip.								

(continued)

TABLE 3: (continued)

		Pollutants (Millions of Pounds/Year)							
SIC	Name	Sulfur Compounds (as S)	Nitrogen Compounds (as N)	Heavy Metals	Organics BOD$_5$	Suspended Solids	Acidity Alkalinity	Phosphorous Compounds (as P)	Radio-activity
37	Transportation Equipment				36	40			
40	Railroad Transportation								
42	Warehousing and Trucking								
44	Water Transportation								
45	Air Transportation								
491	Electric – Power Generation Services						- - -		1,200^e
492	Gas Production and Distribution								
4952	Sanitary Systems – Sewers	960	2,400		8,006	9,651	- - -	540	
4953	Refuse Disposal Systems								
54	Food Stores								
5541	Gasoline Sales								
72	Dry Cleaning								
	Totals	3,200	11,724	18	22,111	23,109	7,000 (p)^d	598	1,200^e
	Totals Without Domestic Sewage	2,240	9,324	18	14,105	13,458	7,000	58	1,200

^{a}Data gap. This sector contributes to this pollutant, but quantitative information is incomplete.

^{b}Partial data. This sector contributes more of this pollutant than shown here but quantified information is incomplete.

^{c}Not elsewhere classified.

^{d}Millions of pounds of $H_2 SO_4$ (sulfuric acid).

^{e}Millions of curies (mega curies).

Source: Report EPA 600/5-73-003

In the assessment of community air pollution, there is a critical need for accurate data on the quantity and characteristics of emissions from the numerous sources that contribute to the problem. The large numbers of these individual sources and the diversity of source types make conducting field measurements of emissions on a source-by-source basis at the point of release impractical. The only feasible method of determining pollutant emissions for a given community is to make generalized estimates of typical emissions from each of the source types.

The emission factor is an average estimate of the rate at which a pollutant is released to the atmosphere as a result of some activity, such as combustion or industrial production, divided by the level of that activity. For example, assume that in the production of 260,000 tons (236,000 metric tons) of ammonia per year, 26,000 tons (23,600 metric tons) of carbon monoxide is emitted to the atmosphere. The emission factor for the production of ammonia would therefore be 200 pounds of CO released per ton (100 kilograms per metric ton) of ammonia produced. The emission factor thus relates the quantity of pollutants emitted to some indicator such as production capacity, quantity of fuel burned, or vehicle miles traveled by autos.

The report *Compilation of Air Pollutant Emission Factors,* Second Ed., Research Triangle Park, N.C., U.S. Environmental Protection Agency (Apr. 1973) reports data available on those atmospheric emissions for which sufficient information exists to establish realistic emission factors. The information contained therein is based on Public Health Service Publication 999-AP-42, *Compilation of Air Pollutant Emission Factors,* by R.L. Duprey, and on a revised and expanded version of *Compilation of Air Pollutant Emission Factors* that was published by the Environmental Protection Agency in February 1972. The scope of this second edition has been broadened to reflect expanding knowledge of emissions.

The emission factors presented in this report were estimated by using the entire spectrum of techniques available for determining such factors. These techniques include: detailed source testing that involve many measurements related to a variety of process variables, single measurements not clearly defined as to their relationship to process operating conditions, process material balances, and engineering appraisals of a given process.

The limitations and applicability of emission factors must be understood. To give some idea of the accuracy of the factors presented for a specific process, each process has been ranked as A, B, C, D, or E. For a process with an A ranking, the emission factor should be considered excellent, i.e., based on field measurements of a large number of sources. A process ranked B should be considered above average, i.e., based on a limited number of field measurements. A ranking of C is considered average; D, below average; and E poor. These rankings are presented below the table titles throughout the present volume where air pollution emission factors are presented.

In general, the emission factors presented are not precise indicators of emissions for a single source. They are more valid when applied to a large number of processes. With this limitation in mind, emission factors are extremely useful when intelligently applied in conducting source inventories as part of community or nationwide air pollution studies.

Another set of symbols used with emission factors in this volume, indicating method of measurement rather than reliability of measurement is shown in Table 4.

TABLE 4: EMISSION FACTOR SYMBOLS

Emission factor symbol	Technique	Estimated accuracy
PV	Plant visits	Unknown.
E	Engineering judgement	Unknown.
Q	Questionnaire surveys	Unknown.
MB	Material balances	Unknown.
Source sampling		
CAA	Flame atomic absorption	Precision is 1, 3, and 10% for minimum detectable levels of 1, 0.5, and 0.1 ppm, respectively, for beryllium and for nickel in oil samples. Minimum detectable level is 0.4 ppm for cadium in oil samples. Minimum detection limit is 0.001 μg/ml of prepared sample for cadmium generally. Less accurate than FAA for mercury.
DM	Dithizone	Comparable to FAA for mercury analysis.
ES	Emission spectroscopy	Semiquantitative.
EST	Emission spectrometry	Semiquantitative.
FAA	Flameless atomic absorption	±5% with minimum resolution of 2 ppb for analysis of mercury.

(continued)

TABLE 4: (continued)

Emission factor symbol	Technique	Estimated accuracy
NA	Neutron activation	± 25% for one sample of particulate analyzed. Arsenic in biological materials had an average deviation of about ± 6.9%. ± 20% precision for mercury in coal. Accuracy of ± 5% for biological tissues analyzed for cadmium.
OES	Optical emission spectrography	Considered to be semiquantitative. Reported to have a precision of ± 25% for samples from cement plants.
S	Saltzman's colorimetric	Unknown
SC	Spectrochemical analyses	Limit of detection is 0.0001% with an accuracy of ± 15% for beryllium analysis in coal.
SSMS	Spark source mass spectrography	Precision ± 100%.
UK	Unknown, reported in literature based on unspecified analytical technique	Unknown.

Source: Report PB 230,894

INTERMEDIA POLLUTION

The major intermedia pollutants emitted initially to either air or water are: sulfur compounds, nitrogen compounds, heavy metals and radioactivity. Those initially discharged to water are: organics, suspended solids, acids and alkalies, and phosphorus compounds; those discharged initially to air are limited to particulates.

Intramedia or lesser intermedia air pollutants are carbon monoxide, hydrocarbons including ethylene, fluorides, hydrogen chloride, arsenic, hydrogen cyanide and ammonia. Intramedia or lesser intermedia water pollutants are: pathogens, pesticides, thermal pollution, metallic salts and oxides, chlorides, surfactants, and liquid hydrocarbons.

The major sources of air pollution and their respective Standard Industrial Classification Codes are: mobile sources, power generation from fuel (491), chemicals manufacture (28), metallurgical processes (33) and refuse incineration (4953). Large contributors of particulates to the atmosphere include: the sand, clay and glass industry (32) and non-metal mining and quarrying (14).

The major sources of water pollution are: agriculture (01, 02), food processing (20), mining (10, 11, 12), paper and allied products (26), chemicals manufacture (28), blast furnaces and basic steel production (331), and sanitary systems and sewers (4952).

Intermedia transfers include direct transfer (removal of a pollutant from one medium and its disposal in another) or indirect (pollution created in another medium and usually in another form by a basic change in a process or industry).

The principal current sources of direct intermedia transfers from water to air are: incineration of sewage sludge or other industrial waste residues including radioactive wastes; ammonia, other gaseous and volatile emissions from wastewater aeration processes, trickling filters, lagoons, stripping towers, sewers, etc.; nitrous oxide emissions from chlorination for ammonia nitrogen control in water; sewage and industrial waste sludge digestion and drying; and removal of radioactive gases from reactor coolant water and their release following insufficient storage time.

Direct intermedia transfers from water to air which are avoidable only with considerable expense are usually of minor significance. They result from wastewater treatment processes such as aeration, trickling filters, lagoons, stripping towers, anaerobic decomposition and chlorination for removal of ammonia nitrogen. The air pollutants produced are nitrogen oxides, hydrogen sulfide, methane, mercaptans and ammonia.

The principal current sources of direct intermedia transfers from air to water are: the use of scrubbers to control gaseous emissions to the atmosphere; flushing with water to remove and carry residues from dry collection equipment such as cyclones; and steam regeneration of activated carbon used to control gaseous emissions, although this depends upon the subsequent treatment of the condensate and the resulting volatilized or oxidized materials.

The principal current sources of indirect intermedia transfers are: [1] Replacement of fossil fuel power generation by nuclear power generation. This eliminates hydrocarbon, particulate, sulfur dioxide, nitrogen oxides and other forms of air pollution from fossil fuel combustion, but it creates possible radioactive pollution of air and water and thermal pollution of water. [2] Waste products created by the manufacture of pollution control equipment. [3] Recycling of water to reduce water usage. This seemingly intramedial alternative may create indirect intermedia transfers by the reduction of production efficiency from the buildup of salinity or scale in either process or cooling equipment. A reduction in efficiency results in an increased new materials and labor input to maintain production rates and the accompanying increased outputs to the environment of more energy, waste materials, people, greater travel and the additional support services required.

Table 5 summarizes intermedia transfers in air treatments and Table 6 summarizes intermedia transfers in wastewater treatment.

TABLE 5: INTERMEDIA TRANSFERS IN AIR TREATMENTS

Treatment	Oxides of Nitrogen	Oxides of Sulfur	Carbon Monoxide	Hydrocarbons - Particulate	Hydrocarbons - Gaseous	Particulates	Radioactivity	Thermal	Air	Water	Land
Water Scrubbing	●	●		●		●				SS, TDS, H_2SO_3, H_2SO_4 Pathogens, Heat	
Electrostatic Precipitator				●		●	●			Residues[a]	Residues
Cyclones-Dry				●		●				Residues	Residues
Settling Chambers				●		●					Residues
Baghouse Filters				●		●	●				Residues
Condensers-Boiler Coolers								●		Heat	
Afterburners	●	●		●	●	●		●	CO2, H_2O Heat		
Adsorbers				●	●	●				Recycled	
Venturi Scrubbers (Wet Cyclone)	●	●		●		●				H_2SO_3, H_2SO_4 Pathogens	

Columns grouped under "AIR POLLUTANTS" (Oxides of Nitrogen through Thermal) and "INTERMEDIA TRANSFERS" (Air, Water, Land).

[a] "Residues" indicates the same or a combined form of the pollutant physically removed.

Source: Report EPA 600/5-73-003

TABLE 6: INTERMEDIA TRANSFERS IN WASTEWATER TREATMENT

WATER POLLUTANTS   INTERMEDIA TRANSFERS

Treatments	Radioactivity	Organics-Soluble	Inorganic – Soluble	Organics-Insoluble	Inorganic-Insoluble	Sulfur Compounds	Phosphorus Compounds	Nitrogen Compounds	Heavy Metals	Pathogens	Thermal	Acidity/Alkalinity	Air	Water	Land
Screening			•	•											Residues[a]
Flotation	•			•											Residues
Coagulation and Sedimentation	•			•	•		•		•	•					Residues
Chemical Addition	•	•	•	•	•	•	•	•	•	•			•		Compounds formed	Compounds formed
Trickling Filter		•	•	•	•							•			formed	Residue
Activated Sludge	•	•	•	•	•	•	•	•	•			•		Residue sludge	Residue & sludge
Lagoons and Stabilization Ponds	•		•	•	•	•								Residue	Residue
Ion Exchange		•			•	•	•	•				•			Residue
Activated Carbon	•	•	•		•		•	•						BOD, SS NO₃ heavy metals	
Reverse Osmosis		•			•	•	•	•						Residues	Residue
Chlorination										•					
Spray Irrigation	•	•	•		•	•	•	•							
Ammonia Stripping								•					NH₃	CaCO₃	CaCO₃
Cooling Towers											•		Heat		
Retention[b]	•													Low-level Radioactivity	Radioactive Drums
Electrodialysis		•			•	•	•	•						Residues	Residues

a. Residues mean the same or a combined chemical term of the pollutant physically removed.

b. Retention by storage as used in nuclear power plants.

Source: Report EPA 600/5-73-003

Reference

(1) Stone, R. and Smallwood, H., "Intermedia Aspects of Air and Water Pollution Control," *Report EPA 600/5-73-003,* Wash., D.C., U.S. Environmental Protection Agency (Aug. 1973).

SOURCES OF SPECIFIC TYPES OF POLLUTANTS

ACIDS AND ALKALIES

In Water

Industrial wastes of many types, and from a variety of processes, exhibit extreme pH values which can greatly influence the quality and aquatic life of receiving waters. Table 7 lists representative industries, and gives waste pH characteristics associated with each industry. Wastes may be released on a batch basis as acid or alkaline processing vats are dumped, or released as a continuous waste stream resulting from a specific industrial process (e.g., rinsing). The volume and composition of a waste stream containing acidic or basic compounds can be quite variable. Dickerson and Brooks (1) have described the waste stream from a cellulose production line of Hercules Powder Corporation as varying in volume from 3,000 to 9,000 gpm with acidity, measured as free sulfuric acid, ranging from 30 to 1,100 mg/l.

TABLE 7: pH CHARACTERISTICS OF INDUSTRIAL WASTES

Industrial Process	Waste pH Characteristics
Food and drugs:	
Pickling	Acidic or alkaline
Soft drinks	High pH
Apparel:	
Textile	Highly alkaline
Leather processing	Variable
Laundry	Alkaline
Chemicals:	
Acids	Low pH
Phosphate and phosphorus	Low pH
Materials:	
Pulp and paper	High or low pH
Photographic products	Alkaline
Steel	Mainly acid, some alkaline
Metal plating	Acids
Oil	Acids
Rubber stores	Variable pH
Naval stores	Acid

(continued)

TABLE 7: (continued)

	Industrial Process	Waste pH Characteristics
	Energy:	
	Coal mining	Acid mine drainage
	Coal processing	Low pH

Source: Report PB 216,162

In addition to the free acids and bases (H_2SO_4, NaOH, etc.) typically encountered in industrial waste streams, some industries produce combined acidic or basic salts, which form weak acids or bases by hydrolysis upon dilution of the waste by receiving water. Typical of such industrial effluents are those originating from steel mills and other metal processing industries, where acid or alkaline cleaning solutions are employed. Control of effluent and stream pH requires treatment not only for the free acids and bases in the waste, but for the acidic and/or basic salts present as well. Waste treatment processes employed for the free acids and bases are effective in controlling acid and basic salts.

Intermedia Aspects

Metallic salts and oxides in water hydrolyze to form acids and alkalies which in turn affect the pH of the water. When wastes containing metallic salts and oxides are discharged in sufficient quantities into natural water bodies, the pH of the latter may also be changed. Like biochemical oxygen demand, pH is not a pollutant, but only an indicator of pollution. Even when volatile acids and bases are dissolved in water at reasonable concentrations, they are not easily transferable to air. On the other hand, water scrubbing of gases containing SO_x and NO_x provides an air-to-water intermedia transfer route. Figure 1 shows the intermedia flow for acidity and alkalinity.

FIGURE 1: INTERMEDIA FLOW CHART: ACIDITY-ALKALINITY

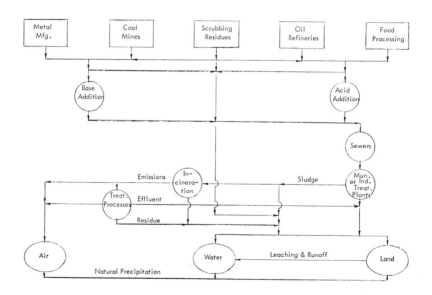

Source: Report EPA 600/5-73-003

A slight change in pH can produce an alteration in the carbonate buffer system on which living organisms rely. Another primary danger which may accompany a change in pH is the synergistic effect of acidity and waterborne substances producing toxicity. For example, a reduction in pH of about 1.5 units can cause a thousandfold increase in the acute toxicity of cyanometallic complex. Not all plant and animal life have the same tolerance to pH changes. Most fish can withstand a variation in pH between 5.0 and 9.0. Beyond this range, replacement communities take over. A high or low pH in livestock watering supplies can be detrimental to the animals.

Industries which either use waters polluted with metallic salts and oxides or have these pollutants in their wastewater, may have serious problems because of the hardness of the water. Industries which use these hard waters generally soften them to prevent the waters leaving a scale deposit on the inside of process tanks, pipes, and boilers. These deposits can decrease the efficiency of the system, shorten its useful life, or even damage it. Industries which discharge these latter pollutants usually have them in large quantities. The metallic salts cause corrosion of pipes, pumps, and other structures made of metal or concrete. Some salts of nontoxic metals (iron and aluminum, for example) react with the natural alkalinity in the water to form stable hydroxides. Some of these are colored and form unsightly deposits in the receiving waters. It is believed that the latter deposits reduce light penetration of the river and interfere with normal, existing ecological systems.

The metallic salts and oxides are likely to be formed by any waste generator which employs metal in its process or piping. This is true of the transportation equipment, mining, primary metals industries, and fabricated metals products. The largest U.S. source is acid drainage from coal mines, the pollutant being sulfuric acid produced by air oxidation of pyrites. These waters may have a pH as low as 2.9. Acid pickling of steel is another source of acidity in waste effluents. At the other end of the pH scale, refinery wastes and food processing wastes are highly alkaline, in part. The most common treatment for acid or alkali waste water is to neutralize it by adding the appropriate basic or acidic solution. The most common alkalies used to neutralize acids are limestone, lime, and soda ash. Sulfuric acid is most commonly used to neutralize basic solutions. Carbonic or nitric acids are also used to a lesser degree.

References

(1) Dickerson, B.W. and Brooks, R.M., "Neutralization of Acid Wastes," *Indus. Engr. Chem.*, 42, 599-605 (1950).

(2) Nemerow, N.L., *Theories and Practices of Industrial Waste Treatment,* Reading, Mass., Addison-Wesley Pub. Co., Inc. (1963).

(3) Stone, R. and Smallwood, H., "Intermedia Aspects of Air and Water Pollution Control," *Report EPA-600/5-73-003,* Washington, D.C., U.S. Environmental Protection Agency (August 1973).

ALDEHYDES

In Air

Aldehydes that pollute the atmosphere result from two main sources: [1] incomplete combustion of organic compounds and [2] atmospheric photochemical reactions involving mainly hydrocarbons and nitrogen oxides. Thus, the highest concentrations of atmospheric aldehydes are expected to be in the populated areas where combustion of fuels and motor-vehicle exhaust emit significant amounts of aldehydes and compounds that form aldehydes through photooxidation. Emission data reported for certain towns, cities, and counties of the United States are summarized (1). The highest reported value for a city is 1,139 tons of aldehydes per year for the city of St. Louis. These data indicate that emission of aldehydes to the atmosphere is primarily due to automobile exhaust, followed by burning of wastes and combustion of fuel.

Natural Occurrence: Natural sources of aldehydes do not appear to be important contributors to air pollution. Acetaldehyde is found in apples and as a by-product of alcoholic fermentation processes. Other lower aliphatic aldehydes are not found in significant quantities in natural products. Olefin and aromatic aldehydes are present in some of the essential oils in fruits and plants. These include citronellal, in rose oil; citral, in oil of lemongrass; benzaldehyde, in oil of bitter almonds; cinnamaldehyde, in oil of cinnamon; anisaldehyde, in anise; and vanillin, in the vanilla bean.

Production Sources: Aldehydes are commercially manufactured by various processes, depending on the particular aldehyde. In general, they are prepared via oxidation reactions of hydrocarbons, hydroformylation of alkenes, dehydrogenation of alcohols, and addition reactions between aldehydes and other compounds.

Product Sources: Aldehydes have a wide variety of uses in numerous industries, such as the chemical, rubber, tanning, paper, perfume, and food industries. The major use is as an intermediate in the synthesis of organic compounds, including alcohols, carboxylic acids, dyes, and medicinals.

Other Sources: Significant amounts of atmospheric aldehydes, particularly formaldehyde, are a result of photooxidation of unsaturated hydrocarbon pollutants. In addition, some important contributing sources of aldehyde air pollution are the burning or heating of organic compounds. These sources include mobile combustion (automobiles, diesel vehicles, and aircraft), stationary combustion (units that burn coal, oil, natural gas, or waste materials), and noncombustion sources (e.g., chemical oxidation processes, and drying and baking in ovens).

Environmental Air Concentrations: In 1967, the National Air Sampling Network began the monitoring of aldehydes. The averages ranged from 3 to 79 $\mu g/m^3$ of aldehyde (calculated as formaldehyde); the maximum values ranged from 5 to 161 $\mu g/m^3$. Other areas reported aldehyde air concentrations before establishment of the National Air Sampling Network program. Until the early 1960's, the analysis method used was the sodium bisulfite method, which is not specific for aldehydes and is sensitive to some ketones as well. The aldehyde (calculated as formaldehyde) concentrations for the ambient air in several cities sampled in 1946 to 1951 ranged from 0 to 324 $\mu g/m^3$ (0 to 0.27 ppm), with the average ranging from 48 to 216 $\mu g/m^3$ (0.04 to 0.18 ppm). It is generally reported that of the aldehydes present in the atmosphere, 50% is accounted for as formaldehyde and 5% as acrolein.

Reference

(1) Stahl, Q.R., "Air Pollution Aspects of Aldehydes," *Report PB 188,081,* Springfield, Va., Nat. Tech. Information Service (September 1969).

AMMONIA

In Air

The major portion of atmospheric ammonia is produced by biological processes in land and sea masses, and the gas then escapes into the atmosphere. Ammonia produced by industry and as a result of urban activities, though of lesser importance, may nevertheless be a factor in air pollution in localized areas. The major source of urban-produced ammonia is the combustion process which occurs in operation of the internal combustion engine, combustion of fuels for heating, and the incineration of wastes. Industrial sources of ammonia are refineries, fertilizer plants, and organic chemical process plants. Other minor sources of ammonia arise from biological degradation in areas where animals are kept, such as stockyards, and from miscellaneous uses of ammonia in cleaning both in industry and in the home.

Natural Occurrence: According to one estimate, 3.7×10^9 tons of ammonia are released into the atmosphere annually. Of this amount, only 4.2 x 10⁶ tons are emitted to the atmosphere as a result of industrial and urban processes; therefore, roughly 99.9% of the atmosphere's ammonia concentration is produced by natural biological processes. The main biological source of ammonia is the decomposition of organic waste material. Approximately 1.0 g of ammonia per man per day is produced metabolically. Ammonia is given off from manure in piggeries and other installations where animals are kept. Ammonia is also generated during treatment of wastewater in sewage plants. No information was found on the quantity of these emissions. Ammonia is also found in sea water and in volcanic gases.

Production Sources: Ammonia is produced commercially in chemical process plants, as a by-product in the manufacture of other chemicals, mainly in making coke from coal, and as a product of combustion, refining of oils, and other processes. The main source of ammonia in oil refineries is from the catalyst regenerators in the catalytic cracking plants. The ammonia releases from oil refineries are given in Table 8.

TABLE 8: AMMONIA RELEASED FROM OIL REFINERIES

Source	Lb/100 Barrels Fresh Feed
Compressor-internal combustion	0.2
Fluid-bed catalytic cracking units	54.0
Thermofor catalytic cracking units	5.0

Source: Report PB 188,082

Table 9 gives the ammonia emissions from regenerator stacks in catalytic cracking units of the Los Angeles area refineries.

TABLE 9: AMMONIA EMISSION FROM CATALYTIC CRACKING

Unit Type	µg/m³	Tons/Day
Fluid bed	47,000-470,000	4.2
Thermofor	20,000-72,000	0.2

Source: Report PB 188,082

Typical exhaust emissions from some metallurgical and ceramic plants are as follows:

From nonferrous foundries: per plant producing 50 tons of castings per day	0.002 tons of ammonia
From gray iron foundries: per plant producing 200 tons of castings per day	0.023 tons of ammonia
From stone, clay, and glass plants: per cement plant producing 4,830 barrels per day	0.17 manufacturing tons

Ammonia is produced as a result of combustion, mainly from the use of fossil fuels and incineration of waste materials. These sources generally result in direct emission of the ammonia into the atmosphere. The emission of ammonia from internal combustion engines has been estimated at 2.0 lb/1,000 gallons burned for gasoline engines and for diesel engines. The rates of emissions of ammonia from various categories of fossil fuels are presented in Table 10.

TABLE 10: AMMONIA EMISSIONS FROM COMBUSTION

Combustion Source	Amount of Emission
Coal	2 lb/ton
Fuel oil	1 lb/1,000 gal
Natural gas	0.3 to 0.56 lb/10^6 ft^3
Bottled gas (butane)	1.7 lb/10^6 ft^3
Propane	1.3 lb/10^6 ft^3
Wood	2.4 lb/ton
Forest fires	0.3 lb/ton

Source: Report PB 188,082

Product Sources: Ammonia is used as a raw material in the production of nitric acid, fertilizers, and the syntheses of hundreds of organic compounds, including many drugs, plastics, and dyes. Approximately 85% of the ammonia is used as anhydrous ammonia fertilizer or as a raw material for other fertilizer production. Very little information is available on ammonia emissions from these plants.

Environmental Air Concentrations: The average concentration of ammonium compounds in the air in urban areas is approximately 20 μg/m^3. The background concentration of ammonium in the lower troposphere is about 6 μg/m^3 in the mid-latitudes and 140 μg/m^3 near the equator. Data on atmospheric concentrations of ammonium for various cities of the United States are presented by Miner (1).

Intermedia Aspects

Where an industrial process emits high concentrations of ammonia, as mentioned above for fertilizer, organic chemicals and nitric acid, its value stimulates recovery for use rather than disposal of the residue, so no intermedia transfer takes place. In the case of agricultural problems, adequate ventilation of enclosed buildings is advised, and controls consist of maintaining dry conditions in the manure to reduce ammonia discharge. Ammonia is, therefore, classified as an intramedia pollutant.

Reference

(1)   Miner, S., "Air Pollution Aspects of Ammonia," *Report PB 188,082,* Springfield, Va., Nat. Tech. Information Service (September 1969).

ARSENIC

In Air

Some estimates of total arsenic emissions by source are given in Table 11.

TABLE 11: ARSENIC EMISSIONS

Source	Amount in Tons	Pollutant, %
Mining	<1	neg
Phosphate rock	neg	neg
Primary Copper		
Roasting	900	10.07
Reverberatory furnaces	400	4.48
Converters	1,150	12.87
Material handling	250	2.80

(continued)

TABLE 11: (continued)

Source	Amount in Tons	Pollutant, %
Primary zinc		
Roasting	1,390	15.55
Primary lead		
Sintering	285	3.19
Blast furnace	80	0.90
Reverberatory furnace	11	0.12
Gray iron foundry	97	1.09
Cotton ginning	19	0.21
Nonferrous alloys	neg	neg
Phosphoric acid	neg	neg
Glass manufacture	638	7.14
Wood preservatives	neg	neg
Miscellaneous arsenic		
chemicals	3.3	0.04
Arsenic pesticide		
production	197	2.20
Pesticide, herbicide,		
fungicide use	2,925	32.72
Power plant boilers		
Pulverized coal	429	4.80
Stoker coal	49	0.55
Cyclone coal	15	0.17
Industrial boilers		
Pulverized coal	19	0.21
Stoker coal	67	0.75
Cyclone coal	9	0.10
All oil	neg	neg
Residential/commercial coal	6	0.07
Total	8,939	100

Source: The MITRE Corporation, Preliminary Results, EPA Contract No. 68-01-0438

There are three major sources of arsenic air pollution: smelting of metals, burning of coal, and use of arsenicals as pesticides as pointed out in a review publication by Sullivan (1). Tables 12 to 16 inclusive summarize emission factors for arsenic from various sources, as presented by Anderson (2).

TABLE 12: EMISSION FACTORS FOR MINING AND INDUSTRIAL SOURCES OF ARSENIC

Source	Emission Factor	Emission Factor Symbol*
Mining	0.1 kg/10^3 kg (0.2 lb/ton) of arsenic in ore	E
Copper smelter	3 kg/10^3 kg (5 lb/ton) of copper	MB, UK
Lead smelter	0.4 kg/10^3 kg (0.8 lb/ton) of lead	MB, UK
Zinc smelter	0.5 kg/10^3 kg (1 lb/ton) of zinc	MB, UK
Cast iron	0.006 to 0.008 kg/10^3 kg (0.01 to 0.02 lb/ton) of metal charged	E
Nonferrous alloys	0.5 kg/10^3 kg (1 lb/ton) of arsenic processed	Q

*Defined in Table 4, page 7.

Source: Report PB 230,894

TABLE 13: EMISSION FACTORS FOR PROCESSING ARSENIC AND ITS COMPOUNDS

Source	Emission Factor	Emission Factor Symbol*
Pesticide production for agricultural uses**	10 kg/10^3 kg (20 lb/ton) of arsenic processed	Q
Glass production	0.08 kg/10^3kg (0.2 lb/ton) of glass produced	OES (1)
Wood preservatives	Negligible	–
Others (paint pigments, pyrotechnics, pharmaceutical, semiconductors, etc.)	2 kg/10^3kg (3 lb/ton) of arsenic processed	Q

*Defined in Table 4, page 7; number in parentheses indicates the number of samples analyzed.
**Controlled emission factor.

Source: Report PB 230,894

TABLE 14: EMISSION FACTORS FOR CONSUMPTIVE USES OF ARSENIC

Source	Emission Factor	Emission Factor Symbol*
Cotton ginning processing	2 kg/10^3 bales (4 lb/10^3 bales) of cotton ginned	UK
Detergents	Negligible	–

*Defined in Table 4, page 7.

Source: Report PB 230,894

TABLE 15: EMISSION FACTORS FOR ARSENIC FROM FUEL COMBUSTION

Fuel and Source	Arsenic Content, ppm	Emission Factor kg/10^3 liters	Emission Factor lb/10^3 gallons	Emission Factor Symbol*
Coal, domestic, controlled	5.44	2**	3**	OES (1)
Foreign residual oil				
No. 6 fuel oil, Mexico	0.1	0.00009	0.0008	NA (1)
No. 6 fuel oil, Virgin Islands	0.1 to 0.2	0.00009 to 0.0002	0.0008 to 0.002	NA (2)
No. 6 fuel oil, Trinidad	0.095	0.00009	0.0007	NA (1)
No. 6 fuel oil, Curacao, N.A.	0.16 to 0.2	0.0002	0.001 to 0.002	NA (2)
Average for foreign residual oil	0.14	0.0001	0.001	NA (8)
Foreign crude oils				
Neutral zone crude No. 24	0.079	0.00007	0.0006	NA (1)
Boscan crude oil, Morovia	0.34	0.0003	0.002	NA (1)
Jabo crude oil, Venezuela	0.053	0.00005	0.0004	NA (1)
Kuwait crude oil, Kuwait	0.01	0.000009	0.00007	NA (1)
Boscan crude oil, Venezuela	0.31	0.0003	0.002	NA (2)
Lagunillas crude oil, Venezuela	0.12	0.0001	0.0009	NA (1)
UMM Farvo, Libya	0.02	0.00002	0.0001	NA (1)
Average for foreign crude oils	0.13	0.0001	0.0009	NA (8)
U.S. crude oils				
St. Tedesa, Illinois	0.61	0.0005	0.004	NA (1)
Maysville, W. Oklahoma	0.031	0.00003	0.0002	NA (1)
Hall-Gurney, Kansas	0.047	0.00004	0.0003	NA (1)
East Texas, Texas	0.007	0.000006	0.00005	NA (1)

(continued)

TABLE 15: (continued)

Fuel and Source	Arsenic Content, ppm	Emission Factor kg/10³ liters	lb/10³ gallons	Emission Factor Symbol*
U.S. crude oils				
Grass Creek, Walker Dome, Wyoming	0.04	0.00003	0.0003	NA (1)
Average for U.S. crude oils	0.15	0.0001	0.001	NA (5)

*Defined in Table 4; numbers in parentheses indicate number of samples analyzed.
**Units for coal are kilograms per 10³ kilograms and pounds per 10³ tons, respectively.

Source: Report PB 230,894

TABLE 16: EMISSION FACTORS FOR ARSENIC FROM SOLID WASTE INCINERATION

Source	Emission Factor	Emission Factor Symbol*
Cotton ginning waste burned	9 kg/10³ bales (17 lb/10³ bales) of cotton ginned	E
Sewage and sludge	0.01 kg/10³ kg (0.02 lb/ton) sewage and sludge	E

*Defined in Table 4.

Source: Report PB 230,894

In Water

Arsenic and arsenical compounds have been reported as waste products of the metallurgical industry, glassware and ceramic production, tannery operations, dye, and pesticide manufacture. Other industrial sources include the organic and inorganic chemicals and petroleum refining industries, and the rare-earth industry. The manufacture of Paris green and calcium meta-arsenate, both insecticides, was reported to produce wastewaters containing 362 mg per liter of arsenious oxide. Although arsenic is widely associated with the manufacture of herbicides and pesticides, a recent study concluded that the inorganic pesticides industry (including arsenic based compounds) is fading. Arsenical wastes from these sources would be expected to assume less importance in that event.

Arsenic wastes from nonferrous smelting have been reported in the literature, although the main form appears to be as extremely small (1 to 3 micron diameter) particles from which arsenic can be profitably recovered. Very little current information is available on arsenic waste treatment in the smelting industry, perhaps because of the highly efficient arsenic recovery processes employed. High arsenic levels have been encountered in raw municipal water supplies, necessitating arsenic removal by the water treatment industry. Ground water supplies in Central and South America are frequently reported to contain excessive arsenic levels. The literature has described deep well waters in Taiwan, China which contain up to 1.1 mg/l arsenic, and are believed to be responsible for an endemic illness in the area called black-foot disease. Deep well waters in many desert areas of the southwest United States contain arsenic at levels exceeding 0.1 mg/l.

References

(1) Sullivan, R.J., "Air Pollution Aspects of Arsenic and Its Compounds," *Report PB 188,071*, Springfield, Va., Nat. Tech. Information Service (September 1969).
(2) Anderson, D., "Emission Factors for Trace Substances," *Report PB 230,894*, Springfield, Va., Nat. Tech. Information Service (December 1973).
(3) Patterson, J.W. and Minear, R.A., "Wastewater Treatment Technology," 2nd Ed., *Report PB 216,162*, Chicago, Institute for Environmental Quality (February 1973).

(4) U.S. Department of the Interior, "Arsenic and Lead in Water — A Bibliography," *Report PB 202,578,* Washington, D.C., Water Resources Scientific Information Center (September 1971).

ASBESTOS

In Air

Some estimates of total asbestos emissions by source are given in Table 17.

TABLE 17: ASBESTOS EMISSIONS

Source	Amount, tons	Pollutant, %
Asbestos mining	5,610	89.6
Kraft pulp mill recovery furnace	15	0.24
Sulfite pulp mill	neg	neg
Asbestos products		
Brake lining production	312	4.98
Shingle and siding production	205	3.27
Asbestos textile production	18	0.29
Installation of asbestos construction material	61	0.97
Spray on steel fire proofing	15	0.24
Insulating cement application	25	0.40
Total	6,261	99.99

Source: The MITRE Corporation, Preliminary Results, EPA Contract No. 68-01-0438

Asbestos fibers enter the atmosphere from a wide variety of sources extending from the weathering and disturbance of natural deposits of asbestos-bearing materials to operations for the ultimate disposal of products containing asbestos. Intermediate emission sources include asbestos mining and milling sites, manufacturing facilities for asbestos-containing products, and construction sites employing asbestos insulating, fireproofing, and structural materials. More than 3,000 products contain commercial asbestos. As these products are used, asbestos is frequently emitted to the atmosphere. Among these products are automotive brake linings and asbestos-asphalt concrete for paving roadways. Asbestos is also present as a natural contaminant in some widely employed materials, such as talc. The asbestos emissions from use of these materials can be significant.

Because asbestos is exceptionally resistant to thermal degradation and chemical attack, settled particles are persistent in the environment and subject to reentrainment into the atmosphere. It can readily be mechanically subdivided into fibers of submicron diameter, which can remain airborne for long periods of time. These factors, coupled with the presence of large numbers of emission sources, as noted above, would indicate the presence of a background level of asbestos in the atmosphere. Semiquantitative data confirm this conjecture and show that urban background concentrations are significantly larger than non-urban ones.

Asbestos emissions are now being controlled to a limited extent, primarily from milling and manufacturing sources to which gas-cleaning devices are readily applicable. The formulation of recommended codes of trade practices governing such operations as the transport, fabrication, application, and disposal of asbestos-containing materials has not proved to be an effective emission control technique. Control of asbestos emissions from some sources, for example, spray-applied asbestos fireproofing, has been made possible by the use of substitute materials for asbestos. It is also true, however, that there is interest in expanding the already vast number of applications for asbestos fibers. A variety of reports has been issued in recent years on the asbestos air pollution problem (1)(2)(3)(4)(5). Tables 18 through 20 (5) summarize asbestos emission factors.

TABLE 18: EMISSION FACTORS FOR ASBESTOS FROM MINING AND MILLING[a]

Source[b]	Emission factor, kg/10^3kg (lb/ton) of asbestos produced	Emission factor symbol[c]
Mining, total	5 (9 to 10)	E
Mining	2 (3)	E
Loading	1 (2)	E
Hauling	1 (2)	E
Unloading	1 (2)	E
Mining, total, 50% control	3 (5)	E
Milling	50 (100)	E
Milling, 50% control	40 (80)	E
Milling, 80% control	10 (20)	E

[a]The emission factors are based on engineering estimates and no source sampling. These factors cannot be used to quantify asbestos emissions.

[b]Uncontrolled unless otherwise specified.

[c]Defined in Table 4.

TABLE 19: EMISSION FACTORS FOR PROCESSING OF ASBESTOS[a]

Source[b]	Emission factor, kg/10^3kg (lb/ton) of asbestos processed	Emission factor symbol[c]
Friction material, controlled	3 (6)	Q
Asbestos cement products, controlled	0.5 (1)	E
Textiles	20 (40)	E
Textiles, controlled	1 (2)	E
Asbestos paper	2 (4)	E
Asbestos paper, controlled	0.5 (1)	E
Floor tile	2 (4)	E
Floor tile, controlled	0.5 (1)	E

[a]The emission factors are based on engineering estimates and no source sampling. These factors cannot be used to quantify asbestos emissions.

[b]Uncontrolled unless otherwise specified.

[c]Defined in Table 4.

Source: Report PB 230,894

TABLE 20: EMISSION FACTORS FOR CONSUMPTIVE USES OF ASBESTOS[a]

Source[b]	Emission factor, kg/10³kg (lb/ton) of asbestos applied	Emission factor symbol[c]
Brake linings	5 (10)[d]	E
Steel fireproofing, controlled	5 (10)	E
Insulating cement, controlled	13 (25)	E
Construction industry	13 (25)	E

[a]The emission factors are based on engineering estimates and no source sampling. These factors cannot be used to quantify asbestos emissions.

[b]Uncontrolled unless otherwise specified.

[c]Defined in Table 4.

[d]A factor of 0.0005 pound per 10^3 vehicle miles can also be used.

Source: Report PB 230,894

In Water

As described by Cook et al (6) asbestiform amphibole minerals, which have been demonstrated to be associated with human health problems, have been detected in substantial quantities in municipal water supplies taken from western Lake Superior water. The total concentration of amphibole minerals in the Duluth, Minnesota, water supply, as measured by x-ray diffraction for daily samples of suspended solids, averages 0.19 milligram per liter with large fluctuations due to seasonal and climatological effects on lake circulation. Electron microscopic examination of these water samples confirms the presence of asbestiform amphibole fibers. A conservative estimate of the fiber count for 1973 Duluth water supply samples is (1 to 30) x 10^6 amphibole fibers identifiable by electron diffraction per liter of water with a mass concentration of 1 to 30 micrograms per liter.

References

(1) Davis, W.E., "National Inventory of Sources and Emissions; Cadmium, Nickel, and Asbestos – Section III, Asbestos" (1968), Leawood, Kansas, W.E. Davis and Associates, *Contract No. CPA 70-128* (February 1970).
(2) U.S. Environmental Protection Agency, "Background Information – Proposed National Emission Standards for Hazardous Air Pollutants," *Report APTD-0753,* Research Triangle Park, N.C., Officer of Air Programs (December 1971).
(3) U.S. Environmental Protection Agency, "Control Techniques for Asbestos Air Pollutants," *Publ. AP-117,* Research Triangle Park, N.C., Office of Air and Water Programs (February 1973).
(4) U.S. Environmental Protection Agency, "Background Information on Development of National Emission Standards for Hazardous Air Pollutants: Asbestos, Beryllium and Mercury," *Publ. PB 222,802,* Springfield, Va., Nat. Tech. Information Service (March 1973).
(5) Anderson, D., "Emission Factors for Trace Substances," *Report PB 230,894,* Springfield, Va., Nat. Tech. Information Service (December 1973).
(6) Cook, P.M., Glass, G.E. and Tucker, J.H., "Asbestiform Amphibole Minerals: Detection and Measurement of High Concentrations in Municipal Water Supplies," *Science 185, No. 4154, 853-855* (September 6, 1974).

BARIUM

In Air

Air pollution aspects of barium and its compounds have been reviewed in some detail by Miner (1). The flow of barium in the United States has been traced and charted for the year 1969. The consumption was 934,000 tons, while domestic production totaled 603,000 tons. Imports and exports were 344,000 and 10,000 tons, respectively. Emissions to the atmosphere during the year were 15,420 tons (Table 21). Nearly 18% of the emissions resulted from the processing of barite, more than 28% from the production of chemicals, 23% from the manufacture of various end products, and about 26% from the combustion of coal. The wear of rubber tires was a relatively minor emission source. Emission estimates for processing, chemical production, and the manufacture of end use products are based on unpublished data obtained from industrial sources.

TABLE 21: EMISSIONS BY SOURCE, 1969

Source Category	Source Group		Emissions - Tons	Emissions %
Mining			30	--
Processing			2,700	17.5
Chemical Production			4,400	28.5
End Product Uses			4,240	27.5
	Well Drilling Mud	70		
	Glass	40		
	Paint	30		
	Rubber	600		
	Miscellaneous	3,500		
Other Emission Sources			4,050	26.5
	Coal	4,000		
	Cast Iron	50		
TOTAL			15,420	100.0

Source: Report PB 210,676

The emission factors presented in Table 22 are the best currently available. They were determined through a combination of methods consisting of: [1] direct observation of emission data and other related plant processing and engineering data; [2] estimates based on information obtained from literature, plant operators, and others knowledgeable in the field; [3] calculations based on experience and personal knowledge of metallurgical processing operations; and [4] specific analytical results where available.

TABLE 22: EMISSION FACTORS

Mining	100 lb/1,000 tons barite mined
Processing	5 lb/ton barite processed
Barium chemicals	50 lb/ton barite processed
End product uses of barium	
Oil and gas well drilling	110 lb/1,000 tons barite used
Glass manufacture	2 lb/ton barium processed
Paint manufacture	2 lb/ton barium processed
Rubber tire wear	0.3 lb/million miles

TABLE 22: (continued)

Other emission sources	
Coal	15 lb/1,000 tons of coal burned
Cast iron	5 lb/1,000 tons of process weight

Source: Report PB 210,676

In Water

Barium as barite (barium sulfate), is used as a white pigment for paint. Barium is employed in metallurgy, glass, ceramics and dyes manufacture, and in vulcanizing of rubber. Barium has been also reported present in explosives manufacturing wastewater. Very little information is available in the literature on levels of barium in industrial wastes, on existing methods of treatment or on levels of barium removal achievable.

References

(1) Miner, S., "Air Pollution Aspects of Barium and Its Compounds," *Report PB 188,083,* Springfield, Va., Nat. Tech. Information Service (September 1969).
(2) Davis, W.E. and Assoc., "National Inventory of Sources and Emissions, Section I: Barium," *Report PB 210,676,* Springfield, Va., Nat. Tech. Information Service (May 1972).
(3) Patterson, J.W. and Minear, R.A., "Wastewater Treatment Technology," 2nd Edition, *Report PB 216,162,* Springfield, Va., Nat. Tech. Information Service (February 1973).

BERYLLIUM

In Air

The air pollution aspects of beryllium and its compounds have been reviewed by Durocher (1). Some estimates of total beryllium emissions by source are given in Table 23. Problems with beryllium air pollution have been reviewed in several publications (1)(2)(3)(4)(5).

TABLE 23: BERYLLIUM EMISSIONS

Source	Amount, tons	Pollutant, %
Mica, feldspar mining	neg	neg
Gray iron foundry		
Cupola	4	2.77
Ceramic coatings	neg	neg
Beryllium alloys and compounds	5	3.64
Beryllium fabrication	neg	neg
Power plant boilers		
Pulverized coal	86	59.62
Stoker coal	10	6.93
Cyclone coal	3	2.08
All oil	2	1.39
Industrial boilers		
Pulverized coal	8	5.55
Stoker coal	13	9.01
Cyclone coal	2	1.39
All oil	2	1.39

(continued)

TABLE 23: (continued)

Source	Amount, tons	Pollutant, %
Residential/commercial boilers		
Coal	1	0.69
Oil	8	5.55
Total	144	100.01

Source: The MITRE Corporation, Preliminary Results, EPA Contract No. 68-01-0438

The adverse effects of airborne beryllium on human health were first recognized in 1940 as a result of the occurrence of lung disease in occupationally exposed workers. Beryllium workers develop two forms of lung disease. One form, an acute chemical pneumonitis, has been observed, with one reported exception, only in workers who were occupationally exposed to beryllium. The chronic form, berylliosis, a progressive, interstitial, granulomatous disease located primarily in the alveolar walls, has been observed in individuals who have never been occupationally exposed to beryllium. Of the 60 people with nonoccupationally incurred disease whose cases are on file with the Beryllium Registry, 27 were exposed to beryllium by washing clothes soiled with beryllium dust. Another 18 were exposed to beryllium in the form of air pollution surrounding beryllium plants, 13 were exposed both to polluted air and contaminated clothing, and the exposure of the remaining 2 was unknown.

Most of the cases of berylliosis involved exposure to beryllium at a time when its hazard was not recognized and its concentration in the air was not measured. Retrospective estimates of the concentrations of beryllium that resulted in some cases of berylliosis from nonoccupational exposure have been made. The report of this work states: "It may therefore be concluded that the lowest concentration which produced disease was greater than 0.01 microgram per cubic meter and probably less than 0.10 microgram per cubic meter."

In 1949, a guideline limit for beryllium concentrations in community air was developed by the Atomic Energy Commission (AEC). The concentration selected was an average of 0.01 microgram of beryllium per cubic meter of air for 30 days. In the period since the implementation of this guideline, no reported cases of chronic beryllium disease have occurred as a result of community exposure. Consequently, the Committee on Toxicology of the National Academy of Sciences has concluded that the average concentration 0.01 $\mu g/m^3$ for 30 days has proved to be a safe level of exposure. Therefore, an average of 0.01 $\mu g/m^3$ for 30 days should be used as a guide in developing emission standards. Tables 24 to 27 summarize emission factors for beryllium as given by Anderson (5).

TABLE 24: EMISSION FACTORS FOR BERYLLIUM FROM INDUSTRIAL AND SOLID WASTE INCINERATION SOURCES

Source[1]	Emission Factor	Emission Factor Symbol[2]
Mining	0.1 kg/10^3 kg (0.2 lb/ton) of Be produced	E
Production of beryllium metal and its compounds, overall value[3]	15 kg/10^3 kg (30 lb/ton) of Be produced	CAA (?)
Cement plants[3]		
Dry process		
Feed to raw mill[4]	0.00002 kg/10^6 kg (0.00003 lb/1,000 tons) of feed	ES (1)
Wet process		
Kiln[5,6]	0.001 kg/10^6 kg (0.002 lb/1,000 tons) of feed	ES (1)
Clinker cooler[5]	0.0004 kg/10^6 kg (0.0008 lb/1,000 tons) of feed	ES (1)
Clinker cooler[7]	0.0005 kg/10^6 kg (0.0009 lb/1,000 tons) of feed	ES (1)

(continued)

TABLE 24: (continued)

Source[1]	Emission Factor	Emission Factor Symbol[2]
Processing or uses of beryllium and its compounds		
Beryllium alloys (stamped and drawn)[8]	0.05 kg/10³ kg (0.1 lb/ton) of Be processed	E
Beryllium alloys (molding)		
Uncontrolled	0.005 kg/10³ kg (0.01 lb/ton) of alloy melted	CAA (4)
After a baghouse	0.0002 kg/10³kg (0.0004 lb/ton) of alloy melted	CAA (2)
Ceramics[3]	0.5 kg/10³ kg (1 lb/ton) of Be processed	Q
Rocket propellants	Negligible	–
Beryllium metal fabrication[3]	0.05 kg/10³ kg (0.1 lb/ton) of Be processed	Q

[1]Uncontrolled unless otherwise specified.
[2]Defined in Table 4; numbers in parentheses indicate number of samples analyzed.
[3]Controlled emission factor.
[4]Exit from baghouse.
[5]Exit from electrostatic precipitator.
[6]At another plant, beryllium emissions from the kiln were below the detection limit of the analytical technique.
[7]Exit from two baghouse collectors in parallel.
[8]It is not known whether this estimate is based on controlled or uncontrolled emissions.

Source: Report PB 230,894

TABLE 25: EMISSION FACTORS FOR BERYLLIUM FROM FUEL COMBUSTION, COAL

Source[1]	Beryllium Content, ppm	- - - Emission Factor - - - kg/10³ kg	lb/ton	Emission Factor Symbol[2]
Power plants				
Kansas[3]	2	0.00002	0.00004	EST (2)
South Carolina[4]	–[5]	0.00001	0.00002	EST (6)
Illinois[4]	1	0.00002	0.00003	EST (2)
Michigan[4]	1	0.00001	0.00002	EST (2)
Coal beds				
Maryland	2.4	0.002	0.003	SC (6)
Ohio	2.6 (0.5 to 5.1)	0.0023	0.0046	SC (137)
Pennsylvania	2.8 (1.8 to 3.7)	0.0013	0.0026	SC (13)
Alabama	2.2 (0.5 to 4.6)	0.0017	0.0033	SC (70)
Georgia	1.6 (0.8 to 2.2)	0.001	0.002	SC (6)
Kentucky-east	4.4 (0.1 to 19)	0.0035	0.0069	SC (87)
Kentucky-west	2.4 (0.5 to 5.0)	0.0022	0.0044	SC (313)
Tennessee	2.8 (0.6 to 11)	0.0012	0.0023	SC (29)
Virginia	1.2 (0.5 to 1.9)	0.0005	0.001	SC (6)
West Virginia	1.3 (0.2 to 3.6)	0.0012	0.0023	SC (25)
Illinois	2.0 (0.5 to 5.7)	0.0018	0.0036	SC (565)
Indiana	4.5 (1.4 to 7.6)	0.0046	0.0091	SC (178)
Arkansas	1.8 (0.1 to 10.2)	0.0010	0.0020	SC (25)
Iowa	3.6 (2 to 5.1)	0.0025	0.0049	SC (22)
Missouri	3.8 (0.5 to 10.4)	0.0039	0.0078	SC (57)
Oklahoma	1.6 (0 to 3.5)[6]	0.0011	0.0022	SC (49)
Montana	2.4 (0 to 9.1)[6]	0.0027	0.0054	SC (118)
North Dakota	0.8 (0 to 4.0)[6]	0.00055	0.0011	SC (56)
Wyoming	1.4 (0 to 12.7)[6]	0.0023	0.0046	SC (75)
Colorado	0.6 (0 to 1.8)[6]	0.0017	0.0033	SC (98)
Utah	0.3 (0 to 0.6)[6]	0.0003	0.0006	SC (33)
U.S. average (coal beds)	2.2	0.0019	0.0037	SC (1875)

[1]Uncontrolled unless otherwise specified.
[2]Defined in Table 4; numbers in parentheses indicate number of samples analyzed.
[3]Exit from limestone wet scrubber.
[4]Exit from electrostatic precipitator.
[5]Not reported.
[6]Zeros indicate below detectable level of analytical technique employed.

Source: Report PB 230,894

TABLE 26: EMISSION FACTORS FOR BERYLLIUM FROM FUEL COMBUSTION, OIL

Source	Beryllium Content, ppm	Emission Factor kg/10³ liters	Emission Factor lb/10³ gallons	Emission Factor Symbol[1]
Residual oil	0.08	0.00008	0.0007	E
Power plant, Connecticut[2]	0.024	0.00002	0.0002	EST (2)
Residual No. 6	0.1	0.00009	0.0008	CAA (2)

[1] Defined in Table 4; numbers in parentheses indicate number of samples analyzed.
[2] Exit from electrostatic precipitator.

TABLE 27: EMISSION FACTORS FOR BERYLLIUM FROM WASTE INCINERATION

Source	Emission Factor kg/10⁶ kg of waste burned	Emission Factor lb/10³ tons of waste burned	Emission Factor Symbol[1]
Sewage sludge incinerator Multiple hearth, after wet scrubber	0.001 (0.0005 to 0.002)	0.002 (0.0009 to 0.003)	OES (3)
Fluidized bed, after wet scrubber	0.001 (0.00005 to 0.001)	0.002 (0.0001 to 0.002)	OES (3)
Municipal incinerator, uncontrolled	0.02	0.03	EST (1)
Municipal incinerator, after electrostatic precipitator	0.02	0.03	EST (1)

[1] Defined in Table 4; numbers in parentheses indicate number of samples analyzed.

Source: Report PB 230,894

References

(1) Durocher, N.L., "Air Pollution Aspects of Beryllium and Its Compounds," *Report PB 188,078,* Springfield, Va., Nat. Tech. Information Service (September 1969).
(2) U.S. Environmental Protection Agency, "Background Information—Proposed National Emission Standards for Hazardous Air Pollutants: Asbestos, Beryllium, Mercury," *Report APTD-0753,* Research Triangle Park, N.C., Office of Air Programs (December 1971).
(3) U.S. Environmental Protection Agency, "Control Techniques for Beryllium Air Pollutants," *Publ. AP-116,* Research Triangle Park, N.C., Office of Air and Water Programs (February 1973).
(4) U.S. Environmental Protection Agency, "Background Information on Development of National Emission Standards for Hazardous Air Pollutants: Asbestos, Beryllium and Mercury," *Publ. APTD-1503,* Research Triangle Park, N.C., Office of Air and Water Programs (March 1973).
(5) Anderson, D., "Emission Factors for Trace Substances," *Report PB 230,894,* Springfield, Va., Nat. Tech. Information Service (December 1973).

BIOLOGICAL MATERIALS

In Air

Microorganisms are ubiquitous in nature. However, all microorganisms found in the air had as their original habitat either soil, water, humans, animals, or plants. Microorganisms become airborne from the soil and from plants by wind disturbances, and from water by wave and wind action. They come from animals through shedding, excreta, and respiratory droplets, and from humans through shedding from skin and clothing and through respiratory droplets produced by speech, coughing, and sneezing. Some types of organisms are more plentiful in the air than others, because of their size (i.e., they are small enough to remain airborne), the magnitude of the emission source, the death rate of organisms suspended in the air, and other factors. Most of the microorganisms found in the air are saprophytic and generally are not pathogenic. Those which are pathogenic, with some exceptions, come from a living host. Since they are usually detrimentally affected by exposure to the atmosphere, they are only found in close proximity to the host. However, certain plant pathogens can have rapid and widespread aerial dispersal over hundreds of miles within a few days.

The survival of biological aerosols has been studied rather intensively in recent years, primarily in laboratory studies. Most of these studies have been concerned with bacteria, and relatively little is known of the behavior of viruses and fungi. No simple relationship has been found between the degree of survival and age of the aerosol. The half-life of the aerosol is affected by such variable factors as the species of microorganisms (spore-former or non-spore-former); metabolic state of the microorganism; the relative humidity, gaseous composition, and temperature of the air; radiation; collection method; and others. Because of the large number of these variables and their interrelationship, both the results and the interpretations of aerosol survival studies are markedly dependent upon the precise technique employed. Bacterial contaminations of air by sewage plants, such as activated sludge treatment plants has been measured and details reported by Finkelstein (1).

Aerosol production in an animal rendering plant has been studied and slurries of harmless tracer bacteria were painted (a spore-former and a non-spore-former) onto the carcasses before the rendering process began, and later air samples were collected at various places inside and outside the plant as processing proceeded. Results showed that the rendering process in use created aerosols of viable microorganisms. Both the vegetative and spore-forming tracer organisms were found in air samples taken inside the plant and at 100 feet downwind from the exhaust stack. These findings supported the suspicion that some of the workers in the plant had become infected with ornithosis at an earlier date when diseased turkeys had been processed.

Other diseases which potentially could have been transmitted by this rendering plant include anthrax, brucellosis, tularemia, glanders, sylvatic plague, Q fever, and virus equine encephalitis. This situation was a health hazard both to the workers within the plant and to the population in surrounding areas, according to Finkelstein (1). Many microorganisms, bacteria, yeasts, and molds, are used in industrial fermentations to produce a number of economically important materials. The latter include butanol, acetone, ethanol, vitamins B_2 and B_{12}, lactic acid, amylase, dextran, diacetyl, acetic acid (vinegar), antibiotics, industrial alcohols, beverage alcohols, citric acid, corticosterone, and gibberellin, as well as dairy products such as butter, cheese, and various fermented milks.

However, even though huge quantities of microorganisms are involved in the production of these materials, no information was found on these fermentations as a source of outdoor or indoor air pollution. Further, it has been stated that no industry has been reported to produce a disease in the general population through air pollution by living organisms. It is not valid to present any one set of values for the aerial microbial concentration of a given area, such as a schoolroom or a playground. Any count is influenced by the temperature, meteorological conditions, vegetation, human and animal population, and time of day, as well as by the inability to determine all types of microorganisms by any one sampling procedure.

However, one study has attempted to evaluate the types and number of viable microorganisms present in the air of an urban area such as Minneapolis–St. Paul. Air samples were obtained at four points (35, 70, 170, and 500 feet) along a 500-foot television tower by means of an Anderson sampler. Sampling was performed at intervals over a 6-month period, and wind, rainfall, humidity, and temperature conditions were recorded with each sample. The mean viable counts were as follows: 58 particles/ft^3 (2,047/m^3) at 35 ft; 38.4 particles/ft^3 (1,355/m^3) at 75 ft; 32.7 particles/ft^3 (1,155/m^3) at 170 ft; and 22.4 particles/ft^3 (790/m^3) at 500 ft. The range of all counts observed was 3.5 particles/ft^3 (123/m^3) to 141 particles per ft^3 (4,977/m^3), with no consistent relationships between the counts and any of the meteorological parameters. Regardless of altitude, molds constituted approximately 70% of the total airborne microflora, bacteria between 19 and 26%, and yeast and actinomycetes the remainder. A significant portion of the viable microorganisms in the air were in the particle size range of 3 to 5μ. Microbial counts in nonurban areas are usually relatively lower than in urban areas, and in both areas are influenced primarily by the degree of activity and dust in the immediate area as well as by seasonal and climatic conditions.

Intermedia Aspects

Figure 2 depicts the intermedial relationships for pathogens. Man-controlled intermedia transfers of pathogens include water or air or land transfers. While humans can affect air contamination they do not directly transfer pathogens from air to water of vice-versa. Human fecal matter is a major source of pathogenic organisms in the environment. Infection of humans can occur by direct contact with contaminated fecal matter or indirectly by contact with water polluted by feces. Air contact with pathogens is also possible, although not as probable as other contact mechanisms. Vectors are another route by which humans can come in contact with pathogens. It is possible that pit toilets, cesspools and septic tanks can contaminate water supplies by percolation and leaching when these sources are located near ground or surface waters.

FIGURE 2: INTERMEDIA FLOW CHART—PATHOGENS

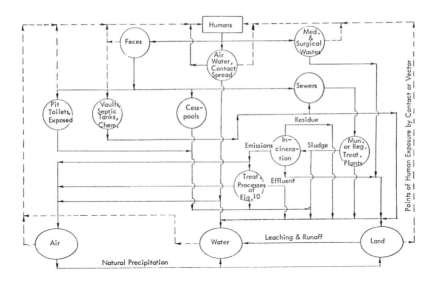

Source: Report EPA 600/5-73-003

Treated sludges from sewage treatment plants contain pathogens and should be treated or disposed in a hygienic manner. Incineration of sludges destroys most pathogens, while unincinerated sludges disposed to the land are potentially capable of contaminating adjacent waters. Exposed sludges may also present a contact source for the various vectors. Aerosols containing pathogens can be formed directly from fecal matter or polluted water. The contaminated aerosols are viable for a short period of time, but the contained pathogens are capable of polluting as the aerosols settle out with natural precipitation.

References

(1) Finkelstein, H., "Air Pollution Aspects of Biological Aerosols (Microorganisms)," *Report PB 188,084,* Springfield, Va., Nat. Tech. Information Service (September 1969).
(2) Stone, R. and Smallwood, H., "Intermedia Aspects of Air and Water Pollution Control," *Report EPA-600/5-73-003,* Washington, D.C., U.S. Environmental Protection Agency (August 1973).

BORON

In Air

The air pollution aspects of boron and its compounds have been reviewed by Durocher (1). The flow of boron in the United States has been traced and charted for the year 1969. The consumption was 85,864 tons, while domestic production totaled 171,361 tons. Imports and exports were 803 and 86,084 tons, respectively. Emissions to the atmosphere during the year were 11,003 tons as shown in Table 28. Nearly 22% of the emissions resulted from the processing of boron compounds, more than 34% from the manufacture and use of various end products, and about 43% from the combustion of coal. Emission estimates for processing and the manufacture of end use products are based on unpublished data obtained from industrial sources.

TABLE 28: EMISSIONS BY SOURCE—1969

Source Category	Source Group	Emissions, Tons		Emissions, %
Mining			100	0.9
Processing			2,400	21.8
End product uses				34.5
	Glass	1,000		
	Ceramic coatings	470		
	Agricultural chemicals	1,800	3,783	
	Soaps and detergents	13		
	Miscellaneous	500		
Other emission sources				42.8
	Coal	4,700	4,720	
	Sewage and sludge	20		
Total			11,003	100.0

Source: Report PB 210,677

The emission factors presented in Table 29 are the best currently available. They were determined through a combination of methods consisting of:

[1] direct observation of emission data and other related plant processing and engineering data;

[2] estimates based on information obtained from literature, plant operators, and others knowledgeable in the field;

[3] calculations based on experience and personal knowledge of metallurgical processing operations; and

[4] specific analytical results where available.

TABLE 29: EMISSION FACTORS

Mining	1 lb/ton boron mined
Processing	28 lb/ton boron processed
End Product Uses of Boron	
Glass Manufacture	70 lb/ton boron processed
Ceramic Coatings	80 lb/ton boron processed
Other Emission Sources	
Coal	18 lb/1,000 tons of coal burned
Sewage and Sludge	55 lb/1,000 tons of sewage and sludge burned

Source: Report PB 210,677

In Water

Despite its widespread industrial use, information on industrial waste levels of boron, and on treatment processes for removal of boron from industrial wastewaters is almost nonexistent in the literature surveyed. Waggott (3) has reported boron levels up to 5.5 mg/liter in raw domestic sewage. A strong correlation between anionic detergent and boron concentrations was observed, indicating that the major source of the boron in this wastewater was sodium perborate, used as bleach in household washing powders.

Fluoborate solutions are sometimes used for plating of cadmium, copper, lead, nickel, tin, zinc and a number of alloys (4). They are used only for special purposes such as deposition of lead on battery bolts and terminals, plating of tin-lead alloys on bearings, or in cases where high-speed deposition is required for continuous wire or strip plating. Lowe has given fluoborate plating solution concentrations (as BF_4) of 125,000 to 265,000 mg/liter (4). This is equivalent to boron concentrations of 15,800 to 33,500 mg/liter.

References

(1)   Durocher, N.L., "Air Pollution Aspects of Boron and Its Compounds," *Report PB 188,085*, Springfield, Va., Nat. Tech. Information Service (September 1969).
(2)   Davis, W.E. and Assoc., "National Inventory of Sources and Emissions, Section II: Boron," *Report PB 210,677*, Springfield, Va., Nat. Tech. Information Service (May 1972).
(3)   Waggott, A., "An Investigation of the Potential Problem of Increasing Boron Concentrations in Rivers and Water Courses," *Water Research*, 3, 749-765 (1969).
(4)   Lowe, W., "The Origin and Characteristics of Toxic Wastes, with Particular Reference to the Metal Industries," *Water Pollution Control*, London, 1970, pp 270-280.

CADMIUM

In Air

The air pollution aspects of cadmium and its compounds have been reviewed by Athanassiadis (1). Some estimates of total cadmium emissions by source are given in Table 29. Emission factors for cadmium from various sources are given in Tables 30 through 35 inclusive as presented by Anderson (2).

TABLE 29: CADMIUM EMISSIONS

Source	Amount in Tons	% This Pollutant
Copper Mining	NEG	NEG
Zinc Mining	<1	0.01
Lead Mining	NEG	NEG
Primary Copper		
Roasting	229	7.59
Reverberatory Furnace	94	3.12
Converters	270	8.95
Material Handling	59	1.96
Primary Zinc		
Roasting	666	22.07
Sintering	284	9.41
Distillation	90	2.98
Material Handling	NEG	NEG
Primary Nickel	NEG	NEG
Primary Lead		
Sintering	66	2.19
Blast Furnace	19	0.63
Reverberatory	3	0.10
Material Handling	NEG	NEG
Secondary Copper		
Sweating Furnace	70	2.32
Blast Furnace	55	1.82
Iron & Steel		
Blast Furnace	1,000	33.14
Non-Ferrous Alloys		
Furnaces	3	0.10
Material Handling	NEG	NEG
Cadmium Paint Pigments	11	0.36
Cadmium-Barium Plastic Stabilizers	3	0.10
Cadmium-Nickel Batteries	<1	0.01
Miscellaneous Cadmium Products	<1	0.02
Use of Pesticides, Herbicides and Fungicides	<1	0.01
Fertilizer Application	<1	0.02
Incinerators	95	3.15
TOTAL	3,018	100.06

Source: The MITRE Corporation, Preliminary Results, EPA Contract No. 68-01-0438

TABLE 30: EMISSION FACTORS FOR CADMIUM FROM INDUSTRIAL SOURCES

Source[a]	Emission factor	Emission factor symbol[b]
Mining of zinc-bearing ore[c]	$0.0005 \ kg/10^3kg$ (0.001 lb/ton) of Zn or $0.1 \ kg/10^3kg$ (0.2 lb/ton) of Cd mined	E
Zinc smelters[d]	$150 \ kg/10^3kg$ (300 lb/ton) of Cd charged or $1.0 \ kg/10^3kg$ (2.0 lb/ton) of Zn produced	PV, MV, Q
Copper smelters	$650 \ kg/10^3kg$ (1300 lb/ton) of Cd charged or $0.4 \ kg/10^3kg$ (0.07 lb/ton) of Cu produced	PV, MV, Q
Lead smelters	$650 \ kg/10^3kg$ (1300 lb/ton) of Cd charged or $0.1 \ kg/10^3kg$ (0.2 lb/ton) of Pb produced	PV, MV, Q
Cadmium refining units	$13 \ kg/10^3kg$ (25 lb/ton) of Cd produced	PV, MV, Q
Secondary copper[c]	$2 \ kg/10^3kg$ (4 lb/ton) of Cu scrap	E
Secondary lead Reverberatory furnace[e]	$0.05 \ kg/10^6kg$ (0.1 lb/10^3 tons) of Pb[i]	ES (3)
Reverberatory furnace[f]	$0.2 \ kg/10^6kg$ (0.4 lb/10^3 tons) of Pb	ES (2)
Steel scrap (galvanized metal)[c]	$0.001 \ kg/10^3kg$ (0.003 lb/ton) of steel	E
Cement plants Dry process Kiln[g]	$0.2 \ kg/10^6kg$ (0.3 lb/10^3 tons) of feed	ES (1)
Raw mill[g]	$0.00005 \ kg/10^6kg$ (0.0001 lb/10^3 tons) of feed	ES (1)
Air separator after raw mill[g]	$0.0005 \ kg/10^6kg$ (0.0009 lb/10^3 tons) of feed	ES (1)
Feed to raw mill[g]	$0.0002 \ kg/10^6kg$ (0.0003 lb/10^3 tons) of feed	ES (1)
Feed to finish mill[g]	$0.003 \ kg/10^6kg$ (0.0005 lb/10^3 tons) of feed	ES (1)
Wet process Kiln[h]	$0.1 \ kg/10^6kg$ (0.2 lb/10^3 tons) of feed	ES (1)
Raw mill[g]	$0.01 \ kg/10^6kg$ (0.02 lb/10^3 tons) of feed	ES (1)
Clinker cooler[g]	$0.00005 \ kg/10^6kg$ (0.0001 lb/10^3 tons) of feed	OES,SSMS (1)

[a] In this table, sources are controlled unless otherwise specified.
[b] Defined in Table 4; numbers in parentheses indicate number of samples analyzed.
[c] Uncontrolled emission factor.
[d] Factors should not be used for electrolytic process.
[e] Exit from a cooling system, three cyclones, and a baghouse (121°C).
[f] Exit from a cooling system, cyclone, manifold, hopper, and baghouse (93°C).
[g] Exit from baghouse.
[h] Exit from electrostatic precipitator. At another plant using an electrostatic precipitator, no cadmium was
    using ES analytical method.
[i] Range, 0.02 to 0.1 $kg/10^6$ kg (0.04 to 0.2 lb/10^3 ton).

Source: Report PB 230,894

TABLE 31: EMISSION FACTORS FOR PROCESSES INVOLVING CADMIUM

Source [a]	Emission factor, kg/10^3kg (lb/ton) of cadmium charged	Emission factor symbol [b]
Pigments [c]	8 (15)	MB
Plastic stabilizers [c]	3 (6)	PV
Alloys and solders [d]	5 (10)	Q
Batteries (Ni-Cd)	1 (2)	Q
Miscellaneous (x-ray screens, cathode ray tubes, nuclear reactor components, etc.)	1 (2)	E

[a] Emission are uncontrolled unless otherwise specified.

[b] Defined in Table 4.

[c] Controlled emissions (bag filters).

[d] Controlled emissions.

Source: Report PB 230,894

TABLE 32: EMISSION FACTORS FOR CONSUMPTIVE USES OF CADMIUM

Source [a]	Emission factors	Emission factor symbol [b]
Rubber tire wear	0.003 kg/10^6km (0.01 lb/10^6 vehicle miles)	E
Fungicides	0.02 kg/10^3 liters (0.05 lb/10^3 gal.) of spray	E
Superphosphate fertilizers	0.0001 kg/10^3kg (0.0002 lb/ton) of fertilizer	E
Motor oil consumption (in vehicle)	0.0006 kg/10^6km (0.002 lb/10^6 vehicle miles)	UK
Cigarettes	16.0 µg/20 cigarettes	S (15)

[a] All sources are uncontrolled.

[b] Defined in Table 4; number in parentheses indicates number of samples analyzed.

Source: Report PB 230,894

TABLE 33: EMISSION FACTORS FOR CADMIUM FROM FUEL COMBUSTION

Source [a]	Cadmium content, ppm	Emission factor	Emission factor symbol [b]
Heating oil (residual assumed)	0.4 to 0.5	0.004 to 0.0005 kg/10^3 liters (0.003 to 0.004 lb/10^3 gal.) of oil	ES (2)
Diesel oil	0.07 to 0.1	0.00007 to 0.00009 kg/10^3 liters (0.0006 to 0.0008 lb/10^3 gal.) of oil	ES (3)
Foreign No. 6 residual fuel oil			
Virgin Islands	5	0.005 kg/10^3 liters (0.04 lb/10^3 gal.) of oil	NA (1)
Virgin Islands (different oil field)	<0.4 [e]	—	CAA (3)
Curacao, N.A.	4	0.004 kg/10^3 liters (0.03 lb/10^3 gal.) of oil	NA (1)
Trinidad	3	0.003 kg/10^3 liters (0.02 lb/10^3 gal.) of oil	NA (1)
Venezuela	<0.4 [e]	—	CAA (10)
Coal, power plant, Kansas [c]	— [f]	0.1 kg/10^6kg (0.2 lb/10^3 tons) of coal burned [g]	EST (2)
Coal, power plant, Michigan [d]	— [e]	0.5 kg/10^6kg (1 lb/10^3 tons) of coal burned [g]	EST (2)

[a] Uncontrolled emissions unless otherwise specified.

[b] Defined in Table 4; numbers in parentheses indicate number of samples analyzed.

[c] Exit from limestone scrubber.

[d] Exit from electrostatic precipitator.

[e] Below detectable limits of techniques.

[f] Not reported.

[g] Based on ash samples.

Source: Report PB 230,894

TABLE 34: EMISSION FACTORS FOR CADMIUM FROM WASTE INCINERATION

Source [a]	Emission factor	Emission factor symbol [b]
Sewage sludge incinerators		
Multiple hearth [c,d]	0.004 kg/10^3kg (0.007 lb/ton) of solid waste incinerated [g]	OES (3)
Fluidized bed [c]	0.0002 kg/10^3kg (0.003 lb/10^3 tons) of solid waste incinerated	OES (2)
Municipal incinerator		
Refuse only [c]	0.4 kg/10^6kg (0.8 lb/10^3 tons) waste incinerated	CAA (3)
Refuse and sludge [c]	0.3 kg/10^6kg (0.6 lb/10^3 tons) waste incinerated	CAA (3)
Overall value for uncontrolled solid waste incineration (municipal) [e]	0.0002 kg/10^3kg (0.003 lb/ton) of solid waste incinerated	E
Lubricating oil [f]	0.0002 kg/10^3 liters (0.002 lb/10^3 gal.) of oil	UK

[a] In this table, emissions are controlled unless otherwise specified.

[b] Defined in Table 4; number in parentheses indicates number of samples analyzed.

[c] Exit from wet scrubber.

[d] Cadmium ratio of metal in particulate to metal in sludge was 2.6.

[e] This emission factor is presented for comparison only; it should not be used for emission estimates.

[f] Emissions of cadmium oxides.

[g] Range, 0.0005 to 0.01 kg/10^3kg (0.001 to 0.02 lb/ton).

Source: Report PB 230,894

In Water

Potential industrial sources of cadmium wastewaters are reported by McKee and Wolf (3) and Santaniello (4) to be: metallurgical alloying, ceramics, electroplating, photography, pigment works, textile printing, chemical industries, and lead mine drainage. Of the total industrial cadmium use, 90% is reported to be utilized in electroplating, pigments, plastics stabilizers, alloying and batteries. Most of the remaining 10% is used for television tube phosphors, golf course fungicides, rubber curing agents and nuclear reactor shields and rods (5).

Considering the industries listed above, only the electroplating industry is discussed to any extent in the literature with regard to cadmium bearing wastes and the treatment thereof.

High concentrations of cadmium in lead mine drainage have been reported, however (3). Cadmium wastewater concentrations reported in the literature are summarized in Table 35. A report on groundwater contamination from plating wastes gives a concentration of 1.2 mg per liter of cadmium in a groundwater recharge basin receiving plant wastes. Private wells in the area reported 0.3 to 0.6 mg per liter, while one test well contained 3.2 mg per liter of cadmium (6).

TABLE 35: CADMIUM CONCENTRATIONS REPORTED FOR INDUSTRIAL WASTE-
WATERS

Process	Cadmium Concentration, mg/l	Reference
Automobile Heating Control Manufacturing	14-22*	7
Plating Rinse Waters Automatic Barrel Zn & Cd Plant Mixed Manual Barrel and Rack	10-15 7-12	8 8
Plating Rinse Waters (large installations)	15 ave, 50 max.	9
Plating Rinse Waters 0.5 gph dragout 2.5 " "	48 240	10 10
Plating Bath	23,000	10
Bright Dip and Passivation Baths	2,000 – 5,000	8
Lead Mine Acid Drainage	1,000	3

* 24 hour composite on days when plating baths were dumped. Cadmium levels much reduced due to dilution when solution dragout was the only contributing source.

Source: Report PB 216,162

Intermedia Aspects

Cadmium is closely related chemically to zinc, and is found with zinc ores in nature. It is obtained as a by-product in the refining of zinc and other metals. Cadmium is emitted to the air and water in mining processes and from metal smelters, especially lead, copper and zinc smelters. Also, industries using cadmium in alloys, paints, and plastics produce cadmium wastes. The burning of oil and scrap metal treatment wastes also contribute to the amount of cadmium entering the air. Cadmium which is emitted to the air is ultimately deposited on the soil and water. High concentrations of cadmium have been found in sewage treatment plant sludges. Soil contamination is possible when these sludges are used as fertilizers. Plants are capable of extracting cadmium from soil and when consumed can be be toxic to man and animals. Certain fertilizers contain cadmium, and the chief route by which cadmium reaches man is believed to be through food grown in soils containing cadmium derived from super-phosphate fertilizers.

References

(1) Athanassiadis, Y.C., "Air Pollution Aspects of Cadmium and Its Compounds," *Report PB 188,086,* Springfield, Va., Nat. Tech. Information Service (September 1969).
(2) Anderson, D., "Emission Factors for Trace Substances," *Report PB 230,894,* Springfield, Va., Nat. Tech. Information Service (December 1973).
(3) McKee, J.E. and Wolf, H.W., "Water Quality Criteria," 2nd ed., California State Water Quality Control Board, *Publication No. 3-A* (1963).
(4) Santaniello, R.M., "Air and Water Pollution Quality Standards, Part 2: Water Quality Criteria and Standards for Industrial Effluents," *Industrial Pollution Control Handbook,* Herbert F. Lund, ed., New York, McGraw-Hill Book Co. (1971).
(5) Anonymous, "Metals Focus Shifts to Cadmium," *Env. Sci. Technol.,* 5, 754-755 (1971).
(6) Lieber, M. and Welsch, W.F., "Contamination of Ground Water by Cadmium," *Jour. Amer. Wat. Works Assoc.,* 46, 541-547 (1954).
(7) Gard, S.M., Snavely, C.A. and Lemon, D.J., "Design and Operation of a Metal Wastes Treatment Plant," *Sew. Ind. Wastes,* 23, 1429-1438 (1951).
(8) Lowe, W., "The Origin and Characteristics of Toxic Wastes with Particular Reference to the Metals Industries," *Water Pollution Control,* London, 1970, pp 270-280.
(9) Pinkerton, H.L., "Waste Disposal: Inorganic Wastes," *Electroplating Engineering Handbook,* 2nd ed., A. Kenneth Graham, ed., New York, Reinhold Publishing Co. (1962).
(10) Nemerow, N.L., *Theories and Practices of Industrial Waste Treatment,* Reading, Mass., Addison Wesley (1963).

CARBON MONOXIDE

In terms of gross amounts, carbon monoxide is the most significant man-made pollutant, with an estimated 147.2 million tons emitted in 1970. Transportation sources account for about 75% of this total, or 111.0 million tons in 1970. Forest fires and other open burning are the second largest contributor, with about 17 million tons, followed by industrial processes and solid waste disposal. In all such calculated estimates of nationwide pollutant emissions the method of calculation is highly important. A different method for calculating CO from automotive exhausts showed an increase from 63.8 to 111.5 million tons from 1968 to 1969. One further complicated estimate claims that 90% of all CO emissions to the global atmosphere, or about 3.5 billion tons per year, are the result of natural processes. Waterways have been cited as CO emitters. Carbon monoxide has been considered a uniquely man-related pollutant, or at least its production from natural sources has been considered relatively unimportant or negligible.

At concentrations of about 1,000 ppm, carbon monoxide is quickly lethal to humans. It kills by oxygen starvation since CO is preferentially chelated by hemoglobin as compared to oxygen. At 100 ppm carbon monoxide induces lassitude, headache and dizziness in humans. The highest concentration of CO recorded at a fixed site in Los Angeles, California was 72 ppm, although instances of concentrations higher than 100 ppm have been found in Los Angeles traffic and in heavy traffic areas of Detroit, Michigan. There is some concern that CO may be a chronic poison, but this position has few adherents in the United States.

There is no evidence for chronic carbon monoxide damage to vegetation, materials, animals, or aesthetics at normal pollution levels. At high concentrations carbon monoxide is known to participate in synergistic toxic reactions with several other gaseous pollutants. However, at CO concentrations generally encountered in the atmosphere, no such synergisms have been established.

In Air

Carbon monoxide is one of the products produced by the incomplete combustion of carbonaceous material; it is formed, for instance, when carbonaceous material is burned in a reducing atmosphere, in which the available oxygen is not sufficient to burn the material completely to carbon dioxide. Because such conditions exist in the cylinders of the gasoline internal combustion engine, and because of the large number of automobiles in use, CO emissions from this source conspicuously exceed those from any combination of other sources. Table 36 is a summary of estimated annual emissions of CO within the United States during 1968. No figures are available for a few operations, such as certain metallurgical operations and the use of explosives. Table 36 shows that the sum of the CO emissions was approximately 100 million tons.

TABLE 36: SUMMARY OF ESTIMATED CARBON MONOXIDE EMISSIONS IN THE UNITED STATES DURING 1968 (tons/year)

Source	Emissions
Mobile fuel combustion	
Motor vehicles	
Gasoline	59,000,000
Diesel	160,000
Aircraft[a]	2,400,000
Railroad	120,000
Vessels	310,000
Non-highway users	1,800,000
Stationary fuel combustion	
Coal	770,000
Fuel oil	50,000
Natural gas	3,000
Wood	1,010,000
Solid waste	
Incineration	800,000
Open burning	3,400,000
Conical burners	3,600,000
Coal-refuse fires	1,200,000
Structural fires	250,000
Forest fires	
Wild fires	4,740,000
Prescribed burning	2,480,000
Agricultural burning	8,250,000
Industrial processes	
Blast-furnace sinter plants	2,400,000
Gray-iron cupolas	3,600,000
Basic oxygen furnaces	100,000
Beehive coke ovens	20,000
Kraft recovery furnaces, lime kilns	830,000
Carbon black	350,000
Petroleum catalytic cracking units	2,200,000
Fluid coking burners	160,000
Methanol	4,000
Formaldehyde	34,000
Subtotal	100,041,000
Ammonia	b
Metallurgical electric furnaces	b
Zinc and lead reduction	b
Aluminum reduction	b
Calcium carbide furnaces	b
Silicon carbide furnaces	b
Phosphorus furnaces	b
Explosions (blasting, etc.)	b

[a]This includes emissions during cruising.
[b]Although these sources are thought to be emitters of CO, no data are available and no emission factors have been developed.

Source: EPA Report AP-65

Large amounts of CO are produced and handled by industry; but, in most cases, this is used as fuel or raw material, and emissions result only from leaks or abnormal operation. Total production of CO by pig-iron blast furnaces, for instance, was estimated to be about 90 million tons in 1968, even more than that generated by the gasoline internal combustion engine, but only a small fraction of this CO escapes from the blast furnace operations. Almost 60% of the CO emissions are due to the motor vehicle; it is also interesting to note the relatively large contribution that still arises from the use of wood as fuel. The small values of CO emissions indicated for burning natural gas must not convey the impression that complete combustion necessarily takes place when gas is burned; actually, copious quantities of CO can be formed if this fuel is burned with too little air. Table 37 is a compilation of available emission factors for CO from various types of sources.

TABLE 37: CARBON MONOXIDE EMISSION FACTORS

Source	Emission factor[a]
Stationary fuel combustion	
Coal	
Less than 10 x 10⁶ Btu/hr capacity	50 lb/ton of coal burned
10 to 100 x 10⁶ Btu/hr capacity	3 lb/ton of coal burned
Greater than 100 x 10⁶ Btu/hr capacity	0.5 lb/ton of coal burned
Fuel oil	
Less than 100 x 10⁶ Btu/hr capacity	2 lb/1,000 gal of oil burned
More than 100 x 10⁶ Btu/hr capacity	0.04 lb/1,000 gal of oil burned
Natural gas	
Less than 100 x 10⁶ Btu/hr capacity	0.4 lb/10⁶ ft³ of gas burned
More than 100 x 10⁶ Btu/hr capacity	Negligible lb/10⁶ ft³ of gas burned
Wood	30–65 lb/ton of wood burned
Solid waste disposal	
Open burning onsite of leaves, brush, paper, etc.	60 lb/ton of waste burned
Open-burning dump	85 lb/ton of waste burned
Municipal incinerator	1 lb/ton of waste burned[d]
Commercial and industrial multiple chamber incinerator	10 lb/ton of waste burned
Commercial and industrial single chamber incinerator	44 lb/ton of waste burned
Flue-fed incinerator	27 lb/ton of waste burned
Domestic incinerator	200 lb/ton of waste burned
Process industries (specific examples)	
Gray iron foundry	
Cupola	
Uncontrolled	250 lb/ton of charge
Controlled with afterburner	8 lb/ton of charge
Iron and steel manufacture	
Blast furnace	1,700–250 lb/ton of pig iron produced[c]
Basic oxygen furnace	124–152 lb/ton of steel produced[d]
Petroleum refinery	
Fluid catalytic unit	13,700 lb/1,000 bbl fresh feed[e]
Moving-bed catalytic cracking unit	3,800 lb/1,000 bbl fresh feed[e]

[a]These emissions are from uncontrolled sources, unless otherwise noted.
[b]This represents excellent design and operation.
[c]Practically all of the CO is burned for heating purposes or in waste gas flares.
[d]Gases emitted from a basic oxygen furnace during the blowing period contain 87 percent CO. After ignition of the gases above the furnace, the CO amounts to 0.0–0.3 percent.
[e]These emissions are completely controlled when CO waste heat boilers are utilized.

Source: EPA Report AP-65

These emission rates represent uncontrolled sources, unless otherwise noted. These emission factors apply best to areas, rather than to specific sources; actual measurements are preferable when calculations are made for specific sources, but emission factors can be used when data are lacking. For a specific source where control equipment is utilized, the listed uncontrolled process source emission rates must be multiplied by 1 minus the percent efficiency of the equipment expressed in hundredths. There are of course, many other process sources of CO emissions. Emission factors for other miscellaneous process sources can be found through literature searches, stack-gas testing, and material balances on the process in question. It must be remembered that the emission factors listed herein are average values and can vary, depending on operating conditions and other factors.

Intermedia Aspects

Because of its low water solubility and low hygroscopicity carbon monoxide is not easily transferred to water or land by physical means (rainwash). Unless oxidized to CO_2, it remains essentially a serious air pollution problem. In spite of the great quantities of man-made CO emitted, if it were all oxidized to CO_2 the result would be only about 1% of all man-made CO_2 emissions. This oxidation pathway is, therefore, insignificant for CO in terms of the intermedia implications of CO_2. Furthermore, the nature of treatment methods for carbon monoxide is such that intermedia transfer from air to water is essentially precluded (Figure 3). As CO from man-related sources results solely from incomplete combustion, which is inherently an air-polluting process, there is very little water pollution by CO. Carbonates discharged to water are reducible to carbon monoxide only with extreme difficulty.

FIGURE 3: INTERMEDIA FLOW CHART—CARBON MONOXIDE

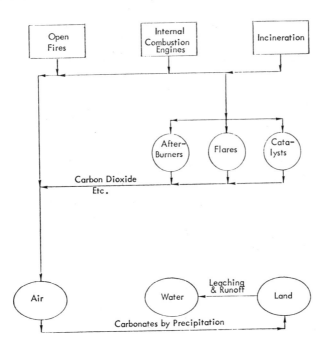

Source: Report EPA 600/5-73-003

References

(1)   U.S. Dept. of Health, Education and Welfare, "Control Techniques for Carbon Monoxide Emissions from Stationery Sources," *Publ. AP-65,* Washington, D.C., Nat. Air Pollution Control Administration (March 1970).
(2)   Stone, R. and Smallwood, S., "Intermedia Aspects of Air and Water Pollution Control," *Report EPA-600/5-73-003,* Washington, D.C., U.S. Environmental Protection Agency (August 1973).

CHLORIDES

In Water

Chlorides may be a problem in such diverse areas as sewage treatment plants and irrigation water. In sewage treatment plants, high chloride concentrations may interfere with plant operation, especially with the activated sludge process. Most agricultural crops will be adversely affected by high salinity before they are affected by chloride per se; however, some fruit crops are harmed by very low concentrations of chlorides. Some crops which are not necessarily damaged by high chloride concentration are damaged by high salt concentration.

Chlorides are present in practically all waters. They may be of natural origin, or derived [1] from sea water contamination of underground supplies; [2] from salts spread on fields for agricultural purposes; [3] from human or animal sewage; or [4] from industrial effluents, such as those from food processing, paper works, galvanizing plants, water softening plants, oil wells, and petroleum refineries (1). Herbert and Berger (2) have reported waste chloride concentrations from a Kraft process paper mill of 350 to 1,760 mg per liter.

An industrial waste study on blast furnaces and steel mills (3) reveals that average size steel mills using advanced production technology produce from their pickling line 9,700 pounds per day of chloride as the ferrous salt, and 4,500 pounds per day as hydrochloric acid. The waste stream from the pickling line contains a total chloride content of 7,000 mg per liter. This is diluted to as little as 15 mg per liter chloride upon mixing of the pickling line waste stream (approximately 0.24 MGD) with the waste flow from other steel processes (approximately 113 MGD).

Oil refineries and petrochemical processes are major producers of high chloride wastes. Crude oil desalting is a significant source of chloride in petroleum refinery wastes. Berger (4) describes a refinery effluent containing total dissolved solids of 1,163 mg per liter, primarily as sodium chloride dissolved out of crude oil by desalting. Petrochemical processes, including chlorination, oxidation, polymerization and alkylation produced chloride wastes, usually associated with calcium or aluminum (5).

The Solvay process, widely used in making soda ash, yields wastewaters high in chloride. Soda ash is a major raw material of many industries, including glass, chemicals, pulp and paper, soap and detergent, aluminum and water treatment. The Solvay process uses sodium chloride brine solution, and yields a substantial amount of waste calcium chloride (6). Jones has reported that the chloride content of a typical soda ash manufacturing waste is 90,000 mg per liter (7). While some calcium chloride is recovered by distillation, this practice is not sufficiently prevalent to prevent the Solvay process having a major waste disposal problem (6).

Food processing industry wastewaters may also be high in chloride content. Burr and Byrne (8) have described a 400,000 gallon per day waste flow from an olive processing facility in California, which contained 3,500 to 6,000 mg per liter of chloride. The chloride represents 35 to 40% of the total dissolved solids concentration of that waste. Another report (9) has cited chloride levels of 36 to 4,280 mg per liter of chloride from olive processing, and 24,220 mg per liter of chloride from cucumber pickling.

Frequently, chloride ion is added to a wastewater as a result of waste treatment required to remove some less desirable constituent. Sodium chloride and hydrochloric acid are often used to regenerate spent ion exchange resins used in water purification. Chloride ions on the regenerated resins then exchange with undesirable anions as the waste water passes through the ion exchange bed. Another example of chloride addition is chlorination treatment of cyanide waste streams.

References

(1) McKee, J.E. and Wolf, H.W., "Water Quality Criteria," California State Water Quality Control Board, *Publication No. 3-A* (1963).
(2) Herbert, A.J. and Berger, H.F., "A Kraft Bleach Waste Color Reduction Process Integrated with the Recovery System," *Proc. 15th Purdue Industrial Waste Conf.,* 15, 49-57 (1962).
(3) Anonymous, *The Cost of Clean Water, Vol. III, Industrial Waste Profiles, No. I - Blast Furnaces and Steel Mills,* Washington, D.C., U.S. Dept. Interior (1967).
(4) Berger, M., "The Disposal of Liquid and Solid Effluents from Oil Refineries," *Proc. 21st Purdue Industrial Waste Conf.,* 21, 759-767 (1966).
(5) Gloyna, E.F. and Ford, D.L., *The Characteristics and Pollutional Problems Associated with Petrochemical Wastes—Summary,* Washington, D.C., U.S. Dept. Interior (1970).
(6) Anonymous, *The Economics of Clean Water, Vol. III, Inorganic Chemistry Industry Profile,* Washington, D.C., U.S. Dept. Interior (1970).
(7) Jones, M.W., "Construction and Operation of a Chloride Holding Basin," *Proc. 16th Purdue Industrial Waste Conf.,* 16, 186-192 (1961).
(8) Burr, D. and Byrne, P.J., "City and Industry Cooperate to Solve Brine and Waste Problems," *Public Works,* January 1971, 46-48.
(9) Anonymous, "Reduction of Salt Content of Food Processing Liquid Waste Effluent," *U.S. EPA Report 12060 DXL 01/71* (1971).

CHLORINE

In Air

Some estimates of total chlorine emissions by source are presented in Table 38.

TABLE 38: CHLORINE EMISSIONS

Source	Amount in Tons	Pollutant, %
Chlorine Fluxing		
Non-Ferrous Metals	100	0.13
Iron and Steel	1,900	2.43
Bleaching, Pulp and Paper	18,000	23.02
Chlorine Industry		
Manufacture	4,000	5.12
Liquefaction and Handling	43,000	54.99
Organic Chlorine Chemicals	8,500	10.87
Hydrochloric Acid Manufacture	800	1.02
Bleach Manufacture	900	1.15
Miscellaneous Chlorine Products	1,000	1.28
TOTAL	78,200	100.01

Source: The MITRE Corporation, Preliminary Results, EPA Contract No. 68-01-0438

Natural Occurrence: Natural occurrences of free chlorine gas are extremely rare, due to the high reactivity of chlorine with many substances. Volcanic gases contain very small amounts of chlorine gas. Low concentrations of chlorine may, however, be formed by atmospheric reactions. For example, chloride compounds and nitrogen dioxide may react to form nitrosyl chloride, which can decompose photochemically to yield free chlorine. Also, chlorides in the presence of strong oxidants (such as ozone) may be oxidized to chlorine.

Production Sources: The major processes for the production of chlorine in the United States are the electrolysis of aqueous alkali chloride via the diaphragm cell or mercury cell, and to a lesser extent, fusion electrolysis of alkali chlorides (Down Process) and the nitric acid process. Diaphragm and mercury cell processes account for over 95% of the chlorine produced. The major sources of atmospheric emissions of chlorine from the production processes are the following: liquefaction process, the filling of containers or transfer of liquid chlorine from one container to another, the cleaning of returned tank cars containing residual chlorine, the improper treating of spent brine solution and of sniff or blow gas (the gas remaining after the final liquefaction step), and occasional equipment failure (for example, a chlorine compressor breakdown).

Product Sources: The largest consumer of chlorine is the chemical industry which uses this substance for the manufacture or preparation of chemical compounds, both organic and inorganic. Nearly 80% of the total chlorine production of this country is consumed in this way with the organic compounds accounting for the greater part, approximately 70% of the total production. The second largest user is the pulp and paper industry (which consumes approximately 16% of total production), followed by water and sewage treatment (approximately 4% of total production).

References

(1) Stahl, Q.R., "Air Pollution Aspects of Chlorine Gas," *Report PB 188,087,* Springfield, Va., Nat. Tech. Information Service (September 1969).
(2) Processes Research, Inc., "Air Pollution from Chlorination Processes," *Report PB 218,048,* Springfield, Va., Nat. Tech. Information Service (March 1972).
(3) U.S. Environmental Protection Agency, "Atmospheric Emissions from Chlor-Alkali Manufacture," *Publ. AP-80,* Research Triangle Park, N.C., Air Pollution Control Office (January 1971).

CHROMIUM

In Air

Air pollution aspects of chromium and its compounds have been reviewed by Sullivan (1). Sources and estimates of chromium-containing emissions as of 1970 have been presented by GCA Corporation (2) as shown in Table 39.

The physical and chemical properties of the emitted particles containing chromium are the direct results of the processes causing the emission, the types of feed materials, and the characteristics of chromium and its compounds. Some of the latter are summarized in Table 40.

Of those listed, the boiling (vaporization) point for chromium of 2665°C may be the most important, in that temperatures below this point may be expected to produce relatively little chromium or chromate fume, while higher temperatures can produce substantial amounts of fume particulate. With this brief background, the emissions from the principal sources will be examined in turn.

TABLE 39: SOURCES AND ESTIMATES OF CHROMIUM-CONTAINING EMISSIONS IN 1970

Source	Uncontrolled Emission Factor		Reliability Code	Production Level (tons/yr)	Percent Cr in Emissions	Reliability Code	Emissions of Cr Before Controls (tons/yr)	Estimated Level of Emission Control	Emissions of Cr After Controls (tons/yr)
	lb/ton	kg/10³ kg							
Mining									
None in U.S.A.							0		0
Refining									
Ferrochromium									
Electric furnace	(200-830) 500ᵃ	(100-415) 250ᵃ	C	375,500ᵉ	22	B	20,600	40%	12,360
Material handling	10	5	B	375,500	65	C	1,220	32%	830
Electrolytic chromium	0.048	0.024	C	9,000	51	A	neg	95%	neg
Refractory									
Noncast	150	75	C	60,300	b		4,500	64%	1,630
Electric cast	225	112	C	6,700	b		754	77%	173
Chemical Processing									
Dichromate	30	15	C	61,000	b		920	90%	92
Other chemicals	–	–	–	–	–	–	–	–	24
Steel and Alloys									
Chromium steels	25	12	C	189,000	b		2,362	78%	520
Cast iron	75	38	C	5,000	b		188	99%	2
Super alloys and alloys	25	12	C	12,000	b		150	78%	33
General steel making	N.A.	N.A.		N.A.	N.A.		N.A.	N.A.	100ᶜ
Inadvertent Sources									
Coal combustion	N.A.	N.A.		33,800,000ᵉ	0.026	A	8,700	82%	1,564
Oil combustion	N.A.	N.A.		287,000ᵈ	0.13	B	370	0	370
Cement production	N.A.	N.A.		934,000ᵈ	0.03	C	N.A.	N.A.	280
Incineration	N.A.	N.A.		931,000ᵈ	0.017	C	N.A.	N.A.	158
Asbestos	N.A.	N.A.		6,579	0.15	C	10	99%	0
Total							39,774	54%	18,136

ᵃIntermediate value
ᵇEmission factor multiplier equal to tons of Cr processed or handled annually.
ᶜNeeds further investigation.
ᵈEmissions, after control.
ᵉEmissions, before control.

NOTE: See page 7 for explanation of Reliability Code.

Source: Report PB 230,034

TABLE 40: SOME PHYSICAL PROPERTIES OF CHROMIUM AND ITS COMPOUNDS

Melting point	1875°C
Boiling point	2665°C (for Cr_2O_3: 4000°C)
Density	7.2 g/cm³
Atomic weight	52.0 awu
Heat of vaporization	73.0 kg cal/g-atom
Mineral hardness of chromite	5.5 (std. minerology scale)

Source: Report PB 230,034

Ferrochromium Production Emissions: Emissions from high-carbon ferrochrome furnaces are reported as follows: 1.07 gr/SCF, before scrubber control at 98.2% efficiency; approximately 1.0 μm diameter particles containing 29.3% Cr_2O_3; dust collected by an electrostatic precipitator containing 4% Cr, and 15% Mg; an unspecified type of ferrochrome furnace emitting particulate varying from 1 to 2 μm in diameter (Control by electrostatic precipitator did not alter the relative concentration of Cr and other metals in the dust, which included 14% Cr and 8.9% Mg.); a maximum of 1.0 μm in diameter, with most particles from 0.1 to 0.4 μm – amorphous, spherical particles, including traces of spinel and quartz; 21% SiO_2, 11% FeO, 15% MgO, 29.3% Cr_2O_3, etc.; fume resistivity – 9.4 x 10^{10} ohm-cm, 200 to 300°F, or with conditioning of gas to 20% moisture, 21 x 10^{10} ohm-cm.

Emissions from ferrochrome silicon furnaces are reported as follows: 1.43 gr/SCF before scrubber control at 92.6% efficiency; 0.245 gr/SCF from a hooded furnace without control equipment. This particulate is described as noncrystalline fused silica and impurities, the fume containing 1.3% Cr and 6.8% Mg, 0.65 μm mass median diameter.

Refractory Production Emissions: No useful description of these was found. The following estimates are made. Most of the emissions are from grinding, screening, mixing, and drying. All except grinding are relatively low-intensity operations which are therefore expected to generate particles and agglomerates roughly 5 μm in diameter and larger, having relatively short distances of travel before disposition. Grinding material having the moderate hardness of chromite is expected to generate moderately small rough-shaped particles, perhaps 1 μm in diameter and larger, of which the smallest may travel distances up to a few miles before disposition. Most of the Cr emitted will be in the chromite form, i.e., $FeO \cdot Cr_2O_3$.

Chrome Steel Production Emissions: Emission characteristics are known to be highly dependent on the material charged to the furnace. Probably some particles consisting almost entirely of Cr_2O_3 will be emitted, while other particles will contain Fe_2O_3 and FeO up to 50%, plus numerous other metallic oxides. Most particles will be spherical, and submicrometer in size, although some will exceed 1 μm in diameter.

Coal Combustion Emissions: Analyses of flyash give average particle diameters from 2 to 30 μm (mass median diameter) depending on the type of feed to the furnace. From 1 to 10% of the particle mass may be submicrometer. The concentration of Cr in flyash as a function of particle size is reported: particles less than 1.7 μm, 0.4% Cr; particles 1.7 to 4.1 μm, 0.05% Cr; particles 4.1 to 30 μm in six increments and over 30 μm, 0.02% Cr. This indication that Cr is concentrated in the smaller particles is not explained by the vaporization temperature of Cr, which is well above the combustion temperature of coal (about 1500°C).

Oil Combustion Emissions: Most particles are between 0.01 and 1 μm in diameter, depending on the atomization and combustion processes. Thus, these particles tend to travel long distances (tens to hundreds of miles) before being removed from the atmosphere by natural processes (settling, washout, agglomeration). The chemical form of the Cr contained in these emissions is not stated, although one report gives "Cr as CrO_2" data of 0.06 and 0.3% of ash, indicating that this may be the chemical form.

Cement Production Emissions: Most emissions come from the kilns from which 5% of the material may be submicrometer. Up to 50% of the particulate can be CaO; silicon, iron, and aluminum oxides represent a large portion of the remainder.

Ambient Chromium and Toxicity: Analyses of the air in and near Cincinnati is reported to contain 0.28 and 0.31 micrograms of chromium per cubic meter, in particles of average size 1.5 and 1.9 micrometers (mass median diameter). Of the particles containing chromium, 45 and 74% were greater than 1 micrometer, somewhat larger than would be predicted from the preceding discussions. Although not a principal source of emissions, chromium plating releases chromic acid particles in the form of chromium trioxide which is said to be poisonous and injurious to the kidneys. Dust from cement plants is said to cause skin problems with some workers, due to the chromium contained in the dust. Chromium-containing chemicals are used as preservatives of wood, slimicides, and herbicides, indicating a potential toxicity to organisms in at least some chemical forms.

In Water

Chromium occurs in nature as both the trivalent (Cr^{+3}) and the hexavalent (Cr^{+6}) ion. Hexavalent chromium present in industrial wastes is primarily in the form of chromate ($CrO_4^=$) and dichromate ($Cr_2O_7^=$). Chromium compounds are added to cooling water to inhibit corrosion. They are employed in manufacture of paint pigments, in chrome tanning, aluminum anodizing and other metal cleaning, plating and electroplating operations. In the metal plating industry, automobile parts manufacturers are one of the largest producers of chromium-plated metal parts. Frequently, the major source of waste chromium is the chromic acid used in such metal plating operations. Table 41 is a compilation of typical sources and concentrations of chromate wastewaters.

TABLE 41: HEXAVALENT CHROMIUM WASTEWATER SOURCES AND TYPICAL CONCENTRATIONS

Industrial Source	Chromium (VI) Concentration mg/l	
	Average	Range
Leather Tanning	40	
Wood Preserving		0.23 - 1.5
Cooling Tower Blowdown	31.4	
Cooling Tower Blowdown		8 - 10.7
Cooling Tower Blowdown		10 - 60
Bright Dip Rinse		1 - 6
Bright Dip Bath		10,000 - 50,000
Bright Dip Bath		20,000 - 75,000
Bright Dip Bath		200 - 600
Anodizing Bath	173	
Anodizing Bath		15,000 - 52,000
Anodizing Rinse	49	
Anodizing Rinse		30 - 100
Plating	1,300	
Plating	600	
Plating		100,000 - 270,000
Plating		60 - 80
Electroplating	140	
Electroplating	41	15 - 70

Source: Report PB 216,162

Hexavalent chromium treatment frequently involves reduction to the trivalent (Cr^{+3}) form prior to removing the chromium from the industrial waste. Thus trivalent chromium in industrial waste may result from one step of the waste treatment process itself, that of chemical reduction of hexavalent chromium. Industries which employ trivalent chromium directly in manufacturing processes include glass, ceramics, photography and textile dyeing.

In addition to the predominant hexavalent form, some few mg/l of trivalent chromium may be encountered in plating wastes, even prior to chemical reduction of Cr^{+6}. For example, Anderson and Iobst (4) report for one process waste a hexavalent chromium concentration ranging from 0.0 to 18.0 mg/l and a total chromium concentration of 2.9 of 31.8 mg/l. By difference, their data indicate a trivalent chromium concentration in this one waste stream of 2.9 to 13.9 mg/l. These same authors report for an acid bath waste stream a hexavalent chromium content of 122 to 270 mg/l and trivalent levels of 37 to 282 mg/l.

Thus, according to these authors, trivalent chromium represents a major component of this acid waste, even before reduction of hexavalent chromium. After reduction, the trivalent content of this latter waste stream would be approximately 160 to 550 mg/l. Evaluation of similar plating rinse data presented by Germain, et al (5) indicates trivalent chromium levels of 2.5 to 15 mg/l prior to treatment.

References

(1) Sullivan, R.J., "Air Pollution Aspects of Chromium and Its Compounds," *Report PB 188,075,* Springfield, Va., Nat. Tech. Information Service (September 1969).
(2) GCA Corp., "National Emissions Inventory of Sources and Emissions of Chromium," *Report PB 230,034,* Springfield, Va., Nat. Tech. Information Service (May 1973).
(3) Patterson, J.W. and Minear, R.A., "Wastewater Treatment Technology," 2nd Edition, *Report PB 216,162,* Springfield, Va., Nat. Tech. Information Service (February 1973).
(4) Anderson, J.S. and Iobst, Jr., E.H., "Case History of Wastewater Treatment in a General Electric Appliance Plant," *Jour. Wat. Poll. Control Fed.,* 10, 1786-1795 (1968).
(5) Germain, J.E., Vath, C.A. and Griffin, C.F., "Solving Complex Waste Disposal Problems in the Metal Finishing Industry," presented at Georgia Wat. Poll. Control Assoc. Meeting, September 1968.

COPPER

The flow of copper in the United States has been traced and charted for the year 1969. The consumption was 3,058,000 tons, while primary and secondary production totaled 3,118,000 tons. Imports and exports were 131,000 tons and 200,000 tons, respectively.

In Air

Emissions to the atmosphere during the year were 13,680 tons as shown in Table 42.

TABLE 42: EMISSIONS BY SOURCE—1969

Source Category	Source Group		Emissions - Tons	Emissions, %
Mining and milling			190	1.4
Metallurgical processing			8,700	63.6
Secondary production			210	1.5
Metal fabrication			neg	
End product uses			230	1.7
	Miscellaneous	230		
Other emission sources			4,350	31.8
	Coal	1,030		
	Oil	50		
	Iron and steel	2,760		
	Foundries	50		
	Incineration	460		
Total			13,680	100.0

Source: Report PB 210,678

About 64% of the emissions resulted from the metallurgical processing of primary copper, and about 20% from the production of iron and steel. The combustion of coal was the only other significant emission source. Emission estimates for mining, production of primary and secondary copper, and reprocessing operations are based on data obtained by personal contact with processing and reprocessing companies.

The emission factors presented in Table 43 are the best currently available. They were determined through a combination of methods consisting of: [1] direct observation of emission data and other related plant processing and engineering data; [2] estimates based on information obtained from literature, plant operators, and others knowledgeable in the field; [3] calculations based on experience and personal knowledge of metallurgical processing operations; and [4] specific analytical results where available.

TABLE 43: EMISSION FACTORS

Mining and milling	200 lb/1,000 tons of copper mined
Metallurgical processing	10 lb/ton of copper produced
Secondary production	300 lb/1,000 tons of copper produced
Metal fabrication	1 lb/1,000 tons of copper fabricated
Other emission sources	
Coal	4 lb/1,000 tons of coal burned
Oil	160 lb/million bbls of oil burned
Blast furnaces	22 lb/1,000 tons of pig iron produced
Open-hearth furnaces	51 lb/1,000 tons of steel produced
Basic oxygen furnaces	2 lb/1,000 tons of steel produced
Electric furnaces	7 lb/1,000 tons of steel produced
Foundries	5 lb/1,000 tons of gray iron produced

Note: The emission factors shown above for mining and milling, metallurgical processing, secondary production, and metal fabrication are based on the averages reported by industry. The degree of accuracy is judged to be plus or minus 30%.

Source: Report PB 210,678

In Water

Primary sources of copper in industrial waste streams are metal process pickling baths and plating baths. Brass and copper metal working requires periodic oxide removal by immersing the metal in strong acid baths. Solution adhering to the metal surface, referred to as drag-out, is rinsed from the metal and contaminates the waste rinse water. Similarly, metal parts undergoing copper or brass plating drag out some of the concentrated plating solution. Copper concentrations in the plating bath depend upon the bath type, and may range from 3,000 to 50,000 mg/l.

For a given bath, the rinse water concentration will be a function of many factors including drainage time over the bath, shape of the parts and their total surface area and the rate of rinse water flow. One author (2) has suggested a typical rinse water level would be 1% of the process bath concentration. Untreated process wastewater concentrations of copper typical of plating and metal processing operations are summarized in Table 43 from Reference (3). Jewelry manufacturers employ copper plating either directly or as a base metal for silver and other precious metal surfaces. Copper is also employed in the alkaline Bemberg rayon process, as cupro-ammonium salts. Copper bearing acid mine drainage also contributes significant quantities of dissolved copper to waste streams. Hill (4) has reported copper concentrations of 3.2 and 3.9 mg/l for two mines, while another discharge contained only 0.12 mg/l.

Wixson, et al (5), have cited mine waters, mine tailings ponds and acid mine drainage (primarily with respect to lead mining) as possible sources of copper. Teer and Russel (6) have reported wood preserving to generate copper containing wastes. Dean, et al (7), indicated additional potential sources of copper bearing wastes; which included pulp and paper mills, paper and paper board mills, fertilizer manufacturing, petroleum refining, basic steel works and foundries, nonferrous metal works and foundries, and motor vehicle and aircraft plating and finishing.

TABLE 43: CONCENTRATIONS OF COPPER IN INDUSTRIAL PROCESS WASTEWATERS

Process	Copper Concentration, mg/l
Plating wash	20 - 120
Plating wash	0 - 7.9
Brass dip	2 - 6
Brass mill rinse	4.4 - 8.5
Copper mill rinse	19 - 74
Metal processing	204 - 370
Brass mill wash	
Tube mill	74
Rod and wire mill	888
Brass mill bichromate pickle	
Tube mill	13.1
Rod and wire mill	27.4
Rolling mill	12.2
Copper rinse	13 - 74
Brass mill rinse	4.5
Appliance manufacturing	
Spent acids	0.6 - 11.0
Alkaline wastes	0 - 1.0
Typical large plater	
Rinse waters	up to 100 (20 avg)
Plating operations	6.4 - 88
Automobile heater production	24 - 33 (28 avg)
Silver plating	
Silver bearing	3 - 900 (12 avg)
Acid wastes	30 - 590 (135 avg)
Alkaline wastes	3.2 - 19 (6.1 avg)
Brass industry	
Pickling bath wastes	4.0 - 23
Bright dip wastes	7.0 - 44
Business machine corp.	
Plating wastes	2.8 - 7.8 (4.5 avg)
Pickling wastes	0.4 - 2.2 (1.0 avg)
Copper plating rinse water	5.2 - 41
Copper tube mill waste	70 (avg)
Copper wire mill waste	800 (avg)
Plating plants	2.0 - 36.0
Brass and copper wire mill	75 - 124
Copper and brass pickle	60 - 69
Copper and brass bright dip	20 - 35
Plating	20 - 30
Plating	10 - 15
Plating	3 - 8
Large plater	11.4

Source: Report PB 216,162

References

(1) Davis, W.E. and Assoc., "National Inventory of Sources and Emissions, Section III: Copper," *Report PB 210,678,* Springfield, Va., Nat. Tech. Information Service (May 1972).

(2) Golomb, A., "Application of Reverse Osmosis to Electroplating Waste Water Treatment, Part II: The Potential of Reverse Osmosis in the Treatment of Some Plating Wastes," *Plating,* 59, 316-319 (1972).

(3) Patterson, J.W. and Minear, J.A., "Wastewater Treatment Technology," 2nd Ed., *Report PB 216,162,* Springfield, Va., Nat. Tech. Information Service (February 1973).

(4) Hill, R.D., "Control and Prevention of Mine Drainage," presented at the Environmental Resources Conference on Cycling and Control of Metals, Battelle Columbus Laboratories (1972).

(5) Wixson, B.G., Bolter, E., Gale, N.L., Jennett, J.C., and Purushothaman, K., "The Lead Industry as a Source of Trace Metals in the Environment," presented at the Environmental Resources Conference on Cycling and Control of Metals, Battelle Columbus Laboratories (1972).

(6) Tear, E.H. and Russell, L.V., "Heavy Metals Removal from Wood Preserving Wastewater," presented at the 27th Industrial Waste Conference, Purdue University (1972).

(7) Dean, J.G., Bosqui, F.L. and Lanouette, K.H., "Removing Heavy Metals from Waste Water," *Env. Sci. Technol.,* 6, 518-522 (1972).

CYANIDES

In Water

Cyanide, as sodium cyanide (NaCN) or hydrocyanic acid (HCN) is a widely used industrial material. Cyanide waste streams result from ore extracting and mining, photographic processing, coke furnaces, synthetics manufacturing, case hardening and pickling of steel, and industrial gas scrubbing. A major source of waste cyanide is the electroplating industry. Electroplaters use cyanide baths to hold metallic ions such as zinc and cadmium in solution. Drag-over of the plating solution, containing cyanide ions and metal-cyanide complexes, contaminates rinsing baths. Table 44 presents typical cyanide levels reported for plating wastewaters (1).

TABLE 44: CONCENTRATIONS OF CYANIDE IN PLATING WASTEWATERS

Process	Average, mg/l	Range, mg/l
Plating rinse	2	0.3 - 4
Plating rinse	700	
Plating rinse		10 - 25
Plating rinse	32.5	
Plating rinse	25	
Plating rinse		60 - 80
Plating rinse		30 - 50
Plating rinse	55.6	1.4 - 256
Bright Dip		15 - 20
General (separate cyanide)	72	9 - 115
General (combined stream)	28	1 - 103
Alkaline cleaning bath		4,000 - 8,000
Plating bath	30,000	
Plating bath		45,000 - 100,000
Plating bath		
Brass		16,000 - 48,000
Bronze		40,000 - 50,000
Cadmium		20,000 - 67,000
Copper		15,000 - 52,000

(continued)

TABLE 44: (continued)

Process	Average, mg/l	Range, mg/l
Silver		12,000 - 60,000
Tin - Zinc		40,000 - 50,000
Zinc		4,000 - 64.000

Source: Report PB 216,162

Reference

(1)   Patterson, J.W. and Minear, R.A., "Wastewater Treatment Technology," 2nd Ed., *Report PB 216,162*, Springfield, Va., Nat. Tech. Information Service (February 1973).

ETHYLENE

In Air

Ethylene in high dilution causes injury to leaves of sensitive plants. As little as 0.1 ppm ethylene in the air causes epinasty in sweet peas and tomatoes, and 0.05 ppm in buckwheat and sunflowers. Injury by ethylene has been observed in greenhouses with leaking gas lines. It is not a major air pollutant nationally, however. Details of the air pollution aspects of ethylene have been presented by Stahl (1).

Ethylene is produced naturally from many, if not all, types of plants. The ethylene can emanate from flowers, fruits, leaves, and stems of a number of plant species. The importance of this biologically derived ethylene in relation to the overall air pollution problem is not known at this time. However, numerous examples are given in the literature in which these emissions caused damage to plants in greenhouses and during storage and shipment, according to Stahl (1). The proportion of ethylene in the atmosphere which results from industrial sources is not known, except in one isolated case where damage to cotton from ethylene emissions from a polyethylene plant was reported.

Ethylene may form as a by-product from incomplete combustion of hydrocarbons and other organic substances. Thus, ethylene has been found to be one of the components of automobile and diesel combustion emissions (exhaust and blow-by emissions), incinerator effluents, and agricultural waste combustion gases.

Automobile and Diesel Emissions: Motor vehicle emissions are believed to be the major source of ethylene pollution in California metropolitan areas. In Los Angeles, the ethylene emission from automobile exhaust has been estimated at 60 tons per day. Several studies have been conducted to determine the ethylene concentration in automobile exhaust in relation to the engine operating mode. The data indicate that the ethylene concentration varies considerably with the operating mode, with the deceleration mode producing the most ethylene. In general, automobile exhaust produces approximately 7.8 pounds of ethylene per ton of gasoline consumed. The composition of the fuel was also found to be an important variable in the amount of ethylene produced. The data indicated that the paraffin content, and to a lesser extent, the olefin content, were the main contributors to ethylene formation in the exhaust.

Incinerator Effluents: Effluents of multiple-chamber incinerators usually contain ethylene in concentrations of less than 11,500 $\mu g/m^3$ (10 ppm). When an afterburner is used, the ethylene concentration becomes nearly zero. Single-chamber incinerators were found to have ethylene concentrations of 23,000 to 31,000 $\mu g/m^3$. However, inadequately designed incinerators may be a major source of pollution, producing concentrations of ethylene as high as 3,565,000 $\mu g/m^3$.

Thus, poor or incomplete combustion of solid wastes may be an important source of ethylene pollution.

Burning of Agricultural Wastes: Laboratory studies have been conducted to determine the amount of ethylene emissions produced from burning of agricultural wastes. Field studies confirm that the laboratory data are probably representative of normal field conditions. The results show that moisture content, type of waste, and location, can be variables that determine the amount of ethylene that is produced. For comparison, automobile exhaust produces approximately 7.8 pounds of ethylene per ton of gasoline, or about 2 to 10 times more ethylene per ton than agricultural wastes.

Environmental Air Concentrations: Only limited data are available on the concentration of ethylene in ambient air. Several studies have been made in California. The results are shown in Table 45.

TABLE 45: CONCENTRATION OF ETHYLENE IN AMBIENT AIR

Location (California)	No. of Samples	Date	Concentration, $\mu g/m^3$ Range	Avg	Max
South Pasadena	4	12/56	290-580	410	580
South Pasadena (at Pasadena Freeway)	1	12/31/56	820		
Los Angeles	125	9/12/61 to 11/14/61	110-150	50	150
Los Angeles*	39	(4-month period, 1967?)		120	820
Oakland*	73	(10-month period, 1967?)		40	
Ventura*	52	(4-month period, 1967?)		63	

*Data given in ppm and converted to µg by dividing by 2 and multiplying by 1.15.

Source: Report PB 188,069

Reference

(1) Stahl, Q.R., "Air Pollution Aspects of Ethylene," *Report PB 188,069,* Springfield, Va., Nat. Tech. Information Service (September 1969).

FLUORIDES

Fluorine is a cumulative poison and the degree of its toxicity is a function of both ingestion level and length of exposure. Fluoride ingestion causes a disturbed calcification of growing teeth. Fluorides are also a protoplasmic poison, a fact which finds its explanation in the blocking of certain enzyme systems. Although there is no evidence to indicate widespread damage at the national level from fluorides, local problem areas do exist. Measurable amounts of fluoride may be found in the atmosphere of any coal burning city in the winter. Agricultural sprays and dusts containing fluorides have caused significant damage in rural areas.

In Air

A study (1) performed by RRI/TRW Systems for the Office of Air Programs of the Environmental Protection Agency, inventoried fluoride emission sources and investigated the technical and economic aspects of implementing soluble fluoride emission controls for major industrial sources. Soluble fluorides are defined as those fluorides with appreciably greater solubility (and ecological impact) than calcium fluoride. Table 46 presents estimates of present and projected (year 2000) emissions of soluble fluorides from the major industrial process sources in the United States. Indications of the relative levels of confidence in the current estimates are included, as are projected emissions assuming that 99% emission control has been attained.

TABLE 46: SOLUBLE FLUORIDE EMISSIONS[A]

	Current			Year 2000 With Current Practice		Year 2000 Assuming 99% Control Efficiency	
Ranking	Process	Emission Thousands Of Tons F Per Year (Yr)	Relative[F] Confidence	Process	Emission Thousands Of Tons F Per Year	Process	Emmission Thousands Of Tons F Per Year
1	Coal Burning For Power	27 (1970)	II	Primary Aluminum Manufacture	141	Primary Aluminum Manufacture	8.1
2	Open Hearth Steelmaking	25 (1968)	II	Coal Burning For Power	86	Coal Burning For Power	0.9
3	Iron Ore Sintering	18 (1968)	II	Iron Ore Pelletizing[B]	39	HF Production	0.7
4	Iron Ore Pelletizing[B]	18 (1968)	II	Expanded Clay Aggregate	25	Electrothermal Phosphorus	0.4
5	Primary Aluminum Production	16 (1970)	II	Wet Process Phosphoric Acid[C]	22	Iron Ore Pelletizing	0.4
6	Heavy Clay Products	10 (1968)	II	HF Alkylation[D]	16	Wet Process Phosphoric Acid[C]	0.3
7	Wet Process Phosphoric Acid[C]	6.4 (1970)	I	Heavy Clay Products	10	Triple Superphosphate[C]	0.3
8	HF Alkylation[D]	5.8 (1971)	I	Triple Superphosphate[C]	7.3	Expanded Clay Aggregate	0.3
9	Expanded Clay Aggregate	5.3 (1968)	II	Electrothermal Phosphorus	6.6	Defluorinated Phosphate Rock[C]	0.2
10	Normal Superphosphate	5.0 (1970)	I	Opal Glass Production	5.5	HF Alkylation	0.2
11	Electrothermal Phosphorus	4.1 (1968)	I	HF Production	5.3	Normal Superphosphate	0.1
12	Triple Superphosphate[C]	3.8 (1970)	I	Iron Ore Sintering	4.8	Heavy Clay Products	0.1
13	Opal Glass Production	3.3 (1968)	II	Defluorinated Phosphate Rock[C]	2.7	Opal Glass Production	<0.1
14	Blast Furnace	2.8 (1968)	II	Blast Furnace[E]	2.6	Iron Ore Sintering	<0.1
15	Defluorinated Phosphate Rock[C]	1.8 (1970)	I	Normal Superphosphate	1.4	Blast Furnace	<0.1
16	HF Production	0.7 (1970)	I	Enamel Frit Production	1.1	Ammonium Phosphate[C]	<0.1
17	Enamel Frit Production	0.7 (1968)	II	Cement Manufacture	0.8	Enamel Frit Production	<0.1
18	Copper Smelting and Refining	0.6 (1967)	III		1.6		<0.1
19	Ammonium Phosphate[C]	0.3 (1970)	I	Ammonium Phosphate[C]	0.8	Cement Manufacture	<0.1
20	Cement Manufacture	0.3 (1964)	II	Open Hearth Steelmaking	0	Open Hearth Steelmaking	0
21	Lead Smelting and Refining	0.2 (1967)	III		2.4		<0.1
22	Zinc Smelting and Refining	0.2 (1967)	III		2.4		<0.1
Total for Processes Considered		155.3			384.3		12.0

(A) Excludes CaF$_2$
(B) Assumes No Fluorspar Addition to Pellet
(C) Includes Prorata Allocation of Gypsum Pond Fluoride Emissions (Estimated as 6,300 Tons for 1970 and 21,000 Tons for 2000)
(D) Assumes 25% of Production Uses Lime Pit Disposal of Acid Sludges
(E) Assumes No Limestone Other Than That in Pellets or Sinter
(F) Relative Confidence Levels: I is Excellent; II is Good; III is Fair to Poor

Source: Report PB 207,506

The evolution and emission factors for each process are presented in Table 47. The evolution factor includes all soluble fluorides leaving the process prior to control. The emission factor corresponds to that portion of the evolved soluble fluorides that eventually enters the atmosphere. The following observations can be made after consideration of the information presented in the body of the report. Five of the first six industries listed typically utilize no fluoride control (the exception is aluminum production). It is technically possible to control soluble fluorides with available devices such as wet scrubbers; the immediate problem lies in implementation of that control, including collection of the evolved fluorides by hoods and similar effluent capture systems for treatment in the abatement devices.

Implementation of control involves a cost which reduces return on investment by varying amounts in different industries. Control of fluoride emissions becomes largely a matter of economics and/or control regulations. The lesser confidence attached to about half of the current emission estimates, including the top four processes in the current rankings of Table 46, indicates the need for more direct experimentally obtained data on both emissions and feed stock compositions for these industries.

The fluorides currently emitted may damage economic crops, farm animals, and materials of decoration and construction. It should be noted, however, that the potential for the observed ambient atmospheric levels to cause fluoride effects in man is negligible. The major problem in measuring the fluoride contents of industrial effluent streams is that of obtaining representative samples which do not change by internal reaction prior to analysis. Current analytical techniques are satisfactory for laboratory analysis. No satisfactory continuous sampling and analytical system for continuous direct monitoring of process streams and stacks has been developed.

Research and development work is required to select control systems from those currently available which minimize the economic impact on a given industry, thereby making implementation of fluoride control as painless as possible. Research and development work on fluoride contents of raw material feed stocks, process streams, and process/plant effluents is required to permit proper control design.

TABLE 47: SOLUBLE FLUORIDE EVOLUTION AND EMISSION FACTORS BY PROCESS

Industry and Process	Evolution Factor LbF/Ton	Product	Emission Factor LbF/Ton	Product
Primary Aluminum Smelting				
Prebaked Anode	46	(Al)	6.94	(Al)
Horizontal Stud Soderberg	46	(Al)	10.12	(Al)
Vertical Stud Soderberg	46	(Al)	9.66	(Al)
Iron and Steel Manufacture				
Iron Ore Sintering	0.73	(Ore)	0.69	(Ore)
Iron Ore Pelletizing	0.73	(Ore)	0.69	(Ore)
Blast Furnace	0.088	(Ore)	0.065	(Ore)
Open Hearth Furnace	0.81	(Steel)	0.77	(Steel)
Basic Oxygen Furnace[a]	0	(Steel	0	(Steel)
Electric Arc Furnace[a]	0	(Steel)	0	(Steel)
Coal Combustion				
Electric Power Generation	0.16	(Coal)	0.16	(Coal)
Cement Ceramic and Glass Mfr.				
Opal Glass Production	21.8	(Glass)	21.8	(Glass)
Enamel Frit Production	3.15	(Dry Frit)	2.64	(Dry Frit)
Heavy (Structural) Clay Products	0.81	(Brick)	0.81	(Brick)
Expanded Clay Products	1.14	(Aggregate)	1.14	(Aggregate)
Portland Cement	0.008	(Cement)	0.008	(Cement)
Phosphate Rock Processing				
Wet Process Phosphoric Acid	4.07	(P_2O_5)[b]	3.36	(P_2O_5)[b]
Diammonium Phosphate	1.3	(P_2O_5)	0.23	(P_2O_5)[b]
Triple Superphosphate	21	(P_2O_5)	5.4	(P_2O_5)[b]
Normal Superphosphate	71	(P_2O_5)	14.2	(P_2O_5)
Electrothermal Phosphorus	30	(P_2O_5)	5.1	(P_2O_5)[b]
Defluorination of Phosphate Rock	210	(P_2O_5)	34	(P_2O_5)[b]

(continued)

TABLE 47: (continued)

Industry and Process	Evolution Factor		Emission Factor	
	LbF/Ton	Product	LbF/Ton	Product
Non-Ferrous Metal Smelting and Refining				
Copper Smelting and Refining	0.78	(blister Cu)	0.78	(blister Cu)
Lead Smelting and Refining	0.34	(Lead)	0.34	(Lead)
Zinc Smelting and Refining	0.46	(Zinc)	0.46	(Zinc)
HF Alkylation Processes	0.18	(bbl alkylate)	0.15	(bbl. alkylate)
HF Production	52	(HF)	4.1	(HF)

(a) Estimated at zero soluble fluoride emission on the basis of thermodynamic equilibria analyses and the assumed total unavailability of hydrogen for conversion of other species to HF.

(b) Includes pro-rata allocation of gypsum pond fluoride emissions.

Source: Report PB 207,506

In Water

Industries which discharge significant quantities of fluorides in process waste streams are: glass manufacturers; electroplating operations; steel and aluminum producers; and pesticide and fertilizer manufacturers. Glass plating wastes typically contain fluoride in the form of hydrogen fluoride (HF) or fluoride ion (F^-), depending upon the pH of the waste. Fluoride discharged from fertilizer manufacturing processes is typically in the form of silicon tetrafluoride (SiF_4), as a result of processing of phosphate rock (2)(3). The aluminum processing industry utilizes the fluoride compound cryolite (Na_3AlF_6) as a catalyst in bauxite ore reduction. Previously, the gaseous fluorides resulting from this process were discharged directly into the atmosphere.

Wet scrubbing of the process fumes, an air pollution abatement procedure, now results in transfer of the fluorides to aqueous waste streams. Average fluoride values for aluminum reduction plants were given as 107 to 145 mg/l in wastewater streams (4). Concentrations an order of magnitude greater have been reported for glass manufacturing, ranging from 1,000 to 3,000 mg/l of fluoride (2).

Intermedia Aspects

Intermedia treatments do exist for some fluorides. Hydrogen fluoride, a common form, can be readily removed by scrubbing processes, thus creating a potential transfer to water.

References

(1)   Robinson, J.M., Gruber, G.I., Lusk, W.D. and Santy, M.J., "Engineering and Cost Effectiveness Study of Fluoride Emissions Control," *Report PB 207,506,* Springfield, Va., Nat. Tech. Information Service (January 1972).
(2)   Zabban, W. and Jewett, H.W., "The Treatment of Fluoride Wastes," *Proc. 22nd Purdue Industrial Waste Conf.,* 22, 706-716 (1967).
(3)   Cherry, J.M., "Fluorine Recovery in Phosphate Manufacture," *Wat. Wastes Eng.,* 7, D-5 (1970).
(4)   Sylvester, R.O., Oglesby, R.T., Carlson, D.A. and Christman, R.F., "Factors Involved in the Location and Operation of an Aluminum Reduction Plant," *Proc. 22nd Purdue Industrial Waste Conf.,* 22, page 441 (1967).

HEAVY METALS

For information on emissions of discharge of heavy metals, the reader is referred to the sections on individual heavy metals such as: cadmium, lead, mercury and nickel.

In Air

Air emissions of 15 metals (or elements such as arsenic and boron generally associated with metallurgical ores) have been estimated for a fair cross section of commercial production-consumption sources. Table 48 summarizes the approximate amounts dispersed from each source.

TABLE 48: ELEMENTAL AIR EMISSIONS FROM U.S. PRODUCTION AND CONSUMPTION OPERATIONS—TONS/YEAR

Element	Production	Consumption	Total
Zn	118,000	33,000	151,000
Mn	16,700	2,300	19,000
V	500	17,500	18,000
Cu	12,100	1,400	13,500
Cr	4,200	7,800	12,000
Ba	7,500	3,300	10,800
B	3,900	5,600	9,500
Pb*	8,100	1,200	9,300
As	5,400	3,600	9,000
Ni	900	5,100	6,000
Cd	2,900	100	3,000
Se	300	600	900
Hg	150	650	800
Sb	250	100	350
Be	10	140	150
	180,910	82,390	263,300

* Composited date for the 1969-1971 period.

Source: Report EPA 600/5-74-005

In total, nearly 264,000 tons/year are emitted. The bulk of the production-related air emissions is derived from: primary and secondary metals processing, e.g., lead, zinc, copper, iron and steel; production of inorganics, e.g., glass and chlorine; and chemical manufacture, e.g., pesticides and petroleum. Application and consumption of coal and oil, agricultural pesticides, herbicides and fungicides, and paint produced the remaining emissions.

Projections of the air emissions of various elements in Table 48 have been made and are shown in Table 49. These projections were derived using: Federal Reserve Board projections of Federal production through 1985; projections and information presented in *U.S. Industrial Outlook–1973*; and figures presented in the *Annual Survey of Manufacturers 1970* and *1971*. The levels shown assume no reductions in pollution due to process changes or additional levels of pollution control. Growth rates for major industry production or consumption source categories for each element were determined, e.g., iron and steel or powerplant boilers. These ranged from 2 to 10%, with most in the 2 to 6% annual range. Were it possible to apply

100% control, these figures would become a measure of pollution control residues available for disposal or recycle.

TABLE 49: PROJECTED AIR EMISSIONS OF SELECTED ELEMENTS

Tons of Pollutant

Element	Base Year*	1978	1983
Zn	151,000	216,700	273,000
Mn	19,000	25,840	31,720
V	18,000	37,240	58,370
Cu	13,500	20,680	24,070
Cr	12,000	14,980	17,800
Ba	10,800	17,290	22,860
B	9,500	14,000	17,690
Pb**	9,300	11,840	14,370
As	9,000	12,750	16,990
Ni	6,000	10,940	17,500
Cd	3,000	4,090	5,050
Se	900	1,240	1,560
Hg	800	1,160	1,560
Sb	350	460	550
Be	150	200	260
Totals	263,300	389,410	503,350

* Data for 1969-71 period

** Excludes automotive sources.

Source: Report EPA 600/5-74-005

Intermedia Aspects

Sludges and solid residues (tailings) generated by industry also constitute a source of trace element contaminants, unless adequate storage or disposal is accomplished. The metal finishing industry has been surveyed in a recent EPA study with respect to sludges from some 15 to 20 thousand facilities known to exist in the U.S. Data extrapolated from questionnaires sent to members of the National Association of Metal Finishers suggest that annual sludge production ranges between 400,000 and 500,000 tons dry basis. These sludges contain potentially valuable amounts of base metals, e.g., of Cu, Ni, Cr, Zn and Cd; precious metals, e.g., Ag, Rh and Au; and other metals, e.g., Fe, Al and Pb.

Typically, concentrations ranged from a few tenths of 1% up to 5% by weight for a single element. Disposal of such sludges in private and public landfills, the principal method used, would likely result in some leaching to the environment of toxic heavy metals. Tailings from mineral and fossil-fuel mining constitute another major source of trace element contamination due to water leaching and airborne dusts. An estimated 1.6 billion tons of tailings per year were generated in 1970, and, with the continuing worldwide exponential growth of demand for raw materials the waste production could double by 1980.

There are other major sources of trace element contaminants to air, water and land receptors, notably automotive exhaust (Pb), leaching from municipal landfills, and more recently identified sources from incineration and land disposal of sewage sludge from municipal wastewater treatment. The concentration of heavy metals in such sewage sludges has been well documented. Table 50 shows a few typical values of selected elements. Finally, applications of agricultural chemicals to millions of acres in the U.S. are a significant source of Hg, Cu, Zn, Cd, Mn, and Cr.

TABLE 50: HEAVY METAL CONCENTRATIONS IN SELECTED SLUDGE SAMPLES*

Location	Zinc	Copper	Nickel	Cadmium
Dayton, Ohio	8,390	6,020	<200	830
Monterey, California	3,400	720	220	<200
Tahoe, California	1,700	1,150	<400	40
Millcreek, Cincinnati Ohio	9,000	4,200	600	<40

*Ppm on dry basis.

Source: Report EPA 600/5-74-005

Figure 4 depicts the intermedia relationships for both water and airborne heavy metal discharges, with the exception of lead, which mainly originates from mobile exhausts.

FIGURE 4: INTERMEDIA FLOW CHART—HEAVY METALS

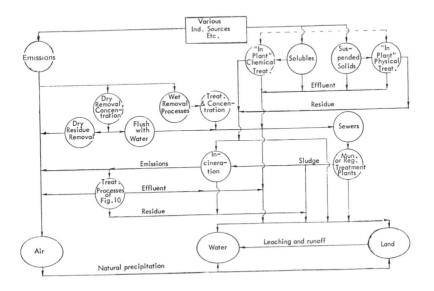

Source: Report EPA 600/5-73-003

Most heavy metal discharges and emissions can be controlled by methods having intermedia implications. For airborne emissions both dry (precipitators, etc.) and wet (water scrubbing, etc.) removal processes can be used. The dry processes may use either a dry process or a water flush to remove the residues. In the latter case, the wash water produces the same problems as the effluent from a scrubber. The residue from dry processes may be incinerated if other combustible materials have been removed along with the heavy metals. Otherwise, they may go directly to land disposal or may be recycled. Recycling is a possibility for both water and air discharges since some heavy metals have a high recovery valve.

Heavy metals in water from either removal processes or primary industrial sources may be either soluble or in a suspended state. Removal of solubles is accomplished by physical-chemical methods, while suspended solids may be removed by either physical or physical-chemical methods. The alternatives for sludge disposal are the same as those discussed for phosphorous compounds. The sludge may be incinerated or may be discharged to receiving water or to the land. Heavy metals may then leach from landfills or be carried by runoff from the land. Most heavy metals precipitate fairly rapidly from the air to water or land and, as with phosphorous compounds, do not transfer readily to the air upon incineration, but tend to remain in the slag or ash.

The principal methods of control of heavy metals in discharges and emissions are pretreatment and the restricted use of heavy metals where substitution can be made. Examples are the elimination of mercury seals in trickling filters, a long standing practice; the restrictions on pesticides containing mercury compounds; and the reduced use of lead-base paints. Several methods have been employed by industry to remove heavy metals from their discharges. For waterborne wastes, physical separation is used to remove suspended solids from effluents, while chemical or biological methods are employed to extract the soluble components. The sludges generated may be dewatered before the ultimate disposal to the land. Heavy metals may appear in industrial discharges to the atmosphere and control measures such as gas scrubbers and electrostatic precipitators may be applied.

Sludges placed to landfills can possibly contaminate underground water, although this is not likely to occur if the landfill is adequately designed and constructed. Heavy metals occur naturally in the environment as part of the earth's crust. Many industrial processes produce pollutants containing heavy metals in various forms which, depending on the dosage received, may be toxic to wildlife, microorganisms and human life. Concentration in the food chain presents increasing hazards to the higher life forms.

Once heavy metals are discharged to the environment, intermedia transfer of the pollutants is possible. Contaminants entering the atmosphere can, after a period of time, settle to the land and water through natural fallout and rainfall. Heavy metal wastes applied to the land and metals settling on land through atmospheric fallout, can further contaminate local surface waters through storm runoff and continental weathering. The heavy metals considered here are lead, mercury, cadmium, and nickel. Quantitative data do exist concerning the intermedia transfer of a few of these metals.

References

(1)   Flinn, J.E. and Reimers, R.S., "Development and Predictions of Future Pollution Problems," *Report EPA 600/5-74-005*, Washington, D.C., U.S. Government Printing Office (March 1974).

(2)   Stone, R. and Smallwood, H., "Intermedia Aspects of Air and Water Pollution Control," *Report EPA 600/5-73-003*, Washington, D.C., U.S. Environmental Protection Agency (August 1973).

HYDROCARBONS

In Air

Atmospheric hydrocarbons and other organic compounds can be thought of as having four origins: petroleum, coal, natural gas, and biological products. Organic gases can escape to the atmosphere at many points during the production, processing, storage, transport, and ultimate use of the originating organic material. Potential sources of hydrocarbon emissions in petroleum production and product processing include leakage from oil field operations, refining, gasoline storage tanks, gasoline loading facilities, blowing of asphalt, blowdown systems, catalyst regenerators, processing vessels, flares, compressors, pumps, vacuum jets, waste-effluent handling equipment, and turnaround operations.

Gasoline distribution and marketing systems emit hydrocarbon vapors from tank-truck loading racks, service station tank-filling operations, and automobile tank-filling operations. Total hydrocarbon emissions to the atmosphere in the United States have been estimated at 34.6 million tons in 1970. Estimates have remained essentially constant at that value for the previous few years. Nearly 20 million tons are attributable to vehicular and other transportation sources, with miscellaneous sources contributing 7.1 million tons and industrial processes 5.5 million tons.

However, there is some controversy over the relative importance of man-related hydrocarbon emissions. It has been reported that natural sources such as vegetation and bacteria are responsible for 85% of the global hydrocarbon emissions and 50% of all United States emissions. However, natural sources are generally well removed from population centers, which may limit their importance as direct impactors on man's environment or reactants with man-made pollutants to form more damaging substances.

Methane from swamps and terpenes from evergreen forests are examples of two predominant natural sources, while gasoline constituents such as branched pentanes and benzene derivatives are the predominant hydrocarbons from man-related sources. Although very low in atmospheric concentrations and total emissions, some high molecular weight fused-ring aromatic hydrocarbons like the benzopyrenes are known carcinogens. They are produced by high-temperature combustion of organic substances as in coal and oil burning, incineration, backyard barbequeing, and cigarette smoking. Sanitary landfills may also be a source of hydrocarbon emissions. Anaerobically decomposing organic wastes produce methane. The potential gross pollution by methane from such point source disposal methods may be far greater than the hydrocarbon emissions during solid waste incineration.

Proposed means of reducing hydrocarbon emissions include propane powered vehicles, improved service station design to lessen evaporation, reduction of hydrocarbon compounds used by industry, minimum reactive hydrocarbons used in degreasing, and lesser use of hydrocarbons in dry cleaning establishments. Hydroelectric and nuclear power plants are presently limited alternatives to fossil-fuel units for abatement of hydrocarbon emissions from power generation.

Gaseous organic compounds in the atmosphere may undergo chemical and physical processes that produce other substances with greatly altered properties. These reactions are photochemical in nature. A description of these processes and the effects of the products are given in *AP-63, Air Quality Criteria for Photochemical Oxidants,* U.S. Department of Health, Education, and Welfare, National Air Pollution Administration, and in *AP-64, Air Quality Criteria for Hydrocarbons,* U.S. Department of Health, Education, and Welfare, National Air Pollution Control Administration.

Estimates for total United States organic emissions from stationary sources are presented in Table 51. These estimates in general were made from emission factors and quantities of fuels consumed, quantities of refuse burned, and quantities of raw materials processed. Although they represent only gross approximations, they are the best estimates currently available.

TABLE 51: SUMMARY OF NATIONWIDE HYDROCARBON EMISSIONS, 1968

$(10^6$ tons/year)

Source	Emissions
Transportation	
Motor vehicles	
Gasoline	15.2
Diesel	0.4
Aircraft	0.3
Railroads	0.3
Vessels	0.1
Nonhighway use, motor fuels	0.3
Fuel combustion–stationary	
Coal	0.2
Fuel oil	0.1
Natural gas	Negligible
Wood	0.4
Industrial processes	4.6
Solid waste disposal	1.6
Miscellaneous	
Forest fires	2.2
Structural fires	0.1
Coal refuse	0.2
Agricultural burning	1.6
Organic solvent evaporation[a]	3.2
Gasoline marketing	1.2
Total	32.0

[a] Includes estimated 25 percent aldehydes, ketones, and esters, and 10 percent chlorinated solvents.

Source: EPA Publication AP-68

Information on hydrocarbon emissions from mobile sources has been presented elsewhere in this volume under such headings as "Aircraft Operation," "Automotive Vehicle Operation" and "Boat and Ship Operation."

Table 52 is a compilation of available emission factors for hydrocarbons from various types of stationary sources. These emissions rates are for uncontrolled sources unless otherwise noted. Except where noted, emission factors were compiled from a Public Health Service publication. For some cases in which a range of emission factors is given, the range reflects the figures given by different sources.

TABLE 52: EMISSION FACTORS FOR HYDROCARBONS

Source	Hydrocarbon emissions, lb/unit given	Unit
Fuel combustion–stationary sources		
Coal combustion unit		
$<10 \times 10^6$ Btu/hr capacity	10	ton of coal
10 to 100 x 10^6 Btu/hr capacity	1	ton of coal
Greater than 100 x 10^6 Btu/hr capacity	0.2	ton of coal
Fuel Oil		
$<10 \times 10^6$ Btu/hr capacity	3	1,000 gal of oil
10 to 100 x 10^6 Btu/hr capacity	2	1,000 gal of oil
Greater than 100 x 10^6 Btu/hr capacity	0.8	1,000 gal of oil

(continued)

TABLE 52: (continued)

Source	Hydrocarbon emissions, lb/unit given	Unit
Solid waste disposal		
Open burning on site of leaves, brush, paper, etc.	12	ton of waste
Open burning dump	30	ton of waste
Municipal incinerator	0.3	ton of waste
Multiple-chamber incinerator	0.5	ton of waste
Single-chamber incinerator	0.8	ton of waste
Flue-fed incinerator	2	ton of waste
Domestic incinerator		
No control	2	ton of waste
Afterburner	1	ton of waste
Process industries		
Phthalic anhydride plant		
Oxidation of napthalene vapors with excess air over catalyst	32	ton of product
Petroleum refinery (as total hydrocarbons)		
Boilers and process heaters	140	1,000 bbl oil
Fluid catalytic unit	220	1,000 bbl fresh feed
Moving-bed catalytic cracking unit	87	1,000 bbl fresh feed
Compressor internal combustion engines	1.2	1,000 ft^3 fuel gas
Blowdown system		
With control	5	1,000 bbl refinery capacity
Without control	300	1,000 bbl refinery capacity
Process drains		
With control	8	1,000 bbl waste water
Without control	210	1,000 bbl waste water
Vacuum jets		
With control	Negligible	1,000 bbl vacuum distillation capacity
Without control	130	1,000 bbl vacuum distillation capacity
Cooling towers	6	10^6 gal cooling water capacity
Pipeline valves and flanges	28	1,000 bbl refinery capacity
Vessel relief valves	11	1,000 bbl refinery capacity
Pump seals	17	1,000 bbl refinery capacity
Compressor seals	5	1,000 bbl refinery capacity
Air blowing, blend changing, and sampling	10	1,000 bbl refinery capacity
Storage		
V.P.$\geq$1.5 psia-fixed roof	47	1,000 bbl refinery capacity
V.P.$\geq$1.5 psia-floating roof	4.8	1,000 bbl storage capacity
V.P.$<$1.5 psia-fixed roof	1.6	1,000 bbl storage capacity
Commercial operations		
Dry cleaning		
Chlor-hydrocarbons	1.7	per capita per year
Hydrocarbon vapors	2.2	per capita per year
Gasoline handling (evaporation)		
Filling tank vehicles		
Splash filling	8.2	1,000 gal throughput
Submerge filling	4.9	1,000 gal throughput
Filling service station tanks		
Splash filling	11.5	1,000 gal throughput
Submerge filling	7.3	1,000 gal throughput
Filling automobile tanks	11.6	1,000 gal throughput

Source: EPA Publication AP-68

Intermedia Aspects

Examination of Figure 5 reveals transfer modes between the fluid media for hydrocarbon pollutants discharged originally to the air. However, it should be emphasized that the physical possibility of transferring hydrocarbon emissions from air to water is very slight, due to their water insolubility and the efficiency of normal intramedial hydrocarbon treatment methods. Volatile hydrocarbons spilled or wasted to waterways will evaporate to air, and incineration of oily spills and sludges may transfer a small amount of unburned fuel. Some hydrocarbons are in particulate form, and may thus be treated by particulate control techniques.

FIGURE 5: INTERMEDIA FLOW CHART—GASEOUS HYDROCARBONS

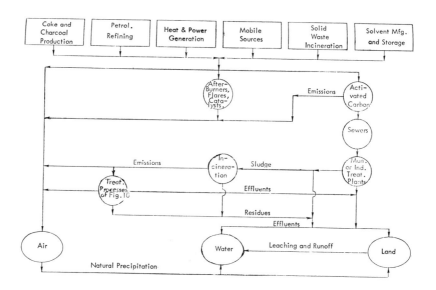

Source: Report EPA 600/5-73-003

Except for fused-ring aromatics, which may be carcinogenic, hydrocarbons are not generally harmful to human or animal life in relatively dilute atmosphere concentrations. In large concentrations they are asphyxiants because of oxygen exclusion. Low-molecular weight hydrocarbons comprise nearly all the gross emissions and are generally colorless at ambient temperatures. They do not generally form aerosols, hence they do not contribute directly to visibility reduction in the atmosphere.

A major impact of hydrocarbon pollutants involves their photochemical reaction with nitrogen oxides to form smog. Olefinic unsaturates are the most reactive, and paraffinic hydrocarbons, the least. Except for the high molecular weight hydrocarbons, which are settleable, hydrocarbon emissions tend to remain airborne. They are water insoluble and do not enter water resources except that the medium molecular weight compounds may enter as flotables. Because hydrocarbons are not easily transferred from the air, they pose a continual hazard as reactants to form such photochemical products as PAN.

Hydrocarbons in liquid form are significant water pollution factors in certain areas of the United States. The intermedia relationships are shown in Figure 6. Liquid hydrocarbon pollution of the sea and of lakes and waterways can severely damage fish, other animal, and plant life. Spills from oil well drilling and operation, from pipeline breaks, from off-shore oil drilling, and from the sinking or washing down of oil tankers are the major sources. Solutions to these problems lie in greater safety precautions, moratoriums or greater controls on offshore oil drilling, and stricter enforcement of laws controlling ocean dumping from ships.

FIGURE 6: INTERMEDIA FLOW CHART—LIQUID HYDROCARBONS

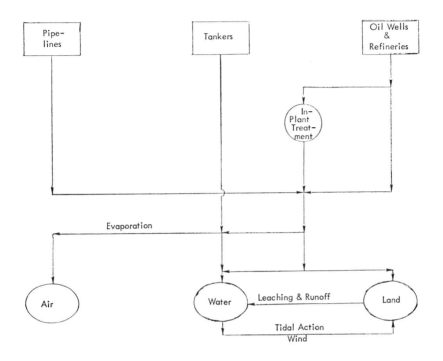

Source: Report EPA 600/5-73-003

References

(1) U.S. Dept. of Health, Education and Welfare, "Control Techniques for Hydrocarbon and Organic Solvent Emissions from Stationary Sources," *Publ. No. AP-68*, Wash., D.C., Nat. Air Pollution Control Administration (March 1970).

(2) Stone, R. and Smallwood, H., "Intermedia Aspects of Air and Water Pollution Control," *Report EPA 600/5-73-003*, Wash., D.C., U.S. Environmental Protection Agency (August 1973).

HYDROGEN CHLORIDE

Hydrogen chloride and other chloride compounds can cause widespread damage to vegetation and property. However, the modern alkali industry, a main source of this pollutant, is based upon the electrolysis of common salt and by-products are usually carefully controlled. Concentrations for U.S. cities are well below commonly-accepted danger levels. Some local problems in rural areas can exist if precautions are not taken in the use of hydrogen chloride as a fumigant. This pollutant creates no major problem nationally nor is it likely to become a major problem.

In Air

The air pollution aspects of hydrogen chloride have been reviewed in detail by Stahl (1). Hydrochloric acid may be emitted from a wide variety of sources. Emissions from some sources may go unnoticed because the hydrochloric acid is generated as an unpredicted product. The sources may be classified as follows: direct manufacturing of hydrochloric acid (e.g., acid-salt and synthesis processes); a predicted by-product of a chlorination process in indirect manufacturing (e.g., by-product process); unpredicted or undesired product of a manufacturing process (e.g., thermal decomposition of chloride-containing reactants or products); use of hydrochloric acid in the production or manufacturing of other products (e.g., pickling of metals); burning or combustion of chloride-containing materials (e.g., fossil fuels, plastics, paper); and heating of chloride-containing materials (e.g., heating of organic matter).

Intermedia Aspects

Hydrogen chloride can be controlled with methods similar to those for sulfur oxides. Some potential controls convert the compound to hydrochloric acid, as a water pollutant, but most control methods recover the gas so that intermedia transfer of hydrogen chloride is not widespread.

Reference

(1) Stahl, Q.R., "Air Pollution Aspects of Hydrochloric Acid," *Report PB 188,067*, Springfield, Va., Nat. Tech. Information Service (Sept. 1969).

HYDROGEN CYANIDE

In Air

Hydrogen cyanide, while extremely lethal, is not a major air pollutant nationally. It can be fatal to animals and humans, and can also injure vegetation, causing surface irritation and root damage. It can cause root injury when leaked into greenhouses from underground gas lines, since it has been found in artificial gas to the extent of 200 to 300 ppm. Hydrogen cyanide is used as a fumigant, and careless handling can cause the damage described.

HYDROGEN SULFIDE

In Air

Hydrogen sulfide is produced in nature primarily through decomposition of proteinaceous material (vegetable and animal) by bacteria. It develops principally in stagnant and insufficiently aerated water such as found in swamps and polluted water. Hydrogen sulfide also occurs naturally as a constituent of natural gas, petroleum, sulfur deposits, and numerous volcanic gases and sulfur springs. The annual worldwide production of hydrogen sulfide in nature is around 90 to 100 million tons, with 60 to 80 million tons coming from land sources and 30 million tons coming from ocean areas. Other estimates range as high as 202 million tons from ocean areas and 82 million tons from land areas. Data on background air concentrations of hydrogen sulfide due to the natural sources are scarce. However, it has been estimated to be between 0.15 and 0.46 $\mu g/m^3$, which is well below the odor threshold or the concentrations at which deleterious effects are known to occur.

Hydrogen sulfide is produced as a by-product in many industrial processes. Production sources include the petroleum industry (refineries and natural gas plants), petrochemical plant complexes, coke oven plants, kraft paper mills, chemical processing industries, dye manufacture, viscose rayon manufacture, sulfur production, manufacture of sulfur-containing chemicals, iron and metal smelters, food processing plants, and tanneries.

Petroleum Industry: Sulfur enters the refinery as a constituent of crude oil and is usually found in combination with hydrogen as hydrogen sulfide and with hydrocarbons as various organic sulfides. Since removal of sulfur from both product and intermediate stocks is necessary to meet low sulfur requirements for fuel oil and to prevent sulfur poisoning of catalysts, processes such as hydrogen treating are employed. During the various steps employed in processing the crude oil, the sulfur compounds are generally converted to hydrogen sulfide and lower molecular weight mercaptans. From each 20,000 barrels of crude oil with high sulfur content processed, approximately 50 tons of hydrogen sulfide are formed.

The main sources of air pollution in refineries are untreated gas stream leaks, vapors from crude oil and raw distillates, and process and condensate sewers. Typical refinery processing systems that have hydrogen sulfide emissions are cracking units, catalytic reforming units, and sulfur recovery units. The cracking process tends to convert sulfur contained in crude oil into hydrogen sulfide in the heavier materials and into mercaptans in the gasoline fractions. Measurements were made in the El Paso, Texas, area of the atmospheric hydrogen sulfide concentration adjacent to an oil refinery. The mean hydrogen sulfide concentration was 6 $\mu g/m^3$. This varied from undetectable amounts to a maximum of 91 $\mu g/m^3$.

Hydrogen sulfide occurs naturally in many areas in association with natural gas. In some areas, such as Alberta, Canada, the sour natural gas can consist of over 50% hydrogen sulfide. In processing, the natural gas stream is treated to remove the hydrogen sulfide, which is generally converted to sulfur. Distributing companies which sell the natural gas for heating and power generation generally require that its hydrogen sulfide content be less than 23,000 $\mu g/m^3$. Another possible source of hydrogen sulfide air pollution in the petroleum industry is the production of asphalt.

Petrochemical Plant Complexes: Hydrogen sulfide is produced in petrochemical plants during cracking and other desulfurization reactions also. Data have been reported on atmospheric hydrogen sulfide concentrations around a petrochemical industrial complex in Russia. The complex consisted of three oil refineries, a synthetic alcohol plant, a chemical plant, and three power plants. Measurements made of air concentrations of hydrogen sulfide showed 17 to 150 $\mu g/m^3$ inside the industrial complex, 8 to 70 $\mu g/m^3$ at 2.5 km from the complex, and 1 to 50 μg at 20 km from the complex.

Kraft Mills: Hydrogen sulfide and organic sulfide are produced and released to the atmosphere in kraft mills in a number of locations. This emission imparts the characteristic odor in the vicinity of kraft paper mills and has been the cause of major air pollution problems. Over 50% of the pulp produced in the United States comes from the kraft or sulfate process. It was estimated that in 1960 about 64,000 tons of hydrogen sulfide were emitted from kraft paper mills throughout the world. In the kraft process, wood chips and a solution of sodium sulfide and sodium hydroxide (white liquor) are cooked in a digester for about 3 hours at elevated temperatures and pressures. The solution dissolves the liquor from the wood. The spent liquor (black liquor) is then separated from the cellulose fiber in the blow tank. The fiber is then washed and processed into paper. The remainder of the process involves the recovery and regeneration of the cooking chemicals from the black liquor. The recovery process is initiated by concentrating the black liquor by evaporation. The concentrated black liquor is then burned in the recovery furnace, and the inorganic chemicals collect on the floor of the furnace in a molten state (smelt). Hot combustion gases from the recovery furnace are used in the direct contact evaporation to concentrate the black liquor.

The major sources of hydrogen sulfide emission in kraft mills are the stack gases from the recovery furnace, including the direct contact evaporator; the stack gases from the lime kilns; and the noncondensibles from the digester relief, the blow tank, and the multieffect evaporator. The quantities of emissions from each source are given in Table 53. The amount of these emissions actually reaching the environment depends upon the efficiency of each of the abatement systems that are installed and operating at each mill.

TABLE 53: HYDROGEN SULFIDE EMISSIONS FROM KRAFT MILL PROCESSORS

Sources	Pounds/Dry Ton Produced	$\mu g/m^3$ Emitted
Digester blow (cooking time 3.75 hr, temp 172°C, sulfidity 22%)	0-0.45	0-600,000
Digester relief and blow (no data on conditions)	0.43	208,000
Digester blow (no data on conditions)	0.9	
Digester relief and blow (cooking time 3.45 hr, temp 172°C, sulfidity 22.5%)	0.66	
Digester relief (no data on conditions)	0.01-0.05	
Recovery furnace	3.6-28	198,000-1,500,000
Recovery-furnace stack gases		750,000-1,150,000
Recovery furnace	9.0	
Multiple-effect evaporators (potential)	1.2	
Multiple-effect evaporators (normal)	0-0.06	
Multiple-effect evaporators	1.2	
Lime kiln		0-900,000

Source: Report PB 188,068

Coke Ovens: Hydrogen sulfide is produced in the coking operation at the rate of about 6.7 pounds per ton of coal charged. The effluent gas from coke ovens contains about 6,000 to 13,000 $\mu g/m^3$ of hydrogen sulfide. During cooling and scrubbing, approximately 50% of the hydrogen sulfide is removed. The remaining gas is either used as is for firing the coke ovens; purified further (partial desulfurization) and then used for firing of coke ovens; or completely desulfurized and used for municipal gas. The hydrogen sulfide content of nonpurified, partially purified, and municipal gas is shown in Table 54.

TABLE 54: HYDROGEN SULFIDE CONTENT OF COKE-OVEN GAS

Type of Gas	$\mu g/m^3$
Nonpurified	5,000–13,000
Partially desulfurized	1,500–5,000
Municipal gas or pipeline gas	None

Source: Report PB 188,068

Hydrogen sulfide emissions can occur throughout the complete coking cycle from coke-oven charging to hydrogen sulfide removal (desulfurization). The sources of these emissions, other than charging and discharging emissions, and their causes are shown in Table 55. No data were found on the magnitude of hydrogen sulfide concentration in the atmosphere in or around coke ovens. However, one reference indicated that it is rarely of sufficient magnitude to create problems or evoke complaints from nearby residents.

TABLE 55: SOURCES OF HYDROGEN SULFIDE EMISSIONS IN COKE PLANTS

Source of Emission	Cause of Emission
Condensation	
Unburnt gases escaping from the gas torches	
In normal operation with torch shut off	Leakage at stop valves
With torch open during operational failures	Failure of ignition device
Gases escaping from water seals	Defective seals
Outflow collectors on coolers; collector and separator tanks	Gas escape from liquids
Ammonia Scrubber	
Outflow collectors and collector tanks	Gas escape from washing of fluid
Secondary coolers for primary-cooler outflow (in semi-direct process)	Escape of hydrogen sulfide with the cooling-tower vapors
Benzol Scrubber and Plant	
Outflow receivers of scrubbers and washing oil tanks	Gas escape from washing fluid
Cooler-ventilating lines	Escape of sulfur-containing compounds with low boiling point, together with ventilating gases
Desulfurization of Gas	
Outflow receivers and tanks for scrubbing fluid	Gas escape from washing fluid

Source: Report PB 188,068

Mining: Burning coal refuse piles have been a continual cause of air pollution, and one of the combustion products emitted to the atmosphere is hydrogen sulfide. Approximately 20% to 50% of the raw anthracite processed in cleaning plants is rejected as refuse. At many operations the refuse discarded amounts to about 33% of the tonnage produced.

This refuse over the years has accumulated in coal refuse piles, some of which contain millions of tons. The piles ignite either through spontaneous combustion, carelessness, or deliberate action. A recent survey indicated that there are approximately 500 burning piles in 15 States.

The hydrogen sulfide generated during combustion disperses into the atmosphere. Significant concentrations of hydrogen sulfide gas have been measured in communities adjacent to burning coal piles. Measurements made in July 1960 adjacent to a large burning anthracite refuse pile showed an hourly maximum average of 600 $\mu g/m^3$. The minimum hourly average was 140 $\mu g/m^3$. Other possible sources of hydrogen sulfide from mines include underground mine fires and sulfide ore mines.

Iron-Steel Industry and Foundries: Small amounts of hydrogen sulfide are given off when blast furnace slag is granulated. The amount of hydrogen sulfide formed is proportional to the amount of hydrogen formed during the quenching process. No information was given on the amount of hydrogen sulfide released to the atmosphere by plant granulation operations. Typical hydrogen sulfide exhaust emissions from foundries are 0.002 tons per nonferrous foundry producing 50 tons of castings per day, and 0.023 tons per gray iron foundry producing 200 tons of castings per day.

Chemical Industry: Hydrogen sulfide is a by-product of many chemical operations. In general, it is formed when sulfur or sulfur compounds are associated with organic materials at a high temperature. For example, it is a by-product in the manufacture of carbon disulfide. The process of producing thiophene by the reaction of sulfur with butane at elevated temperatures also produces hydrogen sulfide.

Other sources of hydrogen sulfide in the chemical industry are the manufacture of sulfur dyes and the production of viscose rayon, ethyl and methyl parathion (pesticides), organic thiophosphate, and many other organic sulfur chemicals. In addition, hydrogen sulfide is evolved from some grease and fatty acid-making processes. Approximately 6 tons of hydrogen sulfide are formed for every 100 tons of viscose rayon produced. Inorganic processes which evolve hydrogen sulfide are zinc smelting and refining, manufacture of barium chloride from barium sulfide, and production of phosphorus compounds, pigments, lithopone, and sodium sulfide. Hydrogen sulfide is also emitted during the manufacture of stove clay and glass. The only data found on the magnitude of hydrogen sulfide emissions were 0.024 tons reported per chemical and allied products plant consuming 10^9 Btu per day, and 0.17 tons per cement plant producing 4,830 barrels per day.

Animal Processing Plants and Tanneries: Hydrogen sulfide is generated in animal processing plants during the decomposition of protein material. Hydrogen sulfide and organic sulfur compounds are produced in an offal cooking plant when hooves and horns of cattle and other animals are treated with high-pressure live steam. Hydrogen sulfide is also produced during the cooking of meat. In stale meat, approximately 0.15 pounds of hydrogen sulfide is formed per ton of raw meat at 100°C. Only a trace is formed during the cooking of fresh meat.

During the 1920's and the early thirties, there were many cases of poisoning among tanners (some fatal) mainly caused by hydrogen sulfide. The air concentrations at those times varied from 1,000 to 540,000 $\mu g/m^3$. However, recent results of a nine year study on tanneries in Russia showed that the hydrogen sulfide has been almost eliminated in modern tanneries. No information was found on United States tanneries, but it can be assumed that the hydrogen sulfide problem has been largely eliminated. Other sources of hydrogen sulfide are stockyards, cheese and dairy plants, and wool scrubbing plants. No information was available on emissions from these sources.

Combustion Processes: Hydrogen sulfide is released when coal, oil, or gas is burned. The amount of hydrogen sulfide depends upon the amount of sulfur in the fuel and the efficiency of the combustion process. In an efficient combustion system, the hydrogen sulfide is oxidized to sulfur dioxide. In a study of sulfur released from domestic boilers, it was found that hydrogen sulfide was given off from open fires during heavy smoke emission, mainly just after refueling. The emission factors for fuel are given in Table 56.

TABLE 56: HYDROGEN SULFIDE EMISSION FACTORS

Combustion Source	Emission Factor
Coal	0.0045*/lb coal
Fuel oil	1 lb/1,000 lb oil
Natural gas	0.13 lb/1,000 lb gas
(density 0.0475)	

*lb sulfur/lb coal

Source: Report PB 188,068

Hydrogen sulfide is also given off by apartment incinerators and sanitary landfills. Estimates of hydrogen sulfide discharged daily from domestic and municipal sources in a metropolitan area of 100,000 persons show that domestic heating would produce 1,000 pounds of hydrogen sulfide daily from coal combustion, 500 pounds from oil, and 0.1 pound from gas; apartment incineration would produce 24 pounds daily, and sanitary landfill would produce trace amounts.

Polluted Water: In some localized situations, air pollution due to natural biological processes in polluted waters includes hydrogen sulfide air concentrations that can produce blackening of paint as well as odors above the hydrogen sulfide odor threshold.

Well Water: Another hydrogen sulfide source is municipal plants for removing hydrogen sulfide from well waters. Water aeration plants in the city of Jacksonville, Fla., emit about 0.15 ton per day of hydrogen sulfide which have been a cause of nuisance complaints.

Sewage Plants and Sewers: Hydrogen sulfide is produced biologically in sewers from organic compounds formed by hydrolysis of materials like cystine and methionine and by reduction of sulfates. Sewage usually contains 1 to 5 ppm of organic sulfur compounds, while some industrial wastes, such as from wool, contain as high as 50 to 100 ppm. Sulfates are present in sewage almost entirely as inorganic sulfates and enter the system in wastewater, saline groundwater, or through industry discharge of tidal or seawater to sewers. The factors that influence hydrogen sulfide generation in sewers include sewage temperature, content of sewage, velocity of flow, age of sewage, pH value of sewage, sulfate concentration, and ventilation of the sewer. Hydrogen sulfide is also generated and released from sewage treatment plants.

The hydrogen sulfide is formed in digesters during anaerobic digestion of the sewage sludge and industrial wastes. Atmospheric measurements made at a sewage treatment plant in El Paso, Tex., in 1958 showed that the hydrogen sulfide concentration varied between 24 $\mu g/m^3$ and 2,120 $\mu g/m^3$, with the average concentration 610 $\mu g/m^3$. At a sampling station 100 yards from the sewage plant, the maximum hydrogen sulfide concentration was 205 $\mu g/m^3$.

Environmental Air Concentrations: Measurements of the concentration of hydrogen sulfide in the environmental air from selected areas for various time periods indicate average levels of hydrogen sulfide in the atmosphere of 1 to 92 $\mu g/m^3$, as indicated in Table 57.

TABLE 57: ATMOSPHERIC HYDROGEN SULFIDE CONCENTRATIONS
(μg/m³)

Location	Average	Maximum
New York City		
1956-61	1	13
1962	1	6
Elizabeth, N. J.		
Aug.-Oct. 1963	1	247
Hamilton Township, N. J.		
May-Oct. 1962	1	49
Woodbridge Township, N. J.		
April-May 1961	1	305
Greater Johnstown Area, Pa.		
1963	3	210
Winston-Salem, N. C.		
Nov.-Dec. 1962	3	011
Lewiston-Clarkston Area, North Lewiston, Idaho, near pulp mill, 1962		37
Great Kanawha-River Valley Industrial Area		
Feb. 1950-Aug. 1951	3-92	410
Camas, Wash.		
1962	0-1	6
Santa Barbara, Calif.		
1949-1954		1,400
St. Louis, Mo.		
1964	2-6	94
Terre Haute, Ind.		
May-June 1964		>460

Source: Report PB 188,068

Reference

(1)   Miner, S., "Air Pollution Aspects of Hydrogen Sulfide," *Report PB 188,068,* Springfield, Va., Nat. Tech. Information Service (Sept. 1969).

IRON

In Air

There are a variety of production sources which introduce iron into the atmosphere as reviewed by Sullivan (1).

Iron and Steel Industry: Measurements of particulate concentrations in the area downwind of iron and steel plants have shown that these emissions can contribute significantly to air pollution. In a report of a study in Ironton, Ohio, during the period September 1965 to August 1967, particulates measured downwind from two iron and steel plants ranged between 190 and 212 μg/m³. The iron content of one sample was 16% by weight. Estimates of the dust concentrations from open-hearth furnaces indicated that the concentration might exceed 1,000 μg/m³ at a distance of 4 km from the source, if the wind conditions were right. Dust emission rates from the Dayton Malleable Iron Plant were 935 tons per year and from the Armco Steel Corporation, 7,926 tons per year. There is a significant difference in the air pollutants when steel mills are shut down during strikes. Comparison of analytical results during and after a strike showed that suspended particulate concentrations were higher by 44 to 171% in the poststrike period, the soiling index was higher by 200%, and the concentration of iron was higher by 2.6 to 10.8 times. In the upper Ohio River Valley where two major steel mills, two large coke plants, and a steam-powered electric-generating plant were located, the suspended particulate concentration was 383 and 186 $\mu g/m^3$ in two nearby cities. A maximum of 1,238 $\mu g/m^3$ was measured in one city. One particulate sample from that city had an iron concentration of 30.8 $\mu g/m^3$. The soiling index averaged 5.5 to 5.3 cohs (coefficient of haze) per 1,000 linear feet.

In a Pennsylvania town where the industry consisted of steel mills with two blast furnaces, 11 open-hearth furnaces, a sintering plant and other equipment, and a zinc plant, the particulate concentration near the furnaces frequently exceeded 500 $\mu g/m^3$ at a distance of 0.25 to 0.5 mile from the furnaces after the strike. In a Michigan community, the iron concentration measured near two coke plants was 5.8 $\mu g/m^3$ compared to a value of 0.6 $\mu g/m^3$ in a residential area. In another small community (500 population) the processing of blast furnace slag was found to be a major source of pollution. Particulate concentrations downwind from the slag processing plant 0.4 and 0.8 miles measured an average of 411,000 and 477,000 $\mu g/m^3$ respectively. While other pollution sources could have contributed to the particulate air pollution, chemical analyses indicated that from 35 to nearly 100% of the dust came from the slag processing plant.

These community studies indicate that the iron and steel industry may be a significant source of iron air pollution. Although the production of iron and steel in the United States is increasing, the air pollution trend by steel production is uncertain. The main reason for the uncertain pollution trend is the remarkable change in steel processing. The Bessemer process has almost entirely been discontinued while the relatively new basic oxygen process has become increasingly important. This has an impact on the air pollution aspects of the iron and steel industry since old furnaces without dust control are being replaced with new furnaces with such control. While every basic oxygen furnace constructed in the U.S. has been equipped with dust control devices, there are several factors which have tended to increase air pollution from oxygen processes:

[1] Fume generation rates in basic oxygen furnaces are approximately six times those of open-hearth furnaces without oxygen lancing.

[2] The oxygen-blowing rates of many basic oxygen furnaces have been increased considerably beyond the original design rate. As a result, the control systems have become overloaded, resulting in the emission of copious amounts of fumes.

[3] Oxygen lancing has been used to increase the production rate in open-hearth furnaces. Because many of the open-hearth furnaces lack dust control, oxygen lancing has resulted in increased emissions from these sources.

Sintering Plants: From U.S. sintering plants, 20 pounds of dust per ton of sinter is likely in a waste gas of 160 scfm. This gas ranges in temperature from 160° to 390°F, and contains 1.1 to 6.8 g/m^3 (0.5 to 3 grains/scf). Dust from a plant in Germany is reported to contain 50% iron.

Blast Furnaces: The average blast furnace produces approximately 1,000 tons of pig iron per day, during which about 100 tons of dust are produced. The input to the furnace is approximately 2,000 tons of ore, 900 tons of coke, 400 tons of limestone and dolomite, and 3,570 tons of air. The normal emissions are 16,000,000 to 22,800,000 $\mu g/m^3$. Approximately 32% of the dust is fine particulates containing 30% iron. Slips are the principal factors in pollution arising from modern blast furnaces having typical air pollution control equipment. A slip occurs when the crust of a furnace charge breaks and slips into a void. This produces a sudden rush of gas which automatically bypasses the control equipment (to avoid high pressures in the system), thus releasing a large black or red cloud of dust to the atmosphere. In recent years automation has helped to reduce the number of slips.

Ferromanganese Blast Furnaces: Uncontrolled emissions from a ferromanganese blast furnace in 1951 produced as much as 8.2 to 15.5 g/m^3 of exhaust gas, with an average of 13.7 g/m^3. Two 350 ton furnaces produced approximately 142 tons of dust per day containing 0.3 to 0.5% iron. The particle size of the fume is extremely small, 0.1 to 1.0μ in diameter.

Open-Hearth Furnaces: For economic reasons, oxygen injection is used to increase the yield of steel from open-hearth furnaces. However, it also increases the air pollution. In 1960, the fume loading was reported to increase from approximately 0.9 g/m^3 (7.5 lb/ton) to 1.4 g/m^3 (9.3 lb/ton) using oxygen injection. This represents about a 50% increase in emissions per unit time, but only about a 25% increase in emissions per ton. More recent evidence indicates the emission rates range from 15 to 40 lb/ton of steel. When the heat time is short (3 hours) the emission rates would be 30 to 40 lb/ton. The average open-hearth furnace in the U.S. has a capacity of about 175 tons of steel per heat. A heat takes about 11 hours without oxygen injection, but only 3 hours (in short heats) with oxygen injection.

Most of the dust is made up of iron oxide, predominantly Fe_2O_3. If the heat contains a large fraction of galvanized steel, zinc oxides may predominate. Studies have shown that the iron oxide content averages about 50% of the total particulates with oxygen lancing. These particles are small, 93% of them less than 40μ in diameter and 46% less than 5μ.

Electric-Arc Furnaces: Dust and fume emissions from an electric-arc furnace average 10.5 lb/ton of steel melted, ranging from 4.5 to 29.4 lb/ton. These particulates contain 40 to 50% Fe_2O_3 with 70% of the particles less than 5μ in diameter.

Basic Oxygen Furnaces: The basic oxygen furnace appears to be the most important furnace for the future. Fortunately, all of the basic oxygen furnaces operating in this country in 1960 were equipped with either wet scrubbers or electrostatic precipitators. Emissions from these control devices range between 55,000 $\mu g/m^3$ and 220,000 $\mu g/m^3$; emission rates of 0.5 to 1 ton/hr have been reduced to about 10 to 20 lb/hr. However, uncontrolled emissions range from 20 to 60 lb/ton of steel with an average of 40 lb/ton.

Gray Iron Cupola: The gray iron cupola is still used to melt over 90% of the gray iron produced in the U.S. There are over 6,000 foundries which use over 3,300 cupolas. These cupolas range in capacity from 1 to over 50 tons/hr. While the quantity and concentration of iron in the dust and fume emitted is dependent on the quality of scrap melted, tests indicate that emissions range from 10 to 45 pounds of dust per ton of melt. From 15 to 55% of the particles from these emissions measure less than 50μ, and 6 to 25% less than 10μ.

Coal: In general, the fly ash in coal contains 2 to 26.8% iron as Fe_2O_3 or Fe_3O_4, while the average concentration in West Virginia coal ash is 15.9%. The fly ash may range from 1 to 10% of the coal burned. Emissions of iron have been reported ranging from 2 to 37 lb/ton (0.1 to 1.8%) before fly-ash collection and 0.09 to 4 lb/ton (0.004 to 0.2%) after collection, depending on the type of equipment. It has been estimated that 6,000,000 tons/year of coal fly ash were produced in the Ohio River Valley in 1971. This fly ash would yield 1,200,000 tons of iron oxide which could make 1,000,000 tons/year of good, high-grade iron pellets for use in blast furnaces. Iron usually occurs in coal as pyrite, FeS_2. Recovery of sulfur and iron from pyrite concentrates requires crushing and grinding, which results in suspension of fine dusts.

Fuel Oil: Fly ash from burning residual fuel oil is most commonly about 69,000 to 80,000 $\mu g/m^3$ or approximately 2 g/lb of oil fired. The concentration of Fe_2O_3 in the fly ash is about 3.5%. This would result in an emission of 50,000 μg of Fe_2O_3 per pound of fuel oil burned. A boiler burning 1,000 pounds of oil per hour would be discharging about 50 grams of Fe_2O_3 per hour into the air.

Incineration: Incineration of municipal wastes may produce some iron pollution in the air. Burning such as this is reported to produce 17 pounds of particulate per ton. Emissions analyses from three incinerators in New York City showed that the collected fly ash contains 6.3% iron, while the fly ash passing through the control equipment contains 2.1% iron.

Welding Rods: Welding rods contribute some iron pollution to the air. In one study, the dust contained 25 to 30% of iron as Fe_3O_4 and gamma-Fe_2O_3.

Environmental Air Concentrations: Air quality data obtained from the National Air Sampling Network show that national average concentration in 1964 was 1.58 $\mu g/m^3$ and the national maximum was 22.0 $\mu g/m^3$. The particle size distribution of iron particulates from samples collected from ambient air in downtown Cincinnati, Ohio, and Fairfax, a suburb of Cincinnati showed that the concentrations and mass median diameters were significantly higher in downtown Cincinnati than in suburban Fairfax. The concentrations were 3.12 and 1.15 $\mu g/m^3$ respectively for Cincinnati and Fairfax, and the mass median diameters were 3.7 and 1.4μ respectively.

In Water

Potential industrial sources of dissolved iron species are reported to be:

> Mining operations
> Ore milling
> Chemical industries (organic,
>   inorganic, petrochemical)
> Dye industries
> Metal processing industries
> Textile mills
> Food canneries
> Tanneries
> Titanium dioxide production
> Petroleum refining
> Fertilizers

Iron exists in the ferric or ferrous form, depending upon conditions of pH and dissolved oxygen concentration. At neutral pH and in the presence of oxygen, soluble ferrous iron (Fe^{+2}) is oxidized to ferric iron (Fe^{+3}), which readily hydrolyzes to form the insoluble precipitate, ferric hydroxide, $Fe(OH)_3$. Therefore, acidic and/or anaerobic conditions are necessary in order for appreciable concentrations of soluble iron to exist. In addition, at high pH values ferric hydroxide will solubilize due to the formation of the $Fe(OH)_4^-$ anion. Ferrous and ferric iron may also be solubilized in the presence of cyanide, due to the formation of ferro-and ferri-cyanide complexes. Such species present considerable problems for both iron and cyanide treatment.

Perhaps the most significant industrial source of soluble iron waste is spent pickling solution. Pickling baths are employed to remove oxides from iron and steel during processing, or prior to plating of other metals on the surface. These baths contain strong acid solutions; usually sulfuric acid of 5 to 20% concentration by weight, although more recently hydrochloric acid has come into wider use. The baths accumulate soluble ferrous and ferric iron until their concentration interferes with product quality. At that time the bath must be replaced. According to Bramer (5), few industrial wastes have had the quantity of research and process development effort directed toward their treatment as has spent pickling solution.

Acid mine drainage waters are also high in soluble iron and represent a formidable treatment problem. It has been reported that 75% of the mines producing these wastes are inactive and abandoned (6). In one area, the flow from abandoned mines represents 90% of the total acid mine drainage of the region (7). Iron concentrations reported for some industrial wastewaters and mine waters are given in Table 58.

TABLE 58: IRON CONCENTRATIONS REPORTED FOR INDUSTRIAL WASTE-
WATERS

Process	Soluble Iron Concentration, mg/l
Mine Waters	
Mine Drainage	360 (36)*
Mine Drainage	10-200, 93 ave.
Mine adit	17-218
Tailings pond	3,200
Acid Mine water	501
Mine waste	67-70 (64-67)*
4 coal mines	122-330 (8-313)*
3 coal mines	40-150 (21-150)*
Motor Vehicle Assembly	
Body assembly	4
Vehicle assembly	3
Steel Processing	
Waste pickle liquor	96,800
Waste pickle liquor	70,000
Pickle bath rinse	200-5,000
Pickle bath rinse	60-1,300, 210 ave.
Steel cold finishing mills	60-150
Metal Processing	
Appliance manufacturer	0.09-1.9
Automobile heating controls	1.5-31
Appliances	
Mixed wastes	0.2-20
Spent acids	25-60
Chrome plating	40
Plating wastes	2-4

* Values in parentheses are for ferrous iron levels.

Source: Report PB 216,162

In addition to dissolved iron species, there is the question of total iron in wastewaters which includes insoluble iron oxides. McKee and Wolf (2) have pointed out that ferric oxides are used as pigments in certain paints. In addition, the authors indicate iron oxide is used as a polishing powder for glass, metal and ceramic materials. Such operations very likely represent a source of iron pollution, due to the colloidal nature of polishing compounds and the fact that water is usually involved as a coolant-lubricant and rinsing agent. In the processing of phosphoric acid, iron phosphates are formed and collected as a sludge, representing a potential waste source in the absence of efficient solids removal (3).

Iron and steel production represents a major source of small iron particles from the furnaces (5)(8), by virtue of air pollution control measures involving flue gas scrubbing. Steel mill wash waters may contain from 1,000 to 2,000 mg/l suspended solids, with their metal composition roughly reflecting the furnace charge. 50% of blast furnace suspended solids may be less than 50 microns in diameter, while open-hearth steel furnace solids may be even smaller (8). Ferro-manganese steel processing produces the highest percentage of semicolloidal particles, with 50% being less than 10 microns in diameter (5).

Milling of iron or steel products yields solids in the quench tank in the form of rust scale. The total iron loss during milling is roughly 2.5% of the finished product weight (5). Much of this scale is of large, readily settleable size, but as metal product thickness decreases so does the scale size, with finishing mill scale at the micron level. Quench water (8) and casting equipment cleaning water (9) in iron foundries represent another source of finely divided iron and iron oxide particles. Dean et al (4), have listed the following industries as potential sources of iron which may in part be particulate, or converted to particulate form during treatment:

> Organic chemicals
> Petrochemicals
> Alkalis
> Chlorine
> Inorganic chemicals
> Fertilizers
> Petroleum refining

References

(1) Sullivan, R.J., "Air Pollution Aspects of Iron and Its Compounds," *Report PB 188,088,* Springfield, Va., Nat. Tech. Information Service (Sept. 1969).

(2) McKee, J.E. and Wolf, H.W., *Water Quality Criteria,* 2nd ed., California State Water Quality Board Publication, No. 3-A (1963).

(3) Anonymous, *The Economics of Clean Water, Vol. III, Inorganic Chemicals Industry Profile,* Washington, D.C., U.S. Dept. Interior (1970).

(4) Dean, J.C., Bosqui, F.L., and Lanouette, K.H., "Removing Heavy Metals from Waste Water," *Env. Sci. Technol.,* 6, 518-522 (1972).

(5) Bramer, H.C., "Iron and Steel," in *Chemical Technology, Vol. 2, Industrial Wastewater Control,* C. Fred Gurnham, ed., New York, Academic Press (1965).

(6) Hill, R.D., "Control and Prevention of Mine Drainage," presented at the Environmental Resources Conference on Cycling and Control of Metals, Battelle Columbus Laboratories, Columbus, Ohio (1972).

(7) Wilmoth, R.C. and Hill, R.D., "Neutralization of High Ferric Iron Acid Mine Drainage," *EPA Report 14010* ETV 8/70 (1970).

(8) Kemmer, F.N., "Pollution Control in the Steel Industry," in *Industrial Pollution Control Handbook,* Herbert F. Lund, ed., New York, McGraw-Hill Book Co. (1971).

(9) Barzler, R.P. and Giffels, D.J., "Pollution Control in Foundry Operations," in *Industrial Pollution Control Handbook,* Herbert F. Lund, ed., New York, McGraw-Hill Book Co. (1971).

LEAD

The flow of lead in the United States has been traced and charted for the year 1970. The consumption was 1,361,000 tons, while primary and secondary production totaled 667,000 and 597,000 tons, respectively. Approximately 20% of the primary lead was recovered from foreign ore.

In Air

Emissions to the atmosphere during the year were 18,050 tons (Table 59). About 18% of the emissions resulted from the burning of waste oil, 13% from municipal incineration, 13% from grey iron foundries, 11% from the processing of gasoline additives, 9% from the production of primary lead, 9% from copper smelting, and 8% from the production of iron and steel. The combustion of coal, the manufacture of storage batteries, and the production of portland cement were also significant emission sources.

Emission estimates for mining, production of primary and secondary lead, lead oxide processing, and the manufacture of end use products are based on unpublished data obtained from industrial sources.

TABLE 59: LEAD EMISSIONS BY SOURCE—1970

Source Category	Source Group	Emissions - Tons	Emissions %
Mining and Milling		60	0.3
Metallurgical Industries		3,860	21.4
	Primary Lead	1,700	
	Primary Copper	1,700	
	Primary Zinc	240	
	Secondary Lead	220	
Lead Oxide		140	0.8
Consumer Product Manufacturing		3,080	17.0
	Storage Batteries	480	
	Gasoline Additives	1,900	
	Pigments	210	
	Solder	110	
	Cable Covering	50	
	Type Metal	200	
	Brass and Bronze	40	
	Metallic Lead Products	90	
Other Emission Sources		10,910	60.5
	Waste Incineration		
	Waste Oil	3,200	
	Municipal Incineration	2,400	
	Sewage & Sludge Incineration	200	
	Coal	650	
	Oil	90	
	Iron & Steel	1,500	
	Grey Iron Foundries	2,300	
	Ferroalleys	70	
	Cement	500	
		18,050	100.0

Source: Report PB 223,652

The emission factors presented in Table 60 are the best currently available. They were
determined through a combination of methods consisting of: direct observation of emission data and other related plant processing and engineering data; estimates based on information obtained from literature, plant operators, and others knowledgeable in the field;
calculations based on experience and personal knowledge of metallurgical processing operations; and, specific analytical results where available.

In Water

Lead is used as an industrial raw material for storage battery manufacture, printing, pigments, fuels, photographic materials, and matches and explosives manufacturing. It is one
of the most widely used nonferrous metals in industry. Despite its wide use, and the in-
dustry-associated lead bearing wastes which must result from its use, there is relatively
little information in the technical literature on the treatability of lead-bearing wastes, or
on associated waste treatment costs.

The storage battery industy is the largest consumer of lead, followed by the petroleum
industry in producing gasoline additives. In a survey of eight battery manufacturing plants,
it was reported that lead losses per battery manufactured ranged from 4.54 to 6,810 mg,
and that water usage ranged from 11 to 77 gallons per battery. Highest lead concentrations originated from the plate forming area. This water has a low pH, which leads to a
high concentration of dissolved lead. The particulate lead is very fine, and is reported to
require long detention periods (above 24 hours) for effective settling. The use of lead (as
sodium plumbite) in petroleum refining produces a lead sludge, from which lead is normally
recovered.

TABLE 60: LEAD EMISSION FACTORS

Source	Factor	Qualifier
Mining and Milling	0. 2 lb/ton lead mined (controlled)	Plant visit
Metallurgical Industries		
Primary Lead Production	5. 0 lb/ton of product (controlled)	Questionnaires
Primary Copper Production	0. 6 lb/ton of Cu concentrates (controlled)	Estimate
Primary Zinc Production	0. 3 lb/ton of Zn concentrates (controlled)	Estimate
Secondary Lead Production	0. 7 lb/ton of product (controlled)	Questionnaires
Lead Oxide Processing	0. 7 lb/ton of lead oxide (controlled)	Questionnaires
Consumer Product Manufacturing		
Storage Batteries	8. 0 lb/ton of lead processed (uncontrolled)	Questionnaires
Storage Batteries	1. 3 lb/ton of lead processed (controlled)	Questionnaires
Gasoline Additives	14. 0 lb/ton of lead processed (controlled)	Questionnaires
Solder	3. 0 lb/ton of lead processed (controlled)	Estimate
Cable Covering	2. 0 lb/ton of lead processed (controlled)	Questionnaires
Type Metal	17. 0 lb/ton of lead processed (controlled)	Questionnaires
Brass and Bronze	4. 0 lb/ton of lead processed (controlled)	Questionnaires
Other Emission Sources		
Waste Oil Combustion	0. 04 lb/bbl of oil burned (controlled)	Estimate
Municipal Incineration	0. 2 lb/ton of charge (controlled)	Stack sampling
Sewage and Sludge Incineration	0. 6 lb/ton of charge (uncontrolled)	Stack sampling
Coal Combustion	2. 2 lb/1,000 tons of coal burned (controlled)	Estimate
Distillate Oil Combustion	0. 1 lb/1,000 bbls of oil burned (uncontrolled)	Stack sampling
Residual Oil Combustion	0. 04 lb/bbl of oil burned (uncontrolled)	Stack sampling
Steel Production		
Open-Hearth	0. 14 lb/ton of steel (uncontrolled)	Atomic absorption
Basic Oxygen	0. 18 lb/ton of steel (uncontrolled)	Atomic absorption
Electric Arc	0. 18 lb/ton of steel (uncontrolled)	Atomic absorption
Grey Iron Foundries	0. 3 lb/ton of iron (uncontrolled)	Estimate
Cement Production	25 lb/1,000 bbl of cement production (uncontrolled)	Stack sampling
Pigments	10 lb/ton of lead processed (uncontrolled)	Questionnaires
Ferroalloys		
Silicomanganese Electric Furnace	0.9 lb/ton of product (uncontrolled)	Estimate
Ferromanganese Electric Furnace	0.4 lb/ton of product (uncontrolled)	Estimate
Blast Furnace	4 lb/ton of product (uncontrolled)	Estimate

Source: Report PB 223,652

Lead is also a common constituent of plating wastes, although not so frequently encountered as copper, zinc, cadmium and chromium. The lead content of wastewater from one engine-parts plating plant was reported as ranging from 2.0 to 140.0 mg/l. The higher concentrations resulted from dumping of spent plating bath solution. The quantity of lead discharged per day might amount to 112 lb (50,848 gr) from a single lead-plating bath. Rinse waters contained 0 to 30 mg/l of lead.

A lead solder-glass frit mixture is used to fuse the front glass panel to the funnel of television tubes. Quality control of color picture tubes may require the salvage and reuse of the glass envelopes. Separation of the two components is accomplished by dissolving the frit in dilute nitric acid, which yields an acidic wastewater containing lead and fluoride. Lead wastewater levels of 400 mg/l have been reported from such a salvage operation.

Anaylysis of lead mine wastewaters, lead levels in wastewater from the milling process, and final values following long-term detention of the wastewater in holding ponds have been reported. Pond detention times were not reported. Initial levels ranging from 0.018 to 0.098 mg/l and effluent levels of 0.022 to 0.055 mg/l were found. With one exception, lead reductions of approximately 45% were observed. This is likely due primarily to sedimentation of the particulate lead contained in the wastewater.

There are significant levels of lead in acid mine drainage. At the reported pH values, the lead would be predominately in the soluble form. Table 61 summarizes reported lead levels in industrial wastewaters.

TABLE 61: REPORTED LEAD LEVELS IN INDUSTRIAL WASTEWATERS

Industry	Lead, mg/l
Battery Manufacture,	
Particulate	5-48
Soluble	0.5-25
Plating	2-140
Plating	0-30
Television Tube Manufacture	400
Mine Drainage	0.02-2.5
Mining Process Water	0.018-0.098
Tetraethyl Lead Manufacture,	
Organic	126.7-144.8
Inorganic	66.1-84.9
Tetraethyl Lead Manufacture	45

Source: Report PB 216,162

Intermedia Aspects

The annual consumption of lead in the United States is one million tons. About 20% of the lead consumed is used for lead alkyls which are the antiknock ingredients in gasolines. Combustion of leaded gasoline is the major source of lead in the atmosphere; about 300,000 tons are added directly to the air annually. This results in an average lead concentration of 0.6 micrograms per cubic meter in urban atmosphere near ground level while in Los Angeles the atmospheric lead concentration is 5 micrograms per cubic meter. The combustion of lead alkyls in gasoline produces aerosol forms of inorganic lead salts such as lead chlorobromide. After emission, lead quickly becomes diluted in the atmosphere and it has been found that about 1,300 feet downwind from a freeway, the average lead concentration reduces to 22% of the roadside value.

Lead aerosols are thought to have a settling half-life of about three hours in urban atmospheres. In rural areas it has been estimated that the quantity of lead fallout is 26,400 tons per year.

Rivers can be contaminated with lead alkyls from atmospheric fallout. The major pathway by which lead alkyls reach surface waters in urban areas probably is by the discharge from storm sewers. About two-thirds of the urban fallout of lead alkyls, which are soluble salts, find their way into storm sewers. Such lead contributions amount to about 8,800 tons per year. The majority of this comes from automotive exhausts.

Surface waters can become contaminated directly from lead fallout. The seepage of lead wastes from scrap heaps and the weathering products of lead paints all contribute lead to surface waters.

References

(1) Davis, W.E. and Assoc., "Emission Study of Industrial Sources of Lead Air Pollutants, 1970," *Report PB 223,652,* Springfield, Va., Nat. Tech. Information Service (April 1973).
(2) Patterson, J.W. and Minear, R.A., "Wastewater Treatment Technology," 2nd ed., *Report PB 216,162,* Springfield, Va., Nat. Tech. Information Service (Feb. 1973).
(3) Stone, R. and Smallwood, H., "Intermedia Aspects of Air and Water Pollution Control," *Report EPA-600/5-73-003,* Wash., D.C., U.S. Environmental Protection Agency (Aug. 1973).

MAGNESIUM

In Air

Table 62 shows the principal sources of emissions of magnesium.

Reference

(1) GCA Corp., "National Emissions Inventory of Sources and Emissions of Magnesium," *Report-450/3-74-010,* Research Triangle Park, N.C., U.S. Environmental Protection Agency (May 1973).

MANGANESE

Manganese is one of the more abundant metals of the earth's crust. It is widely distributed in rocks, soils, fresh waters, seawaters, and sediments, and it is present in all or nearly all organisms. Its concentration is variable within each of these categories, often measured in parts per million, or in parts per billion in the case of waters. In rocks and soils, it ranges up to ores containing 50% manganese; sea-floor deposits have reported individual samples that high in manganese.

There are numerous valence states for manganese, with the divalent form giving the most stable salts, and the tetravalent, the most stable oxide. At least 100 minerals contain

TABLE 62: SOURCES AND ESTIMATES OF MAGNESIUM-CONTAINING EMISSIONS

Source	Uncontrolled Particulate Emission Factor (lb/ton)	(kg/kgx10³)	Reliability Code	Production Level (tons/yr)	% Mg in Emissions	Mg Emissions Before Controls (tons/yr)	Estimated Level of Emission Control	Mg Emissions After Controls (tons/yr)
MINING & ORE PROCESSING								
Mining, crushing, and drying of dolomite	465	233.0	(C)	2,200,000	13 (A)	66,500	95.0	3,320
Mining, crushing, drying, and briquetting of magnesite	475	238.0	(D)	631,000	40 (B)	59,950	90.3	5,790
OXIDE PRODUCTION								
Dolomite - vertical kilns	7	3.5	(B)	137,300	17 (C)	82	39.0	50
Dolomite - rotary kilns	180	90.0	(B)	1,236,000	17 (C)	18,911	81.0	3,593
Magnesite - rotary kilns	180	90.0	(C)	125,000	50 (C)	5,620	80.0	1,130
Hydroxide - rotary kilns	180	90.0	(C)	625,618	60 (B)	33,700	95.0	1,690
METALLURGY								
Magnesium production	0	0	(D)	136,500	60 (D)	0	0	0
Alloying and refining	4	2	(C)	124,000		149	90.0	15
REFRACTORIES								
Grinding and mixing	150	75.0	(C)	483,000	See Note a.	36,200	80.0	7,250
Electric casting	75	37.5	(C)	44,000	See Note a.	1,650	85.0	250
INADVERTENT SOURCES								
Iron and Steel								
Sinter Process	20 lb/ton sinter	10.0ᵈ	(B)	51,000,000	0.6 → 6.0 (C)	(3,060 → 30,600) 15,300ᵇ	90.0	(306 → 3,060) 1,530ᵇ
Blast Furnace	130 lb/ton pig iron	65.0	(B)	88,800,000	0.1 → 3.6 (C)	(5,770 → 207,900) 69,300ᵇ	99.0	(58 → 2,079) 693ᵇ
Open Hearth	17 lb/ton steel	8.5	(B)	65,800,000	0.2 → 0.7 (C)	(1,130 → 3,970) 1,700ᵇ	40.0	(670 → 2,360) 1,000ᵇ
Basic Oxygen	40 lb/ton steel	20.0	(B)	48,000,000	0.4 → 0.7 (C)	(3,840 → 6,700) 4,800ᵇ	99.0	(38 → 67) 48ᵇ
Electric Arc	10 lb/ton steel	5.0	(B)	16,800,000	0.2 → 9.2 (C)	(168 → 7,700) 4,200ᵇ	78.0	(37 → 1,700) 924ᵇ
Coal	N/A	N/A		33,800,000ᶜ	0.70 (A)	236,000	82.0	42,600
Oil	N/A	N/A		287,000ᶜ	0.15 (C)	430	0	430
Asbestos	N/A	N/A		6,579ᶜ	19.0 (A)	1,250	0	1,250
Cement	N/A	N/A		7,790,000ᶜ	0.07 → 1.15(C)	(5,400 → 89,000) 31,100ᵇ	88.0	(650 → 10,700) 3,730ᵇ
Totals						586,842	87.0%	75,293

NOTES:
a. Emission factor multiplier equal to tons of Mg processed or handled annually. See page 7 for explanation of Reliability Code
b. Intermediate value
c. Particulate generated, before control.   N/A - Not applicable.

Source: Report EPA 450/3-74-010

manganese as an essential element, and perhaps 200 others contain it as an accessory element. Its ores have wide distribution over the tropical, subtropical, and warmer temperate zones of the earth.

Manganese occurs in nature in constantly changing complex relations involving physical, chemical, and biologic activity, with the oxides being the dominant form, although carbonates and silicates occur commonly as minerals. Any study of the data must cope not only with the actual variations in content but also with variations resulting from different methods of sampling and analysis, which are of uncertain reliability and interpretation. This is particularly true with respect to water, with its complexity, low concentrations, and problems of sampling and analysis. Two samples of water taken at the same place at different times cannot necessarily be expected to show the same composition.

Man's principal use of manganese is in the production of steels and cast irons to nullify the harmful effects of sulfur, and approximately 90% (or more) of its consumption is by the iron and steel industry for this purpose or as an alloying agent. Steel cannot be made in quantity without it. It is commonly added in the form of ferromanganese or silicomanganese and less commonly as manganese metal or spiegeleisen. Manganese is important to the production of aluminum, magnesium, and some copper alloys. High-carbon ferromanganese and spiegeleisen are made in either blast or electric furnaces; the other manganese ferroalloys are made, with few exceptions, in electric furnaces; and manganese metal is most commonly the product of electrolysis.

Dioxide ores, either as particular ores or as a refined chemical, are essential constituents of the common dry-cell battery. The same or other ores have a variety of chemical and miscellaneous uses. In considering manganese emissions to the atmosphere as a result of various human activities, one does not find much in the way of good recent data. The data that are available must be used with care, taking into account various qualifying factors, such as geographic location, atmospheric conditions, degree of control, and nature of the sampling and analysis.

In Air

Table 63 summarizes the sources and quantities of manganese emissions to the atmosphere.

The air pollution aspects of manganese and its compounds have been reviewed in some detail by Sullivan (1).

In the past, the blast and the electric manganese ferroalloy furnaces have been the worst offenders. Examples of efficient cleaning systems are reported, but there does not appear to be any documentation of their extent of application or of the degree of their overall efficiency today. Although total volumes on an uncontrolled basis are much greater, emissions from steel furnaces and pig-iron blast furnaces have much lower manganese concentrations, are more easily controlled, and apparently are being efficiently controlled for the most part.

Emission of dusts arising from the handling of manganese ores and other manganiferous materials (including recovered dusts) and emission of fumes and dusts from foundries and miscellaneous uses can be controlled and probably are to be considered essentially local problems where they are not controlled. There appears to be little or no published information on such emission to the general atmosphere.

Manganese emission from the burning of coals and fuel oils apparently does not normally present a serious problem, particularly when one considers the cleaning practices in use. The U.S. Public Health Service initiated an air sampling program in 1953, which has grown into the National Air Surveillance Networks of the Environmental Protection Agency. Analyses for manganese that have been obtained in the program range from below the minimum detectable, for several nonurban western locations, to more than 10 $\mu g/m^3$, for some industrial urban sampling locations.

TABLE 63: MANGANESE EMISSIONS

Source	Amount in Tons	% This Pollutant
Manganese Mining	5	0.03
Primary Manganese Preparation	325	1.71
Iron and Steel		
Blast Furnace	1,000	5.27
Open Hearth Furnace	1,660	8.74
Basic Oxygen Furnace	1,060	5.58
Electric Arc Furnace	620	3.26
Gray Iron Foundry		
Cupola	2,770	14.58
Ferro-Alloy Preparation		
Blast Furnace	1,113	5.86
Electric Furnace	3,669	19.32
Non-Ferrous Alloy Preparation		
Furnaces	60	0.32
Material Handling	NEG	NEG
Silico Manganese Preparation		
Electric Furnace	4,164	21.92
Manganese Chemical Preparation	300	1.58
Dry Storage Battery Production	90	0.47
Welding Rod Production	24	0.13
Sewage and Sludge Burning	175	0.92
Power Plant Boilers		
Pulverized Coal	1,409	7.42
Stoker Coal	162	0.85
Cyclone Coal	49	0.26
All Oil	2	0.01
Industrial Boilers		
Pulverized Coal	62	0.33
Stoker Coal	218	1.15
Cyclone Coal	31	0.16
All Oil	2	0.01
Residential/Commercial Boilers		
Coal	20	0.11
Oil	3	0.02
TOTAL	18,993	100.10

Source: The MITRE Corporation, Preliminary Results, EPA Contract No. 68-01-0438

The average manganese concentration for urban air over the United States from 1953 through 1965, the last year for which sample results have been published, was approximately 0.10 $\mu g/m^3$ (2).

Tables 64 to Table 67 inclusive show manganese emission factors from various sources as taken from Anderson (3).

TABLE 64: EMISSION FACTORS FOR MANGANESE FROM INDUSTRIAL SOURCES

Source[a]	Emission factor	Emission factor symbol[b]
Mining	0.1 kg/10^3kg (0.2 lb/ton) of Mn mined	Q
Manganese metal production	13 kg/10^3kg (25 lb/ton) of Mn processed	Q
Production of manganese alloys		
Ferromanganese		
Blast furnace[c]	41 kg/10^3kg (82 lb/ton) of ferromanganese produced	E
Electric furnace	12 kg/10^3kg (24 lb/ton) of ferromanganese produced	E
Siliconmanganese, electric furnace	35 kg/10^3kg (70 lb/ton) of siliconmanganese produced	E
Processing of manganese and its compounds		

(continued)

TABLE 64: (continued)

Source[a]	Emission factor	Emission factor symbol[b]
Steel production (carbon and alloy steels)		
Blast furnace	0.4 kg/10^3kg (0.8 lb/ton) of pig iron produced	E
Open hearth furnace[d]		
Oxygen lance	0.06 kg/10^3kg (0.1 lb/ton) of steel produced	OES (1)
No lancing	0.3 kg/10^3kg (0.6 lb/ton) of steel produced	OES (1)
Basic oxygen furnace[e]	0.7 kg/10^3kg (1 lb/ton) of steel produced	E
Electric furnace		
Oxygen lance	0.2 kg/10^3kg (0.3 lb/ton) of steel produced	E
No lancing	0.1 kg/10^3kg (0.2 lb/ton) of steel produced	E
Cast iron	0.2 kg/10^3kg (0.3 lb/ton) of metal charged	E
Welding rods	8 kg/10^3kg (16 lb/ton) of Mn processed	Q
Nonferrous alloys	6 kg/10^3kg (12 lb/ton) of Mn processed	Q
Batteries	5 kg/10^3kg (10 lb/ton) of Mn processed	Q
Chemicals	5 kg/10^3kg (10 lb/ton) of Mn processed	Q
Cement plants		
Dry process		
Kiln[f]	0.04 kg/10^6kg (0.07 lb/10^3 tons) of feed	ES (1)
Feed to raw mill[f]	0.005 kg/10^6kg (0.01 lb/10^3 tons) of feed	ES (1)
Air separator after raw mill[f]	0.01 kg/10^6kg (0.02 lb/10^3 tons) of feed	ES (1)
Raw mill[f]	0.004 kg/10^6kg (0.008 lb/10^3 tons) of feed	ES (1)
Feed to finish mill[f]	0.004 kg/10^6kg (0.008 lb/10^3 tons) of feed	ES(1)
Air separator after finish mill[f]	0.005 kg/10^6kg (0.01 lb/10^3 tons) of feed	ES (1)
Wet process		
Kiln[g]	0.005 kg/10^6kg (0.01 lb/10^3 tons) of feed	ES (2)
Clinker cooler[g]	0.02 kg/10^6kg (0.03 lb/10^3 tons) of feed	ES (1)
Clinker cooler[h]	0.1 kg/10^6kg (0.2 lb/10^3 tons) of feed	ES(1)
Clinker cooler[f]	0.0002 kg/10^6kg (0.0004 lb/10^3 tons) of feed	OES, SSMS (1)
Air separator after finish mill[f]	0.00005 kg/10^6kg (0.0001 lb/10^3 tons) of feed	OES, SSMS (1)

[a] Emissions are uncontrolled unless otherwise specified.

· [b] Defined in Table 4 ; numbers in parentheses indicate number of samples analyzed.

[c] Average particle size of particulate matter is 0.3 micron.[1]

[d] Mean particle size of dust is 0.5 micron.[1]

[e] Mean particle size of particulate matter is 0.7 micron.[1]

· Exit from baghouse.

[g] Exit from electrostatic precipitator.

[h] Exit from two baghouses in parallel.

Source: Report PB 230,894

TABLE 65: EMISSION FACTORS FOR MANGANESE FROM FUEL COMBUSTION, COAL

Source	Manganese content, ppm	Emission factor	Emission factor symbol [a]
Power plant study			
South Carolina [b]	—[d]	0.0002 (0.0002 to 0.0003 kg/10^3 kg) 0.0004 (0.0003 to 0.0005 lb/ton)	EST (6)
Michigan [b]	5	0.0002 to 0.005 kg/10^3kg (0.0003 to 0.01 lb/ton)	EST (2)
Illinois [b]	30 to 40	0.0004 to 0.0005 kg/10^3kg (0.0008 to 0.0009 lb/ton)	EST (3)
Average [b]		0.0005 kg/10^3kg (0.001 lb/ton)	EST (11)
Kansas [c]	200	0.001 kg/10^3kg (0.002 lb/ton)	EST (3)
Six-boiler study	—[d]	0.04 kg/10^3kg (0.08 lb/ton)	EST (6)

[a] Defined in Table 4 ; numbers in parentheses indicate number of samples analyzed.

[b] Exit from electrostatic precipitator.

[c] Exit from limestone wet scrubber.

[d] Not reported.

TABLE 66: EMISSION FACTORS FOR MANGANESE FROM FUEL COMBUSTION, OIL

Source	Manganese content, ppm	Emission factor kg/10^3 liters	Emission factor lb/10^3 gal.	Emission factor symbol [a]
U. S. crude oil				
Arkansas	0.12	0.0001	0.0009	UK
California	0.138	0.0001	0.001	UK
Colorado	0.208	0.0002	0.001	UK
Kansas	0.013	0.00001	0.00009	UK
Montana	0.005	0.000004	0.00004	UK
New Mexico	0.021	0.00002	0.0001	UK
Oklahoma	0.030	0.00003	0.0002	UK
Texas	0.029	0.00003	0.0002	UK
Utah	1.45	0.001	0.01	UK
Wyoming	0.044	0.00004	0.0003	UK
U. S. crude oil, average	0.21	0.0002	0.001	UK
U. S. residual fuel oils, average	0.16	0.0002	0.001	UK
Residential units (distillate)	—[b]	0.00001 to 0.00004	0.0001 to 0.0003	EST (2)
Commercial units (residential No. 4)	—[b]	0.00002 to 0.00004	0.0002 to 0.0003	EST (2)
Commercial units (residential No. 5)	—[b]	0.00005 to 0.00007	0.0004 to 0.0006	EST(2)
Commercial units (residential No. 6)	—[b,c]	0.00005 to 0.00006	0.0004 to 0.0005	EST (2)

[a] Defined in Table 4 ; numbers in parentheses indicate number of samples analyzed.

[b] Not reported.

[c] Particle size range for particulate collected with a cascade impactor was 25 percent by weight less than 0.21 micron and 80 percent less than 7.4 microns. Mass median particle size was approximately 1.2 microns.

Source: Report PB 230,894

TABLE 67: EMISSION FACTORS FOR MANGANESE FROM WASTE INCINERATION

Source [a]	Emission factor, kg/10³kg (lb/ton) of waste burned	Emission factor symbol [b]
Sewage sludge incinerators		
Multiple hearth [c]	0.0005 (0.001) [f]	OES (4)
Fluidized bed [c]	0.0003 (0.0005) [g]	OES (3)
Solid waste incinerator [d]	0.02 (0.03)	EST(1)
Solid waste incinerator [e]	0.004 (0.007)	EST (1)

[a] In this table, all sources are controlled.

[b] Defined in Table 4 ; numbers in parentheses indicate number of samples analyzed.

[c] Exit from wet scrubber.

[d] Inlet to electrostatic precipitator (could be used as an uncontrolled emission factor).

[e] Exit from electrostatic precipitator.

[f] Range, 0.001 to 0.003 kg/10³kg (0.002 to 0.006 lb/ton).

[g] Range, 0.0002 to 0.0004 kg/10³kg (0.0003 to 0.0007 lb/ton).

Source: Report PB 230,894

In Water

From a study of atmospheric precipitation sampling for six metals at 32 stations distributed over the United States, it was concluded that manganese in atmospheric precipitation was primarily the result of man's activities. The average manganese concentration was 0.012 ppm in the precipitate. From a 1962 U.S. Geological Survey study of public water supplies of the 100 largest U.S. cities, it was calculated that the average manganese concentration of untreated surface waters was 0.070 ppm. This compares with the Public Health Service Drinking Water Standard's recommended upper limit of 0.05 ppm, which was established in 1962 primarily for esthetic and economic reasons, in view of the fact that no concentration was known to be dangerous to health.

This and other surveys disclosed mining and industrial pollution of waters, brought out the fact that the introduced manganese concentration was soon dissipated or diluted in rivers, and showed great differences in manganese concentration between samples taken at the same station at different times.

McKee and Wolf (4) have reported that manganese and its salts are used in the following industries:

> Steel alloy
> Dry cell battery
> Glass and ceramics
> Paint and varnish
> Ink and dye
> Match and fireworks

Among the many forms and compounds of manganese, only the manganous salts and the highly oxidized permanganate anion are appreciably soluble. The latter is a strong oxidant which is reduced under normal circumstances to insoluble manganese dioxide. The highly soluble manganous chloride is used in dyeing operations, linseed oil driers and electric batteries, while the equally soluble manganous sulfate is used in porcelain glazing, varnishes and production of specialized fertilizers (4).

Hill (5) reports manganese concentrations of 1.2 to 8.0 mg/l for mine waters. Kunin and Downing (6) cite a value as high as 50 mg/l for an acid mine water. Since the chemistry of manganese is similar to that of iron (7) it would be expected that any pickling operation involving manganese steel alloy results in dissolution of manganous ion and the presence of this ion in pickling and rinse solutions. In contrast to iron, the divalent (manganous) form is not readily oxidized to the insoluble manganic form, other than at elevated pH (7).

Intermedia Aspects

Mine effluents, fly ash, and impure fertilizers and soil additives are the chief sources of contamination of soils by manganese, but there are few reports of such contamination. The manganese content of soils depends primarily on the manganese content of the parent rocks as modified by physical, chemical, and biologic factors, such as weathering, drainage, and transportation. Manganese in the ecosystem has well-established lines of movement from rocks to soils to plants to animals, from soils to water to organisms, and back to water and soils. Marine organisms can concentrate manganese in their bodies to many times the concentration in seawater, and man retains manganese at a concentration three or four times higher than that in his food.

There are, in addition, less apparent movements of manganese among the components of the ecosystem; e.g., from atmosphere to water to soils, from water to sediments and vice versa, from soils to atmosphere with or without man's assistance, and even from soils to rocks (if one thinks in terms of geologic time), as pointed out in the National Academy of Sciences publication, "Manganese."

References

(1) Sullivan, R.J., "Air Pollution Aspects of Manganese and Its Compounds," *Report PB 188,079,* Springfield, Va., Nat. Tech. Information Service (Sept. 1969).
(2) National Academy of Sciences, "Manganese," a volume in the Medical and Biologic Effects of Environmental Pollutants, Series, Wash., D.C. (1973).
(3) Anderson, D., "Emission Factors for Trace Substances," *Report PB 230,894,* Springfield, Va., Nat. Tech. Information Service (Dec. 1973).
(4) McKee, J.E. and Wolf, H.W., *Water Quality Criteria,* California State Water Quality Control Board Publication No. 3-A (1963).
(5) Hill, R.D., "Control and Prevention of Mine Drainage," presented at the Environmental Resources Conference on Cycling and Control of Metals, Battelle Columbus Laboratories, Columbus, Ohio (1972).
(6) Kunin, R. and Downing, D.G., "Ion Exchange System Boasts More Pulling Power," *Chem. Eng.,* 78, 67-69 (June 28, 1971).
(7) Cotton, F.A. and Wilkinson, G., *Advanced Inorganic Chemistry,* New York, Interscience Publishers (1962).

MERCURY

Because of its high volatility, emissions of mercury emanate from any source where mercury is exposed to the atmosphere or where any material bearing mercury is processed. Some major sources of atmospheric mercury are considered to be coal-fired power plants, paint, primary nonferrous smelters, incinerators, mercury-cell chlor-alkali plants, primary and secondary mercury-processing plants, and general laboratories and hospitals.

Mercury and its compounds enter the atmosphere through emission in the form of vapors and particulates from industries and also by evaporation from soil and water. Once in the atmosphere, mercury is widely transported by wind currents. Eventually, some of the atmospheric mercury returns to the earth's surface as settleable particulates, but most of it returns with rainfall. Because the mercury that falls on soil does not penetrate deeply, it can reenter the atmosphere by evaporation or wash off into an aqueous system. Other ways by which mercury can gain entrance into aquatic systems are through settleable particulates, rainfall, and soil erosion. After entering the hydrosphere, all forms of mercury appear to be directly or indirectly capable of being converted by bacteria to highly toxic methyl-and dimethylmercury. The solubility of methylmercury in water causes it to be incorporated into the body tissues of aquatic life forms and, ultimately, into human food chain. Dimethyl-mercury can evaporate from the water system and reenter the atmosphere.

As is readily seen from the foregoing discussion, mercury is extremely mobile in the environment. Natural processes such as methylation, evaporation, and solution provide means for mercury compounds to cycle between air, water, and land for an indefinite time. Atmospheric mercury is not only a local inhalation hazard, but can contribute to contamination of food and drinking water or produce hazards in other ecological systems. Currently there are few existing data concerning atmospheric concentrations of mercury. Those data that do exist, however, indicate that a concentration of 1 μg Hg/m^3 may be approached, on a 24 hour basis, in large industrial cities. The measurement of mercury and its compounds in ambient air and from industrial-plant effluents has only recently received attention; as a result, measurement methodology is in a state of evolution.

In Air

Table 68 summarizes estimates of mercury emissions to the atmosphere by source. It should be noted that primary mercury is expected to decrease to 5.5 and electrolytic chlorine to 26.9 after implementation of the new emission standards.

TABLE 68: EMISSIONS OF MERCURY TO THE ATMOSPHERE*

| | Emissions (Short Tons) | |
	1968	1971
Mining	2.6	2.5
Processing		
Primary mercury**	55.0	33.5
Secondary mercury	0.5	0.5
Nonferrous: Copper	31.0	35.0
Zinc	9.7	11.0
Lead	4.4	5.0
Reprocessing		
Paint	0.8	0.8
Electrical apparatus	2.6	2.6
Consumptive uses		
Paint	215.0	229.0
Agricultural	18.8	8.1
Pharmaceuticals	2.6	4.2
Electrolytic chlorine (mercury cell)**	185.4	150.0
Instruments	2.6	2.6
Dental preparations	1.2	0.9
General laboratory use	4.8	6.9
Other	3.0	
Coal: Power plants	57.5	59.3
Other	34.5	31.8
Oil: Power plants	3.4	10.2
Incineration and other disposal		
Incineration	10.8	10.8
Sewage and sludge	4.4	4.4
Other	124.0	77.5
TOTAL	777.9	686.6

*Does not include estimates of crustal mercury emissions
**Sources covered by the mercury emission standard (NESHAPS)

Source: Report PB 222,802

Few industries currently control atmospheric mercury emissions solely for the sake of protecting public health. In general, economic reasons have dictated the use of those mercury emission controls that are employed. In the primary and secondary mercury industries, process efficiency improves with lower mercury emissions, thereby making reduction of mercury emissions profitable. Some primary nonferrous smelters collect mercury from their gaseous effluents as a by-product. The basic control method employs condensation to remove mercury from a gas stream. The amount of cooling accomplished depends on the temperature of the ambient air and available cooling water. In the mercury-cell chlor-alkali industry, the hydrogen stream is cooled to collect valuable mercury that must be replaced if atmospheric losses occur.

In certain cases, the hydrogen stream is either treated further with impregnated activated carbon or cooled to very low temperatures and sold as a by-product to the chemical industry. The mercury concentration in the chlor-alkali cell room is controlled to less than 100 $\mu g/m^3$, as recommended by the National Conference of Governmental Industrial Hygienists (NCGIH), to protect operating personnel from mercury poisoning. The control is maintained by dilution of the cell-room air with large ventilation flow rates, resulting in sizable atmospheric emissions from the cell rooms.

Tables 69 to Table 74 inclusive show emission factors for mercury from various sources.

TABLE 69: EMISSION FACTORS FOR PROCESSING AND UTILIZATION OF MERCURY AND ITS COMPOUNDS

Source[a]	Emission factor	Emission factor symbol[b]
Instruments	9 kg/10^3kg (17 lb/ton) of contained Hg	Q
Electrolytic production of chlorine Hydrogen stream Uncontrolled	0.001 to 0.005 kg/10^3kg (0.002 to 0.01 lb/ton) of Cl_2 produced	FAA (5)
After one carbon adsorber unit	0.0005 kg/10^3kg (0.001 lb/ton) of Cl_2 produced	FAA (2)
End-box ventilation	0.005 kg/10^3kg (0.01 lb/ton) of Cl_2 produced[c]	FAA (6)
Cell room ventilation Ridge vent Fan ventilation (1 fan)	0.002 kg/10^3kg (0.003 to 0.004 lb/ton) of Cl_2 0.0003 kg/10^3kg (0.0005 lb/ton) of Cl_2 produced	FAA (4) FAA (2)
Loss in hydrogen stream	0.01 kg/10^3kg (0.02 lb/ton) of Cl_2 produced	MB, Q
Loss in ventilation	0.02 kg/10^3kg (0.04 lb/ton) of Cl_2 produced	MB, Q
Comparative data Loss in hydrogen stream End-box ventilation Cell room ventilation	0.1 kg/10^3kg (0.2 lb/ton) of Cl_2 produced[d] 0.04 kg/10^3kg (0.075 lb/ton) of Cl_2 produced 0.02 kg/10^3kg (0.03 lb/ton) of Cl_2 produced[e]	UK UK UK
Paints	Negligible	—
Pharmaceuticals	Negligible	—
Pulp and paper	Negligible	—
Amalgamation	Negligible	—
Electrical apparatus	4 kg/10^3kg (8 lb/ton) of Hg used	E

[a]Uncontrolled unless otherwise specified.
[b]Defined in Table 4 ; number in parentheses indicates number of samples analyzed.
[c]Range, 0.003 to 0.01 kg/10^3kg (0.006 to 0.02 lb/ton).
[d]Range, 0.003 to 7.5 kg/10^3kg (0.006 to 15 lb/ton).
[e]Range, 0.005 to 0.027 kg/10^3kg (0.01 to 0.054 lb/ton).

Source: Report PB 230,894

TABLE 70: EMISSION FACTORS FOR MERCURY FROM SOLID WASTE INCINERATION

Source	Emission factor	Emission factor symbol [a]
Solid waste incineration	0.7 kg/10^6 kg (1 l5/10^3 tons) of refuse burned	E
Sewage sludge incineration Uncontrolled	0.02 kg/10^3 kg (0.03 lb/ton) sewage and sludge burned	E
After wet scrubber	Negligible	FAA (9)

[a] Defined in Table 4 ; number in parentheses indicates number of samples analyzed.

Source: Report PB 230,894

TABLE 71: EMISSION FACTORS FROM FUEL COMBUSTION FOR MERCURY, COAL

Source [a]	Mercury content, ppm	Emission factor kg/10^6 kg	Emission factor lb/10^3 tons	Emission factor symbol [b]
Eastern states coals				
Ohio	0.13	0.1	0.3	NA (3)
Belmont County	0.15 ± 0.03	0.15	0.30	NA, FAA, CAA, (32)
Harrison County	0.41 ± 0.06	0.41	0.82	NA, FAA, CAA (28)
Jefferson County	0.24 ± 0.04	0.24	0.48	NA, FAA, CAA (30)
West Virginia				
Kanawha County	0.07 ± 0.02	0.07	0.14	NA, FAA, CAA (27)
Pennsylvania				
Washington County	0.12 ± 0.04	0.12	0.24	NA, FAA, CAA (29)
Pittsburgh bed	0.24 ± 0.02	0.24	0.48	NA, FAA, DM (?)
Lower Kittanning	0.31 ± 0.03	0.31	0.62	NA, FAA, DM (?)
Washington County				
100-g/hr combustor	0.15 ± 0.02	—	—	FAA (12)
Fly ash	0.83 to 0.97 ± 0.13	0.09	0.18	FAA (12)
Flue gas	—	0.06	0.12	FAA (12)
227-kg/hr combustor	0.18 ± 0.04	—	—	FAA (23)
Fly ash	0.22 ± 0.04	0.05	0.09	FAA (17)
Flue gas	—	0.2	0.3	E (17)
Missouri				
Henry County				
100-g/hr combustor	0.24 ± 0.05	—	—	—
Fly ash	0.31 to 0.37 ± 0.06	0.08	0.16	FAA (21)
Flue gas	—	0.16	0.32	E (21)

(continued)

TABLE 71: (continued)

Source [a]	Mercury content, ppm	Emission factor kg/10⁶kg	Emission factor lb/10³ tons	Emission factor symbol [b]
Kentucky				
Muhlenberg County	0.19 ± 0.03	0.19	0.38	NA, FAA, CAA (30)
Indiana				
Clay County	0.08 ± 0.02	0.08	0.16	NA, FAA (23)
Illinois	0.18	0.18	0.36	NA (55)
Power plants				
Fly ash	0.10 to 0.26 ± 0.04	0.02	0.04	FAA (32)
Flue gas	—	0.2	0.3	E
New York (fly ash only)	0.13	0.2	0.3	FAA (10)
Western states coals				
Montana	0.08	0.08	0.2	NA (2)
Rosebud County	0.061 ± 0.007	0.061	0.122	NA, FAA, CAA (22)
Colorado	0.02	0.02	0.04	NA (2)
Montrose County	0.05 ± 0.01	0.049	0.098	NA, FAA, CAA (29)
Arizona	0.02	0.02	0.04	NA (1)
Navago County	0.06 ± 0.01	0.06	0.12	NA, FAA, CAA (26)
Utah	0.04	0.04	0.08	NA (1)
Average U.S. coals [c]	0.20	0.20	0.40	NA, FAA, CAA (246)

[a] All sources uncontrolled.

[b] Defined in Table 4 ; number in parentheses indicates number of samples analyzed.

[c] Based on results from Reference 11.

TABLE 72: EMISSION FACTORS FOR MERCURY FROM MINING, PRIMARY AND SECONDARY SOURCES

Source [a]	Emission factor	Emission factor symbol [b]
Mining	0.005 kg/10³kg (0.01 lb/ton) of ore mined	Q
Primary ore processing		
Smelter stack	0.16 kg/10³kg (0.031 lb/ton) of ore processed [c]	FAA (18)
Hoeing operations	0.01 kg/10³kg (0.02 lb/ton) of ore processed	FAA (3)
Retort operation	0.001 kg/10³kg (0.002 lb/ton) of ore processed [c]	FAA (3)
Secondary production	20 kg/10³kg (40 ob/ton) Hg processed	E

[a] All sources are uncontrolled.

[b] Defined in Table 4 ; numbers in parentheses indicate number of samples analyzed.

[c] Range, 0.09 to 0.22 kg/10³kg (0.18 to 0.44 lb/ton).

Source: Report PB 230,894

TABLE 73: EMISSION FACTORS FOR CONSUMPTIVE USES OF MERCURY AND ITS COMPOUNDS

Source	Emission factor	Emission factor symbol[a]
Paint	650 kg/10³kg (1300 lb/ton) of contained Hg	E
Agricultural spraying	500 kg/10³kg (1000 lb/ton) of contained Hg	E
Pharmaceuticals	200 kg/10³kg (400 lb/ton) of Hg applied	E
Dental preparations	10 kg/10³kg (20 lb/ton) of Hg handled	E
General laboratory handling	40 kg/10³kg (80 lb/ton) of Hg used	E

[a] Defined in Table 4 .

Source: Report PB 230,894

TABLE 74: EMISSION FACTORS FOR MERCURY FROM FUEL COMBUSTION, OIL

Source [a]	Mercury content, ppm	Emission factor		Emission factor symbol [b]
		kg/10³ liters	lb/10³ gal.	
Oils				
Imported residual oils				
No. 6 fuel oil, Mexico	0.30	0.0003	0.002	NA (1)
No. 6 fuel oil, Virgin Islands	0.22	0.0002	0.002	NA (1)
No. 6 fuel oil, Trinidad	0.10	0.00009	0.0008	NA (1)
No. 6 fuel oil, Curacao, N.A.	0.13	0.0001	0.001	NA (1)
No. 6 fuel oil, St. Croix, V.I.	0.007	0.000007	0.00006	NA (1)
Bunker "C" fuel oil, Venezuela	0.005	0.000005	0.00004	NA (1)
Average value for imported residual oils	0.13	0.0001	0.001	NA (6)
Imported No. 6 low-sulfur fuel oils				
Virgin Islands	0.007	0.000007	0.00006	NA (2)
Curacao, N.A.	0.02	0.0002	0.0002	NA (1)
Freeport, Bahamas	0.001	0.0000009	0.000008	NA (1)
Average value for imported low-sulfur fuel oils	0.009	0.000008	0.00007	NA (4)
Foreign crude oils				
Neutral Zone-Crude No. 24, Nevada Zone	0.20	0.0002	0.001	NA (1)
Mesa crude oil, Venezuela	0.05	0.00004	0.0004	NA (1)
Monogas crude oil, Venezuela	0.025	0.00002	0.0002	NA (1)

(continued)

TABLE 74: (continued)

Source[a]	Mercury content, ppm	Emission factor		Emission factor symbol[b]
		kg/10^3 liters	lb/10^3 gal.	
Jobo crude oil, Venezuela	0.016	0.00001	0.0001	NA (1)
Bosean crude oil, Venezuela	0.02	0.00002	0.0001	NA (1)
Kenai Peninsula, Venezuela	0.006	0.000005	0.00004	NA (1)
UMM Tarvo, Libya	0.01	0.000009	0.00007	NA (1)
Gamba, Gabon	0.03	0.00003	0.0002	NA (1)
Tia Juanna, Venezuela	0.05	0.00004	0.0004	NA (1)
Ral Al Khafti, Kuwait	0.09	0.00008	0.0006	NA (1)
Durl, Sumatra	0.04	0.00003	0.0003	NA (1)
Darius, Iran	0.01	0.000009	0.00007	NA (1)
Average value for foreign crude oils	0.05	0.00004	0.0004	NA (12)
U.S. crude oils				
Wesson, Arkansas	0.03	0.00003	0.0002	NA (1)
Midland Farms, Texas	0.08	0.00007	0.0006	NA (1)
East Texas, Texas	0.007	0.000006	0.00005	NA (1)
Yates, Texas	0.06	0.00005	0.0004	NA (1)
Vacuum, New Mexico	0.20	0.0002	0.001	NA (1)
St. Tedesa, Illinois	0.076	0.00006	0.0005	NA (1)
Maysville, Oklahoma	0.002	0.000002	0.00001	NA (1)
Hall-Gurney, Kansas	0.006	0.000005	0.00004	NA (1)
Huntington Beach, California	0.11	0.00009	0.0008	NA (1)
Main Pass, Louisiana	0.06	0.00005	0.0004	NA (1)
Average value for U.S. crude oils	0.05	0.00004	0.0004	NA (10)

[a] All sources uncontrolled.

[b] Defined in Table 4 ; numbers in parentheses indicate number of samples analyzed.

Source: Report PB 230,894

In Water

The major consumptive user of mercury is the chlor-alkali industry, using the DeNora or similar electrolytic cells (7)(8)(9)(10)(11)(12), whose losses to waste discharge without process modification or wastewater treatment have been variously reported as 0.1 to 0.5 pound mercury per ton of chlorine produced (8)(9)(11). The second largest user is the electrical and electronics industry (8)(9)(11). Other users of various mercury forms are: explosives manufacturing, photographic industry, and the pesticide and preservative industry (7)(8)(10)(11)(12).

Within the chemical and petrochemical industry, mercury compounds have found use as catalysts in plastics production, hydrogenation and dehydrogenation, sulfonation, oxidation, chlorination and acidolysis (7)(10). Two of the most widely publicized uses are probably the use of mercuric chloride in vinyl chloride production and mercuric oxide in acetaldehyde production from acetylene. Both processes were involved in the Minamata Bay mercury pollution incident (7), although the latter has been assigned responsibility for the mercury input to the bay.

Both the paint and pharmaceutical-cosmetic industries have been users of mercury compounds as preservatives and antifouling or mildew proofing agents (7)(8)(10). Pulp and paper operations use mercurials as fungicides, slimicides and biocides (7)(8)(10), but recently use has been curtailed. As of August 1970, 48 algacides, slimicides and mildew controlling agents had been removed from the market, with indications of additional cancellations to follow (8).

Other sources of mercury wastes are seed preservatives, research and hospital labs, sealants used in engineering, thermometers and manometers, and film production and processing (7)(10). Power generation is a large source of mercury release into the environment, through combustion of fossil fuel (10)(11). Scrubber devices installed on thermal power plant stacks for sulfur dioxide removal could conceivably accumulate mercury if extensive recycle is practiced. Coal has been reported to contain 0.010 to 300 mg mercury/kg (12).

Limited data is available regarding actual wastewater mercury levels. Wastewater emanating from an acetaldehyde production unit was reported to contain 20 mg/l of mercury and the combined wastewater, prior to treatment, contained mercury at 18 mg/l (13). An unidentified wastewater was reported to contain 400 to 600 mg/l of mercury at discharge, and mercury levels of 20 mg/l were observed 1 mile downstream (14). Effluent mercury from one chlor-alkali plant is estimated to have been 1.4 to 2.8 mg/l, based on 600 gpm waste discharge and 10 to 20 lb/day mercury loss (11).

Cranston and Buckley (15) report the mercury effluent data shown in Table 75. They also reported river water mercury concentrations of 0.134 to 0.380 mg/l near a municipal outfall. In conjunction with these values it was mentioned that commercial bleach preparations alone contain up to 0.200 mg/l of mercury. In contrast, D'Itri (12) reports normal background levels of mercury in natural waters to range from 0.02 to 0.7 μg/l. Minamata Bay water concentrations have been reported to be 0.36 to 1.6 μg/l, and it has been estimated that 600 tons of mercury have entered the bay (12).

TABLE 75: LEVELS OF MERCURY IN INDUSTRIAL EFFLUENTS

Waste	Dissolved Mercury, mg/l	Suspended Solids Mercury, mg/Kg dry weight of solids
Paper Mill Settling Pond	.00008	10
Paper Mill Effluent	.002-.0034	5.6
Fertilizer Mill	.00026-.004	32.0
Smelting Plant	.002-.004	-
Chlor-Alkali Plant	.080-2.0	14.0

Source: R.E. Cranston and D.E. Buckley

Another source of mercury is reported by Morrison (16). A particular natural gas containing mercury at 180 $\mu g/m^3$ was reduced to 40 $\mu g/m^3$ by condensing the water and organic vapors. Although a major portion of the mercury remained in the condensed organic fraction, calculations indicated that 17 to 27 pounds of mercury had been discharged into the nearby river over a six year period.

Intermedia Aspects

There is a natural intermedia transfer of mercury in the environment which is increased because of the additional mercury contributed by industrial processes. Mercury can enter the atmosphere in both the gaseous and particulate forms. Its mobility is enhanced by its physical properties, which are unique among metals. As a metallic liquid its vapor pressure is relatively high for a metal and makes possible natural land to air transfer.

Gaseous and particulate mercury compounds are commonly contained in the emissions from various industrial processes. Mercury has been found in 36 United States' coals. A conservative estimate of the average mercury concentration in all United States' coals is around 1 mg/l. Approximately 2 billion tons of coal are burned annually in the United States, resulting in a release of 3,000 tons of mercury to the environment.

Mercury vapor discharges from some coal burning power plants, municipal incinerators, and several types of industrial plants may range from 100 to 5,400 pounds per year at individual sites. The estimated annual rate of mercury vapor discharge from 12 locations in Missouri and Illinois exceeds the rate of mercury discharge into waterways by the nation's 50 major mercury polluters. Mercury vapors can also escape to the atmosphere as dust from open pit mining.

Mercury is continually being removed from the atmosphere and deposited on the earth's surface. Estimates of the input rate of mercury from the atmosphere to the entire global surface from fallout are between 5.50×10^7 and 9.68×10^8 pounds per year. The fallout is expected to be higher in the more industrial areas.

About 5,000 tons of mercury per year are released to rivers by continental weathering. Mine drainage contributes significant amounts of mercury to streams. Soluble mercury introduced into streams is rapidly reduced to its metallic mercury form by various natural chemical processes. Approximately 8,800 pounds per year of mercury is discharged into Southern California coastal waters via sewage effluents. The average mercury concentration of the effluent is 0.003 mg/l.

Localized mercury inputs from sewage outfalls result in mercury concentrations near the outfalls which are 50 times larger than natural concentrations. Ocean sediment samples collected near Los Angeles sewage outfalls contained about 1 mg/l mercury on a dry weight basis as compared to a control area concentration of 0.02 mg/l. No attempt has yet been made to study air-water interactions of mercury, but this mode of transfer has potential significance.

References

(1) Stahl, Q.R., "Air Pollution Aspects of Mercury and Its Compounds," *Report PB 188,074*, Springfield, Va., Nat. Tech. Information Service (Sept 1969).
(2) Jones, H.R., *Mercury Pollution Control,* Park Ridge, N.J., Noyes Data Corp. (1971).
(3) U.S. Environmental Protection Agency, "Background Information—Proposed National Emission Standards for Hazardous Air Pollutants: Asbestos, Beryllium, Mercury, *Publ. No. APTD-0753,* Research Triangle Park, N.C., Office of Air Programs (Dec. 1971).
(4) U.S. Environmental Protection Agency, *Mercury and Air Pollution—A Bibliography with Abstracts,* Research Triangle Park, N.C., Office of Air Programs (Oct. 1972).
(5) U.S. Environmental Protection Agency, "Background Information on Development of National Emission Standards for Hazardous Air Pollutants: Asbestos, Beryllium

and Mercury," *Report PB 222,802,* Research Triangle Park, N.C., Office of Air and Water Programs (March 1973).

(6) Anderson, D., "Emission Factors for Trace Substances," *Report PB 230,894,* Springfield, Va., Nat. Tech. Information Service (Dec. 1973).

(7) Anonymous, "Mercury Pollution: Interest in it Now on the Upswing," *Wat. Wastes Eng.,* 8, A13-A16 (January, 1971).

(8) National Industrial Pollution Control Council, *Mercury,* Staff Report, Wash., D.C., U.S. Government Printing Office 0-409-779 (October, 1970).

(9) Zugger, P.D. and Ghosh, M.M., "The Effect of Mercury on the Activated Sludge Process," presented at the 27th Indust. Waste Conf., Purdue Univ. (1972).

(10) Cheremisinoff, P.N. and Habib, Y.H., "Cadmium, Chromium, Lead, Mercury: A Plenary Account for Water Pollution, Part 1, Occurrence, Toxicity and Detection," *Wat. Sew. Works,* 119, 73-85 (1972).

(11) Turney, W.G., "Mercury Pollution: Michigan's Action Program," *Jour. Wat. Poll. Control Fed.,* 43, 1427-1438 (1971).

(12) D'Itri, F.M., "Mercury in the Aquatic Ecosystem," *Technical Report No. 23,* Institute of Water Research, East Lansing, Michigan, Michigan State University (1972).

(13) Irukayama, K., "The Pollution of Minamata Bay and Minamata Disease," Proc. 3rd Int'l Conf. Wat. Poll. Res., Munich, 3, 153-180 (1966).

(14) Krenkle, P.A., "Report International Conference on Environmental Mercury Contamination-Summary," *Wat. Res.,* 5, 1121-1122 (1971).

(15) Cranston, R.E. and Buckley, D.E., "Mercury Pathways in a River and Estuary," *Env. Sci. Techn.,* 6, 274-278 (1972).

(16) Morrison, J., "NAM Recovers Mercury Produced with Dutch Natural Gas," *Oil Gas Jour.,* 70, 72-73 (April, 1972).

(17) Stone, R. and Smallwood, H., "Intemedia Aspects of Air and Water Pollution Control," *Report EPA-600/5-73-003,* Wash., D.C., U.S. Environmental Protection Agency (Aug. 1973).

MOLYBDENUM

In Air

Table 76 summarizes information on sources and emissions of molybdenum.

Reference

(1) GCA Corp., "National Emissions Inventory of Sources and Emissions of Molybdenum," *Report PB 230,035,* Springfield, Va., Nat. Tech. Information Service (May 1973).

NICKEL

In Air

Nickel air pollution usually occurs as particulate emissions and may be controlled with the particulate by the usual dust handling equipment such as bag filters, precipitators, and scrubbers. Nickel poses no unique control problems except with gaseous nickel carbonyl. Control of nickel carbonyl is accomplished by taking advantage of the fact that nickel carbonyl decomposes at temperatures above 60°C. For this reason gases containing nickel carbonyl are usually passed through a furnace, where the nickel carbonyl decomposes into its two components, nickel and carbon monoxide. The resulting nickel metal can then be removed as particulate.

TABLE 76: SOURCES AND ESTIMATES OF MOLYBDENUM-CONTAINING EMISSIONS

Source	Uncontrolled Particulate Emission Factor (lb/ton)	(kg/kg×10³)	Reliability Code	Production Level (tons/yr)	% Mo in Emissions	Mo Emissions Before Controls (tons/yr)	Estimated Level of Emissions Control	Mo Estimated Emissions Following Controls (tons/yr)
MINING								
Open Pit	10	15	D	13,000	*	65	0	65
Underground	0.5	0.25	D	25,986	*	6	0	6
Copper Open Pit	10	5	D	16,690	*	83	0	83
BENEFICIATION	52	26	D	34,821	*	906	98.8%	11
ROASTING	100	50	C	34,821	*	1,740	98%	35
METALLURGICAL								
Ferromolybdenum-Electric Arc	200	100	D	6,221	*	622	90%	62
Molybdenum Metal	2000	1000	B	803	*	803	99.9%	1.0
CHEMICAL PRODUCTION	Nil	Nil		1,148	*	Nil	Nil	Nil
STEEL & ALLOY PRODUCTION								
Steel	25	12.5	C	14,712	*	184	78%	41
Cast Iron	15	7.5	C	1,958	*	15	99%	0.1
Super Alloys	200	100	C	1,254	*	126	78%	28
Alloys	25	12.5	C	395	*	5	78%	1
Mill Products	0.5	0.25	C	858	*	0	-	0

(continued)

TABLE 76: (continued)

Source	Uncontrolled Particulate Emission Factor (lb/ton)	Uncontrolled Particulate Emission Factor (kg/kg×10³)	Reliability Code	Production Level (tons/Yr)	% Mo in Emissions	Mo Emissions Before Controls (tons/yr)	Estimated Level of Emissions Control	Mo Estimated Emissions Following Controls (tons/yr)
NON-METALLURGICAL USES	2	1	D	1,861	*	2	0	2
INADVERTENT SOURCES								
Coal burning	NA	NA		33,800,000 (uncontrolled emissions)	.01 (B)	3,380	82%	610
Oil combustion	NA	NA		287,000 (uncontrolled emissions)	.01 (C)	29	0	29
Mineral Processing	NA	NA		6.9 x 10⁶ (controlled emissions)	.00023 (D)	16	NA	16
TOTALS						7,179	86%	990

NOTE: NA = Not Applicable

See page 7 for explanation of Reliability Code

*Emission factor multiplier equal to tons of Mo processed annually

Source: Report PB 230,035

The air pollution aspects of nickel and its compounds have been reviewed by Sullivan (1).

Table 77 summarizes the quantities and sources of nickel emissions to the atmosphere. Tables 78 to Table 81 inclusive show emission factors for nickel from various sources.

TABLE 77: NICKEL EMISSIONS

Source	Amount in Tons	% This Pollutant
Nickel Mining	2	0.03
Iron and Steel		
Blast Furnace	100	1.67
Gray Iron Foundry		
Cupola	79	1.32
Ferro-Alloys		
Blast Furnace	491	8.20
Electric Furnace	98	1.64
Non-Ferrous Alloys		
Furnaces	64	1.07
Material Handling	NEG	NEG
Power Plant Boilers		
Pulverized Coal	87	1.45
Stoker Coal	10	0.17
Cyclone Coal	3	0.05
All Oil	1,441	24.08
Industrial Boilers		
Pulverized Coal	7	0.12
Stoker Coal	23	0.38
Cyclone Coal	3	0.05
All Oil	1,139	19.03
Residencial/Commercial		
Coal	3	0.05
Oil	2,435	40.69
TOTAL	5,985	100.00

Source: The MITRE Corporation, Preliminary Results, EPA Contract No. 68-01-0438

TABLE 78: EMISSION FACTORS FOR NICKEL FROM FUEL COMBUSTION, COAL

Source [a]	Nickel content, ppm	Emission factor kg/10³kg	Emission factor lb/ton	Emission factor symbol [b]
Coal-fired boilers	— [e]	0.0002	0.003	EST (6)
Power plant study				
South Carolina [c]	— [e]	0.0003 (0.0002 to 0.0005)	0.0007 (0.0004 to 0.0009)	EST (6)
Michigan [c]	10	0.0003 (0.0002 to 0.0005)	0.0006 (0.0003 to 0.0009)	EST (2)
Illinois [c]	10 to 20	0.0002	0.0003	EST (2)
Average [c]	12	0.0003	0.0006	EST (10)
Kansas [d]	30	0.0003 (0.0002 to 0.0005)	0.0006 (0.0003 to 0.0009)	EST (2)

[a] Uncontrolled unless otherwise specified.

[b] Defined in Table 4 ; numbers in parentheses indicate number of samples analyzed.

[c] Exit from an electrostatic precipitator.

[d] Exit from wet scrubber.

[e] Not reported.

Source: Report PB 230,894

TABLE 79: EMISSION FACTORS FOR NICKEL FROM INDUSTRIAL SOURCES

Source [a]	Emission factor	Emission factor symbol [b]
Mining and metallurgical	9 kg/10^3kg (17 lb/ton) of Ni produced	MB
Processing of nickel and its compounds		
Stainless steel	5 kg/10^3kg (10 lb/ton) of Ni charged or 0.3 kg/10^3 kg (0.6 lb/ton) of stainless steel produced	Q
Nickel alloy steels [c]	5 kg/10^3kg (10 lb/ton) of Ni charged	MB
Iron and steel scrap	0.0008 kg/10^3 kg (0.0015 lb/ton) of steel and iron	MB
Nickel alloys (other) [d]	1 kg/10^3kg (2 lb/ton) of Ni charged	E
Copper base alloys [d]	1 kg/10^3kg (2 lb/ton) of Ni charged	E
Electrical alloys [d]	1 kg/10^3kg (2 lb/ton) of Ni charged	E
Cast iron	10 kg/10^3kg (20 lb/ton) of Ni charged	E
Electroplating	Negligible	—
Batteries	4 kg/10^3kg (8 lb/ton) of Ni processed	Q
Catalysts	Negligible	—
Cement plants		
Dry process		
Kiln [d]	0.2 kg/10^6kg (0.3 lb/10^3 tons) of feed	ES (1)
Feed to raw mill [d]	0.005 kg/10^6kg (0.01 lb/10^3 tons) of feed	ES (1)
Air separator after raw mill [d]	0.0005 kg/10^6kg (0.001 lb 10^3 tons) of feed	ES (1)
Raw mill [d]	0.0003 kg/10^6kg (0.0006 lb/10^3 tons) of feed	ES (1)
Air separator after finish mill [d]	0.002 kg/10^6kg (0.003 lb/10^3 tons) of feed	ES (1)
Feed to finish mill [d]	0.005 kg/10^6kg (0.01 lb/10^3 tons) of feed	ES (1)
Wet process		
Kiln [e] (2 different plants)	0.1 to 1 kg/10^6kg (0.2 to 2 lb/10^3 tons) of feed	ES (2)
Clinker cooler [d]	0.002 kg/10^6kg (0.004 lb/10^3 tons) of feed	SSMS, OES (2)
Clinker cooler [e]	0.05 kg/10^6kg (0.1 lb/10^3 tons) of feed	ES (2)
Clinker cooler [f]	0.1 kg/10^6kg (0.2 lb/10^3 tons) of feed	ES (2)
Finishing mill after air separator [d]	0.002 kg/10^6kg (0.004 lb/10^3 tons) of feed	SSMS, OES (1)

[a] Uncontrolled unless otherwise specified.

[b] Defined in Table 4 ; numbers in parentheses indicate number of samples analyzed.

[c] Considered controlled.

[d] Exit from baghouse.

[e] Exit from electrostatic precipitator.

[f] Exit from two baghouses (in parallel).

Source: Report PB 230,894

TABLE 80: EMISSION FACTORS FOR NICKEL FROM FUEL COMBUSTION, OIL

Source [a]	Nickel content, ppm	Emission factor		Emission factor symbol [b]
		kg/10^3 liters	lb/10^3 gal.	
U.S. crude oils				
Texas	2.4 (1.4 to 4.3)	0.002 (0.001 to 0.004)	0.02 (0.01 to 0.03)	UK (3)
Louisiana	2.1	0.002	0.01	UK
Kansas	6.3	0.005	0.04	UK
Wyoming	2.7	0.002	0.02	UK
California	64	0.05	0.5	UK
Alaska	11.3	0.01	0.08	UK
Average for U.S. crude oils	15 (1.4 to 64)	0.01	0.1	UK
Average for imported crude oils	10 (0.3 to 28.9)	0.009	0.07	UK
Foreign crude oils				
Monogas crude, Venezuela	44.1	0.04	0.3	CAA (1)
Crude oil, Venezuela	7	0.006	0.05	CAA (1)
Nigerian crude oil	8.8	0.007	0.06	CAA (1)
Jobo crude, Puerto Ordaz, Venezuela	59	0.05	0.4	CAA (1)
Aguasay crude oil, Venezuela	1.8	0.002	0.01	CAA (1)
Average for foreign crude oils	25.6	0.02	0.2	CAA (5)
Foreign residual oils				
No. 6 fuel oil, Virgin Islands	31.8 to 33.8	0.03	0.3	CAA (2)
No. 6 fuel oil, Venezuela	40.5 (4 to 61.2)	0.04 (0.004 to 0.06)	0.3 (0.03 to 0.5)	CAA (7)
No. 6 fuel oil, Aruba, N.A.	35.2	0.03	0.3	CAA (1)
No. 6 fuel oil, St. Croix, V.I.	15.5	0.01	0.1	CAA (1)
No. 6 fuel oil, El Patito, Venezuela	59.7	0.06	0.5	CAA (1)
No. 6 fuel oil, Puerto La Auz Venezuela	48	0.05	0.4	CAA (1)
No. 6 fuel oil, Spain	17.9	0.02	0.1	CAA (1)
No. 6 fuel oil, Trinidad	29 to 35	0.03	0.2 to 0.3	CAA (2)
No. 6 fuel oil, Canada	22.7	0.02	0.2	CAA (1)
Average for foreign fuel oils	36.3 (4 to 61.2)	0.03 (0.004 to 0.06)	0.3 (0.03 to 0.5)	CAA (17)
U.S. boilers				
Residential units (distillate)	— d	0.00005 to 0.00006	0.0004 to 0.0005	EST (2)
Commercial units (residual No. 6)	48.5	0.01	0.08	EST (1)
Commercial units (residual No. 5)	31	0.007	0.06	EST (1)
Commercial units (residual No. 4)	18	0.006	0.05	EST (1)

[a] Uncontrolled unless otherwise specified.
[b] Defined in Table 4 ; numbers in parentheses indicate number of samples analyzed.
[c] Particle size distribution (particulate collected with a cascade impactor): 25 percent (by weight) less than 0.21 micron,
—80 percent less than 7.4 microns, mass mean particle size approximately 1.2 microns.
[d] Not reported.

Source: Report PB 230,894

TABLE 81: EMISSION FACTORS FOR NICKEL FROM WASTE INCINERATION

Source [a]	Emission factor	Emission factor symbol [b]
Sewage sludge incinerators		
Multiple hearth [c]	0.002 kg/10^3kg (0.003 lb/ton) of solid waste incinerated [d]	OES (4)
Fluidized bed [c]	0.0002 kg/10^3kg (0.0003 lb/ton) of solid waste incinerated [e]	OES (2)
Municipal incinerator		
Refuse only [c]	0.002 kg/10^3kg (0.003 lb/ton) of solid waste incinerated	CAA (3)
Refuse and sludge [c]	0.003 kg/10^3kg (0.005 lb/ton) of solid waste incinerated	CAA (3)
Lubricating oil	0.008 kg/10^3 liters (0.07 lb/10^3 gal.) of lubricating oil [f]	UK

[a] In this table, all sources except lubricating oil are controlled.

[b] Defined in Table 4 ; numbers in parentheses indicate number of samples analyzed.

[c] Emission from wet scrubber.

[d] Range, 0.0003 to 0.004 kg/10^3kg (0.0006 to 0.008 lb/ton).

[e] Range, 0.0001 to 0.0002 kg/10^3kg (0.0002 to 0.0003 lb/ton).

[f] Emission factor for nickel oxides.

Source: Report PB 230,894

In Water

Wastewaters containing nickel originate primarily from metal industries; particularly plating operations. High levels of nickel have also been reported in wastes from silver refineries (3). In addition, basic steel works and foundaries, motor vehicle and aircraft industries, and printing have been mentioned as potential sources of nickel (4).

Golomb (5) states that generally, rinse waters may be considered to contain 1% of the plating bath concentration. Lowe (6) has reported that acid and sulfamate nickel plating baths generally contain nickel at 40,000 to 90,000 mg/kg of solution, while electrodeless process baths are of weaker concentration, containing 7,000 to 7,500 mg/kg of solution.

Industries concerned primarily with copper or brass plating or processing may also contain nickel in plant wastewaters, but generally at low levels, 1 mg/l or less, and are therefore of little concern. In addition, Hill (7) has reported nickel concentrations ranging from 0.46 to 3.4 mg/l in acid mine drainage.

Nickel exists in waste streams as the soluble ion. In the presence of complexing agents such as cyanide, nickel may exist in a soluble complexed form. The presence of nickel cyanide complexes interferes with both cyanide and nickel treatment, and the presence of this species may be responsible for increased levels of both cyanide and nickel in treated wastewater effluents.

Nickel concentrations typically found in plating and metal processing wastes are summarized in Table 82 (8).

TABLE 82: SUMMARY OF NICKEL CONCENTRATIONS IN METAL PROCESSING
AND PLATING WASTEWATERS

Industry	Nickel Concentration, mg/l	
	Range	Average
Tableware Plating		
Silver bearing waste	0-30	5
Acid Waste	10-130	33
Alkaline waste	0.4-3.2	1.9
Metal Finishing		
Mixed wastes	17-51	-
Acid wastes	12-48	-
Alkaline wastes	2-21	-
Small parts fabrication	179-184	181
Combined degreasing, pickling and Ni dipping of sheet steel	3-5	-
Business Machine Manufacture		
Plating wastes	5-35	11
Pickling wastes	6-32	17
Plating Plants		
4 different plants	2-205	-
Rinse waters	2-900	-
Large plants	up to 200	25
5 different plants	5-58	24
Large plating plant	88 (single waste stream)	-
	46 (combined flow)	-
Automatic plating of Zinc base castings	45-55	-
Automatic plating of ABS type plastics	30-40	-
Manual barrel and rack	15-25	-

Source: Report PB 216,162

Intermedia Aspects

Nickel has many industrial applications, but nickel alloys used in many food-processing operations are considered to be the major sources of nickel in water and soil and which ultimately reach man. Nickel carbonyl is considered to be the most serious environmental hazard of all the nickel compounds studied. Nickel carbonyl is formed when inorganic nickel in the air reacts with carbon monoxide. The use of nickel-based gasoline additives such as nickel isodecyl orthophosphate should, therefore, be discouraged, according to Stone and Smallwood (9).

References

(1) Sullivan, R.J., "Air Pollution Aspects of Nickel and Its Compounds," *Report PB 188,070,* Springfield, Va., Nat. Tech. Information Service (Sept. 1969).

(2) Anderson, D., "Emission Factors for Trace Substances," *Report PB 230,894,* Springfield, Va., Nat. Tech. Information Service (Dec. 1973).

(3) Banerjee, N.G. and Banerjee, T., "Recovery of Nickel and Zinc from Refinery Waste Liquors: Part I—Recovery of Nickel by Electrodeposition," *Jour. Sci. Industr. Res.,* 11B, 77-78 (1952).

(4) Dean, J.G., Bosqui, F.L., and Lanouette, K.H., "Removing Heavy Metals from Waste Water," *Env. Sci. Tech.,* 6, 518-522 (1972).

(5) Golomb, A., "Application of Reverse Osmosis to Electroplating Waste Water Treatment, Part II, The Potential of Reverse Osmosis in the Treatment of Some Plating Wastes," *Plating,* 59, 316-319, (1972).

(6) Lowe, W.,"The Origin and Characteristics of Toxic Wastes with Particular Reference to the Metals Industries," *Wat. Poll. Control* (London), 1970, 270-280.

(7) Hill, R.D., "Control and Prevention of Mine Drainage," presented at the Environmental Resources Conference on Cycling and Control of Metals, October 31-November 1-2, 1972, Battelle Columbus Laboratories, Columbus, Ohio.

(8) Patterson, J.W. and Minear, R.A., "Wastewater Treatment Technology," 2nd Ed., *Report PB 216,162,* Springfield, Va., Nat. Tech. Information Service (Feb. 1973).

(9) Stone, R. and Smallwood, H., "Intermedia Aspects of Air and Water Pollution Control," *Report EPA-600/5-73-003,* Wash., D.C., U.S. Environmental Protection Agency (Aug. 1973).

NITROGEN COMPOUNDS

Intermedia Aspects

Intermedia flows of nitrogen compounds artificially produced are shown in Figure 7 and Figure 8. Man-controlled transfers of these compounds between the fluid media are few and inefficient: incineration of nitrogenous sludges is an effective mode of transfer to air of compounds in which nitrogen is in a highly reduced state (as ammonia), but ineffective for nitrites and nitrates; and nitrogen oxides are not efficiently scrubbed from exhaust stacks. As a result of man's inability efficiently to effect intermedia transfer of nitrogen compounds, the important control methods for abatement of nitrogen-compound pollution are process-centered rather than treatment-centered and are distinctly nonintermedial in character.

Corrosion of materials can occur as a result of reactions with atmospheric nitric acid, from NO_2 via N_2O_5 and water. A unique effect of atmospheric NO_2 is sky discoloration. Due to its absorption in the blue-green region of the spectrum, NO_2 imparts a brownish-red color to the atmosphere, thus creating a visible smog.

In bodies of water, the effects of nitrogen may encourage rapid eutrophication, and aid the development of sludge deposits. A high nitrogen concentration serves as a nutrient building material for algae. As the algae grow, they use nutrients in the water until the nutrients are consumed. When the algae begin to die, bacteria decompose the organic material, using up the dissolved oxygen in the water.

The large amounts of living and dead algae which result from the nitrogen compounds cause turbidity, disagreeable color, taste, and odor, sludge solids deposition, and other nuisances. The cost of removal of the nitrogen nutrient must be viewed in terms of the alternate uses of the receiving waters. Ammonia, nitrates, nitrites and organic nitrogen are common nutrients.

Found in water, nitrogen and its compounds, particularly ammonia, are formed as the undesirable by-products of such industries as agricultural production, food and kindred products, chemicals and allied products and sanitary services.

FIGURE 7: INTERMEDIA FLOW CHART NITROGEN OXIDES

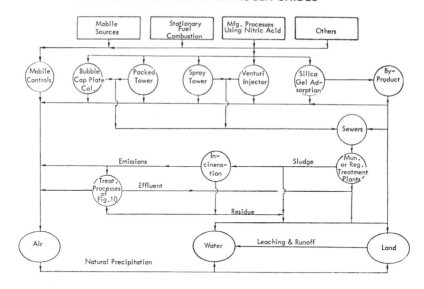

Source: Report EPA 600/5-73-003

FIGURE 8: INTERMEDIA FLOW CHART NITROGEN COMPOUNDS

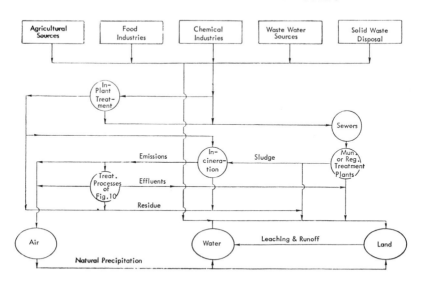

Source: Report EPA 600/5-73-003

Reference

(1) Stone, R. and Smallwood, H., "Intermedia Aspects of Air and Water Pollution Control," *Report EPA 600/5-73-003,* Wash., D.C., U.S. Environmental Protection Agency (Aug. 1973).

NITROGEN OXIDES

In Air

Of the various oxides of nitrogen, the most important as air pollutants are nitric oxide (NO) and nitrogen dioxide (NO_2), generally grouped together and referred to as nitrogen oxides (NO_x). Other oxides of nitrogen, N_2O_3, N_2O_4, N_2O_5, and NO_3 are present in very low concentrations, and although they may participate in photochemical reactions they can be neglected for present purposes. Nitrous oxide or laughing gas (N_2O) is present in the atmosphere in concentrations of about 450 $\mu g/m^3$ (0.25 ppm) but is regarded as being harmless physiologically, and there is no evidence that it participates in photochemical reactions in the lower atmosphere.

Most NO_x is produced biologically; fixation by lightning seems to be relatively unimportant. Natural sources produce on the order of 50×10^7 tons per year (about 90% of the total) of NO_x, worldwide, while man-made sources emit about 5×10^7 tons per year (about 10% of the total). Naturally occurring sources of NO_x have been found to cause nonurban concentrations, usually less than 10 $\mu g/m^3$. However, urban concentrations are generally 10 to 100 times higher, which indicates the importance of the man-made sources even though, overall, they are a small fraction of the natural sources. Nationwide, the man-made sources that are the major cause of NO_x emissions are fuel combustion in furnaces and in engines. Industrial chemical processes are responsible for high, localized emissions, but are not significant on a large scale. A summary of estimated man-made NO_x emissions in the U.S. in 1969 is given in Table 83.

TABLE 83: SUMMARY OF ESTIMATED MAN-MADE NITROGEN OXIDES EMISSIONS IN THE UNITED STATES, 1969

Source	10^6 Tons/Year
Transportation	11.1
Motor vehicles, gasoline	7.6
Motor vehicles, diesel	1.1
Aircraft	0.3
Railroads	0.1
Vessels	0.2
Nonhighway	1.8
Fuel Combustion in Stationary Sources	10.0
Coal	3.8
Fuel oil	1.3
Natural gas	4.7
Wood	0.2
Industrial Process	0.2
Solid waste	0.4
Miscellaneous	2.0
Forest fires	1.6
Structural fires	–
Coal refuse	0.1
Agricultural	0.3
TOTAL	23.7

Source: Environmental Protection Agency, May 1971

Table 84 is a compilation of average emission factors for NO_x from various types of sources. These emission rates represent uncontrolled sources unless otherwise noted. For an operation in which control equipment is utilized, the emission rate given for an uncontrolled source must be multiplied by: 1 minus the percent efficiency of the equipment, expressed in hundredths.

TABLE 84: EMISSION FACTORS FOR NITROGEN OXIDES DURING COMBUSTION OF FUELS AND OTHER MATERIALS

Source	Average emission factor
Fuels	
Coal	
Household and commercial	8 lb/ton
Industry	20 lb/ton
Utility	20 lb/ton
Fuel oil	
Household and commercial	12-72 lb/10^3 gal
Industry	72 lb/10^3 gal
Utility	104 lb/10^3 gal
Natural gas	
Household and commercial	116 lb/10^6 ft^3
Industry	214 lb/10^6 ft^3
Utility	390 lb/10^6 ft^3
Wood	11 lb/ton
Combustion sources	
Gas Engines	
Oil and gas production	770 lb/10^6 ft^3
Gas plant	4,300 lb/10^6 ft^3
Pipeline	7,300 lb/10^6 ft^3
Refinery	4,400 lb/10^6 ft^3
Gas Turbines	
Gas plant	200 lb/10^6 ft^3
Pipeline	200 lb/10^6 ft^3
Refinery	200 lb/10^6 ft^3
Waste Disposal	
Open burning	11 lb/ton
Conical incinerator	0.65 lb/ton
Municipal incinerator	2 lb/ton
On-site incinerator	2.5 lb/ton
Other combustion	
Coal refuse banks	8 lb/ton
Forest burning	11 lb/ton
Agricultural burning	2 lb/ton
Structural fires	11 lb/ton
Chemical industries	
Nitric acid manufacture	57 lb/ton HNO$_3$ product
Adipic acid	12 lb/ton product
Terephthalic acid	13 lb/ton product
Nitrations	
large operations	0.2-14 lb/ton HNO$_3$ used
small batches	2-260 lb/ton HNO$_3$ used

Source: EPA Publication Number AP-67

References

(1) U.S. Dept. of Health, Education and Welfare, "Control Techniques for Nitrogen Oxide Emissions from Stationary Sources," *Publ. No. AP-67,* Wash., D.C., Nat. Air Pollution Control Admin. (Mar. 1970).
(2) National Academy of Engineering, "Abatement of Nitrogen Oxide Emission from Stationary Sources," Wash., D.C., Nat. Research Council (1972).

ODOROUS COMPOUNDS

In Air

The air pollution aspects of ordorous compounds have been reviewed in some detail by Sullivan (1), and are the subject of an extensive EPA bibliography (2).

Table 85 summarizes the concentration and quantity of emissions from some typical processes as reviewed by Goldberg (3).

TABLE 85: ODOR EMISSIONS FROM TYPICAL INDUSTRIAL EQUIPMENT AND ODOR CONTROL DEVICES

Type of Equipment or Operation	Odor Levels and Emission Rates, Uncontrolled		Odor Levels and Emission Rates, Controlled			
	Vent Gas Odor Concentration Range, ou/scf[a]	Modal Odor Emission Rate, ou/min[b]	Type of Odor Control Equipment	Vent Gas Odor Concentration, ou/scf[a]	Odor Emission Rate, ou/min[b]	Temperature[c] and Efficiency[d]
Rendering cooker	5,000 to 500,000[e] (Mode 50,000)	25,000,000	Direct-Fired (DF)*	100 to 150 (Mode 120)	90,000	1200°F 99+%
			Surface condenser**	100,000 to 10,000,000[f] (Mode 500,000)	12,000,000	80°F Negative[f]
			Jet condenser followed by a DF after-burner*	20 to 50 (Mode 25)	2,000	1200°F 99+%
			Surface condenser followed by a DF after-burner*	50 to 100 (Mode 75)	6,000	1200°F 99+%
			Jet (or contact condenser)**	2,000 to 20,000 (Mode 10,000)	70,000	80°F 80%
Rendering cooker (blood drying) Dry batch type	10,000 to 1,000,000[g]	Not measured				
Rendering cooker (edible charge) Dry batch type / Wet batch type / Continuous type	2,500[h] 350[h] 650 to 7,000[h,i]	70,000[h]				
Fish-meal drier	1,000 to 5,000 (Mode 2,000)	50,000,000	Packed column type scrubber**	200 to 1,000 (Mode 400)	10,000,000	70°F 80%
			Chlorination[k] plus packed column scrubber**	30 to 50 (Mode 40)	1,000,000	70°F 98%

(continued)

TABLE 85: (continued)

Type of Equipment or Operation	Odor Levels and Emission Rates, Uncontrolled		Odor Levels and Emission Rates, Controlled			
	Vent Gas Odor Concentration Range, ou/scf[b]	Modal Odor Emission Rate, ou/min[b]	Type of Odor Control Equipment	Vent Gas Odor Concentration, ou/scf[g]	Odor Emission Rate, ou/min[b]	Temperature[c] and Efficiency[d]
Air blowing of fish oils	10,000 to 70,000 (Mode 50,000) (Estimated)	30,000,000	DF afterburner*	25 to 75 (Mode 50) (Estimated)	50,000[l]	1200°F 99+%
Air blowing of linseed oil	120,000[h]	Not measured	DF afterburner*	2,000	Not measured	1200°F 97.5%
Varnish cooker Batch type	10,000 to 200,000[e] (Mode 25,000)	10,000,000	Recirculating spray contact scrubber followed by a DF afterburner*	10 to 25 (Mode 20)	10,000	1200°F 99+%
			Recirculating spray (contact) scrubber**	20,000	Not measured	
			DF afterburner*	100 to 400 (Mode 250)	100,000	1200°F 99%
			Recirculating spray (contact) scrubber**	100,000[h]	Not measured	
Lithographing oven metal decorating	700 to 10,000 (Mode 3,000)	15,000,000	DF afterburner*	50 to 500 (Mode 200)	1,200,000	1200°F 95%
			Catalytic afterburner*	450[h]	2,300,000	1000°Fm
Coffee roaster Batch type	300 to 30,000[e]	3,000,000[h] (Estimated)	DF afterburner	3,000[h]	1,700,000[h] (Estimated)	800°F
				150 to 15,000[n]		1100°F 50%
Coffee roaster Continuous type	500 to 1,000[i] (Mode 1,000) (Estimated)	3,000,000[l]	DF afterburner*	300 to 1,000 (Mode 350) (Estimated)	1,200,000[l]	900°F 65%

(continued)

TABLE 85: (continued)

Type of Equipment or Operation	Odor Levels and Emission Rates, Uncontrolled		Type of Odor Control Equipment	Odor Levels and Emission Rates, Controlled		
	Vent Gas Odor Concentration Range, ou/scf[a]	Modal Odor Emission Rate, ou/min[b]		Vent Gas Odor Concentration, ou/scf[a]	Odor Emission Rate, ou/min[b]	Temperature[c] and Efficiency[d]
Bread baking oven	1,000[h]	Not measured				940°F
Tallow hydrolyzer	Not measured	Not measured	Surface condenser[o] followed by a DF afterburner*	2,000,000 2,000 750 150 70	Not measured	1100°F 1200°F 1300°F 1400°F
			Surface condenser**	6,000	Not measured	
Phthalic anhydride manufacturing unit	1,800 to 3,500[j] (Mode 2,500)	15,000,000	DF afterburner*	45 to 120 (Mode 75)	500,000	1200°F 97%
			Catalytic afterburner*	1,800	11,000,000	745°F 27%
			Catalytic afterburner*	180	1,100,000	810°F[m] 93%

*Afterburner odor control devices.

**Nonafterburner odor control devices.

[a] Odor units per standard cubic foot (at 70°F and 14.7 psia).

[b] Odor units discharged per minute, based on average volumetric discharge rate and modal odor concentration.

[c] Temperature of gases after leaving flame-contact zone (afterburners); temperature of vent gases in other cases.

[d] Odor control efficiency, on a modal odor concentration basis.

[e] Odor concentrations in batch processes vary with materials charged and phase of operation.

[f] Surface condensers increase odor concentrations in the vent gases but reduce total odor emission rates.

[g] Hundred-fold increase from beginning to end of cycle.

[h] One test only.

[i] Samples collected from several points of odor emissions.

[j] In continuous processes, odor concentrations vary with temperatures maintained and materials charged.

[k] Chlorine (20 ppm) mixed with drier off-gases, which are then scrubbed. More or less chlorine increases odor concentrations.

[l] Estimated from two tests only.

[m] Maximum temperature at which this catalytic unit can operate.

[n] Outlet odor concentration rises and falls with inlet odor concentration.

[o] The surface condenser is an integral part of the hydrolyzing unit; low temperature incineration increases odor concentration above condenser vent level.

Source: Report PB 223,568

Reference

(1) Sullivan, R.J., "Air Pollution Aspects of Odorous Compounds," *Report PB 188,089,* Springfield, Va., Nat. Tech. Information Service (Sept. 1969).
(2) U.S. Environmental Protection Agency, "Odors and Air Pollution: A Bibliography with Abstracts," *Publ. No. AP-113,* Research Triangle Park, N.C., Office of Air Programs (Oct. 1972).
(3) Goldberg, A.J., "A Survey of Emissions and Controls for Hazardous and Other Pollutants," *Report PB 223,568,* Springfield, Va., Nat. Tech. Information Service (Feb. 1973).

OILY WASTES

In Water

Oily waste materials are measured in terms of their hexane solubility for purposes of pollution evaluation. Hexane is an organic solvent employed to separate oily organic compounds from wastewaters. Oily wastes include greases, as well as many types of oils. Grease is not a specific chemical compound, but a rather general group of semiliquid materials which may include fatty acids, soaps, fats, waxes and other similar materials extractable into hexane. Unlike some industrial oils which represent precise chemical composition, greases are, in effect, defined by the analytical method employed to separate them from the water phase of the waste (1). A waste treatment manual recently published by the American Petroleum Institute (2) suggests the following classification for types of oily wastes:

[1] Light hydrocarbons—including light fuels such as gasoline, kerosine and jet fuel, and miscellaneous solvents used for industrial processing, degreasing or cleaning purposes. The presence of waste light hydrocarbons may make removal of other, heavier oily wastes more difficult.

[2] Heavy hydrocarbons, fuels and tars—includes the crude oils, diesel oils, Number 6 fuel oil, residual oils, slop oil, asphalt and road tar.

[3] Lubricants and cutting fluids—oil lubricants generally fall into two classes; nonemulsifiable oils such as lubricating oils and greases, and emulsifiable oils such as water-soluble oils, rolling oils, cutting oils and drawing compounds. Emulsifiable oils may contain fat, soap, or various other additives.

[4] Fats and fatty oils—these materials originate primarily from processing of foods and natural products. Fats result from processing of animal flesh. Fatty oils for the most part originate from the plant kingdom. Quantities of these oils result from processing soybeans, cottonseed, linseed and corn.

There are many industrial sources of oily wastes. Table 86 presents the major categories of industry producing oil and grease-laden waste streams, and lists characteristic types and sources of oily wastes associated with each category. Table 87 summarizes reported oily waste concentrations for many of these industries. By far the three major industrial sources of oily waste are petroleum refineries, metals manufacture and machining, and food processors. Petroleum refineries produce large quantities of oil and oily emulsion wastes.

Because of the long history of pollution problems associated with petroleum refining, the American Petroleum Institute (API) has exerted a good deal of effort in developing and publishing methods of oily waste control [e.g., (2)(3)]. Nevertheless, a recently published comprehensive federal industrial waste profile on the petroleum refining industry reported no values for oil content of refinery wastewater, and stated that . . . "data concerning the amounts of oil . . . (in wastewater) . . . are not complete enough to justify inclusion . . . " in that report (4)! Indicative of the oil content of refinery waste, however, is a reported value of 154 mg/l of oil, after preliminary skimming to partially remove floating oils (5).

TABLE 86: INDUSTRIAL SOURCES OF OILY WASTES

Industry	Waste Character
Petroleum	Light and heavy oils resulting from producing, refining, storage, transporting and retailing of petroleum and petroleum products.
Metals	Grinding, lubricating and cutting oils employed in metal-working operations, and rinsed from metal parts in clean-up processes.
Food Processing	Natural fats and oils resulting from animal and plant processing, including slaughtering, cleaning and by-product processing.
Textiles	Oils and grease resulting from scouring of natural fibers (e.g., wool, cotton).
Cooling and Heating	Dilute oil-containing cooling water, oil having leaked from pumps, condensers, heat exchangers, etc.

Source: API (2)

TABLE 87: CONCENTRATIONS OF OILY MATERIAL IN INDUSTRIAL WASTE-WATERS

Industrial Source	Oily Waste Conc., mg/l
Petroleum Refinery*	40-154
Petroleum Refinery*	35-178
Petroleum Refinery*	20
Steel Mill	
Hot Rolling	20
Cold Rolling	700 (500 free oil)
Cold Rolling	60-500
Cold Rolling Coolant	2,088-48,742 (2,035-36,664 free oil)
Cold Rolling Rinse	113-3,034 (83-2,284 free oil)
Food Processing	3830
Food Processing (Fish)	520-13,700
Metal Finishing	100-5,000
Metal Finishing	665
Oil Field Brine	25-50
Paint Manufacture	1900
Aircraft Maintenance	500-1200 (250-500 free oil)

* API oil separator effluent.

Source: API (2)

In the metals industry the two major sources of oily wastes are steel manufacture and metal working. Oily wastes include both emulsified and nonemulsified or floating oils. In steel manufacture, steel ingots are rolled into desired shapes in either hot or cold rolling mills. Oily wastes from hot rolling mills contain primarily lubricating and hydraulic pressure fluids. In cold strip rolling, however, the steel ingot is usually oiled prior to rolling, to lubricate and to reduce rusting. Additional oil-water emulsions are sprayed during rolling to act as coolants. After shaping, the steel is rinsed to remove the adhering oil. Rinse and coolant waters from the cold rolling mills may contain several thousand mg/l of oil, of which 25% or more may be emulsified and thus difficult to separate from the wastewater. Emulsified oil in hot rolling mill effluents rarely exceeds 20 mg/l (2). More concentrated oily wastes, such as from batch dumps of spent coolant or lubricating fluid must also be treated.

Metal working produces shaped metal pieces such as pistons and other machine parts. Oily wastewaters from metal working processes contain grinding oils, cutting oils, and lubrication fluids. Coolant oil-water emulsions are also employed in many metal working processes. Soluble and emulsified oil content of wastewaters may vary from 100 to 5,000 mg/l (6).

The third major source of oily wastes, and particularly greases, is the food processing industry. In the processing of meat, fish and poultry, oily and fatty materials are produced primarily during slaughtering, cleaning, and by-product processing. The major grease sources are the rendering areas, in particular from the wet (or steam) rendering process which gives the highest levels, pound per pound of scrap processed, of hexane extractables in food processing waste streams (2). Grease content in meat packinghouse waste streams may run several thousand mg/l (7), and it has been reported that waste from a fish processing plant contained 520 to 13,700 mg/l of fish oil (8).

A recent volume (9) reviews techniques for oil spill prevention and removal which is allied to the question of emissions of oily wastes.

References

(1) Amer. Public Health Assoc., *Standard Methods for the Examination of Water and Wastewater,* 13th ed., APHA, New York (1971).
(2) Amer. Petroleum Instit., *Industrial Oily Waste Control,* API, New York (undated).
(3) Amer. Petroleum Instit., *Manual on Disposal of Refinery Wastes,* 7th ed., API, New York (1963).
(4) Anonymous, *The Cost of Clean Water: Vol. III, Industrial Waste Profiles, No. 5—Petroleum Refining,* Washington, D.C., U.S. Dept. Interior (1967).
(5) Wigren, A.A. and Burton, F.L., "Refinery Wastewater Control," *Jour. Wat. Poll. Control Fed.,* 44, 117-128 (1972).
(6) Brink, R.J., "Operating Costs of Waste Treatment in General Motors," Proc. 19th Purdue Indust. Waste Conf., 19, 12-16 (1964).
(7) Garrison, V.M. and Geppert, R.J., "Packinghouse Waste Processing Applied Improvement of Conventional Methods," Proc. 15th Purdue Indust. Waste Conf., 15, 207-217 (1960).
(8) Chun, M.J., Young, R.H.F., and Burbank, N.C., Jr., "A Characterization of Tuna Packing Waste," Proc. 23rd Purdue Indust. Waste Conf., 23, 786-805 (1968).
(9) Sittig, M., *Handbook of Oil Spill Prevention and Removal,* Park Ridge, N.J., Noyes Data Corp. (1974).

ORGANICS

Intermedia Aspects

Organics in water most commonly are reported as measured biochemical oxygen demand.

The latter is not a physical pollutant but an indicator. Other common measures of organics content are chemical oxygen demand, and total organic carbon. Five day biochemical oxygen demand (BOD_5) indicates the amount of oxygen consumed in the decomposition of organic matter by bacteria in a given sample volume over a period of 5 days at 20°C. Thus BOD is a measure of the oxygen-depleting effects of the contained organic matter. The oxygen consumed may ultimately be incorporated into water, carbon dioxide and compounds of nitrogen and sulfur, principally. Not all organic matter is rapidly biodegradable since organic substances may have widely different carbon-hydrogen ratios and refractory characteristics.

There are three methods by which organics may be removed from their primary medium, water, and transfered to an alternate medium. One of these methods transfers the pollutant to air and the other two transfer it to the land. All three transfers result from residue disposal as shown in Figure 9.

FIGURE 9: INTERMEDIA FLOW CHART ORGANICS AND SUSPENDED SOLIDS IN WATER

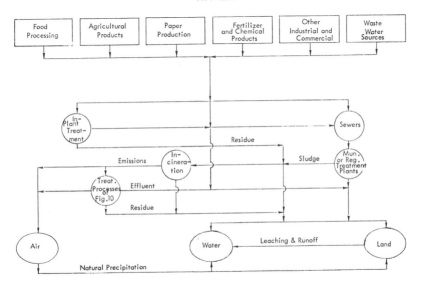

Source: Report EPA 600/5-73-003

The water to air transfer is achieved by incineration of the sludge produced by any of the treatments mentioned below. If this sludge is disposed to a landfill, the pollutant is transfered to the land. A more positive transfer of the pollutant to the land occurs if the treated or partially treated wastewater effluents and/or sludges are discharged on crop or forage lands. This transfer makes the nutrients available for plant growth and utilizes the solids for their nutrient and soil conditioning values. Biochemical oxygen demand is not then a significant consideration provided that runoff to adjacent watercourses is prevented or controlled.

When organic material is discharged into receiving waters, its biodegradation consumes the oxygen in the water. As the biochemical oxygen demand increases, the dissolved oxygen is rapidly depleted, depriving the fish and other aerobic organisms of their needed oxygen. When the dissolved oxygen drops to about 46% of saturation, fish will not enter the area. At the same time, high organic concentration encourages eutrophication with the rapid growth of both algae and bacteria. This combined symbiotic activity and the resulting floatable by-products can produce undesirable scum, suspended solids, and bottom sludge deposits in the water body. Aside from the environmental and esthetic impacts, high social costs, economic costs of treatment and the costs of resource destruction are all related to the degradation of the receiving waters.

Table 88 shows the reported quantities of industrial wastewaters discharged in 1963 and Federal Water Pollution Control Agency (FWPCA) estimates of the quantities of organics measured as BOD_5 and the settleable and suspended solids contained in these waters. Waste-load estimates indicate that industries included in the categories of chemicals (SIC 28), paper (SIC 26) and food and kindred products (SIC 20) generated about 90% of all BOD_5 in industrial wastewater before treatment.

TABLE 88: SELECTED ESTIMATED VOLUME OF INDUSTRIAL WASTES BEFORE TREATMENT, 1963[a]

PPB[b] Code	SIC[c] Code	Industry Group(s)	Total[d] Wastewater, (Billion Gallons)	Process Water Intake, (Billion Gallons)	Standard Biochem. Oxygen Demand[i] (Million Pounds)	Settleable and Suspended Solids (Million Pounds)
1201	33,34	Metal & Metal Products	> 4,300	1,000	> 480	> 4,700
1202	28	Chemical & Allied Products	3,700	560	9,700	1,900
1203	--	Power Production	N.A e, f	N.A.[f]	N.A.[f]	N.A.[f]
1204	26	Paper & Allied Products	1,900	1,300	5,900	3,000
1205	29	Petroleum & Coal	1,300	88	500	460
1206	20	Food & Kindred Products	690	260	4,300	6,600
1207	35,36, 37	Machinery & Transportation Equipment	> 481	109	> 250	> 70
1208	32	Stone, Clay, and Glass Products	218[e]	88	N.A.	N.A.
1209	22	Textile Mill Products	140	110	890	N.A.
1210	24,25	Lumber & Wood Products	126[e]	57	N.A.	N.A.
1211	30	Rubber & Plastics	160	19	40	50
1212	12,19, 21,27, 31,38, 39,72	Miscellaneous Industrial Sources	450	190	> 390	> 930
1200		All Manufacturing	≥ 13,100	≥ 3,700	≥ 22,000	≥ 18,000
		For Comparison: Sewered Population of U.S. 5,300[g]		N.A.	7,300[h]	8,800[k]

[a]Ref: Volume II–The Cost of Clean Waters, 1968. [b]Program Planning & Budget. [c]Standard Industrial Classification. [d]Includes Cooling Water & Steam Production Waters. [e]Included in Total for all Mfg. [f]Not Available or Not Applicable. [g]120,000,000 persons x 120 gal x 365 days. [h]120,000,000 persons x 1/6 lb x 365 days. [k]120,000,000 persons x 0.2 lb x 365 days. [i]BOD_5 at 20°C.

Source: Report EPA 600/5-73-003

A typical process change to reduce the discharge of organic wastes would be the substitution of dry processes for treating and cleaning in industries. Recirculation and reuse of process waters and the treatment of wastewaters for product recovery are other changes that may be used. Better housekeeping to reduce discharge of wastes to the sewers and better utilization of materials will also help.

A number of treatment processes have been proven effective in the stabilization of organics; many are in common use today. The processes range in efficiency from about 40% to more than 98% removal of BOD_5. These methods include: fine-screening (5 to 10%), settling and flotation (5 to 40%), chemical precipitation (78%), activated sludge (85 to 95%), trickling filter (80 to 95%), stabilization ponds (70%), and carbon adsorption (85 to 98%). All of these methods generate gaseous and solid residues which present the ultimate disposal requirement with intermedial effects mentioned above.

Reference

(1)   Stone, R. and Smallwood, H., "Intermedia Aspects of Air and Water Pollution Con-
      trol," *Report EPA 600/5-73-003* (Aug. 1973).

PARTICULATES

In Air

The emission of particulate pollutants to the atmosphere can create numerous problems
related to health, esthetics, and/or economics. The severity of these problems is related
to the total rate of emission, the physical and chemical characteristics of the emissions,
and the environment surrounding the emission source. A source particulate pollutant may,
therefore, be important because of the total amount emitted or because of the objectionable properties of the material emitted. Deleterious aspects of particulate pollution include
the following.

Health problems include: [1] respiratory, created by large concentrations of particles less
than 1μ; [2] toxic reaction, created by toxic particles such as fluoride salts, beryllium compounds, etc.; and [3] eye or skin irritation, created by acid mists or small particles. Esthetic problems include: [1] soiling, created by large mass emission rates, tarry or sticky
emissions, etc.; [2] turbidity (lowering of visibility), created by emission of large concentrations of small particles (i.e., less than 1μ); and [3] odor, created by numerous compounds emitted by various sources. Economic problems include: [1] material (living and
nonliving) damage, created by corrosive particles; [2] soiling, cleaning costs related to soiling by particulate pollutants; and [3] secondary economic problems, created by neighborhood deterioration.

To alleviate or avert these problems requires control at the source. Adequate and economic
control requires a knowledge of the sources and the properties of the effluents from the
sources. Some sources are extremely difficult to control because of incomplete information
about sources and properties or unusual properties, and research and development will be
required to establish successful control techniques.

Table 89 shows the sources of particulate emissions by industry as given by A.E. Vandegrift
et al in Report PB 203,128. It also shows possible emissions based on 90% of that collection efficiency. Incineration (municipal, industrial and open burning) was not handled in
the same manner as the other industrial sources and, in fact, was not considered an industrial source. It is, however, an important source. The total particulate emitted per year
from incineration was estimated at 931,000 tons.

In developing the ranking of stationary sources represented in Table 89, more attention
was directed to calculating the emissions from the primary pieces of processing equipment
such as, kilns, furnaces, reactors, and dryers. In several cases, we have included emissions
for secondary sources which include crushing, materials handling, grinding, stock piles, etc.
The calculations involving these secondary sources are in general much less accurate than
those involving the primary processing equipment because data on secondary sources are
meager or nonexistent.

The emission quantities listed for these secondary sources are at best order of magnitude
calculations, and it is possible that secondary sources may emit as much or more particulate matter than the primary sources.

TABLE 89: MAJOR INDUSTRIAL SOURCES OF PARTICULATE POLLUTION

Source	Emissions (tons/year)	Possible Emissions Based on 90% Collection Efficiency (tons/year)
Fuel Combustion	5,953,000	8,624,000
Crushed Stone, Sand and Gravel	4,600,000	4,714,000
Agricultural Operations (grain elevators, feed mills and cotton gins)	1,817,000	1,843,000
Iron and Steel	1,442,000	2,183,000
Cement	934,000	1,629,000
Forest Products	580,000	1,002,000
Lime	573,000	708,000
Clay	467,000	542,000
Primary Nonferrous (copper, aluminum, zinc and lead)	476,000	654,000
Fertilizer and Phosphate Rock	328,000	337,000
Asphalt (batch plants and roofing)	218,000	721,000
Ferroalloy	160,000	180,000
Iron Foundries	143,000	148,000
Secondary Nonferrous (copper, aluminum, zinc and lead)	127,000	134,000
Coal Cleaning	94,000	94,000
Carbon Black	93,000	93,000
Petroleum	45,000	45,000
Acids (sulfuric and phosphoric)	16,000	28,000
Total	18,066,000	23,679,000

Source: Report PB 203,128

Nonindustrial sources of particulate pollutants with their corresponding emission estimates are shown in Table 90. The amount of natural dust listed, 63,000,000 tons, was calculated from data in an article written by personnel working in soil conservation (see R.M. Smith and P.C. Twiss, *Trans. Kansas Acad. Sci.* 68, No. 2, 311-21 (1965). The article describes a continuing project on dust deposition at 15 stations located from New Jersey to Texas and Montana. The stations are located in rural areas at sites protected for at least several hundred feet by surrounding ground cover that effectively prevents surface soil from blowing.

At the present time, yearly averages at each of the stations (data from 13 stations) range from 19 to 280 lb of dust deposited/acre/month. The highest record for a single month is 3,616 lb/acre at Tribune, Kansas, on the Colorado border, the center of the nation's dust bowl. If the median of the range of yearly averages is taken at 150 and multiplied by the number of acres subject to wind erosion, estimated at 70,000,000, the tons of natural dust settling out of the atmosphere each year are calculated to be 63,000,000.

$$70,000,000 \text{ acres} \times \frac{150 \text{ lb dust}}{\text{acre-month}} \times \frac{12 \text{ months}}{\text{year}} \times \frac{1 \text{ ton}}{2,000 \text{ lb}} = 63,000,000 \text{ tpy}$$

The estimates of particulate emissions from forest fires were obtained from the Forest Service of the U.S. Department of Agriculture. These estimates were made on the basis of research data on products of combustion and statistics on acres of forests burned. Data have been obtained for the amount of gas produced (CO_2, hydrocarbons, etc.) from a given amount of fuel in full-scale fires.

The amounts of fuel required to produce only the measured amounts of gas were calculated. It was then assumed that if fuel left the fire in other than gaseous form, it left as a solid. If it left as a solid, it was particulate. No data have been obtained on size distribution, but, in intense fires, the particulate would include firebrands of quite significant size. Also listed as miscellaneous significant sources are transportation and incineration. Estimates of particulate emissions from these sources were obtained from the National Air Pollution Control Administration. Other minor sources have been included for completeness. The total emissions from stationary industrial sources amount to over 18 million tons/year, and the miscellaneous source total is nearly 123 million tons/year. The grand total for particulate pollutant emissions in the continental United States is thus estimated at some 141 million tons/year.

TABLE 90: MISCELLANEOUS SIGNIFICANT SOURCES OF PARTICULATE POLLUTION

Source	Emissions, tons/year	
Natural dusts		63,000,000
Forest fires		
Wildfire	37,000,000	
Controlled fire		
Slash burning	6,000,000	
Accumulated litter	11,000,000	
Agricultural burning	2,400,000	56,400,000
Transportation		
Motor vehicles		
Gasoline	420,000	
Diesel	260,000	
Aircraft	30,000	
Railroads	220,000	
Water transport	150,000	
Nonhighway use		
Agriculture	79,000	
Commercial	12,000	
Construction	3,000	
Other	26,000	1,200,000
Incineration		
Municipal incineration	98,000	
On-site incineration	185,000	
Wigwam burners (excluding Forest Products disposal)	35,000	
Open dump	613,000	931,000
Other minor sources		
Rubber from tires	300,000	
Cigarette smoke	230,000	
Cosmic dust	24,000	
Aerosols from spray cans	390,000	
Ocean salt spray	340,000	1,284,000
TOTAL		122,815,000

Source: Report PB 203,128

Much of today's atmospheric pollution is smoke and dust of such a size that it falls out or is washed out of the atmosphere within a day or so. A certain portion of the pollution is, however, composed of particles a micron or smaller in diameter. These particles may exist in the atmosphere for long periods of time. The presence in the air of submicron particles of various chemical compositions from man-made sources is of increasing interest to many governmental and scientific groups. The interests of these groups range from upper atmosphere physics and air pollution control engineering to the protection of human health.

Fine particulates emitted from man's activity contribute significantly to all the major adverse aspects of air pollution. Fine particles can initiate or contribute to problems related to human health, atmospheric physical properties, and/or economics. The effects of particulate matter on human health are, for the most part, related to injury to the surfaces of the respiratory system. Such injury may be permanent or temporary. It may be confined to the surface, or it may extend beyond, sometimes producing functional or other alterations.

Particle size, together with specific gravity and chemical nature, largely determine: where particles are deposited in the respiratory system, the fate of the particles after deposition, and to a considerable extent their physiological action. Particulate material in the respiratory tract may produce injury itself, or it may act in conjunction with gases, altering their sites or their modes of action. A combination of particulates and gases may produce an effect that is greater than the sum of the effects caused by either individually (i.e., synergistic effects).

Fine particulate pollutants also affect the physical properties of the atmosphere. The chemical and physical properties of the atmosphere affected include: its electrical properties; its ability to transmit radiant energy; its ability to convert water vapor to fog, cloud, rain and snow; its ability to damage and to soil surfaces.

Table 91 presents a source priority ranking, based on the mass and number of fine particles emitted, of the industrial sources for which adequate information was available to estimate emissions. The priority ranking indicates those sources where initial efforts should be focused to control the emission of fine particulate pollutants. Improved control of fine particle emissions from the high priority sources would result in a significant reduction in the mass and/or number of fine particulates emitted from stationary sources.

TABLE 91: PRIORITY RANKING OF INDUSTRIAL SOURCES OF FINE PARTICULATE POLLUTANTS

(1) Ferroalloy furnaces	(6) Municipal incinerators
(2) Steel-making furnaces	(7) Iron foundry cupolas
(3) Coal-fired power plants	(8) Crushed stone plants
(4) Lime kilns	(9) Hot mix asphalt plants
(5) Kraft pulp mill recovery furnaces	(10) Cement kilns

Source: Report PB 203,521

To indicate the relative magnitude of the fine particulate problem from stationary sources, fine particle emissions were also estimated for mobile and miscellaneous sources. Particulate emissions from mobile sources were assumed to be uncontrolled and to consist of all $<2\mu$ particulates. Table 92 summarizes the estimate of fine particle emissions from these sources. Table 93 presents an estimate of fine particle emissions from miscellaneous sources for which adequate particle size data were available. Fine particle emissions from natural dusts, forest fires, and agricultural burning could not be estimated because of lack of reliable particle size data.

TABLE 92: FINE PARTICLE EMISSIONS FROM MOBILE SOURCES

Source	Emissions, tons/year
Motor vehicles	
Gasoline	420,000
Diesel	260,000
Aircraft	30,000
Railroads	220,000
Water transport	150,000
Nonhighway use	
Agriculture	79,000

(continued)

TABLE 92: (continued)

Source	Emissions, tons/year
Commercial	12,000
Construction	3,000
Other	26,000
TOTAL	1,200,000

Source: Report PB 203,521

TABLE 93: FINE PARTICLE EMISSIONS FROM MISCELLANEOUS SOURCES

Source	Emissions, tons/year
Rubber from tires	300,000
Cigarette smoke	230,000
Ocean salt spray	340,000
Aerosol from spray cans	390,000
TOTAL	1,260,000

Source: Report PB 203,521

Table 94 presents a ranking of important stationary industrial sources of particulate pollutants. The stationary sources represented were ranked by calculating the emissions from
the primary pieces of processing equipment such as kilns, furnaces, reactors, and dryers.
In several cases, emissions from secondary sources which include crushing operations, materials handling, stockpiles, etc. have been included. Total emissions for an industry were
obtained as a sum of the emissions from primary and secondary sources.

The leading sources are stationary combustion processes, crushed stone, agriculture and
related operations, iron and steel, and cement. Emissions from residential and commercial combustion sources, field burning, and slash burning are not included in the totals
shown for the stationary combustion processes, agricultural operations, and forest products
categories.

TABLE 94: MAJOR INDUSTRIAL SOURCES OF PARTICULATE POLLUTANTS
BASED ON 1968 PRODUCTION DATA

Source	Emissions, tons/year
Fuel combustion:	
Coal	
Electric utility	
Pulverized	2,710,000
Stoker	217,000
Cyclone	182,000
Total from electric utility coal	3,109,000
Industrial boilers	
Pulverized	322,000
Stoker	2,234,000
Cyclone	39,000
Total from industrial coal	2,595,000
Fuel oil	
Electric utility	36,000
Industrial	
Residual	87,000
Distillate	18,000
Total from fuel oil	141,000

(continued)

TABLE 94: (continued)

Source	Emissions, tons/year
Natural gas & LPG	
Electric utility	24,000
Industrial	84,000
Total from natural gas & LPG	108,000
Total from utility and industrial fuel combustion	5,953,000
Crushed stone, sand & gravel:	
Crushed stone	4,554,000
Sand & gravel	46,000
Total from crushed stone, sand & gravel	4,600,000
Operations related to agriculture:	
Grain elevators	1,700,000
Cotton gins	45,000
Feed mills	
Alfalfa mills	23,000
Mills other than alfalfa	49,000
Total from listed agricultural operations	1,817,000
Iron and steel:	
Ore crushing	82,000
Materials handling	446,000
Pellet plants	80,000
Sinter plants	
Sintering process	51,000
Crushing, screening, etc.	56,000
Coke manufacture	
Beehive	130,000
By-product	90,000
Pushing and quenching	21,000
Blast furnace	58,000
Steel furnaces	
Open hearth	337,000
Basic oxygen	10,000
Electric arc	18,000
Scarfing	63,000
Total from iron and steel	1,442,000
Cement:	
Wet process	
Kilns	435,000
Grinders, dryers, etc.	65,000
Dry process	
Kilns	310,000
Grinders, dryers, etc.	124,000
Total from cement	934,000
Forest products:	
Wigwam burners	132,000
Pulp mills	
Kraft process	
Recovery furnace	164,000
Lime kilns	33,000
Dissolving tanks	42,000
Sulfite process	
Recovery furnace	10,000
NSSC process	
Recovery furnace	1,000
Fluid-bed reactor	42,000
Bark boilers	82,000
Particleboard, etc.	74,000
Total from forest products	580,000

(continued)

TABLE 94: (continued)

Source	Emissions, tons/year
Lime:	
Crushing, screening	264,000
Rotary kilns	294,000
Vertical kilns	4,000
Materials handling	11,000
Total from lime	573,000
Primary nonferrous metals:	
Aluminum	
Grinding of bauxite	8,000
Calcining of hydroxide	58,000
Reduction cells	
H.S. Soderberg	35,000
V.S. Soderberg	10,000
Prebake	20,000
Materials handling	11,000
Total from primary aluminum	142,000
Copper	
Ore crushing	170,000
Roasting	7,000
Reverberatory furnace	28,000
Converters	33,000
Materials handling	5,000
Total from primary copper	243,000
Zinc	
Ore crushing	18,000
Roasting	
Fluid-bed	15,000
Ropp, multihearth	4,000
Sintering	3,000
Distillation	15,000
Materials handling	2,000
Total from primary zinc	57,000
Lead	
Ore crushing	4,000
Sintering	17,000
Blast furnace	10,000
Dross reverberatory furnace	2,000
Materials handling	1,000
Total from primary lead	34,000
Total from primary nonferrous	476,000
Clay:	
Ceramic	
Grinding	72,000
Drying	110,000
Refractories	
Kiln-fired	
Calcining	25,000
Drying	13,000
Grinding	47,000
Castable	14,000
Magnesite	7,000
Mortars	
Grinding	2,000
Drying	2,000
Mixes	4,000

(continued)

TABLE 94: (continued)

Source	Emissions, tons/year
Heavy clay products	
Grinding	72,000
Drying	99,000
Total from clay	467,000
Fertilizer and phosphate rock:	
Phosphate rock	
Drying	14,000
Grinding	1,000
Materials handling	30,000
Calcining	8,000
Fertilizers	
Ammonium nitrate	28,000
Urea	10,000
Phosphates	
Rock pulverizing	10,000
Acid-rock reaction	9,000
Granulation and drying, etc.	169,000
Materials handling	18,000
Bagging	4,000
Ammonium sulfate	27,000
Total from fertilizers and phosphate rock	328,000
Asphalt:	
Paving material	
Dryers	161,000
Secondary sources	40,000
Roofing material	
Blowing	3,000
Saturator	14,000
Total from asphalt	218,000
Ferroalloys:	
Blast furnace	1,000
Electric furnace	150,000
Materials handling	9,000
Total from ferroalloys	160,000
Iron foundries:	
Furnaces	105,000
Materials handling	
Coke, limestone, etc.	37,000
Sand	1,000
Total from iron foundries	143,000
Secondary nonferrous metals:	
Copper	
Material preparation	
Wire burning	41,000
Sweating furnaces	–
Blast furnaces	2,000
Smelting and refining	17,000
Total from secondary copper	60,000
Aluminum	
Sweating furnaces	6,000
Refining furnaces	1,000
Chlorine fluxing	51,000
Total from secondary aluminum	58,000
Lead	
Pot furnaces	–

(continued)

TABLE 94: (continued)

Source	Emissions, tons/year
Blast furnaces	1,000
Reverberatory furnaces	3,000
Total from secondary lead	4,000
Zinc	
Sweating furnaces	
Metallic scrap	–
Residual scrap	3,000
Distillation furnace	2,000
Total from secondary zinc	5,000
Total from secondary nonferrous metals	127,000
Coal cleaning: Thermal dryers	94,000
Carbon black:	
Channel process	82,000
Furnace process	
Gas	5,000
Oil	6,000
Total from carbon black	93,000
Petroleum: FCC units	45,000
Acids:	
Sulfuric	
New acid	
Chamber	2,000
Contact	4,000
Spent-acid concentrators	8,000
Phosphoric	
Thermal process	2,000
Total from acids	16,000
Total from major industrial sources	**18,066,000**

Source: Report PB 203,128

In addition to the data compiled on major industrial particulate pollutant sources and tabulated above, information of various types have been accumulated on several minor sources. Table 95 presents emission factors and other assorted facts for these sources.

TABLE 95: EMISSION FACTORS FOR MINOR SOURCES

Source of Particulate	Production tons/year	Control Device†	Emission Factor Range lb./ton	Average Emission Factor lb./ton
Chemical Process Industry:				
Phenolic resin handling				48
Pigment process	–			
Grinder		water spray and oil filters		0.33*
Blender		water spray and oil filters		1.9*
PVC manufacturing	1.1×10^6	FF		0.35*
Polypropylene mfg.	3.2×10^5			3.0
Rubber process	2.7×10^6			
Mixer		baghouse		0.13*
Grinder		cyclones		~20
				1.4*

(continued)

TABLE 95: (continued)

Source of Particulate	Production tons/year	Control Device†	Emission Factor Range lb./ton	Average Emission Factor lb./ton
Food Industry:				
Coffee roasting	1.5 x 10⁶			
Roaster				
Direct fired		cyclone		7.6
				2.2*
Indirect fired		cyclone		4.2
				1.2*
Stoner and cooler		cyclone		1.4
				0.4*
Instant coffee				
Spray dryer		cyclone + wet scrubber		1.4*
Roaster		cyclone		1.9*
Coffee/tea processing	7 x 10⁴**			
Roaster		cyclone		3.1*
Citrus plants				
Peel dryer			1.8–9.6	5.0*
Orange pulp dryer				75.3
Mineral Products Industry:				
Abrasive grit manufacturing		baffle sedimentation chamber		0.11*
Blast grit and roofing				
Granule manufacturing				
Dryer		cyclone		0.42*
Screens		cyclone		4.1*
Calcium carbide mfg.	9.2 x 10⁵			
Coke dryer		cyclone + spray dryer		0.2*
Electric furnace hood				1.7
Furnace room vents				2.6
Main stack		impingement scrubber		2.0*
Concrete batching				5.2
Fiberglas manufacturing				
Melting furnace				
Regenerative			1–4	3
Recuperative			0.5–1	1
Forming line			23.5–100	62
Curing oven			3.4–14	6.9
Frit manufacturing				
Rotary melter				17
Reverberatory furnace			5.9–45.5	16.5
Glass manufacturing furnace				2
Reverberatory furnace				3
Furnace			1.5–7.9	3
Furnaces			0.02–1.1	0.31
Ground brick plant				120
Gypsum board manufacturing				
Grinder		ESP	0.14–0.24	0.19*
Trim saw			lb./350 ft.²	
Gypsum manufacturing				
Dryer		FF		22
		cyclone + ESP	4–40	0*, 0.2*
Grinder		FF		1, neg.*
Calciner		FF		90, 0.1*
Conveyer		FF		0.7, neg.*
Magnesite plant				430
Perlite manufacturing	4.1 x 10⁵			
Furnace				21
Plaster of paris mfg.		ESP		0.9*
Quartz plant			7.8–9.5	8.6*
Rock wool manufacturing				
Cupola			16–28	21.6
Reverberatory furnace				4.8
Blow chamber			4–56	21.6
Curing oven			1.5–5.9	3.6
Cooler			0.4–5.5	2.4
Cupola				32.4

(continued)

TABLE 95: (continued)

Source of Particulate	Production tons/year	Control Device†	Emission Factor Range lb./ton	Average Emission Factor lb./ton
Miscellaneous:				
Electric arc welding				0.01-0.02 lb./lb. electrode
Soldering				0.005 lb./lb. solder
Plywood manufacturing dryer				0.51
Mercury smelting				40.0

*Controlled emission factor based on indicated control device.
**Tea production only.
†FF = fabric filter and ESP = electrostatic precipitator.

Source: Report PB 203,522

Intermedia Aspects

Particulates are perhaps the most prevalent of the intermedia pollutants, transferring readily from one medium to another, often with little change in form or character. Many are quite capable of change either as a cause or result of the transfer process, and bring about quite different effects than caused by the mere presence of solid particles.

Metallic salts and oxides, which may be significant as particulates in the atmosphere, when collected and discharged to receiving waters can affect the pH. Organics, when removed from wastewaters and incinerated, can easily be transferred to the atmosphere as particulates. Other examples are so numerous that they are best defined by following their paths through the intermedia flow chart, Figure 10, as shown on the following page.

The transfer of particulates from the atmosphere to either land or water may take place in three separate ways: the particulates may merely settle out, they may be washed out by the impact or rain drops, or they may rain out. In the latter case the particle serves as a nucleus for the formation of the rain drop.

Disposal of residues from particulate collection treatments can represent a major intermedia pollution problem which may necessitate extensive wastewater treatment facilities. The latter treated residues should be ultimately disposed to land or landfill sites even though they may have received intermediate sludge or residue treatments.

References

(1) U.S. Dept. of Health, Education and Welfare, "Control Techniques for Particulate Air Pollutants," *Publ. No. AP-51,* Wash. DC, National Air Pollution Control Administration (Jan. 1969).
(2) Midwest Research Institute, "Particulate Pollutant System Study," *Reports PB 203,128, PB 203,521 and PB 203,522,* Springfield, Va., National Tech. Information Service (1971).
(3) Jones, H.R., *Fine Dust and Particulates Removal,* Park Ridge, N.J., Noyes Data Corp. (1972).
(4) Stone, R. and Smallwood, H., "Intermedia Aspects of Air and Water Pollution Control," *Report EPA 600/5-73-003,* Wash., D.C., U.S. Environmental Protection Agency (Aug. 1973).

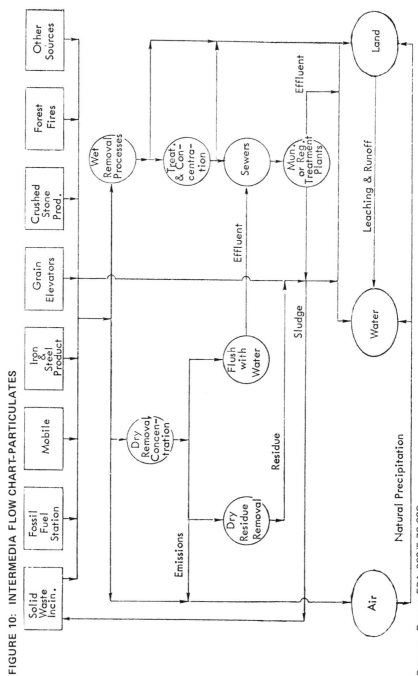

FIGURE 10: INTERMEDIA FLOW CHART-PARTICULATES

Source: Report EPA 600/5-73-003

PESTICIDES

In Air

The manufacture, formulation, and packaging of pesticides present possible air pollution hazards. The pesticides are generally manufactured in closed systems of a continuous-process nature. The process systems are normally maintained at a slightly negative pressure to avoid leakage. No data were found on the emission rates of pesticides from production plants according to Finkelstein (1). Sampling of air in a community (Fort Valley, Georgia) in which a formulating plant was operating revealed that DDT concentrations in ambient air ranged up to 0.007 $\mu g/m^3$ during the spraying season (May to June) and 0.004 $\mu g/m^3$ when spraying season was over (September). It was concluded that most of the DDT found in the air during September came from the formulating plant.

The primary source of pesticide contamination of the environment is the process of application. Since much of this application is by spraying or dusting, some part of the quantity dispensed can remain in the atmosphere and be diluted and dispersed. However, under certain meteorological conditions, dilution to an ineffectual concentration may not occur, and the chemical can drift in the air mass and cause adverse effects to nontarget forms of life some distance from the original application site.

In addition to the production and large-scale agricultural uses of pesticides, there are other sources of pesticide air contamination. Pesticides are released into the atmosphere during application in public areas, buildings, and homes. Pesticides may be inhaled in dusts from treated soils, from house dust contaminated by applications for household pests, or from moth-proofed rugs, blankets, and clothes. However, it is difficult to estimate the degree of importance to be placed on any of these sources as contributors to the concentration of pesticides in the ambient air, either locally or over a wide area. Storage and handling of pesticides presents a possible air pollution problem. Pesticides are usually stored in bags; glass, plastic, or metal bottles; or cans or drums. Contamination of the air may occur during handling, storage, or transit from breakage, leakage or spillage.

In Water

For a discussion of the problems presented by pesticides and pesticide intermediates and by-products in water, the reader is referred to the process section in the latter part of this volume on "Pesticide Manufacture."

Intermedia Aspects

Figure 11 illustrates the intermedial relationships for pesticides. Most intermedial flows stem from agricultural application of pesticides and are natural processes. Waste effluents from pesticide manufacturing operations are not the major source of pesticide pollution. These effluents, however, can be treated effectively with activated carbon. Discharge is frequently directly into sewer systems and sometimes, unfortunately, to nearby surface waters. Sewage treatment plants and the activated carbon treatment both create residues.

The carbon can be regenerated, incinerated or disposed to the land. Sewage sludges may be either incinerated or disposed to the land. Incineration of these materials can produce emissions containing pesticides. While agricultural and domestic application of pesticides are directed primarily toward the land, air application of pesticides creates uncontrollable aerosols. Winds and other climatic conditions affect whether air-applied pesticides will fall as intended or drift to adjacent lands and waters.

Pesticide residues on land can be transported to adjacent waters via leaching, irrigation and storm runoff. Chlorinated hydrocarbons are highly volatile and readily transfer to the air through evaporation. Pesticides in the air resulting from industrial emissions, sewage sludge incineration, industrial, agricultural, and domestic applications eventually return to the land or waterways through natural fallout and precipitation. Herbicides are normally classed

FIGURE 11: INTERMEDIA FLOW CHART PESTICIDES (HERBICIDES)

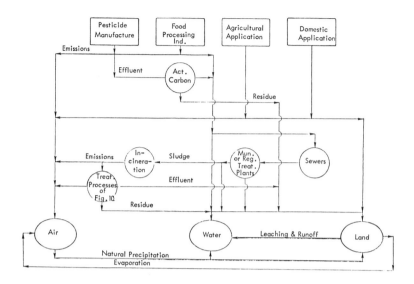

Source: EPA 600/5-73-003

with pesticides. Many types of pesticides are used for such purposes as control of insects, weeds, fungi and rodents. After application the most persistent of the pesticides are the chlorinated hydrocarbons, also known as the organochlorine pesticides. Attention is given here to the chlorinated hydrocarbon pesticides (DDT, chlordane, aldrin, dieldrin, endrin, heptachlor, toxaphene, methoxychlor) because of their reluctance to undergo chemical and biological degradation. Because of this persistence, the occurrence of the chlorinated hydrocarbon pesticides has the greatest impact on the environment.

Since these compounds persist a long time in the environment, they may be transferred by wind, water, animals and food to places far from where they were applied. This mobility of pesticides tends to contaminate nontarget areas and living species. The result is that localized areas may be treated with pesticides, but subsequent spreading of these small amounts may spread to much larger areas and affect wildlife species which are sensitive to low concentrations of pesticides.

Pesticides are a unique source of pollution since usually they are intentionally introduced into the natural environment. Pesticides reach the environment by direct application to the land for agricultural purposes. Also pesticides inadvertently enter the environment from industrial discharges, accidental spills, and from domestic sources such as home garbage disposals. Herbicides have the similar impacts as pesticides.

Industrial waste effluents can be treated to remove large concentrations of pesticides. Pesticide adsorption by activated carbon has been shown to be the most effective treatment for reducing high concentrations of pesticides from water. The removal of low level pesticide contamination is much more difficult to control, and evidence indicates that current conventional water treatment methods are not effective. Sludges from pesticide related industries and municipal sewage treatment plants can contain significant amounts of pesticides. Incineration of these sludges will produce pesticide emissions which require further treatment to avoid discharge to the atmosphere.

Pesticides in the Soil: Pesticides are applied directly to the soil for agricultural purposes. Repeated applications may create accumulations. These pesticide residues in the soil are of concern since they may reach man and wildlife through uptake from soil by consumable crops, by leaching into water supplies, by volatilization into the air and by direct contact with the soil. The factors which affect pesticide persistence in the soil are: pesticide molecular configuration, pesticide adsorption, organic content of the soil, soil moisture and temperature, uptake by plants, and leaching of pesticides from soil by water.

The chlorinated hydrocarbon pesticides are extremely hydrophilic, making them highly insoluble in water. The solubility of a substance is inversely proportional to its affinity for adsorption. Chlorinated hydrocarbon pesticides are therefore highly capable of being adsorbed and concentrated on soils and finely divided clays. The adsorbed pesticides can then be carried with the soil and clay particles into natural waters. Surface runoff after either rainfall or irrigation may transport particles to which pesticides adhere or the water may leach the pesticide from the soil particles.

Chlorinated hydrocarbon pesticides are more persistent in soils where the organic content of the soil is high. Chlorinated hydrocarbons are highly resistant to biological attack so their retention in soil is not affected appreciably by the microorganism concentration. Adsorption rates and soil microbial activity are both affected by soil temperature and moisture. High moisture content and temperature enhance the degradation process and increases the amount of volatilization which occurs. Volatilization is a major pathway of loss for the chlorinated hydrocarbons.

The process involves the desorption of the pesticide from the soil, diffusion upward to the soil surface, and then volatilization of the compounds into the atmosphere. Rates of loss by volatilization are related to the vapor pressures which, for chlorinated hydrocarbons are relatively low. However, the degradation products of lindane and DDT have much higher vapor pressures than their parent compounds, which means that the presence of these degradation products in significant amounts is an indication that volatilization of degradation products provides a major pathway for loss of some organochlorine insecticides from soil.

Plants can absorb pesticides from the soil, concentrating the residue within their structure. This mechanism constitutes a potential exposure hazard for man and animals when the absorbing plants are edible or forage crops. Experiments in Great Britain have shown that plants can also absorb organochlorine residues from the surrounding air.

Pesticides in the Water: The major pathways by which pesticides reach the water environment are through direct application to surface waters, indirect application during treatment of adjacent areas, percolation and runoff from treated agricultural or forested lands, and by the discharge of certain wastewaters. Most chlorinated hydrocarbon pesticides reach the aquatic environment attached to soil or clay particles because of the hydropholic nature of these compounds.

Usually these particles settle to form the bottom sediments of streams and lakes. Under certain conditions, a portion of the adsorbed pesticides can be desorbed directly into the water where they are maintained by a dynamic adsorption-desorption equilibrium system. Consequently, pesticide desorption provides a continuous supply of toxic material to water and creates many serious water pollution problems.

Pesticide residues are concentrated by soil and clay particles and also by microorganisms. It is possible through these associations for pesticides to reach ground and surface waters although the extent which the quality of ground water is threatened is not as well established as that of surface waters. Factors of paramount importance in the consideration of ground water pollution by pesticide residuals in soil are the amount of residue, the solubility of the pesticide in water, the amount of infiltrating water, and the adsorptive rate and capacity of the soil. One study has concluded that dieldrin could not be transported through soils into subsurface waters in significant amounts by infiltrating waters. A period of several hundred years was determined to be the time required for dieldrin to be transported in solution at a residual concentration of 20 ppb to a depth of 1 ft in natural soils. It appeared from this study that residues of dieldrin applied on the upper layers of soil do not threaten the quality of ground water at the assumed permissible concentration of 20 ppb. Studies have shown that dieldrin residues in soils have been detected up to 7 yr after application, and 72 to 90% of the residues remain in the top 3" of the soil. Trace quantities of dieldrin have been found as deep as 9" in soil. Dieldrin is used in experimental studies because it is considered to be one of the most persistent of the pesticides in soil.

Another study revealed that after 10 yr, 60 to 75% of residual DDT remained in the top 12" of the soil. The movement of DDT to lower soil depths was attributed to top soil being washed by rainfall into large vertical cracks in the ground. Even though these studies show that pesticides do not usually migrate to great depths in soils, incidents of pesticide contamination of ground waters have been documented.

During one incident in 1951, crops were damaged when irrigated with well water contaminated with the herbicide 2,4-D. A nearby 2,4-D manufacturing plant had discharged its wastes into lagoons from 1943 to 1957. It had taken between 7 and 8 yr for the pesticide to migrate 3.5 miles and eventually contaminate an area of 6.5 square miles. The herbicide 2,4-D was also reported to have been inadvertently dumped into a sewer. The waste reached underground strata which supplied well water to Montebello, California. The taste and odor of the herbicide was evident for over 5 yr.

Pesticides in the Air: The application of pesticides to land for agricultural uses is most generally accomplished by air. About 80% of the pesticides are applied by aircraft. An understanding of the ways in which air and pesticide particle size influence pesticide applications is necessary to apply pesticides without affecting nontarget areas. Studies have been made to determine the correlation between pesticide particle size and drift from the intended target area. As expected, these studies indicate the greatest potential nontarget contamination hazard resulting from drift occurs with smaller diameter particles.

The control of drop size to provide larger drops and reduce the drift potential, results in a decrease in coverage by spraying. Coverage increases as drop size decreases. Although large drops hit the target area more frequently, the extent of coverage is less. A compromise is necessary to minimize drift and obtain a good coverage. A wide distribution of drop size is inevitable with commonly used spray equipment, and the measurement of wind and atmospheric conditions is therefore important in determining the safety of a pesticide application. In the case of chlorinated hydrocarbon, it is not unusual to find 50% or more of the applied pesticide unaccounted for where a material balance of the treated area is made immediately after application. Most of the unaccounted portion is dispersed in the air as fine particles or aerosols, or carried to adjacent areas.

Pesticides can also enter the air when soils contaminated with pesticides are subjected to erosion by wind. With appropriate conditions of soil, moisture, humidity and wind, pesticide residues from soils can enter the air and be transported great distances. In the air DDT can be transported as vapor, tiny crystals or a mixture with dust particles. One study traced DDT and other pesticides in a dust storm from West Texas to Cincinnati. The mobility of pesticides in air is also demonstrated by the fact that Antarctic ice and snow contain thousands of tons of DDT residues transported there through the air.

Pesticides in Industrial Wastes: The industrial wastes from the pesticide manufacturing and food processing industries usually may not be safely discharged directly to the environment. The pesticides in liquid effluents require treatment to remove the danger to aquatic life. Settling basins are used to allow time for gravity removal of some solids; solid and liquid sludge wastes can be incinerated, but the scrubbing of stack gases is needed to remove contained pesticides. The deep well disposal of pesticides is only practical when the geological characteristics of the area are sufficient to protect against ground water contamination. There is the possibility that pesticide wastes in the disposal areas of pesticide manufacturers will leach from these sites into waters and soils for hundreds of years.

References

(1)   Finkelstein, H., "Air Pollution Aspects of Pesticides," *Report PB 188,091,* Springfield,
      Virginia, Nat. Tech. Information Service (September 1969).
(2)   Stone, R. and Smallwood, S., "Intermedia Aspects of Air and Water Pollution Control,"
      Report EPA-600/5-73-003, Washington, D.C., U.S. Environmental Protection Agency,
      (August 1973).

PHENOLS

In Air

Phenolic air pollutants result from the production of glass fiber products as noted by Sittig
(1), but quantitative data on such emissions were not available. Such phenols may be removed by aqueous alkaline scrubbing processes, transferring the problem from an air pollution problem to a water pollution problem.

In Water

The following industries are characteristic sources of phenolic pollutants: gas works (production), wood distillation, oil refineries, sheep and cattle dip, chemical plants, photographic
developers, explosives, coal tar distilling, mine flotation wastes, insecticides, resin manufacture, and coke ovens. In addition, aircraft maintenance, foundry operations, Orlon manufacture, caustic air scrubbers in paper processing plants, rubber reclamation plants, nitrogen
works, fiberboard factories, plastic factories, glass production, stocking factories and fiber
glass manufacturing have been reported as contributing phenols to wastewaters. Table 96
summarizes the levels of phenol found in wastes of various industries (2).

TABLE 96: LEVELS OF PHENOL REPORTED IN INDUSTRIAL WASTEWATERS

Industrial Source	Phenol Concentration, mg/l
Coke Ovens	
Weak ammonia liquor, without dephenolization	3,350–3,900
	1,400–2,500
	2,500–3,600
	3,000–10,000
	580–2,100
	700–12,000
	600–800
Weak ammonia liquor, after dephenolization	28–332
	10
	10–30
	4.5–100
Wash oil still wastes	30–150
Oil Refineries	
Sour water	80–185 (140 ave.)
General waste stream	40–80
Post-stripping	80
General (catalytic cracker)	40–50
Mineral oil wastewater	100
API Separator effluent	0.35–6.8 (2.7 ave.)

(continued)

TABLE 96: (continued)

Industrial Source	Phenol Concentration, mg/l
General wastewater	30
General wastewater	10–70
General wastewater	10–100
Petrochemical	
General petrochemical	50–600
Benzene refineries	210
Nitrogen works	250
Tar distilling plants	300
Aircraft maintenance	200–400
Herbicide manufacturing (includes chloro derivaties and Phenoxy acids).	210 239–524
Other	
Rubber reclamation	3–10
Orlon manufacturing	100–150
Plastics factory	600–2,000
Fiberboard factory	150
Wood carbonizing	500
Phenolic resin production	1,600
Stocking factory	6,000
Synthetic phenol, plastics, resins	12–18 369 (after cooling water separation)
Fiberglass manufacturing	40–400

Source: Report PB 216,162

Although described in the technical literature simply as phenols, this waste category may include a variety of similar chemical compounds among which are various phenols, chlorophenols and phenoxyacids. In terms of pollution control, reported concentrations of phenol are thus the result of a standard analytical methodology which measures a general group of similar compounds rather than being based upon specific identification of the single compound, phenol (hydroxybenzene).

References

(1) Sittig, M., *Pollutant Removal Handbook,* Park Ridge, N.J., Noyes Data Corp. (1973).
(2) Patterson, J.W. and Minear, R.A., "Wastewater Treatment Technology," 2nd edition, *Report PB 216,162,* Springfield, Virginia, Nat. Tech. Information Service (Feb. 1973).

PHOSPHORUS (ELEMENTAL)

In Air

Elemental phosphorus is produced as a final product for the chemical industry and as an intermediate product in the production of phosphoric acid and fertilizers. Analysis of the flue dust samples obtained from the phosphorus furnace operations of the Monsanto Chemical Company (Monsanto, Tenn.) showed that phosphorus pentoxide (P_2O_5) as particulate constituted 18.89% of the flue dust. Data on gases evolved from the flue dust during heating showed that phosphine and phosphorus pentoxide fumes (collected in water) amount to 0.09 and 0.67% by weight of the total charge. Flue dust is carried by the furnace gases along with the vaporized elemental phosphorus and is collected, while hot, by an electrostatic precipitator before the gases reach the phosphorus condenser.

It should be noted that during the industrial production of phosphorus, it is possible that particles of elemental phosphorus may be emitted. Free phosphorus can exist in air as a vapor because there are upper and lower limits on the ratios in which oxygen and phosphorus react. Consequently, it cannot be assumed that all of the free phosphorus reacts to form phosphorus oxides and suboxides.

Reference

(1)   Athanassiadis, Y.C., "Air Pollution Aspects of Phosphorus and Its Compounds," *Report PB 188,073,* Springfield, Virginia, Nat. Tech. Information Service (Sept. 1969).

PHOSPHORUS COMPOUNDS

In Air

Air pollution aspects of phosphorus and its compounds have been reviewed in some detail by Athanassiadis (1). Table 97 gives estimates as to sources and quantities of phosphorus-containing emissions as provided in a recent EPA sponsored survey (2).

In Water

As pointed out by Sittig (3), phosphates occur in rivers and wastewaters from a variety of sources: detergent laden wastes where phosphate builders were present in the detergent formulation; agricultural runoff from lands fertilized with phosphate-containing fertilizers; and miscellaneous industrial and municipal wastewaters.

Intermedia Aspects

Figure 12 illustrates the intermedial relationships for phosphorus compounds. The main sources are agricultural runoff, urban drainage, food processing, chemicals and allied industries, and wastewater from other sources. Agricultural runoff and urban drainage may flow directly to receiving waters, although urban drainage may be directed through combined sewers to treatment plants where it may or may not be bypassed.

The wastes from the food processing industry and chemicals and allied products industry may or may not receive in-plant treatment, either partial or complete. The effluent from in-plant treatment may either be discharged to sewer systems or directly to the water or land. The residues from treatments may be incinerated, discharged to receiving waters or deposited in landfills. The incineration process creates emissions which may or may not be treated. These possible treatments were discussed under "Particulates" and are shown in Figure 10. The residues from these treatments may be disposed to either the land or water.

Figure 12 also illustrates the natural intermedia flows. Phosphorus compounds may leach from landfills or runoff from the land to receiving waters, where they provide nutrients to plant growth. They may also precipitate from the air to water and land. It should be mentioned that incineration does not transfer a large amount of gaseous phosphorus compounds (such as phosphine) to the air. This intermedia route, then, is less significant for phosphorus than it is for sulfur compounds. The major impact of phosphorus in water is similar to that of nitrogen. As a nutrient, phosphorus promotes algal growth in the same manner and with the same consequences and can be removed for approximately the same cost.

TABLE 97: SOURCES AND ESTIMATES OF PHOSPHORUS-CONTAINING EMISSIONS

	Uncontrolled Particulate Emission Factor		Reliability Code	Production Level (10⁶ tons yr)	% P₂O₅ in Emissions	Reliability Code	P₂O₅ Before Controls (10³ tons/yr)	Estimated Level of Emission Control	P₂O₅ Emissions After Controls (10³ tons/yr)
	(lb/ton)	(Kg/10³kg)							
1. PHOSPHATE MINING									
a. Land-Pebble	0.	0.	(D)	115.	10	(B)	.0	-	.0
b. Hard-Rock	.5	.25	(D)	10.2	26	(B)	.7	0%	.7
c. Underground	.3	.15	(D)	.5	28	(B)			.0
2. PHOSPHATE ROCK PROCESSING									
Drying	15.	7.5	(B)	35.	30	(B)	79.	94%	4.7
Grinding	20.	10.	(C)	35.	30	(B)	105.	97%	3.2
Calcining-Cooling	40.	20.	(C)	4.0	30	(B)	24.	95%	1.2
Material Handling	1.	0.5	(C)	39.	30	(B)	5.8	50%	2.9
3. PHOSPHORIC ACID (WET PROCESS)									
Grinding	20.	10.	(C)	12.	30	(B)	36.	97%	1.1
Material Handling (ore)	2.0	1.	(C)	12.	30	(B)	3.6	52%	1.8
Reactor, Filter, Absorber	0.25**	.12**	(B)	3.78	30	(B)	-	**	0.5

(continued)

TABLE 97: (continued)

	Uncontrolled Particulate Emission Factor (lb/ton)	(Kg/10³kg)	Reliability Code	Production Level (10⁶ tons/yr)	% P2O5 in Emissions	Reliability Code	P2O5 Before Controls (10³ tons/yr)	Estimated Level of Emission Control	P2O5 Emissions After Controls (10³ tons/yr)
4. FERTILIZERS - (GENERAL)									
Rock Pulverizing	6.0	3.	(C)	7.24	30%	(B)	6.6	30%	1.3
Material Handling	2.0	1.	(C)	7.24	30%	(B)	2.2	0%	2.2
Screening	2.0	1.	(C)	6.5	*		6.5	85%	1.0
Bagging	1.0**	0.5**	(C)	3.2	*		-	**	1.6
Bulk Loading	1.0	0.5	(C)	3.2	*		1.6	0%	1.6
a. NORMAL SUPERPHOS.									
Drying	105.	52.5	(C)	1.37	*		72.0	90%	7.2
Cooling	90.	45.	(C)	1.37	*		61.7	90%	6.2
Product Grinding	0.5	0.25	(C)	1.37	*		.3	80%	.1
b. TRIPLE SUPERPHOS.									
Product Grinding	0.5	0.25	(C)	3.25	*		.8	80%	.2
Drying	105.	52.5	(C)	3.25	*		171.	90%	17.1
Cooling	90.	45.	(C)	3.25	*		146.	90%	14.6

(continued)

TABLE 97: (continued)

	Uncontrolled Particulate Emission Factor (lb/ton)	(Kg/10³Kg)	Reliability Code	Production Level (10⁶ tons/yr)	% P₂O₅ in Emissions	Reliability Code	P₂O₅ Before Controls (10³ tons/yr)	Estimated Level of Emission Control	P₂O₅ Emissions After Controls (10³ tons/yr)
c. AMMONIUM PHOS.									
Ammoniator-Granulator	30.	15.	(B)	3.5	*		52.5	85%	7.9
Drying	55.	27.5	(D)	3.5	*		96.0	85%	14.4
Cooling	25.	12.5	(C)	3.5	*		43.8	85%	6.6
5. FERTILIZER APPLICATION									
Bagging-Bulk Loading	1.0	0.5	(C)	6.5	*		3.2	50%	1.6
Field Spreading	4.0	2.	(D)	6.5	*		13.0	0%	13.0
6. ELEMENTAL PHOSPHORUS									
Ore Handling	2.0	1.	(C)	6.46	25%	(B)	1.6	0%	1.6
Grinding	20.	10.	(C)	6.46	25%	(B)	16.2	97%	.5
Briquetting, Sintering or Nodulizing	1.3	0.65	(B)	6.46	***		4.2	60%	1.7
Furnace Operations	1.0	0.5	(B)	6.46	***		3.2	90%	.3
Flares	1.8	0.9	(B)	6.46	***		5.8	0%	5.8

(continued)

TABLE 97: (continued)

	Uncontrolled Particulate Emission Factor (lb/ton)	(Kg/10³kg)	Reliability Code	Production Level (10⁶ tons/yr)	% P₂O₅ in Emissions	Reliability Code	P₂O₅ Before Controls (10³ tons/yr)	Estimated Level of Emission Control	P₂O₅ Emissions After Controls (10³ tons/yr)
7. PHOSPHORIC ACID (THERMAL)									
Hydrator-Absorber	2.6**	1.3**	(C)	1.2	*	-	-	**	1.6
8. SODIUM PHOSPHATE									
Drying-Cooling & Calcining-Cooling	60.	30.	(D)	.8	*	-	24.	90%	2.4
Material Handling	2.	1.	(C)	.8	*	-	0.8	0%	.8
Grinding	20.	10.	(C)	.8	*	-	8.0	97%	.2
Bagging	1.0**	0.5**	(C)	.8	*	-	-	**	.4
9. SOAPS & DETERGENTS									
Mixing	40.	20.		.6	*	-	12.	98%	.2
10. CALCIUM PHOSPHATE									
Drying-Cooling	60.	30.	(D)	.2	*	-	6.0	90%	.6
Material Handling	2.	1.	(C)	.2	*	-	.2	0%	.2
Grinding	20.	10.	(C)	.2	*	-	2.0	97%	.1
Bagging	1.0**	0.5**	(C)	.2	*	-	-	**	.1
Material Handling in Feeding	2.0	1.	(C)	.2	*	-	.2	0%	.2

(continued)

TABLE 97: (continued)

	Uncontrolled Particulate Emission Factor		Reliability Code	Production Level (10⁶ tons/yr)	% P₂O₅ in Emissions	Reliability Code	P₂O₅ Before Controls (10³ tons/yr)	Estimated Level Emission Control	P₂O₅ Emissions After Controls (10³ tons/yr)
	(lb/ton)	(Kg/10³ Kg)							
11. PLATING, POLISHING, & MISC.									
Phos. Acid Bath	0.25**	0.12	(D)	.25	*		-	**	.03
12. INADVERTENT SOURCES									
a. Refuse Incineration	NA	NA	(C)	0.15 (uncontrolled Emissions)	0.23%	(D)	.3	37%	.2
b. Iron Mfg.	150.	75.	(C)	245.	0.2-0.5% ****.35%	(C)	64.	9C%	6.4
c. Steel Mfg.	25.	12.5	(C)	145.	0.2-0.5% ****.35%	(C)	6.3	9C%	.6
d. Cement Mfg.	NA	NA	(D)	7.8 (uncontrolled Emissions)	.04%	(D)	3.1	88%	0.4
e. Fuel Oil Combustion	NA	NA	(C)	.287 (uncontrolled Emissions)	.9%	(C)	2.6	0%	2.6
f. Coal Combustion	NA	NA	(A)	33.8 (uncontrolled Emissions)	0.34%	(A)	115.	87%	20.7
TOTALS							1206.8	86.7%	160.3

* Emission factor multiplier includes % P₂O₅
** Emission Factor After Control
*** Emission Factor includes % P₂O₅
**** Weighted Average
NA Not Applicable

Note: See page 7 for explanation of Reliability Code.

Source: Report PB 231,670

FIGURE 12: INTERMEDIA FLOW CHART–PHOSPHORUS COMPOUNDS

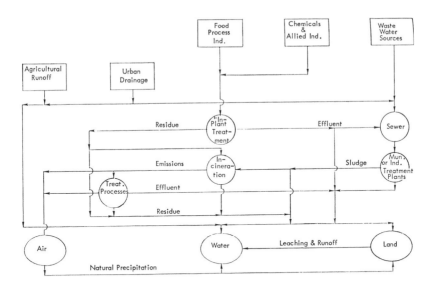

Source: Report EPA 600/5-73-003

Phosphine (PH_3), the final compound in the biological breakdown of phosphates, is toxic to certain fish, and has been detected in some polluted waters in concentrations exceeding 3.6 mg/l. Phosphorus has received much attention because of its use in detergents and the resulting increase in the phosphate content of sewage. Other sources of phosphorus in wastewater are agricultural production, food and kindred products, chemicals and allied products, and urban drainage. Controls on the discharge of phosphorus compounds to receiving waters are limited primarily to use of substitute materials and interception of surface drainage. Nonphosphate detergents are an example of the former, while use of drainage ditches, impoundment and spreading are frequently employed for the latter.

References

(1) Athanassiadis, Y.C., "Air Pollution Aspects of Phosphorus and Its Compounds," *Report PB 188,073,* Springfield, Virginia, Nat. Tech. Information Service (Sept. 1969).
(2) GCA Corp., "National Emissions Inventory of Sources and Emissions of Phosphorus," *Report PB 231,670,* Springfield, Virginia, Nat. Tech. Information Service (May 1973).
(3) Sittig, M., *Pollutant Removal Processes,* Park Ridge, N.J., Noyes Data Corp. (1973).
(4) Stone, R. and Smallwood, H., "Intermedia Aspects of Air and Water Pollution Control," *Report EPA-600/5-73-003,* Washington, D.C., U.S. Environmental Protection Agency (August 1973).

POLLENS

In Air

The aeroallergens encompass a wide variety of materials, but the pollens are the most important member of this group. The plants which produce allergenic pollens are widely distributed geographically, and their distribution and seasonal growth characteristics in part determine their importance as aeroallergens. Ragweed has been found in all 50 states; it produces large quantities of pollen, and the pollen grains are especially adapted for aerial dissemination by virtue of their size (20μ), shape, and density. In addition, because of its allergenic property for a large percent of the population, it has been the most studied, as noted by Finkelstein (1).

Approximately 40 species of ragweed are known at this time, the majority of which occur in North America. Ragweed is most prevalent in the North Central and Northeastern states followed by the Southern, Great Plains, Intermountain and Pacific Coast states. The weed has also been introduced into Hawaii. Six species are sufficiently widespread and abundant within the United States to be of importance as pollen sources on the state or national level. These ragweed species characteristically establish themselves quickly in freshly turned soil, and their special ability to grow as weeds enables them to flourish in cultivated grain fields, where they are most abundant today. Pollination is by aerial dissemination. Because pollination by wind is very inefficient, many thousands of pollen grains must be liberated and disseminated for effective pollination.

Ragweed tends to be crowded out by other vegetation if the soil is not disturbed. Therefore, the plowing of fields, especially cereal grain fields, is responsible for the growth of a major portion of the ragweed in many areas of the country. The ragweed seedlings develop during the ripening stage of the cereal grain and grow rapidly in the stubble after the grain has been harvested in late summer. An estimated one-third of the 60 million acres of wheat stubble is infested with ragweed. In fields of winter wheat, oats, and barley, with no cover crop, about 172,000 ragweed plants per acre have been observed, which is more than 300 times the plant density in pastures. Ragweed is found also in urban areas where soil has been disturbed.

For example, in a new subdivision where the soil had been overturned but untouched during the spring and summer, ragweed concentration amounted to 56,500 plants per acre. Along railroads, one count gave 13,000 plants per acre. It has been estimated that an acre of giant ragweed may produce as much as 50 lb of pollen during a single season. A ragweed survey was conducted within parts of the Ann Arbor and Superior Townships of Washtenau County, Michigan. Land along selected roads and highways was classified by land use, and the density of ragweed was then determined. The results of the survey showed that the general pattern of ragweed distribution prevailed: high density in croplands and low density in marshes and woodlands.

The fungi (molds and yeasts) are an important group of aeroallergens. These microorganisms are ubiquitous and saprophytic in nature. Their usual habitats are soil and dust, and they become airborne by means of local air disturbances. Their concentration in the air is dependent upon the magnitude of the source, their death rate in the air, humidity, temperature, and other factors. The largest percent of the airborne fungi are found in the air up to 5,000 ft, and their number decreases rapidly above that level. However, viable mold spores have been recovered up to 90,000 ft.

Danders, which include feathers of fowl and hair of animals (including humans), are found in the air close to their source point. Their concentration in the air is limited, and they are allergenic to humans when the source is in close proximity to the susceptible individuals. House dust consists of the small particulate organic materials (fragments of animal hair, wool, cotton linters, kapok, feathers, pollen which has seeped in from outdoors, etc.) found in the home.

Molds are also frequently found in the home, especially in old, damp dwellings (e.g., basements), and can be included in house dust. The house dust which is of concern in this connection does not include sand, soil, and powdered rock, which are not allergenic. The mite *Dermatophagoides pteronysissimus* has been determined to be the allergenic agent in house dust. Skin reactions to extracts of this mite in sensitive individuals are both quantitatively and qualitatively indistinguishable from those obtained with house dust.

In addition, the mite has a worldwide distribution, as well as seasonal variation of occurrence similar to the frequency of house dust allergenicity (peak in the autumn). Although a non-specific house dust extract is available for skin testing, some clinicians believe that house dust is specific for each home. Because it is found in every indoor environment, house dust is probably the most common aeroallergen after pollens. The miscellaneous aeroallergens, which include vegetable fibers and dust, cosmetics, paints, and varnishes, are limited in the air and affect susceptible individuals only when in close proximity to the source. They constitute a minor problem in terms of the total number of people affected.

Reference

(1)   Finkelstein, H., "Air Pollution Aspects of Aeroallergens (Pollens)," *Report PB 188,076,* Springfield, Virginia, Nat. Tech. Information Service (September 1969).

POLYNUCLEAR AROMATICS

In Air

The air pollution aspects of the carcinogenic polynuclear aromatic hydrocarbons (PAH) and of benzo[a]pyrene (BAP) in particular have been reviewed in some detail by Olsen and Haynes (1). These materials are also referred to as polynuclear organic materials (POM) and a tabulation of POM emissions and sources is presented in Table 98. The total emissions of benzo[a]pyrene (BAP) and some emission factors for BAP are given in Table 99 as presented by Goldberg (2).

TABLE 98: HAZARDOUS POLLUTANT SOURCES

Source	POM EMISSIONS	
	Amount in Tons	% This Pollutant
Iron and Steel		
Metallurgical Coke	43,380	0.90
Asphalt Industry		
Paving Material Preparation	2,800	0.06
Roofing Material Preparation	23,230	0.48
Petroleum Refining	2,170	0.05
Incineration		
Industrial	2,228	0.05
Domestic	730	0.02
Auto Body	14,602	0.30
Conical Burner	212,211	4.42
Open Burning	526,843	10.98
Agricultural Burning	2,161,142	45.05
Natural Fires, Forest	1,433,712	29.89
Natural Fires, Urban	6,060	0.13
Municipal	682	0.01
Coal Refuse	193,500	4.03
Power Plant Boilers		
Pulverized Coal	8,980	0.19
Stoker Coal	1,032	0.02
Cyclone Coal	310	0.01
All Oil	7,675	0.16
All Gas	6,151	0.13

(continued)

TABLE 98: (continued)

	POM EMISSIONS	
Source	Amount in Tons	% This Pollutant
Industrial Boilers		
Pulverized Coal	1,896	0.04
Stoker Coal	6,635	0.14
Cyclone Coal	948	0.02
All Oil	10,001	0.21
All Gas	20,220	0.42
Residential/Commercial		
Coal	66,796	1.39
Oil	33,105	0.69
Gas	10,065	0.21
TOTAL	4,797,104	100.00

Source: The MITRE Corporation. Preliminary Results, EPA Contract No. 68-01-0438

TABLE 99: ESTIMATED ANNUAL BENZO[a] PYRENE (BaP) EMISSIONS FOR THE UNITED STATES

Source	Estimated BaP Emission Rate	Estimated Annual Consumption or Production	Estimated Annual BaP Emission (tons)
Heat generation	$(\mu g/10^6$ Btu$)$	$(10^{15}$ Btu$)$	
Coal			
Residential			
(i) hand-stoked	1,400,000	0.26	400
(ii) underfeed	44,000	0.20	9.7
Commercial	5,000	0.51	2.8
Industrial	2,700	1.95	5.8
Electric generation	90	6.19	0.6
Oil	200	6.79	1.5
Gas	100	10.57	1.2
Total			421.6
Refuse burning	$(\mu g/ton)$	$(10^6$ tons$)$	
Incineration			
Municipal	5,300	18	0.1
Commercial	310,000	14	4.8
Open burning			
Municipal refuse	310,000	14	4.8
Grass, leaves	310,000	14	4.8
Auto components	26,000,000	0.20	5.7
Total			20.2
Industries	$(\mu g/bl)$	$(10^6$ bl$)$	
Petroleum catalytic			
Cracking (catalyst regeneration)			
FCC[a]			
(i) no Co boiler[b]	240	790	0.21
(ii) with CO boiler	14	790	0.012
HCC[c]			
(i) no CO boiler	218,000	23.3	5.6
(ii) with CO boiler	45	43.3	0.0024
TCC[d] (air lift)	$(\mu g/bl)$	$(10^6$ bl$)$	
(i) no CO boiler	90,000	131	13.0
(ii) with CO boiler	< 45	59	< 0.0029
CC (bucket lift)			
(i) no CO boiler		119	0.0041
(ii) with CO boiler	< 31	0	0
Asphalt road mix	50 μg/ton	187,000 tons	0.000010
Asphalt air blowing	< 10,000 μg/ton	4,400 tons	< 0.000048
Carbon-black manufacturing	Atmospheric samples indicate that BaP Emissions		
Steel & Coke manufacturing	from these processes are not		
Chemical complex	extremely high		
Total			18.8

TABLE 99: (continued)

Source	Estimated BaP Emission Rate (µg/gal)	Estimated Annual Consumption or Production (10^{10} gal)	Estimated Annual BaP Emission (tons)
Motor vehicles			
Gasoline			
Automobiles	170	4.61	8.6
Trucks	>460	2.01	>10
Diesel	690	0.257	2.0
Total			> 20.6
Total (all sources tested)			481

[a]FCC: fluid catalytic cracker.

[b]CO boiler: carbon monoxide waste heat boiler.

[c]HCC: Houdriflow catalytic cracker.

[d]TCC: Thermo for catalytic cracker.

Source: Report PB 223,568

Natural Occurrence: Organic carcinogens are primarily unwanted by-products of imperfect combustion. However, a few sources of organic carcinogens might be defined as naturally occurring. Bituminous coal contains BAP, benz[a] anthracene, and other PAH. However, studies in England reveal a lower incidence of lung cancer among coal miners than in the population at large. Consequently, although carcinogenic agents are present in raw coal, they have no apparent biological effect in that form.

Asbestos miners, in comparison to coal miners, have a greater potential for developing malignancy. Two of the three types of asbestos used industrially contain appreciable quantities of natural oils. These oils, which adhere to asbestos particles, contain BAP. The quantities (1 to 5 µg/100 g asbestos) are small enough that their presence may play no more than a minor carcinogenic role. Another potential naturally derived source of airborne carcinogens has been reported by Troll, namely, various molds in the environment that possibly produce toxic compounds, some of which may be carcinogenic.

Production Sources: Prior to the introduction of gasoline and diesel engines, PAH were emitted primarily from the burning of coal and other solid fuels used for home and industrial heating as well as for the production of industrial products and power. However, in recent times the use of gasoline and diesel engine fuels has resulted in a new major source of PAH.

The U.S. Public Health Service, recognizing the need for specific emission data, has obtained information concerning those sources that involve burning conventional fuels and certain commercial and municipal solid wastes. The sources of heat generation that were tested ranged in size from residential heaters to heavy industrial power plant boilers and employed the following firing methods: pulverized coal burners, chain grate stokers, spreader stokers, underfeed stokers, and hand stoked coal burners; steam-atomized, centrifugal-atomized, and vaporized-oil burners; and premix gas burners.

The incineration sources tested ranged from small commercial incinerators to large municipal incinerators. Open-burning sources and several industrial processes were also tested. From such studies, in conjunction with the annual consumption and production figures, sufficient emission data were collected to reveal the probable major sources of carcinogenic agents as indicated in Table 99.

In interpreting the significance of this information, it should be remembered that the calculation of total emissions involved a considerable amount of estimation, as well as a number of assumptions, with respect to both emission rates and annual consumption or production figures. Also, the aggregate emissions from a number of small sources not considered, although probably small, cannot be estimated. Recommended limits of 200 $\mu g/m^3$ for an 8 hr workday have been adopted by the American Conference of Governmental Industrial Hygienists for coal tar pitch volatiles (benzene fractions containing anthracene, BAP, phenanthrene, acridine, chrysene, pyrene, etc.).

References

(1) Olsen, D.A. and Haynes, J.L., "Air Pollution Aspects of Organic Carcinogens," *Report PB 188,090,* Springfield, Virginia, Nat. Tech. Information Service (September 1969).
(2) Goldberg, A.J., "A Survey of Emissions and Controls for Hazardous and Other Pollutants," *Report PB 223,568,* Springfield, Virginia, Nat. Tech. Information Service (February 1973).

RADIOACTIVE MATERIALS

In Air

The air pollution aspect of radioactive materials have been reviewed in some detail by Miner (1). The two major sources of natural radioactivity are the gases which emanate from minerals in the earth's crust and the interaction of cosmic radiation with gases in the atmosphere.

Radioactive Dusts: Soils and rocks contain naturally radioactive minerals such as radium-226 and radium-229 in variable amounts. The radioactive progeny of two nuclides, the noble gases radon-222 and radon-220 (thoron), emanate from the earth's crust and contribute greatly to atmospheric radioactivity. Their concentration is higher in areas where there are substantial amounts of uranium and thorium ores. Therefore, these gases may occur as air pollutants in the vicinity of uranium mines, mills and refineries, or where radium and its ores and by-products are processed.

Radon, with a half-life of 3.8 days, has a much higher probability of emanating from the earth's crust before it decays than thoron, which has a half-life of 54 sec. The atmospheric concentration of these noble gases and their daughter products also depends on many geological and meteorological factors. The daughter products of thoron and radon attach themselves to the inert dust in the atmosphere, endowing these dusts with apparent radioactivity. In addition, some dust particles from naturally radioactive minerals and soils also find their way into the atmosphere, but they contribute very little to natural radiation.

Cosmic Rays: Interactions of cosmic rays with atmospheric gases produce a number of radioactive species, the most important of which are tritium, carbon-14, and beryllium-7. Of lesser importance are beryllium-10, sodium-22, phosphorus-32, phosphorus-33, sulfur-35, and chlorine-39. These interactions produce electrons, gamma rays, nucleons, and muons. At low radiation levels the muons account for 70% of the cosmic radiation.

Combustion Emissions: Fossil fuels contain radioactive materials that escape into the atmosphere when the fuel is burned. Coal ash contains a number of radionuclides which originate from traces of uranium-238 and thorium-232. It has been estimated that uranium-238 and thorium-232 are present in coal in concentrations of 1.1 and 2.0 ppm, respectively. Fly ash released from the stack when coal is burned contains 10.8 μCi of radium-235 and 17.2 μCi of radium-226 per electrical megawatt per year. Oil-burning plants normally discharge nearly all of their combustion products into the atmosphere, a 1,000 Mw station which consumes 460 million gallons of oil per year will discharge about 0.5 μCi of radium-226 and radium-228. A joint study of natural gas from northwestern New Mexico and southwestern Colorado by the U.S. Public Health Service and the El Paso Natural Gas Company showed that radon-222 (a daughter of radium-226) is present in natural gas at concentrations ranging from 0.2 to 158.8 pCi/liter. There is a lack of data concerning concentrations of radon-222 in the stack effluent of natural gas power plants, but it can be assumed to be minimal because of the short (3.8 day) half-life of radon-222 and the relatively long time required for transit of the gas from the well to the plant where it is burned, as well as for storage. There will be some activity from the longer-lived daughter products of radon, but these are hard to determine since the daughter products occur as particulates and are subject to many removal forces.

Natural Radioactivity: Measurements have been made of ground level atmospheric radioactivity at a number of places throughout the world. The radon concentration is inferred from the lead-214 measurements, since radon and its daughter products lead-214 are in radioactive equilibrium when the radon laden air and dust coexist for 2 hr. The thoron concentration is inferred from measurements of lead-212. The thoron series has no long-lived daughters and its secular equilibrium is determined by the 10.6 hr half-life of lead-212.

The meteorological factors related to an air mass for several days prior to its observation influences its radon and thoron content. Both passage of the air over oceans and precipitation tend to reduce the concentration of these gases, whereas periods of temperature inversion cause them to increase. Washington, D.C., which is some distance from the ocean, had the highest thoron (lead-214) concentration of any coastal area studied, followed by seaports, midocean islands, and finally Antarctica. The doses of ionizing radiation that originate from cosmic rays and from gamma-emitting radionuclides in the earth's crust received in populated areas vary from 75 to 175 mrad/yr.

Radiation emissions associated with the burning of fossil fuels are distributed generally throughout the country. The majority of the emissions will be concentrated in areas where large power plants are located. Therefore, the distribution of radioactive materials in the atmosphere from this source will follow distribution patterns similar to those of other fossil fuel combustion products (for example, sulfur dioxide).

The radioactive nuclides from production sources originate either as fission products or activation products; the ones encountered in atmospheric pollution are the same regardless of whether they are produced by nuclear reactor, a nuclear or thermonuclear bomb, or a plant reprocessing spent reactor fuel. The potential for radioactive contamination of the environment exists in all phases of processing radioactive materials. This processing involves mainly heavy industries, such as the uranium and thorium mines, metallurgical factories, nuclear reactors, and chemical plants.

Radioactive pollution of the atmosphere can occur by the release of airborne radioactive materials in routine industrial operations or as the consequence of an accidental release of airborne contaminants. Nuclear reactor operations and nuclear spent fuel processing are the principal sources of radioactive gases.

Production of Nuclear Fuel: The production of nuclear fuel for use in reactors or for nuclear explosions involves the mining of crude uranium or thorium ore, washing and concentrating the ore in processing plants adjacent to the mines, producing ingots of refined uranium or thorium, and physically separating the different isotopes of uranium and thorium. All these operations use only the naturally occurring radioactive elements belonging to the uranium and thorium families.

Mining, Milling, and Refining of Uranium: Uranium mining gives rise to the usual dust problems associated with conventional ore mining. The presence of radium in particles is not considered as important as the presence of radon gas daughter products. Adequate ventilation at the working faces of the mine must therefore be provided. The release of mine ventilation air to the atmosphere and subsequent dispersion provide large dilution factors.

In addition, mines are frequently in remote areas at significant distances from population centers. Therefore, this is mainly an occupational problem rather than a general air pollution problem. Ore concentration begins with crushing and pulverizing the ore. These operations yield dusts containing a small concentration of radioactive materials, but the other toxic materials present (such as silica, vanadium, arsenic, and selenium) pose greater problems than the radioactive materials present. Adequate filtering of the ventilation air prevents the release of pollutants to the atmosphere. The large tailing piles that have accumulated around uranium mills have recently become an area of public concern from the air pollution standpoint. It is feared that radon gas emanating from these piles may be an air pollution hazard to the general public in the surrounding areas.

The concentrates, consisting of impure U_3O_8 (pitchblende), are further processed for isolation and purification of uranium. Solvent extraction and fluoride volatilization are the principal methods used to produce pure compounds for reduction to metal or for the production of uranium hexafluoride, which is used in the gas diffusion process for producing uranium-235. The airborne radioactive products released from these processes are dilute, volatile uranium fluorides and uranium-containing dusts.

During the feed preparation step, less than 3 μCi/day of uranium are released as the hexafluoride. Uranium dust and other uranium compounds are controlled at the diffusion plants so that downwind concentrations are consistently less than the MPC. Uranium ore processing plants are located in the states of Colorado, Utah, New Mexico, South Dakota, Texas, and Wyoming. Most of these locations are remote from populated areas. The milling plants are in general located close to the mines.

Fuel Fabrication: The fabrication of fuel elements for power reactors involves the metalworking processes of rolling, extruding, heat treating, machining, and cladding the uranium. Experience to date has shown that the potential for radioactive airborne pollution from these processes is minor. Since the development of the breeder reactors, there has been much interest in plutonium and plutonium alloy fuels. Plutonium metal is pyropheric and extremely toxic, hence, great care must be exercised in its loading.

To minimize the release of plutonium during fabrication, leak-tight enclosures are used for all work, and all exhaust heat is produced by nuclear fission. Radioactive wastes formed by nuclear fission are of two types: fission products, which remain incorporated in the nuclear fuels; and activation products, found mainly in the coolant. Both the fuel elements and the coolants are thus potential sources of radioactive atmospheric pollution.

The pollution may come about through release into the atmosphere of radioactive gases, such as xenon and krypton (fission products); through the induced activity of atmospheric argon; through the formation of radioactive aerosols containing fuels (uranium, thorium, plutonium); through the release of fission products (strontium-90, cerium-144, barium-140, zirconium-95, and others); or through induced activity of other kinds.

The civilian nuclear power reactors (built for generation of electricity) operating at this time are generally located around the Great Lakes and in the Eastern portion of the country. The new plants planned for the near future are concentrated in the same areas as well as in the Southeast, the upper Mississippi and Missouri regions, and the Pacific Coast.

In addition to the power reactors, a number of research and test reactors are located throughout the United States, as well as plutonium production reactors. The gases are filtered at least twice through high efficiency filters. Operating experience at Hanford, Oak Ridge, Argonne, and Los Alamos has shown that intricate operations with all forms of plutonium can be carried out without significant release of airborne plutonium. Despite this fact, there have been some serious fires and explosions in plutonium-handling facilities, decontamination costs and equipment damage have been the most serious results. No serious releases to the atmosphere have occurred. From experience to date, the airborne radioactive contamination from uranium mining, milling, refining and fuel fabrication processes is considered to be minor.

In processing plutonium into fuels, great care is exercised in the design of equipment and control features to insure that negligible quantities are released to the atmosphere in day-to-day operations and in fires and other serious accidents. Fuel fabrication plants are located in a number of areas throughout the United States.

Nuclear Reactors: Nuclear fuels are introduced into reactors. The largest concentrations of test reactors are at the National Reactor Testing Station at Idaho Falls and Oak Ridge National Laboratories. Large plutonium production reactors are located at Hanford, Wash. and Aiken, S.C. In addition, there are now commercial power-producing reactors at many locations.

Normal Reactor Operation: The quantity and nature of the gaseous effluents will be influenced by the type of reactor used. The air-cooled reactor at Brookhaven Laboratories releases large quantities of argon-41, an isotope with a 112 min-half-life. Each operating day, some 14,000 Ci are released from a 300 ft stack. The waste gases released from the water reactors at Dresden 1, Big Rock Point, Humboldt Bay, Elk River, Yankee, and Indian Point 1 have been studied in some detail. Dresden 1, Big Rock Point, Humboldt Bay and Elk River are boiling water reactors; Yankee and Indian Point 1 are pressurized water types.

The power ratings of the stations vary from 24 Mw(e) (megawatts of electrical energy) for Elk River to 200 Mw(e) for Dresden 1, and the periods of operation range from 4 to 7 yr. All these plants have operated within the limits authorized by the AEC for release of radioactive wastes to the environment. The maximum annual average releases of gaseous activation products and noble gases have ranged from 22 Ci/yr (0.7 μCi/sec) at Yankee, to 35,000 μCi/sec at Big Rock Point. The releases varied from a maximum of 0.002% of the limit at Indian Point to as much as 28% of the limit at Humboldt Bay.

Releases of halogens and particulates in the gaseous wastes ranged from 2×10^{-8} μCi/sec at Indian Point to nearly 1.2 μCi/sec at Big Rock Point, corresponding to about 0.00001 and 30% of the respective limits. The maximum annual average releases of 0.07 μCi/sec of halogens and particulates at Humboldt Bay corresponded to 38% of that station's licensed limit. The maximum offsite dosage measured above background at Humboldt Bay (integrated over 12 consecutive months) was only 50 mrems. Offsite air monitoring at other sites has yielded measurements at or very near the background level in all cases.

Tritium is produced in nuclear reactors by fissioning of uranium, neutron capture in boron and lithium added to the coolant, neutron capture reaction with boron in control rods, activation of deuterium in water, and high energy capture reactions with structural materials. In light water reactors the main sources of tritium in the primary coolant are leaking of fission produced tritium through cladding defects and boron and lithium reactions. In heavy water reactors, neutron activation of the deuterium moderator and coolant is the major source of tritium.

The majority of the tritium released from the coolant reaches the environment as liquid waste. Only about 1% of the total tritium entering the atmosphere is released as gaseous waste. Measurements made by the Bureau of Radiological Health's Nuclear Engineering Laboratory at a boiling water reactor indicate that the gaseous tritium release may be less than 0.5 Ci/yr. Gaseous tritium releases from the Yankee pressurized water reactor are reported to be less than 100 Ci/yr. In heavy water reactors, only limited loss of heavy water can be tolerated for economy considerations, a consideration which effectively limits the release of tritium from this source. In addition, the rather high permissible concentration of tritium in ambient air also reduces this isotope's significance as an air pollutant from reactors.

Very short-lived nitrogen and oxygen isotopes are formed in large quantities from activation but do not pose an air pollution problem because of their rapid decay. Several incidents in the past 20 yr have occurred during fuel discharge of the Hanford reactors that resulted in temporary off-standard releases of airborne material. An estimated 4 Ci was released in one episode, yet only minor contamination was found in the environment.

Filters and charcoal beds were installed in 1960, and since that time releases have been entirely insignificant. A serious incident occurred in 1958 at the NRU reactor, a heavy water moderated Canadian experimental reactor, yet recovery was possible and only minor releases to the environment resulted. Due to a faulty mechanism, a highly irradiated fuel assembly was caught and could not be inserted into the discharge cask. A 3 ft portion melted and burned. A detectable level of contamination was found at a distance of 1,000 ft from the reactor building. Decontamination of the reactor required about 3 months.

Reactor Accidents: Reactor accidents which result in melting of a large fraction of the highly irradiated fuel are highly unlikely, although credible. Upon melting, the core could release to the reactor building the noble gas fission isotopes and a fraction of the halogens and other volatile isotopes. The postulated accident which could cause this is called the design basis accident, and the reactor system is designed to preclude such an event. In addition, special designs (such as for a containment vessel) are required which ensure confinement or containment to a very high degree in the event of a serious accident. The AEC reviews all reactor designs prior to licensing to ensure the safety of the general public in case of an accident.

Some serious reactor accidents in the Western world have occurred in the United States, Canada, and England. The most recent of these, which resulted in the death of three military personnel, occurred in 1961 at the Army low power (SL-1) reactor at the National Reactor Testing Station in Idaho. Through inadvertent withdrawal of a safety rod, the reactor went critical and the nuclear excursion resulted in a violent chemical explosion. Even though the reactor building was conventionally constructed, the radioactivity released from the core was substantially confined within the reactor building. An estimated 10 Ci of iodine (about twice the background radiation) was released and was detectable about 80 miles downwind.

Through a series of compounding events in Canada at the NRX reactor in Chalk River, Ontario, a power surge melted about 10% of the uranium fuel rods in 1952. Some 10,000 curies of fission products were carried below the reactor and spread through auxiliary equipment. Evacuation of the area for a weekend was required because of airborne gases.

The only accident which caused any generalized environmental contamination occurred at Windscale, England in 1957. The accident followed an attempt to anneal graphite by nuclear heat. The uranium elements in 150 fuel channels rose to such temperatures that cladding failed and the elements reached a glowing red heat. After carbon dioxide was found ineffectual, water was used to quench the uranium. The reactor cooling air was released to the atmosphere through a 410 ft stack, at the top of which was a low efficiency filter. Some 20,000 Ci of iodine-131, 600 Ci of cesium-137, 80 Ci of strontium-89, and 9 Ci of strontium-90 were released.

Milk was contaminated by iodine-131 in a 200 square mile area to greater than the permissible level, and sale of milk from this area was forbidden for 3 to 6 wk. The largest thyroid dose recorded among the inhabitants was 19 rads in one child. The reactor was never put back into operation. Even though the foregoing represent the worst accidents to date involving radioactive air pollution, the consequences in respect to air pollution were much less serious than some of the documented nonradioactive air pollution incidents on record.

Fuel Reprocessing: The highly radioactive fuels taken from power reactors are reprocessed to separate the uranium and plutonium from the many curies of fission products. Many processes have been developed for removing the cladding material, dissolving the fuel, and extracting the uranium and plutonium. Radioactive airborne contamination from a reprocessing plant is a potential problem, since all the highly radioactive fission products are released from the fuel during the dissolution step. Unless they are deliberately recovered, all noble gas isotopes in the fuel at the time of dissolution are swept out of the dissolver into the atmosphere. Krypton-85 releases are not now an air pollution problem; however, based on projected nuclear power expansion and population growth by the year 2060, it is estimated that the radiation dose from krypton-85 would be of the order of 50 mrads per year, and may be as high as 100 mrads per year. From the public health standpoint, 50 mrads per year may be acceptable if other sources of exposure are adequately controlled. Of greater concern is the potential for day-to-day emission of radioactive particles and volatile isotopes. The most critical volatile isotope is iodine-131, which can be reduced to negligible quantities by allowing a long storage time after the fuel is removed from the reactor. In addition, good processes are available for removing iodine-131 from exhaust air. Experience has shown that on a long term basis and with adequate fuel-cooling and iodine-131 removal facilities, the routine iodine-131 emissions can be kept well below 1 Ci/day from a large separations plant. Another isotope of iodine whose emissions from fuel reprocessing plants may be significant is iodine-129.

The fuel processing plants at Savannah River and at Hanford have experienced momentary releases of iodine-131 on occasion, due to equipment failure or inadvertent processing of fuel which had cooled less than 4 months. For example, the Savannah River plant released 153 Ci of iodine-131 during a 5 day period in 1961. The levels reached in the environment did not require withholding milk from consumption or any precautions other than monitoring action. A very similar incident occurred at Hanford in September 1963 when about 60 Ci of iodine-131 were released. The maximum off-project grass level reached about 1.3×10^{-5} μCi/g. Increases in milk were detectable, but not dangerous.

Another isotope which forms volatile compounds is ruthenium, prominently present in the fission product mixture as ruthenium-103 (with a half-life of 40 days) and ruthenium-106 (with a half-life of 1 yr). Ruthenium is relatively easily oxidized to the tetroxide, volatilized, and trapped in a caustic scrubber. Radioactive particles are generated at almost every point in the process where a liquid is boiled, sprayed, agitated, or pumped. The very fine sprays may be carried out through the vessel vents or through very small leaks. The liquid evaporates, leaving a very small solid residue that carries with it the radioactive material. Very efficient filters are utilized for all air leaving the operation.

Isolation and purification of plutonium during fuel processing is accomplished through precipitation, fluorination, and eventual reduction to metal. Plutonium aerosols are generated from droplets and dry powders. Each enclosure where the work is performed is exhausted through a high efficiency fire resistant filter. The air is again filtered before release to the atmosphere through a tall stack.

The uranium stream from the fuel separations process becomes the feed for a calcining operation which converts the nitrate to oxide. The calcining yields airborne uranium oxide particles, practically all of which are retained on high efficiency filters in the ventilation air exhaust. Radioactive air pollution due to fuel reprocessing plants to date has been minor.

Nuclear Tests: Testing of nuclear explosives is another source of atmospheric pollution. The nuclear explosives are either based on fission processes employing uranium-235 or plutonium-239 or fusion reactions employing light nuclei (hydrogen or lithium). The explosion takes the form of a nonmoderated chain reaction which produces large neutron fluxes that activate the surrounding material. The radioactive products released in a nuclear explosion are the fission products strontium-90, cesium-137, iodine-131, and others, and the activation products calcium-45 and sodium-24. After some time has elapsed, the principal contaminants remaining are strontium-90 and cesium-137.

The force of the explosion and the accompanying rise in temperature convert these radioactive materials into gases or else eject fine particles high into the atmosphere. The immediate result is thus a primary pollution of the atmosphere at the site of the explosion. This is followed by a secondary pollution due to radioactive fallout. The distance covered by the particles of radioactive material will vary with the height to which they are ejected and with their size. They will eventually settle out or be carried down by rain and become dispersed over the surface of the ground. In this way, pollution is produced at points remote from the site of the explosion, the distance depending upon the size of the explosion, the prevailing meteorological conditions, and the latitude at which ejection into the stratosphere takes place. Examples of this remote type of pollution are illustrated in reports of fission product fallout in the United States from the Chinese nuclear tests in 1964 and 1965. It is estimated that from World War II until the end of 1962, the total explosive yield of all nuclear detonations by the United States, the United Kingdom, and the Soviet Union was equivalent to 511 megatons of TNT.

In 1963, a moratorium on open air testing was adopted by the United States, the United Kingdom and Russia. Since then, there has been a small amount of venting from underground tests conducted by the United States and Russia, but this has not added a significant amount of radioactivity to the total atmospheric inventory. Moreover, China, France, and India have tested nuclear weapons, but these tests have not added appreciably to the radioactivity totals made prior to 1962. During tests prior to 1963, it is estimated that about 30% of the radioactivity produced by the nuclear explosions was deposited in the immediate vicinity of the test sites.

The hazard to man arises primarily from fallout since most of the debris is carried to the earth's surface in rainfall. The greatest source of human exposure is the radionuclides absorbed by man via the food chain (for example, the contamination of grass by iodine-131 fallout, with subsequent ingestion by cows and concentration in their milk). The majority of the radiation received from inhaled radioactive debris from weapons testing originated from zirconium-95 and cerium-144. During the heavy weapons testing in 1962 and 1963, doses to the lung amounted to only a few mrad per year, which is small in comparison with the normal background dose. The total radiation from nuclear testing has added only about 10 to 15% to the normal natural radiation background dose.

Product Sources: Radionuclides are used as tracers in industry, biology, and agriculture and for internal irradiation in medicine. Another application of radionuclides is as sealed sources for gammagraphy and for massive external irradiation (sterilization). Radioactive wastes result only from the first type of application. These wastes may be either the unused remains of the radionuclides employed, or products of transformation or excretion. The quantities involved are small, and atmospheric pollution from this source is generally of little significance.

Other Sources: Other sources of radioactivity are pilot plants, research laboratories, and laundries for washing contaminated clothes, as well as metallurgical examination of fuels, and incineration of slightly contaminated clothing and radioactivity filters. In such cases, the release of particulate activity is easily controlled by absolute filtration of all air from active laboratories, the levels of gaseous activity are invariable so low that no significant air pollution occurs.

Another potential source of radioactivity can result from the peaceful use of nuclear explosions underground to stimulate gas production, provide gas storage basins, enhance the production of oil from oil shale, and facilitate solution mining of copper. Many such projects have been proposed and a few have been conducted, but without notable success.

Typical of these was a project called Gas Buggy which was conducted in December 1967 to stimulate natural gas production. The potential for release of radioactivity occurs mainly during the production and use of the end products, for example, fuel gases produced or stored can become contaminated with radioactive materials that are released when the gas is burned. To minimize the possibility of such contamination, the area is not used for a period of time afterwards to allow the radioactivity produced during the explosion to decay.

In Water

Some of the aspects of aqueous emissions from mining and milling operations and from nuclear fuel reprocessing have been touched upon by a publication from the University of California (2).

Intermedia Aspects

Figure 13 illustrates the intermedial relationships for radioactive wastes. This is definitely an intermedial pollutant problem and is rapidly becoming a national problem. In the fission process, by which nuclear reactors product energy, both liquid and gaseous wastes are created. The two main sources of these wastes are from the spent fuels and from leaks into the primary cooling water. Radioactive gas wastes require continual removal from the primary coolant and are then normally stored temporarily on a batch basis to reduce radioactivity before release to the atmosphere.

Several in-plant treatments can be employed to remove radioactivity from liquid waste effluents. These treatments result in the radioactivity being low in the treated effluent and fairly high in the treatment residues. The treated effluents can then be discharged into local sewers or adjacent waterways while the residues can be incinerated if the radioactivity level is below a certain level. The various levels which have been established will be discussed later. Incinerator emissions can be controlled as shown in Figure 13.

FIGURE 13: INTERMEDIA FLOW CHART–RADIOACTIVE MATERIALS

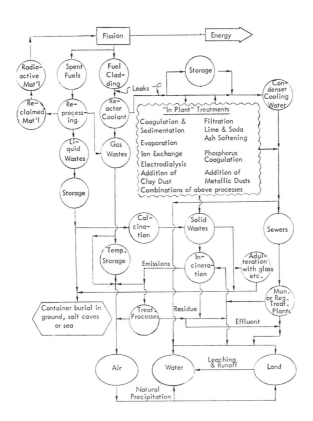

Source: Report EPA 600/5-73-003.

If the level of radioactivity is above that established as a limit for incineration, the residues may be sealed in containers for burial in specified locations. The reprocessing of spent uranium fuel elements results in liquid wastes of high radioactivity. These wastes are usually permanently stored underground, in salt caves, or at sea. An alternative to direct liquid waste disposal is the transformation of highly radioactive liquid wastes to solid wastes by calcination.

These solid wastes can be incorporated into a vitrified solid mass containing ferrous and glass aggregates prior to ultimate disposal underground or to the sea. Such adulteration enhances the heat-rejecting capability of the resulting solid mass. Radioactive residues applied to the land can be transmitted to adjacent waters by leaching and runoff. Radioactive exhaust emissions from nuclear reactors and incinerators can return to water and land by natural fallout and precipitation.

All radioactive wastes discharged to the environment should be evaluated in terms of their potential contribution to radiation exposure of the surrounding community so that the total radiation dose from the waste and from existing radiation sources can be determined. Radiation wastes may be classified quantitatively as low-level, if the activity can be measured in microcuries/liter or per gallon; intermediate-level, if the activity is measured in millicuries/liter or per gallon; and high-level, if the activity is measured in curies/liter or per gallon.

These criteria for measuring radioactive wastes are very simple, but not completely adequate, because the radiotoxicity of various isotopes in nuclear waste is not considered. Radionuclides contained in the water must first be identified before knowledge of the radiotoxicity can be acquired. The radiotoxicity of specific nuclides may be obtained by referring to the MPC values for air and water recommended by the International Commission on Radiation Protection (ICRP) and by the National Committee on Radiation Protection (NCRP).

When assessing the possible effects of environmental radioactivity on living matter, it is important to ascertain: the population affected by a particular isotope, the toxicity of the isotope, the organs within the specie affected by the isotope, and the mechanism of exposure to the isotope. For example, in sufficient doses I-131 is harmful (toxicity) to the thyroid gland (target organ) of man (population). The isotope I-131 is commonly used in medical research (exposure mechansim).

The sources of radioactivity in the environment are the fallout from nuclear testing and the discharge or disposal of radioactive wastes from nuclear power plants and other research facilities. The main radionuclides of concern are Sr-90, Cs-137 and I-131. Artificial radioactivity is present in natural waters mainly because of controlled waste disposal into rivers and lakes. Radioactivity can also contaminate surface waters from wastes released to the atmosphere or to the ground.

Another source of radioactivity affects a limited population living adjacent to uranium mines where contaminated water is discharged after being used for milling and from tailings. Water containing radioactivity can result in the exposure of people using the water for drinking, working, and recreational purposes. Exposure can also occur if contaminated water is used to irrigate plants or by the radioactive pollution of an aquatic environment used for cultivating marine life for human consumption. If radioactive wastes are discharged into the ocean, biological accumulation of radionuclides can result in considerable contamination of marine organisms.

When fuel is fissioned in a nuclear reactor to sustain a chain reaction, large quantities of radioactivity and heat are created. Approximately 2.68×10^4 curies of radioactivity are formed for each megawatt of thermal power. Since the conversion of thermal power to electrical power is only about 33% efficient, about 8×10^4 curies of radioactivity are formed for each megawatt of electrical power. Fission products from the fuel are usually contained within a metal clad solid matrix, but occasionally tiny pinhole leaks or fractures develop in the cladding and certain fission products are released to the reactor coolant.

In general, before discharge to the environment, radioactive materials produced in reactor fuels must pass through several barriers which have large retention factors. The first is the solid matrix where the fission products are usually formed; second is the cladding of the fuel element; third is the coolant and its confinement system; and the last is the containment vessel of the power reactor. Of public health concern are the long lived nuclides such as Sr-90 and Ce-137 which are fission products. Some of the other fission products are gases and include radioactive isotopes of iodine, krypton, and xenon.

Figure 13 illustrates the pollution source points and controls possible in nuclear power plants as well as the intermedia flows.

Three concepts of radioactive waste handling are widely used in industrial activities: [1] Dilute and Disperse—typical of this method is the dispersion of radioactive gases for dilution in the atmosphere and the controlled release of radioactive liquids into surface waterways. [2] Concentrate and Contain—this method applies to the concentration and containment of nuclear fuel reprocessing wastes, evaporator residues, and incinerator solids. [3] Delay and Decay—the radioactivity of an element is lost only through decay. Temporary retention of short lived radionuclides results in sufficiently low levels of radioactivity which may be released to the environment. An application of the delay and decay concept has found use in the introduction of radioactive material into selected soils, where slow ground water movement and the opportunity for the exhange of radioactive ions in solution with soil cations provide the delay necessary for decay.

Low and intermediate level wastes are released to the environment. The major practice in the treatment and handling of high level wastes consists of underground tank storage with no direct release to the environment. Other methods of handling high level radioactive wastes are being studied which may supersede tank storage. These methods include calcination, incorporation of glass, deep well disposal, and salt dome disposal.

References

(1)   Miner, S., "Air Pollution Aspects of Radioactive Substances," *Report PB 188,092,* Springfield, Virginia, Nat. Tech. Information Service (September 1969).
(2)   Lawrence Berkeley Laboratory, "Instrumentation of Environmental Monitoring: Radiation," *Report LBL-1,* volume 3, Berkeley, University of California (Oct. 1973).
(3)   Stone, R. and Smallwood, H., "Intermedia Aspects of Air and Water Pollution Control," *EPA-600/5-73-003,* Washington, D.C., U.S. Environmental Protection Agency (August 1973).

SELENIUM

In Air

The air pollution aspects of selenium and its compounds have been reviewed in some detail by Stahl (1). The production and use of selenium in the United States has been traced and charted for the year 1969. The consumption was 728 tons, while primary and secondary production was only 630 tons. Imports, principally from Canada, totaled 273 tons.

Emissions to the atmosphere during the year 1969 were 986 tons as shown in Table 100 (2). The emissions that resulted from the combustion of coal were about 65% of total emissions, and those due to the manufacture of glass were nearly 21%. Emissions from metallurgical processing of nonferrous metals and the burning of fuel oil were 9 and 7% respectively, while all other emissions were less than 1% of the total. Emission estimates for the manufacture of glass and metallurgical processing are based solely on data obtained from industry.

Emissions due to the combustion of coal and fuel oil are based on the analytical results obtained from relatively few samples. The emission factors presented in Table 101 are the best currently available.

TABLE 100: SELENIUM EMISSIONS BY SOURCE (1969)

Source Category	Source Group	Emissions - Tons	Emissions %
Mining and Milling		N	
Metallurgical Processing		85	8. 6
Secondary Production		1	0. 1
End Product Uses		204	20. 7
	Electronic and Electrical	N	
	Glass	203	
	Duplicating Machines	N	
	Pigments	1	
	Miscellaneous	N	
Other Emission Sources		696	70. 6
	Coal	630	
	Oil	65	
	Incineration	1	
TOTAL		986	100. 0

N - Negligible

TABLE 101: EMISSION FACTORS

Mining and Milling	2 lb/million tons of copper ore mined
Metallurgical Processing	277 lb/ton of selenium produced
Secondary Production	100 lb/ton of selenium recovered
End Product Uses	
Glass	2. 8 lb/1, 000 sq. ft. of colored structural plate glass
Pigments	15 lb/ton of selenium processed
Other Emission Sources	
Coal	2. 5 lb/1, 000 tons of coal burned
Oil	0. 21 lb/1, 000 bbls of oil burned
Incineration	20 lb/million tons of refuse burned

Source: Report PB 210,679.

They were determined through a combination of methods consisting of: direct observation of emission data, and other related plant processing and engineering data where available; estimates based on information obtained from literature, plant operators, and others knowledgeable in the field; calculations based on experience and personal knowledge of metallurgical processing operations; and specific analytical results (in the case of coal and oil) where available.

In Water

Essentially no information is available in the water pollution literature, the industrial waste literature, or the technical literature of the industries which use selenium, on levels of selenium in industrial wastewaters, treatment methods for selenium wastes or costs associated with removal of selenium from industrial wastewaters. Industries using selenium include paint, pigment and dye producers, electronics, glass manufacturers, and insecticide industries.

Johnson (3) has reported that selenium is present in almost all conceivable types of paper, and it might be concluded from this information that pulp and paper mill wastes could contain selenium. Johnson's primary concern was with the release of selenium as an air pollutant, upon incineration of paper products. However, he did report selenium wastewater levels as follows. For incinerator fly ash quench water, soluble selenium was measured at 5 to 23 $\mu g/l$, while incinerator residue quench water contained 3 $\mu g/l$ of soluble selenium. The incinerator handled a solid waste containing approximately 55 to 69% paper.

Hutchinson (4) reports that selenium occurs as an impurity in the form of selenide, $Se^=$, in metallic sulfide ores. When the sulfide ores are roasted in air (to convert the sulfide and drive off SO_2) the selenide is oxidized to selenium dioxide and is released in the flue gas. If the flue gas is quenched, selenium dioxide reacts with the quench water to form selenious acid, H_2SeO_3. The selenious ion appears to be the most common form of selenium in wastewater except for pigment and dye wastes, which contain the selenide (e.g., yellow cadmium selenide). Other forms decompose to yield the $SeO_3^=$ ion. Hutchinson also reports that selenium dioxide is readily reduced, to precipitate finely divided elementary selenium. It is probable that this same reaction might occur for selenious ion at acidic pH. Selenide may be precipitated as the metallic salt, which is highly insoluble. (4).

Secondary municipal sewage treatment plant effluents containing 2 to 9 $\mu g/l$ of selenium have been reported (5). A tertiary sequence of treatment, which included lime treatment to pH 11, sedimentation, mixed media filtration, activated carbon adsorption and chlorination yielded selenium removal of 0 to 89%. Best removal was obtained at higher initial selenium concentrations. Raible (6) has noted the presence of selenium in a pharmaceutical wastewater, but did not report concentrations. Linstedt, et al (7), measured 2.3 $\mu g/l$ of selenium in the effluent from a secondary sewage treatment plant. These workers investigated the efficiencies of removal of selenium, as $SeO_3^=$, by several advanced wastewater treatment processes and reported the results (8).

References

(1) Stahl, Q.R., "Air Pollution Aspects of Selenium and Its Compounds," *Report PB 188,077,* Springfield, Virginia, Nat. Tech. Information Service (September 1969).

(2) Davis, W.E. and Associates, "National Inventory of Sources and Emissions: Barium, Boron, Copper, Selenium and Zinc, Section IV-Selenium," *Report PB 210,679;* Springfield, Virginia, Nat. Tech. Information Service (May 1972).

(3) Johnson, H., "Determination of Selenium in Solid Waste," *Envir. Sci. Tech.,* 4, 850-853 (1970).

(4) Hutchinson, E., *Chemistry: The Elements and Their Reactions,* Philadelphia, W.B. Saunders Co. (1959).

(5) Argo, D.G. and Culp, G.L., "Heavy Metals Removal in Wastewater Treatment Processes: Part 2—Pilot Plant Operation," *Water Sew. Wks.,* 119, 28-132 (1972).

(6) Raible, J.A., "Fluorometric Determination of Selenium in Effluent Streams with 2,3-Diaminonaphthalene," *Envir. Sci. Tech.,* 6, 621-622 (1972).

(7) Linstedt, K.D., Houck, C.P., and O'Connor, J.T., "Trace Element Removals in Advanced Wastewater Treatment Processes," *Jour. Water Pollution Control Fed.*, 43, 1507-1513 (1971).

(8) Ahlgren, R.M., "Membrane vs Resinous Ion Exchange Demineralization," *Indust. Water Engrs.*, 8, 12-14 (1969).

SILVER

In Air

Table 102 summarizes sources and estimates of silver emissions from those sources (1).

TABLE 102: SOURCES AND ESTIMATES OF SILVER-CONTAINING EMISSIONS

Source	Uncontrolled Particulate Emission Factor lb/ton	kg/10³kg	Reliability Code	Production Level (tons/yr)	% Ag in Emissions	Reliability Code	Ag Emissions Before Controls tons/yr	10⁶oz/yr	Estimated Level of Emission Control (%)	Ag Emissions After Controls tons/yr	10⁶oz/yr
Iron and Steel											
Sinter Process	20 Sinter	10	C	51,000,000	0.01	D	51	1.5	90	5.1	0.15
Blast Furnace	130 Pig Iron	65	C	88,800,000	0.01	D	578	16.8	99	5.8	0.17
Open Hearth	17 Steel	8.5	C	65,800,000	0.05	C	280	8.2	40	168.0	4.90
Basic Oxygen	40 Steel	20	C	48,000,000	0.004	D	38	1.1	99	0.38	0.01
Electric Arc	10 Steel	9	C	16,800,000	0.05	D	42	1.2	78	9.2	0.27
Iron Foundry	20	10	C	22,000,000	0.008	D	18	0.5	75	4.5	0.13
Coal	NA	NA		33,800,000[a]	0.0007	A	236	6.9	82	42.5	1.24
Oil	NA	NA		287,000[a]	0.005	A	14	0.41	0	14.0	0.41
Gasoline, etc.	NA	NA		450,000,000	0.000001[b]	B	4.5	0.13	0	4.5	0.13
Incineration											
Municipal	NA	NA		100,000[a]	0.037	B	37	1.1	37	23.0	0.68
Commercial	--	--		---	---		220	6.5	99	2.2	0.07
Cement	NA	NA		7,790,000[c]	(0.008-0.015) 0.01[e]	B	896[d]	26.0[d]	83	107.0	3.1[d]
Non-ferrous Metal Processes											
Mining & milling	0.2	0.1	D	45,000,000 oz Ag	NA		0.15	0.0045	0	0.15	0.0045
Primary Smelting	(3-1800)	(1.5-900)									
Copper	1000[d]	500[d].	C	1,765,100 tons Cu	0.035	B	308	9.0	99	3.0	0.09
Other	25	12.5	D	1,460,000	0.175	D	32	0.9	95	1.5	0.045
Secondary Refining											
Copper	NA	NA		210[f]	25 oz Ag/ ton Cu	C	6	0.167	97	0.18	0.0053
Other	32	16	C	2,760 tons Ag	100	-	44	1.28	98.5	0.66	0.019
Consumption											
Brazing & Soldering	1%	---	D	14,000,000 oz Ag/yr	100	-	4.8	0.14	50	2.4	0.070
Electrical contacts	5%	---	D	12,000,000 oz Ag	100	-	21	0.60	0	21.0	0.60
All Other	---	---					1.5	0.043	--	1.5	0.043
Total							2832	82.5	85	416.6	12.1

NA - Not applicable
[a] Uncontrolled emission
[b] % Ag in fuel
[c] Particulate emissions before control
[d] Intermediate value, see text
[e] GCA Estimate
[f] Controlled Cu emissions
Note: See page 7 for explanation of Reliability Code

Source: Report PB 231,368

In Water

McKee and Wolf (2) have reported that silver, as the solid metal, is used in the jewelry, silverware, metal alloy, and food and beverage processing industries. Little soluble silver waste would be expected to result from use of the solid metal. The only appreciably soluble common silver salt, silver nitrate, is used in the following industries: porcelain, photographic, electroplating, and ink manufacture, and has also found application as an antiseptic (2). The two major sources of soluble silver wastes are the photographic and electroplating industries.

References

(1) GCA Corp., "National Emissions Inventory of Sources and Emissions of Silver," *Report PB 231,368,* Springfield, Virginia, Nat. Tech. Information Service (May 1973).
(2) McKee, J.E. and Wolf, H.W., *Water Quality Criteria,* 2nd edition, California State Water Quality Control Board Publication No. 3A (1963).

SOLID WASTE INCINERATION PRODUCTS

As defined in the Solid Waste Disposal Act of 1965, the term "solid waste" means garbage, refuse, and other discarded solid materials, including solid waste materials resulting from industrial, commercial, and agricultural operations, and from community activities. It includes both combustibles and noncombustibles.

Solid wastes may be classified into four general categories: urban, industrial, mineral and agricultural. Although urban wastes represent only a relatively small part of the total solid wastes produced, this category has a large potential for air pollution since in heavily populated areas solid waste is often burned to reduce the bulk of material requiring final disposal. The following discussion will be limited to the urban and industrial waste categories.

An average of 5.5 lb (2.5 kg) of urban refuse and garbage is collected per capita per day in the United States. This figure does not include uncollected urban and industrial wastes that are disposed of by other means. Together, uncollected urban and industrial wastes contribute at least 4.5 lb (2.0 kg) per capita per day. The total gives a conservative per capita generation rate of 10 lb (4.5 kg) per day of urban and industrial wastes. Approximately 50% of all the urban and industrial waste generated in the United States is burned, using a wide variety of combustion methods with both enclosed and open burning.

Atmospheric emissions, both gaseous and particulate, result from refuse disposal operations that use combustion to reduce the quantity of refuse. Emissions from these combustion processes cover a wide range because of their dependence upon the refuse burned, the method of combustion or incineration, and other factors. Because of the large number of variables involved, it is not possible, in general, to delineate when a higher or lower emission factor, or an intermediate value should be used. For this reason, an average emission factor has been presented in the section on Incineration of Municipal Refuse, page 312, for example.

SULFUR OXIDES

In Air

Sulfur oxides, primarily SO_2, are generated during the combustion of any sulfur-bearing fuel, and by many industrial processes that use sulfur-bearing raw materials.

In 1966, about 28.6 million tons of SO_2 were emitted in the United States. The various sources of SO_2 are shown in Figure 14. The distribution of emissions by source category in any particular city or specific location may differ markedly from that shown.

FIGURE 14: SOURCES OF SULFUR DIOXIDE EMISSIONS IN THE UNITED STATES, 1966

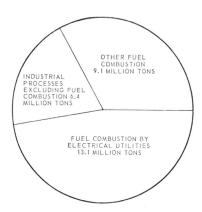

Source: EPA Publication AP-52

The main sources of airborne sulfur oxides are fossil fuel combustion and sulfide ore refining. Table 103 lists world gaseous emissions of sulfur dioxide.

TABLE 103: SULFUR DIOXIDE WORLDWIDE GASEOUS EMISSIONS

Source	Total SO_2 (10^6 tons)
Coal	102
Petroleum combustion & refining	28.5
Smelting	
Copper	12.9
Lead	1.5
Zinc	1.3
Total	146.2

Source: Report EPA 600/5-73-003

Combustion Sources: Combustion of fuels accounts for 77% of all SO_2 emitted. This is due to the relatively high sulfur content of some bituminous coals and residual fuel oils, and to the very large amounts of these fuels consumed in this country. Bituminous coal and residual fuel oil usually contain from 1 to 3% sulfur by weight.

Combustion of these fuels produces about 2 lb of SO_2 and about 0.03 lb of SO_3 for each pound of sulfur in the fuel. Data on SO_2 emissions from fuel combustion in 1966 are presented in Table 104.

TABLE 104: SO_2 EMISSIONS FROM FUEL COMBUSTION IN 1966

Sources	SO_2 Emissions, tons
Utility coal	11,925,000
Utility oil	1,218,000
Other coal	4,700,000
Other oil	4,386,000
Natural gas	3,500
Total	22,232,500

Source: EPA Publication AP-52

Industrial Process Sources: Smelting of metallic ores and oil refinery operations are the major industrial process sources of SO_2 emissions. Increased demand for sulfur and sulfuric acid should result in a more profitable recovery market for these emissions, tending to prevent any large, future increase of SO_2 emissions from these sources. Sulfur dioxide emissions from industrial process sources in 1966 are given in Table 105.

TABLE 105: SO_2 EMISSIONS FROM INDUSTRIAL PROCESS SOURCES IN 1966

Sources	SO_2 Emissions, tons
Ore smelting	3,500,000
Petroleum	1,583,000
Sulfuric acid manufacturing	550,000
Coke processing	500,000
Refuse burning	200,000
Miscellaneous*	75,000
Total industrial process	6,408,000

*Includes chemical manufacturing and pulp and paper products.

Intermedia Aspects

In Figures 15 and 16 the sources, treatment and fate of sulfurous wastes are summarized to show their intermedial relationships. Intermedia transfer of sulfur compounds between the water and air may occur directly (water scrubbing of gaseous exhausts) or indirectly (leaching and runoff from residues deposited on land, and incineration of sulfur-containing sludges).

Regardless of the mode by which it is accomplished, transfer to water from air generally is much more easily accomplished than transfer to air from water. The net result of this is that watercourses tend to become the ultimate sink for sulfur emissions in the absence of biodegradation to volatile products (H_2S) or environmentally isolated deposit on land.

Natural processes are estimated to be responsible for a great portion of incidental intermedia transfer of sulfur compounds (biological decomposition and sea spray), but their relative contribution tends to be inversely proportional to population density. Sulfur dioxide may easily be oxidized in the air to SO_3, and both compounds have destructive environmental impacts. Exposures to SO_2 concentrations of 3 to 4 ppm may occur in urban atmospheres but no persistent effect on humans has been detected for these levels. At 5 ppm human exposure for an hour may cause choking.

FIGURE 15: INTERMEDIA FLOW CHART–SULFUR OXIDES (GASEOUS)

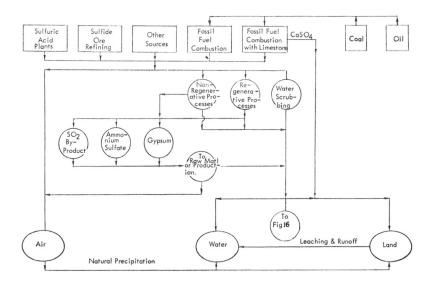

FIGURE 16: INTERMEDIA FLOW CHART–SULFUR COMPOUNDS IN WATER

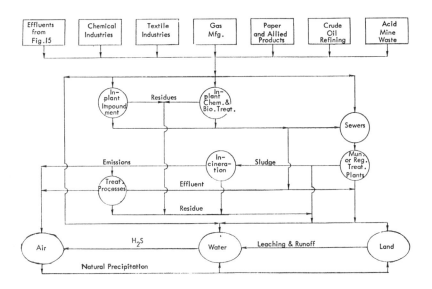

Source: Report EPA 600/5-73-003

Some evidence exists that sulfur dioxide and certain sulfur compound aerosols produce a toxicity synergism. The hydrate of SO_3 is sulfuric acid, which is more toxic than SO_2 or its hydrate H_2SO_3. Sulfuric acid is 4 to 20 times as physiologically damaging to animals as SO_2. The Air Conservation Commission of the American Association for the Advancement of Science states: "Sulfuric acid must have been the principle cause of air pollution disasters in the Meuse Valley, Germany; Donora, Pennsylvania; and London, England." As a result of the conversion of SO_2 to H_2SO_4, sulfur dioxide is especially injurious to plant life, being phytotoxic to some species in concentrations as low as 0.1 to 0.2 ppm. Necrotic blotching and streaking are the chief symptoms, and photosynthesis may be inhibited or terminated.

By the action of H_2SO_3 or H_2SO_4, metals and other materials may be corroded or degraded, especially when moisture is present. Sulfur dioxide may be a major contributor to visibility reduction in urban atmospheres, as its great hygroscopicity allows the formation of aerosol droplets in the size range of less than one micron; this is most effective in scattering visible light. Bluish-white plumes from industrial and power plant stacks, as well as hazes in industrial regions, are often attributable to the presence of SO_3 and H_2SO_4. Oxides of sulfur may easily be washed by rainfall from the atmosphere as sulfite salts and are oxidized to sulfates.

Sulfate compounds are negligibly degradable and they will tend to leach into surface and subsurface water supplies after deposition onto the land. These effects were noted near Ducktown, Tennessee, 30 yr after initiation of the recovery of sulfur dioxide formerly discharged into the atmosphere from copper smelting operations. Waterborne sulfur compounds also can be damaging. In addition to their presence in liquid wastes from SO_x scrubbers, several primary sources of sulfur compounds in water are known. Chemicals and allied products, textile mill products, gas manufacture, and paper and allied products are major contributors of sulfur compounds to wastewater.

Effluents with excessive contents of sulfur compounds are objectionable because they may anaerobically generate sulfides which are malodorous and even toxic. The recommended maximum sulfate concentration for drinking water in the United States, where no other source of water is available, is 250 mg/l. Above that level gastrointestinal irritation may result. Various sulfur compounds in water have been shown to have a toxic effect on fish.

For example, the toxic concentration of sodium sulfide for fish is about 3.2 mg/l and of dissolved hydrogen sulfide is about 0.5 to 1.0 mg/l. Hydrogen sulfide is toxic and has caused the death of many sewer workers. It has an offensive rotten egg odor, blackens lead paints, copper and brass, and causes corrosion of concrete. A concentration of 200 milligrams per liter sulfur compounds renders water undesirable for irrigational use, and a 500 mg/l concentration is excessive if the water is used to water cattle or other stock.

Sulfur dioxide pollution may be reduced by substitution of alternate processes, such as nuclear, solar, or geothermal energy sources for electric power production or fossil fuel combustion. Retaining fossil fuel combustion, but switching to fuels with low sulfur content, such as natural gas, is currently (1973) a popular control method that is widely applied. Precleaning of fossil fuels to remove sulfur is possible but relatively costly. Increased combustion efficiency may reduce SO_2 emissions, but it increases conversion to SO_3.

Some of the treatment processes enable recovery of valuable products. They are already in common use in applications where the concentrations are high and the potential recovery value is significant. Current problems lie more with sources that generate emissions significant in their impact on pollution but not of a sufficiently high level that internal economics make removal treatment attractive. Regulation of emissions is changing this pattern. The processes that recover the product in usable form are applicable mainly to the high-concentration emissions, while the nonrecovery systems usually are more efficient for low-concentration emissions. Metallurgical processes that handle sulfur-containing ores are the primary high concentration sources. Fuel combustion and sulfuric acid manufacture generate low concentration emissions.

Treatment by dilution utilizes stacks which may be as high as 800 ft or more. This can be somewhat effective in reducing nearby ground level SO_x concentrations. The method takes advantage of the dilution capacities of air and is not an intermedia treatment. It has been shown that the maximum downwind ground level concentration is inversely proportional to the square of the effective stack height, which is the physical stack height plus the plume rise as influenced by the exit velocity of the gas, the difference in density between the gas and the atmosphere, and the meteorological conditions.

References

(1) U.S. Department of Health, Education and Welfare, "Control Techniques for Sulfur Dioxide Air Pollutants," *Publication Number AP-52,* Washington, D.C., National Air Pollution Control Administration (January 1969).

(2) Stone, R. and Smallwood, H., "Intermedia Aspects of Air and Water Pollution Control," *Report EPA-600/5-73-003,* Washington, D.C., U.S. Environmental Protection Agency (August 1973).

SURFACTANTS

In Water

Surfactants are not as great a problem as they once were when the detergents used were largely nonbiodegradable. In sewage treatment it has been found that surfactants interfere with anaerobic sludge digestion. Also, there may be a synergistic action between these substances and certain pesticides such as DDT.

Intermedia Aspects

A number of treatments have been proposed for the removal of surfactants from water. These include foaming, aeration, flotation, coagulation, flocculation, adsorption on carbon and other inert matter, biological treatment, and treatment by resins. These treatments do not create water-to-air transfers of pollutants and are not intermedial.

Reference

(1) Jones, H.R., *Detergents and Pollution—Problems and Technological Solutions,* Park Ridge, N.J., Noyes Data Corp. (1972).

SUSPENDED SOLIDS

In Water

As with particulates in the atmosphere, suspended solids (SS) in the liquid medium are made up of many different types of materials. The majority of suspended solids are organic in nature and therefore exert a biochemical oxygen demand on the receiving waters. Suspended solids are determined as the residue that can be removed by filtration, as opposed to dissolved solids, which are determined by evaporation.

In addition to the oxygen demand problem caused by the organic fraction of suspended solids, other problems also result. Among these are aesthetic degradation and interference with the growth, survival, and propagation of algae, plants, fish and shellfish. Aside from any toxicity which may exist, suspended solids may kill fish, shellfish, and other aquatic life through abrasive injuries, by clogging gills and respiratory passages, by smothering eggs, young, food chain organisms, and by destroying spawning beds. In concentrations over 750 mg/l, the development of eggs and larvae of the venus clam is decreased. In a river containing 6,000 mg/l of suspended solids, trout population was one-seventh and that of invertebrates one-nineteenth the comparable densities in a control river source.

Controls for the limitation of suspended solids are essentially the same as for organics. Wastewater treatment processes have efficiencies of suspended solids removal ranging from 50 to more than 98%. These processes, even when not intended primarily for suspended solids removal, have the following approximate efficiencies: screening, 0 to 80%; flotation, 70 to 95%; chemical precipitation, 70 to 90%; primary sedimentation, 50 to 90%; activated sludge, 85 to 95%; trickling filter, 70 to 92%; carbon adsorption, 90%; sand filtration, 70 to 90%; and coagulation, sedimentation, sand filtration, 90 to 99+%.

Intermedia Aspects

Because of the ease with which solids can be transferred between media, suspended solids are included in the flow chart for particulates (Figure 10, page 128). The major difference between particulates in the air and suspended solids in the water is the amount of natural transfer which takes place between media. In the section on particulates, natural transfer was shown to be fairly extensive. Figure 10 shows that transfer out of the water medium by natural processes is far less so. Some man controlled processes such as residue disposal from treatment processes do effect intermedia transfers. This may be through incineration or land disposal.

Reference

(1) Stone, R. and Smallwood, H., "Intermedia Aspects of Air and Water Pollution Control," *Report EPA-650/5-73-003,* Washington, D.C., U.S. Environmental Protection Agency (August 1973).

TIN

In Air

Table 106 shows estimates of sources and quantities of tin emissions to the atmosphere.

TABLE 106: TIN EMISSIONS

Source	Amount in Tons	% This Pollutant
Iron and Steel		
Open hearth	260	73.03
Power Plant Boilers		
Pulverized Coal	68	19.10
Stoker Coal	8	2.25
Cyclone Coal	2	0.56
Industrial Boilers		
Pulverized Coal	3	0.84
Stoker Coal	11	3.09
Cyclone Coal	2	0.56

(continued)

TABLE 106: (continued)

Source	Amount in Tons	% This Pollutant
Residential/Commercial Coal	1	0.28
All Boilers, Oil	1	0.28
Total	356	99.99

Source: The MITRE Corporation, Preliminary Results, EPA Contract No. 68-01-0438.

TITANIUM DIOXIDE

In Air

Table 107 summarizes sources and estimates of titanium dioxide emissions from those sources to the atmosphere (1).

Reference

(1)   GCA Corporation, "National Emissions Inventory of Sources and Emissions of Titanium," *Report PB 229,634,* Springfield, Virginia, Nat. Tech. Information Service (May 1973).

TOTAL DISSOLVED SOLIDS

In Water

The most significant impact of metallic salts and oxides is the salinity produced in water. Hydroxides can, however, produce an impact through their influence on acidity. Most metallic oxides are in a transitory state in water although a few do precipitate to add to the suspended solids. Most discussions of dissolved solids in the literature relate only to salinity.

The most important effect of salinity or total dissolved salts in irrigation water is the toxicity to plants. Of course, different plants have different tolerances to salt concentration. A prime example of the effects of salinity is shown in the Monterey area where the intrusion of salt water from the ocean has forced the change from lettuce, as a major crop, to artichokes, which have a greater salt tolerance. For human consumption, the recommendation of the U.S. Department of Health, Education and Welfare is that drinking water does not contain more than 500 mg/l and preferably less than 200 mg/l of total dissolved solids. Salt buildups also affect the functioning of industrial water reuse systems and cause maintenance expenditures.

Although dissolved solids, to some extent are present in any water, natural or waste, most industrial processes and, in fact, many industrial waste treatment processes, result in increased quantities of TDS. In general, TDS of natural waters consist mainly of carbonates, bicarbonates, chlorides, sulfates, phosphates and nitrates of calcium, magnesium, sodium and potassium, with traces of iron, manganese and other substances (1).

TABLE 107: SOURCES AND ESTIMATES OF TITANIUM-CONTAINING EMISSIONS

	Uncontrolled Particulate Emission Factor (lb/ton) (kg/kg x 10³)		Reliability Code	Production Level (tons/yr)	% TiO₂ in Emissions	Reliability Code	TiO₂ Emissions Before Controls	Estimated Level of Emission Control	TiO₂ Emissions After Controls (tons/yr)
1. MINING & BENEFICIATION									
Open Pit Mining	.2	.1	(D)	1,850,000	15	(A)	30	0	30
Beneficiation									
Open Pit	38	19	(B)	152,000	*	(B)	2,890	90	289
Dredging	9.5	4.75	(D)	308,000	*	(B)	1,460	90	146
2. METAL PROCESSING									
Metal Ingots	0	0		–	–		0	–	0
Carbides	0	0		–	–		0	–	0
Alloys:									
Titanium-Base	0	0		–	–		0	–	0
Ferrotitanium	150	75	(D)	5,872 (tons of alloy)	45%	(C)	200	40%	120
Steel Production	25	12.5	(C)	2,590 (TiO₂ consumed as Ferrotit. in steel mfg.)	*	(-)	31	78%	7
3. OTHER MINERAL USES									
Welding Rod Coating	0	0	(D)	–	–	(B)	0	–	0
Ceramics	16	8	(C)	370	*	(A)	3	67%	1
Fiberglass	2	1	(C)	27,000	*	(A)	27	0% (controls generally not used)	27
Pigment Production									
Sulfate Calcining	150	75	(B)	344,000	100%	(B)	25,000	97%	750
Grinding	.35	.175	(C)	655,000	100%	(A)	115	10%	104
Drying & Sacking	NA	NA	(C)	655,000	100%	(A)	NA	99.5%	3,275

(continued)

TABLE 107: (continued)

	Uncontrolled Particulate Emission Factor (lb/ton) (kg/kg x 10³)		Reliability Code	Production Level (tons/yr)	% TiO₂ in Emissions	Reliability Code	TiO₂ Emissions Before Controls	Estimated Level of Emission Control	TiO₂ Emissions After Controls (tons/yr)
4. PIGMENT CONSUMPTION									
Paints, etc.	15	7.5	(D)	385,000	*		2,880	10%	2,600
Other Uses	15	7.5	(D)	270,000	*		2,020	5%	1,920
5. INADVERTENT EMISSIONS									
Coal Burning	NA	NA		33,800,000 U.E.	1.2%	(A)	405,000	82%	73,000
Coal Cleaning	NA	NA		2,000,000 U.E.	0.12%	(A)	2,400	85%	360
6. Oil Burning									
Residual	NA	NA		123,000 U.E.	.05%	(A)	62	0	62
Distillates	NA	NA		146,000 U.E.	.0065%	(B)	10	0	10
7. INCINERATION	NA	NA		100,000 U.E.	4.2%	(B)	4,200	37%	2,650
8. Non Metallic Minerals	NA	NA		6,900,000 (controlled emissions)	**	(D)	**	**	3,000 **
U.S. TOTALS							446,328†		83,351

NA = Not Applicable

U.E. = Uncontrolled Emission

* Emission factor multiplier equal to tons of TiO₂ processed or handled annually

** See Text

† Does not include pigment drying and non-metallic minerals

Note: See page 7 for explanation of Reliability Code

Source: Report PB 229,634

The mineral content of natural waters is normally raised by the addition of chemical wastes, dissolved salts, acids, and alkalis. Zievers, et al (2), have reported, for example, the TDS in metal plating effluent after waste treatment for acid neutralization and reduction of cyanide and heavy metals. Their data are presented in Table 108.

TABLE 108: TOTAL DISSOLVED SOLIDS CONCENTRATION IN TREATED METAL PLATING WASTES

Plant No.	Location	TDS, mg/l
1	Illinois	555
2	Michigan	490
3	Illinois	15,100
4	Virginia	906
5	Illinois	1,065
6	California	625
7	Kentucky	810
8	California	435
9	New York	610
10	Indiana	166
11	New York	1,580
12	Arkansas	393
13	Illinois	2,260
14	Ohio	4,066
15	Pennsylvania	793

Source: J.F. Sievers, P.W. Crain, and F.G. Barclay (2)

A recent government publication (3) has reported that TDS in inorganic chemical industry effluents may range from 1,000 to 150,000 mg/l. Benger (4) reports an effluent TDS of 1,163 mg/l from a petroleum refinery, and Stone (5) has reported TDS in a carpet mill waste to 5,850 mg/l, with a raw water supply TDS of 410 mg/l. Evans (6) has pointed out that source water TDS is variable, depending upon geographical location and surface versus ground water source, and therefore can influence wastewater TDS.

The nature of nonprecipitable dissolved solids, and the treatment processes available to re- move or reduce such solids are such that there are, practically speaking, no intermediate levels of treatment. That is to say, treatment for removal of dissolved solids, if applied at all, generally produces a water containing very low residual TDS (3). Partial treatment of a high TDS waste stream thus may involve splitting the stream, treating one portion to re- move essentially all solids, and recombining the treated and untreated streams in the pro- portion necessary to achieve the desired final solids concentration.

The major processes employed for reduction of TDS are reverse osmosis, electrodialysis, distillation, and ion exchange. Ion exchange and electrodialysis are generally applicable for TDS concentrations up to 5,000 mg/l, while reverse osmosis and distillation are employed for TDS concentrations to 50,000 mg/l (3). Evaporative ponds may be required at extremely high TDS concentrations above 50,000 mg/l.

Intermedia Aspects

The main controls for salt compounds in water are evaporation, dialysis, ion exchange, and some other miscellaneous methods, all of which are intramedial. These methods all separate the salts from the water, leaving salt in a solid form. These residues can then be used or disposed to land. In either case no air-to-water or water-to-air transfer results. In some cases salts can result from an intermedia transfer from air such as from the precipitation of airborne metallic oxides. Since further treatments are intramedial, metallic salts and oxides are classed as intramedial pollutants.

References

(1)   McKee, J.E. and Wolf, H.W., *Water Quality Criteria,* 2nd edition, California State Water Quality Control Board Publication No. 3-A (1963).

(2)   Zievers, J.F., Crain, R.W. and Barclay, F.G., "Waste Treatment Metal Finishing: U.S. and European Practices," *Plating,* 55, 1171-1179 (1968).

(3)   "The Economics of Clean Water," Volume 3, *Inorganic Chemicals Industry Profile,* U.S. Department of the Interior, Washington, D.C. (1970).

(4)   Benger, M., "The Disposal of Liquid and Solid Effluents from Oil Refineries," Proc. 21st Purdue Industrial Waste Conference, 21, 759-767 (1966).

(5)   Stone, R., "Walter Carpet Mill Industrial Waste System," *Journal Water Pollution Control Federation,* 44, 470-478 (1972).

(6)   Evans, R.L., "Addition of Common Ions from Domestic Use of Water," *Journal American Water Works Association,* 60, 315-320 (1968).

VANADIUM

In Air

The air pollution aspects of vanadium and its compounds have been reviewed in some detail by Athanassiadis (1). The production and use of vanadium in the United States has been traced and charted for the year 1968 by W.E. Davis and Assoc. (2). The consumption was 5,495 tons, exports were 741 tons and imports were 652 tons. About 80% (4,350 tons) was used in the production of steel.

Emissions to the atmosphere during the year 1968 totalled 19,219 tons, as shown in Table 109. Emissions due to the combustion of fuel oil and coal were 17,000 tons and 1,750 tons respectively. Emissions resulting from the production of ferrovanadium were 144 tons and those from the production of steel were 236 tons.

TABLE 109: VANADIUM EMISSIONS BY SOURCE 1968

Source Category	Source Group		Short Tons
Mining and Processing			81
Metallurgical Processing			144
Reprocessing			244
	Steel		
	Blast Furnace	63	
	Open-Hearth Furnace	166	
	Basic Oxygen Furnace	7	
	Electric Arc Furnace	N	
	Cast Iron	1	

(continued)

TABLE 109: (continued)

Source Category	Source Group	Short Tons	
	Nonferrous Alloys	3	
	Chemicals and Ceramics		
	Catalysts	2	
	Glass and Ceramics	N	
	Miscellaneous	2	
Consumptive Uses			18,750
	Coal	1,750	
	Oil	17,000	
	TOTAL		19,219

N - Negligible (less than 1 ton)

Source: Report PB 221,655

Emissions estimates made for coal and fuel oil combustion were considered to be reasonably accurate. They were based on the average vanadium content of many coal and oil samples. Emission factors for vanadium from various sources are given in Tables 110 through 112 inclusive as presented by Anderson (3).

TABLE 110: EMISSION FACTORS FOR VANADIUM FROM INDUSTRIAL SOURCES

Source [a]	Emission factor	Emission factor symbol [b]
Mining and processing	13 kg/10^3kg (25 lb/ton) of vanadium processed	Q
Metallurgical processing Ferrovanadium, electric furnaces [c]	25 kg/10^3kg (50 lb/ton) of vanadium processed	UK
Handling losses	5 kg/10^3kg (10 lb/ton) of vanadium processed	E
Steel production Blast furnace	0.02 kg/10^3kg (0.03 lb/ton) of pig iron produced	E
Open-hearth furnace No oxygen lance Oxygen lance	0.003 kg/10^3kg (0.006 lb/ton) of steel produced 0.006 kg/10^3kg (0.01 lb/ton) of steel produced	OES (1) OES (1)
Basic oxygen furnace [d]	0.001 kg/10^6kg (0.002 lb/ton) of steel produced	OES (1)
Basic oxygen furnace	0.005 kg/10^3kg (0.009 lb/ton) of steel produced	E
Electric furnace	Negligible	—
Cast iron production	0.1 kg/10^6kg (0.2 lb/10^3 ton) of charge	Q
Cement plants Dry process (for all processes)	— h	ES (6)
Wet process Kiln (average of two) [e] Clinker cooler [f] Clinker cooler [e] Clinker cooler [g] Finishing mill after air separator [f]	0.05 kg/10^6kg (0.1 lb/10^3 ton) of feed 0.0003 kg/10^6kg (0.0005 lb/10^3 ton) of feed 0.01 kg/10^6kg (0.02 lb/10^3 ton) of feed — h 0.0001 kg/10^6kg (0.0002 lb/10^3 ton) of feed	ES (2) OES, SSMS (1) ES (1) ES (1) OES, SSMS (1)
Processing of vanadium and its compounds Nonferrous alloys	6 kg/10^3kg (12 lb/ton) of vanadium processed	Q

(continued)

TABLE 110: (continued)

Source [a]	Emission factor	Emission factor symbol [b]
Catalysts	10 kg/10³kg (20 lb/ton) of vanadium processed	Q
Glass and ceramics	Negligible	—
Miscellaneous (steel alloying, magnetic alloys, paint oils, color film)	5 kg/10³kg (10 lb/ton) of vanadium processed	Q

[a] Uncontrolled unless otherwise specified.
[b] Defined in Table 4 ; numbers in parentheses indicate number of samples analyzed.
[c] Particle size, 0.1 to 1 micron. [i]
[d] Exit from two venturi scrubbers.
[o] Exit from electrostatic precipitator.
[f] Exit from baghouse.
[g] Exit from two baghouse collectors (in parallel).
[h] Emission below detection limit of analytical technique.

TABLE 111: EMISSION FACTORS FOR VANADIUM FROM FUEL COMBUSTION, COAL

Source [a]	Vandium content, ppm	Emission factor kg/10³kg	Emission factor lb/ton	Emission factor symbol [b]
Illinois [c]	20	0.0003 (0.0003 to 0.0004)	0.0006 (0.0005 to 0.0007)	EST (3)
South Carolina [c]	—[e]	0.0002 (0.0002 to 0.0003)	0.0004 (0.0003 to 0.0006)	EST (6)
Michigan [c]	10	0.0003 (0.0002 to 0.0003)	0.0005 (0.0003 to 0.0006)	EST (2)
Average [c]	15	0.0003	0.0006	EST (11)
Kansas [d]	10 to 20	0.002	0.003 (0.003 to 0.004)	EST (2)
Based on vanadium content in coal	22.5	15	30	UK

[a] Controlled unless otherwise specified.

[b] Defined in Table 4 ; numbers in parentheses indicate number of samples analyzed.

[c] Exit from electrostatic precipitator.

[d] Exit from limestone wet scrubber.

[e] Not reported.

Source: Report PB 230,894

TABLE 112: EMISSION FACTORS FOR VANADIUM

From Fuel Combustion, Oil

Source[a]	Vanadium content, ppm	Emission factor		Emission factor symbol[b]
		kg/10^3liters	lb/10^3 gal.	
Crude oils, U.S.				
Arkansas	9.3	0.008	0.07	UK
California	50.0	0.04	0.4	UK
Colorado	00.44	0.0004	0.003	UK
Kansas	15.1	0.01	0.1	UK
Louisiana	0.5	0.0004	0.004	UK
Montana	78	0.07	0.6	UK
New Mexico	0.1	0.00009	0.0007	UK
Oklahoma	4.0	0.003	0.03	UK
Texas	2.6	0.002	0.02	UK
Utah	4.6	0.004	0.03	UK
Wyoming	49.7	0.04	0.4	UK
Average for U.S. crude oils	19.5	0.02	0.1	UK
Average, residual oil - United States	30	0.03	0.2	UK
Foreign crude and residual oil				
Average, crude oil -				
Western Venezuela	356	0.3	3	UK
Average, crude oils -				
Eastern Venezuela	116	0.1	0.8	UK
Average, residual oil - Venezuela	280	0.3	2	UK
Average, residual oil - Middle East	50	0.05	0.4	UK
U.S. boilers				
Residual units (distillate)	_[d]	0.000008	0.00007	ES(2)
Commercial units (residual No. 6)[c]	223	0.04	0.3	ES(1)
Commercial units (residual No. 5)	88	0.01	0.09	ES(1)
Commercial units (residual No. 4)	86	0.02	0.2	ES(1)

[a] Uncontrolled unless otherwise specified.
[b] Defined in Table 4 ; numbers in parentheses indicate number of samples analyzed.
[c] Particle size range (particulate collected with a cascade impactor): 20 percent (by weight) less than 0.21 micron, 80 percent less than 7.4 microns, mass mean particle size 1.2 microns.
[d] Not reported.

From Solid Waste Incineration

Source	Emission factor	Emission factor symbol[a]
Uncontrolled	0.0005 kg/10^3kg (0.001 lb/ton) of waste burned	EST (1)
After electrostatic precipitation	_[b]	EST (1)

[a] Defined in Table 4 ; numbers in parentheses indicate number of samples analyzed.

[b] Emissions below detection limit of analytical technique.

Source: Report PB 230,894

References

(1) Athanassiadis, Y.C., "Air Pollution Aspects of Vanadium and Its Compounds," *Report PB 188,093,* Springfield, Virginia, Nat. Tech. Information Service (September 1969).
(2) Davis, W.E. and Associates, "National Inventory of Sources and Emissions: Vanadium–1968," *Report PB 221,655,* Springfield, Virginia, Nat. Tech. Information Service (June 1971).
(3) Anderson, D., "Emission Factors for Trace Substances," *Report PB 230,894,* Springfield, Virginia, Nat. Tech. Information Service (December 1973).

ZINC

In Air

The air pollution aspects of zinc and its compounds have been reviewed in some detail by Athanassiadis (1). The flow of zinc in the United States has been traced and charted for the year 1969 by W.E. Davis and Associates (2). The consumption was 1,797,000 tons, while primary and secondary production totaled 1,417,000 tons. Imports and exports were 354,000 and 43,000 tons, respectively. Ore used directly in processing was 127,000 tons. Emissions to the atmosphere during the year were 159,922 tons (Table 113). About 31% of the emissions resulted from the metallurgical processing of zinc, more than 30% from the production of iron and steel, and nearly 18% from the incineration of refuse.

TABLE 113: EMISSIONS BY SOURCE (1969)

Source Category	Source Group		Emissions - Tons	Emissions %
Mining and Milling			72	--
Metallurgical Processing			50,000	31.3
Secondary Production			3,800	2.4
End Product Uses			23,270	14.6
	Zinc-Base Alloys	3,000		
	Zinc Coatings	950		
	Brass and Bronze	180		
	Zinc Oxide			
	Zinc Oxide Production	8,100		
	Rubber Tires	8,400		
	Photocopying	1,500		
	Paint	10		
	Other	1,000		
	Zinc Sulfate	30		
	Miscellaneous	100		
Other Emission Sources			82,780	51.7
	Coal	4,310		
	Oil	450		
	Iron and Steel			
	Blast Furnace	1,070		
	Open-Hearth Furnace	39,000		
	Basic Oxygen Furnace	900		
	Electric Furnace	7,400		
	Foundries	1,700		
	Incineration	27,950		
TOTAL			159,922	100.0

Source: Report PB 210,680

The production of zinc oxide, the wear of rubber tires, and the combustion of coal were also significant emission sources. Emission estimates for mining, production of primary and secondary zinc, manufacture of zinc-base alloy products, and the production of zinc oxide are based on unpublished data obtained from industrial sources. There are large variations in the concentrations of zinc fumes that are discharged to the atmosphere during the production of secondary zinc. Emissions from brass melting operations may vary from less than 0.5 to 6% or more of the total metal charge, and 2 to 15% of the zinc content. Uncontrolled particulate emissions have been reported to range from 1 to 70 lb/ton of charge, whereas controlled emissions range from 0.1 to 1 lb/ton of charge.

Chemical analyses of dust collected in a brass and bronze smelter baghouse show the zinc content varies from 45 to 77%. Another report on a series of tests in Los Angeles County indicated the zinc oxide content of tume from representative red and yellow brass furnaces averaged 59%. Data obtained from industry during this study indicates that zinc emissions to the atmosphere during the production of secondary zinc vary from 0.1 to 124 lb/ton of product, averaging 20 lb/ton. During 1969 zinc emissions to the atmosphere totaled 3,800 tons. The emission factors presented for zinc in Table 114 are the best currently available.

TABLE 114: EMISSION FACTORS

Mining and Milling	0. 2 lb/ton zinc mined
Metallurgical Processing	
Electrolytic Plants	60. 0 lb/ton of product
Vertical-Retort Plants	80. 0 lb/ton of product
Horizontal-Retort Plants	170. 0 lb/ton of product
Secondary Production	20. 0 lb/ton of zinc produced
End Product Uses of Zinc	
Zinc-Base Alloys	10. 0 lb/ton of zinc processed
Zinc Coatings	4. 0 lb/ton of zinc processed
Brass and Bronze	2. 0 lb/ton of zinc content
Zinc Oxide Production	60. 0 lb/ton of zinc oxide
Rubber Tire Wear	4. 2 lb/million miles
Other Emission Sources	
Coal	17. 0 lb/1,000 tons of coal burned
Oil	1. 4 lb/1,000 bbls of oil burned
Blast Furnaces	0. 02 lb/ton of pig iron produced
Open-Hearth Furnaces	1. 3 lb/ton of steel produced
Basic Oxygen Furnaces	0. 03 lb/ton of steel produced
Electric Furnaces	0. 74 lb/ton of steel produced
Foundries	0. 18 lb/ton of process weight

Source: Report PB 210,680

The emission factors presented in Table 114 were determined through a combination of methods consisting of: direct observation of emission data and other related plant processing and engineering data; estimates based on information obtained from literature, plant operators, and others knowledgeable in the field; calculations based on experience and personal knowledge of metallurgical processing operations; and specific analytical results where available.

In Water

Industries discharging waste streams which carry significant quantities of zinc include the following: steel works with galvanizing lines, zinc and brass metal works, zinc and brass plating, silver and stainless steel tableware manufacturing, viscose rayon yarn and fiber production, groundwood pulp production, and newsprint paper production. In addition, recirculating cooling water systems employing a commercial process, Cathanodic Treatment, contain zinc, which is discharged during blowdown. Perhaps 10,000 tons of waste zinc per year are discharged from viscose rayon production, and 5 tons of zinc may be used per day in a large groundwood pulp mill. Similar quantities were estimated for a typical brass company.

The primary source of zinc in wastewaters from plating and metal processing industries is the solution adhering to the metal product after removal from pickling or plating baths. The metal is washed free of this solution, referred to as dragout, and the contaminants are thus transferred to the rinse water. The pickling process consists of immersing the metal (zinc or brass) in a strong acid bath to remove oxides from the metal surface. Finished metals are brightened by submergence in a bright dip bath containing strong chromate concentrations in addition to acid.

Plating solutions typically contain 5,000 to 34,000 mg/l of zinc. These concentrated solutions may be discharged periodically, due to contamination. The concentration of zinc in the rinse water will be a function of the bath zinc concentration, drainage time over the bath, and the volume of rinse water used. Zinc and brass plating and rinse solutions generally also contain cyanide. Waste concentrations of zinc range from less than 1 to more than 1,000 mg/l in various waste streams described in the literature. Average values, however, seem to be between 10 and 100 mg/l. Table 115 summarizes values reported for various zinc bearing wastewaters.

Treatment processes employed for wastewater zinc removal may involve either destruction or recovery. The destruction process is essentially chemical precipitation; with disposal of the resultant sludge. Recovery processes include ion exchange and evaporative recovery, but may also be precipitation processes, where a relatively pure zinc sludge is recovered. Recovery of plating wastes frequently proves to be more economical on an overall basis than conventional destructive treatment.

TABLE 115: CONCENTRATIONS OF ZINC IN PROCESS WASTEWATERS

Industrial Process	Zinc Concentration, mg/l Range	Average
Metal Processing		
Bright dip wastes	0.2-37.0	
Bright mill wastes	40-1,463	
Brass mill wastes	8-10	
Pickle bath	4.3-41.4	
Pickle bath	0.5-37	
Pickle bath	20-35	

(continued)

TABLE 115: (continued)

Industrial Process	Zinc Concentration, mg/l Range	Average
Metal Processing		
Aqua fortis and CN dip	10-15	
Wire mill pickle	36-374	
Plating		
General	2.4-13.8	8.2
General	55-120	
General	15-20	15
General	5-10	
Zinc	20-30	
Zinc	70-150	
Zinc	70-350	
Brass	11-55	
Brass	10-60	
General	7.0-215	46.3
Plating on zinc castings	3-8	
Galvanizing of cold rolled sttel	2-88	
Silver Plating		
Silver bearing wastes	0-25	9
Acid waste	5-220	65
Alkaline	0.5-5.1	2.2
Rayon Wastes		
General	250-1000	
General	20*	
General	20-120	
Other		
Vulcanized fiber	100-300	
Cooling tower blowdown	6	

* After process recovery of zinc by ion exchange

Source: Report PB 216,162

References

(1)   Athanassiadis, Y.C., "Air Pollution Aspects of Zinc and Its Compounds," *Report PB 188,072,* Springfield, Virginia, Nat. Tech. Information Service (September 1969).
(2)   Davis, W.E. and Associates, "National Inventory of Sources and Emissions: Barium, Boron, Copper, Selenium, Zinc; Section V—Zinc," *Report PB 210,680,* Springfield, Virginia, Nat. Tech. Information Service (May 1972).
(3)   Patterson, J.W. and Minear, R.A., "Wastewater Treatment Technology," 2nd edition, *Report PB 216,162,* Springfield, Virginia, Nat. Tech. Information Service (Feb. 1973).

EMISSIONS FROM SPECIFIC PROCESSES

ADIPIC ACID MANUFACTURE

Adipic acid, $HOOC(CH_2)_4COOH$, is a dibasic acid used in the manufacture of synthetic fibers. The acid is made in a continuous two-step process. In the first step, cyclohexane is oxidized by air over a catalyst to a mixture of cyclohexanol and cyclohexanone. In the second step, adipic acid is made by the catalytic oxidation of the cyclohexanol/cyclohexanone mixture using 45 to 55% nitric acid. The final product is then purified by crystallization(1).

Air Pollution

The only significant airborne emissions from the manufacture of adipic acid are nitrogen oxides. In oxidizing the cyclohexanol/cyclohexanone, nitric acid is reduced to unrecoverable N_2O and potentially recoverable NO and NO_2. This NO and NO_2 can be emitted into the atmosphere. Table 116 shows typical emissions of NO and NO_2 from an adipic acid plant(2).

TABLE 116: EMISSION FACTORS FOR AN ADIPIC ACID PLANT WITHOUT CONTROL EQUIPMENT, EMISSION FACTOR RATING: D

Source	Nitrogen oxides (NO,NO_2)	
	lb/ton	kg/MT
Oxidation of cyclohexanol/cyclohexanone(3)	12	6

Source: EPA Publication AP-42, 2nd Edition

Water Pollution

Wastewater from adipic acid manufacture results from the overhead stream from evaporation as noted by Campbell et al(4).

References

(1) M. Goldbeck, Jr. and F.C. Johnson; U.S. Patent 2,703,331; March 1, 1955; assigned to E.I. du Pont de Nemours and Company.
(2) U.S. Environmental Protection Agency, "Compilations of Air Pollutant Emission

Factors," 2nd Edition, *Publication Number AP-42,* Research Triangle Park, N.C., Office of Air and Water Programs, April 1973.

(3)   "Control Techniques for Nitrogen Oxides from Stationary Sources," U.S. DHEW, PHS, EHS, National Air Pollution Control Administration, Washington, D.C., *Publication Number AP-67,* March 1970, pages 7-12, 7-13.

(4)   C.R. Campbell, D.E. Danly and M.J. Matthews; U.S. Patent 3,267,029; August 16, 1966; assigned to Monsanto Company.

AIRCRAFT OPERATION

Aircraft engines are of two major categories; reciprocating (piston) and gas turbine. The basic element in the aircraft piston engine is the combustion chamber, or cylinder, in which mixtures of fuel and air are burned and from which energy is extracted through a piston and crank mechanism that drives a propeller. The majority of aircraft piston engines have two or more cylinders and are generally classified according to their cylinder arrangement, either "opposed" or "radial". Opposed engines are installed in most light or utility aircraft; radial engines are used mainly in large transport aircraft.

The gas turbine engine in general consists of a compressor, a combustion chamber, and a turbine. Air entering the forward end of the engine is compressed and then heated by burning fuel in the combustion chamber. The major portion of the energy in the heated air stream is used for aircraft propulsion. Part of the energy is expended in driving the turbine, which in turn drives the compressor. Turbofan and turboshaft engines use energy from the turbine for propulsion; turbojet engines use only the expanding exhaust stream for propulsion.

The aircraft classification system used is listed in Table 117. Both turbine aircraft and piston engine aircraft have been further divided into subclasses depending on the size of the aircraft and the most commonly used engine for that class. Jumbo jets normally have approximately 40,000 pounds maximum thrust per engine, and medium-range jets have about 14,000 pounds maximum thrust per engine. For piston engines, this division is more pronounced. The large transport piston engines are in the 500 to 3,000 horsepower range, whereas the small piston engines develop less than 500 horsepower.

TABLE 117: AIRCRAFT CLASSIFICATION

Aircraft class	Representative aircraft	Engines per aircraft	Engine commonly used
Jumbo jet	Boeing 747	4	Pratt & Whitney
	Lockheed L-1011	3	JT-9D
	McDonald Douglas DC-10	3	
Long-range jet	Boeing 707	4	Pratt & Whitney
	McDonald Douglas DC-8	4	JT-3D
Medium-range jet	Boeing 727	3	Pratt & Whitney
	Boeing 737	2	JT-8D
	McDonald Douglas DC-9	2	
Air carrier turboprop	Convair 580	2	Allison 501-D13
	Electra L-188	4	
	Fairchild Hiller FH-227	2	

(continued)

TABLE 117: (continued)

Aircraft class	Representative aircraft	Engines per aircraft	Engine commonly used
Business jet	Gates Learjet Lockheed Jetstar	2 4	General Electric CJ610 Pratt & Whitney JT-12A
General aviation turboprop	—	—	Pratt & Whitney PT-6A
General aviation piston	Cessna 210 Piper 32-300	1 1	Teledyne-Continen- tal Ø-200 Lycoming Ø-320
Piston transport	Douglas DC-6	4	Pratt & Whitney R-2800
Helicopter	Sikorsky S-61 Vertol 107	2 2	General Electric CT-58
Military transport			Allison T56A7
Military jet			General Electric J-79 Continental J-69
Military piston			Curtiss-Wright R-1820

Source: EPA Publication AP-42, 2nd Edition

A landing-takeoff (LTO) cycle includes all normal operation modes performed by an aircraft between the time it descends through an altitude of 3,500 feet (1,100 meters) on its approach and the time it subsequently reaches the 3,500 foot altitude after takeoff. It should be made clear that the term "operation" used by the Federal Aviation Administration to describe either a landing or a takeoff is not the same as the LTO cycle. Two operations are involved in one LTO cycle. The LTO cycle incorporates the ground operations of idle, taxi, landing run, and takeoff run and the flight operations of takeoff and climbout to 3,500 feet and approach from 3,500 feet to touchdown.

Each class of aircraft has its own typical LTO cycle. In order to determine emissions, the LTO cycle is separated into five distinct modes: (1) taxi-idle, (2) takeoff, (3) climbout, (4) approach and landing, and (5) taxi-idle. Each of these modes has its share of time in the LTO cycle. Table 118 shows typical operating time in each mode for the various types of aircraft classes during periods of heavy activity at a large metropolitan airport.

TABLE 118: TYPICAL TIME IN MODE FOR LANDING-TAKEOFF CYCLE AT A METROPOLITAN AIRPORT[a]

Aircraft	Time in mode, minutes				
	Taxi-idle	Takeoff	Climbout	Approach	Taxi-idle
Jumbo jet	19.00	0.70	2.20	4.00	7.00
Long range jet	19.00	0.70	2.20	4.00	7.00

(continued)

TABLE 118: (continued)

Aircraft	Time in mode, minutes				
	Taxi-idle	Takeoff	Climbout	Approach	Taxi-idle
Medium range jet	19.00	0.70	2.20	4.00	7.00
Air carrier turboprop	19.00	0.50	2.50	4.50	7.00
Business jet	6.50	0.40	0.50	1.60	6.50
General avia- tion turboprop	19.00	0.50	2.50	4.50	7.00
General aviation piston	12.00	0.30	4.98	6.00	4.00
Piston transport	6.50	0.60	5.00	4.60	6.50
Helicopter	3.50	0	6.50	6.50	3.50
Military transport	19.00	0.50	2.50	4.50	7.00
Military jet	6.50	0.40	0.50	1.60	6.50
Military piston	6.50	0.60	5.00	4.60	6.50

[a]References 1 and 2.

Source: EPA Publication AP-42, 2nd Edition

Air Pollution

Emissions factors for the complete LTO cycle presented in Table 119 were determined using the typical times shown in Table 118.

TABLE 119: EMISSION FACTORS PER AIRCRAFT LANDING-TAKEOFF CYCLE[a, b]
(lb/engine and kg/engine)
EMISSION FACTOR RATING: B

Aircraft	Solid particulates[a]		Sulfur oxides[d]		Carbon monoxide[e]		Hydrocarbons[e]		Nitrogen oxides[d] (NO_x as NO_2)	
	lb	kg	lb	kg	lb	kg	lb	kg	lb	kg
Jumbo jet	1.30	0.59	1.82	0.83	46.8	21.2	12.2	5.5	31.4	14.2
Long range jet	1.21	0.55	1.56	0.71	47.4	21.5	41.2	18.7	7.9	3.6
Medium range jet	0.41	0.19	1.01	0.46	17.0	7.71	4.9	2.2	10.2	4.6
Air carrier turboprop	1.1	0.49	0.40	0.18	6.6	3.0	2.9	1.3	2.5	1.1
Business jet	0.11	0.05	0.37	0.17	15.8	7.17	3.6	1.6	1.6	0.73
General aviation turboprop	0.20	0.09	0.18	0.08	3.1	1.4	1.1	0.5	1.2	0.54
General aviation piston	0.02	0.01	0.014	0.006	12.2	5.5	0.40	0.18	0.047	0.021
Piston transport	0.56	0.25	0.28	0.13	304.0	138.0	40.7	18.5	0.40	0.18
Helicopter	0.25	0.11	0.18	0.08	5.7	2.6	0.52	0.24	0.57	0.26
Military transport	1.1	0.49	0.41	0.19	5.7	2.6	2.7	1.2	2.2	1.0
Military jet	0.31	0.14	0.76	0.35	15.1	6.85	9.93	4.5	3.29	1.49
Military piston[f]	0.28	0.13	0.14	0.04	152.0	69.0	20.4	9.3	0.20	0.09

[a] References 1 through 5.
[b] Emission factors based on typical times in mode shown in Table 118.
[c] References 1 and 5.
[d] Based on 0.05 percent sulfur content fuel.
[e] References 1, 2, and 4.
[f] Engine emissions based on Pratt & Whitney R-2800 engine scaled down two times.

Source: EPA Publication AP-42, 2nd Edition

References

(1) "Nature and Control of Aircraft Engine Exhaust Emissions," Northern Research and Engineering Corporation, Cambridge, Mass. Prepared for National Air Pollution Control Administration, Durham, N.C., under Contract Number PH22-68-27. November 1968.

(2)   "The Potential Impact of Aircraft Emissions upon Air Quality," Northern Research
      and Engineering Corporation, Cambridge, Mass. Prepared for the Environmental
      Protection Agency, Research Triangle Park, N.C., under Contract Number 68-02-0085.
      December 1971.
(3)   "Assessment of Aircraft Emission Control Technology," Northern Research and En-
      gineering Corporation, Cambridge, Mass. Prepared for the Environmental Protection
      Agency, Research Triangle Park, N.C., under Contract Number 68-04-0011. Septem-
      ber 1971.
(4)   "Analysis of Aircraft Exhaust Emission Measurements," Cornell Aeronautical Labora-
      tory Inc., Buffalo, N.Y. Prepared for the Environmental Protection Agency, Re-
      search Triangle Park, N.C., under Contract Number 68-04-0040. October 1971.
(5)   EPA communication with Dr. E.K. Bastress, IKOR Incorporated, Burlington, Mass.
      November 1972.
(6)   U.S. Environmental Protection Agency, "Compilation of Air Pollutant Emission
      Factors," 2nd Edition, *Publication Number AP-42,* Research Triangle Park, N.C.
      Office of Air & Water Programs, April 1973.

ALFALFA DEHYDRATING

An alfalfa-dehydrating plant produces an animal feed from alfalfa. The dehydration and
grinding of alfalfa that produces alfalfa meal is a dusty operation most commonly carried
out in rural areas. Wet, chopped alfalfa is fed into a direct-fired rotary drier. The dried
alfalfa particles are conveyed to a primary cyclone and sometimes a secondary cyclone in
series to settle out the product from air flow and products of combustion. The settled
material is discharged to the grinding equipment, which is usually a hammer mill. The
ground material is collected in an air-meal separator and is either conveyed directly to
bagging or storage, or blended with other ingredients.

Air Pollution

Sources of dust emissions are the primary cyclone, the grinders, and the air-meal separator(1).
Over-all dust losses have been reported as high as 7%(2), but average losses are around 3%
by weight of the meal produced(3). The use of a baghouse as a secondary collection system can greatly reduce emissions. Emission factors for alfalfa dehydration are presented
in Table 120.

TABLE 120: PARTICULATE EMISSION FACTORS FOR ALFALFA DEHYDRATION[a], EMISSION FACTOR RATING: E

	Particulate emissions	
Type of operation	lb/ton of meal produced	kg/MT of meal produced
Uncontrolled	60	30
Baghouse collector	3	1.5

[a]Reference 3.

Source: EPA Publication AP-42, 2nd Edition

References

(1)   U.S. Environmental Protection Agency, "Compilation of Air Pollutant Emission
      Factors," 2nd Edition, *Publication Number AP-42,* Research Triangle Park, N.C.,
      Office of Air & Water Programs, April 1973.

(2) Stern, A., ed. *Air Pollution, Volume III, Sources of Air Pollution and Their Control,* 2nd Edition, New York, Academic Press, 1968.

(3) "Process Flow Sheets and Air Pollution Controls," American Conference of Governmental Industrial Hygienists, Committee on Air Pollution, Cincinnati, Ohio, 1961.

ALUMINUM CHLORIDE MANUFACTURE

Air Pollution

Aluminum chloride is made by reaction of chlorine with molten aluminum. The aluminum chloride formed vaporizes and is collected on air-cooled condensers. The tail gases leaving the condensers are the only source of wastes downstream of the reaction zone. In this process, there are two sources of waste, uncondensed aluminum chloride and chlorine in tail gases and unreacted aluminum metal. At the exemplary facility, the first waste is utilized to manufacture another product and the unreacted aluminum is disposed of as a solid waste. The raw waste loads are shown below.

Waste Product	Source	kg/kkg of Product Average	Range (lb/ton)
AlCl$_3$	Tail gases	80 (160)	64-96 (128-192)
Unreacted aluminum	Reactor	22 (44)	

At the exemplary plant there is an integrated blower system to exhaust the plant, packing station, condensers, etc. All blower exhaust is treated in an absorption tower where the aluminum chloride and chlorine vapors are absorbed into a recycling scrubber system. From this scrubber, about 121 liters of solution per kkg of product (29 gallons per ton) are drawn off, filtered and further treated to produce a 28% aluminum chloride solution which is sold. There are no waste streams. The water input and use for the scrubber system is an equivalent volume. This water is supplied from a well for makeup to the system. None of this is recycled. It is used to make 28% solution product.

There are three types of aluminum chloride manufactured, all from the same process.

[1] Yellow, this product is made using a slight excess of chloride (0.0005%) and may contain some iron due to reaction of the chlorine with the vessel.
[2] White, this product has a stoichiometric aluminum/chloride ratio.
[3] Grey, this product contains 0.01% excess aluminum. The unreacted aluminum raw waste load is higher for the grey material.

Industrially, it generally makes little difference which of the above grades is employed. In some pigment and dye intermediate applications, however, the yellow material is preferred as it is free of elemental aluminum. There is no waterborne effluent for this facility. The only wastes are airborne.

Reference

(1) U.S. Environmental Protection Agency, "Development Document for Proposed Effluent Limitations Guidelines & New Source Performance Standards for the Major Inorganic Products Segment of the Inorganic Chemicals Manufacturing Point Source Category," *Report EPA 440/1-73/007,* Washington, D.C., August 1973.

ALUMINUM PRODUCTION, PRIMARY

Air Pollution

Emissions from aluminum reduction processes consist primarily of gaseous hydrogen fluoride and particulate fluorides, alumina, hydrocarbons or organics, sulfur dioxide from the reduction cells and the anode baking furnaces. Large amounts of particulates are also generated during the calcining of aluminum hydroxide, but the economic value of this dust is such that extensive controls have been employed to reduce emissions to relatively small quantities. Finally, small amounts of particulates are emitted from the bauxite grinding and materials handling processes.

The source of fluoride emissions from reduction cells is the fluoride electrolyte, which contains cryolite, aluminum fluoride (AlF_3), and fluorspar (CaF_2). For normal operation, the weight or "bath" ratio of sodium fluoride (NaF) to AlF_3 is maintained between 1.36 and 1.43 by the addition of Na_2CO_3, NaF, and AlF_3. Experience has shown that increasing this ratio has the effect of decreasing total fluoride effluents. Cell fluoride emissions are also decreased by lowering the operating temperature and increasing the alumina content in the bath. Specifically, the ratio of gaseous (mainly hydrogen fluoride) to particulate fluorides varies from 1.2 to 1.7 with horizontal stud Soderberg (HSS) and prebaked (PB) cells, but attains a value of approximately 3.0 with vertical stud Soderberg (VSS) cells.

Particulate emissions from reduction cells consist of alumina and carbon from anode dusting, cryolite, aluminum fluoride, calcium fluoride, chiolite ($Na_5Al_3F_{14}$), and ferric oxide. Representative size distributions for PB and HSS particulate effluents are presented in Table 121. Particulates less than 1 micron in diameter represent the largest percentage (35 to 44% by weight) of uncontrolled effluents.

TABLE 121: REPRESENTATIVE PARTICLE SIZE DISTRIBUTIONS OF UNCONTROLLED EFFLUENTS FROM PREBAKED AND HORIZONTAL STUD SODERBERG CELLS

Size range, μm	Particles within size range, wt%	
	Prebaked	Horizontal-stud Soderberg
<1	35	44
1 to 5	25	26
5 to 10	8	8
10 to 20	5	6
20 to 44	5	4
>44	22	12

Source: EPA Publication AP-42, 2nd Edition

Moderate amounts of hydrocarbons derived from the anode paste are emitted from horizontal and vertical Soderberg pots. In vertical cells these compounds are removed by combustion via integral gas burners before the off-gases are released.

Because many different kinds of gases and particulates are emitted from reduction cells, many kinds of control devices have been employed. To abate both gaseous and particulate emissions, one or more types of wet scrubbers, spray tower and chambers, quench towers, floating beds, packed beds, venturis, and self-induced sprays, are used on all three cells and on anode baking furnaces. In addition, particulate control methods, such as electrostatic precipitators (wet and dry), multiple cyclones, and dry scrubbers (fluid-bed and coated-filter types), are employed with baking furnaces on PB and VSS cells. Dry alumina adsorption has been used at several PB and VSS installations in foreign countries. In this technique, both gaseous and particulate fluorides are controlled by passing the pot off-gases through the entering alumina feed, on which the fluorides are absorbed; the technique has an over-all control efficiency of 98%.

In the aluminum hydroxide calcining, bauxite grinding, and materials handling operations, various dry dust collection devices, such as centrifugal collectors, multiple cyclones, or electrostatic precipitators, and wet scrubbers or both may be used. Controlled and uncontrolled emission factors for fluorides and total particulates are presented in Table 122.

TABLE 122: EMISSION FACTORS FOR PRIMARY ALUMINUM PRODUCTION PROCESSES[a], EMISSION FACTOR RATING: A

Type of operation	Total particulates[b]		Gaseous fluorides (HF)		Particulate fluorides (F)	
	lb/ton	kg/MT	lb/ton	kg/MT	lb/ton	kg/MT
Bauxite grinding[a,c]						
Uncontrolled	6.0	3.0	Neg	Neg	NA[d]	NA
Spray tower	1.8[e]	0.90	Neg	Neg	NA	NA
Floating-bed scrubber	1.7	0.85	Neg	Neg	NA	NA
Quench tower and spray screen	1.0	0.50	Neg	Neg	NA	NA
Electrostatic precipitator	0.12	0.060	Neg	Neg	NA	NA
Calcining of aluminum hydroxide[a,c]						
Uncontrolled	200.0	100.0	Neg	Neg	NA	NA
Spray tower	60.0	30.0	Neg	Neg	NA	NA
Floating-bed scrubber	56.0	28.0	Neg	Neg	NA	NA
Quench tower and spray screen	34.0	17.0	Neg	Neg	NA	NA
Electrostatic precipitator	4.0	2.0	Neg	Neg	NA	NA
Anode baking furnace[f]						
Uncontrolled	3.0 (1.0 to 5.0)[g]	1.5 (0.5 to 2.5)	0.93	0.47	Neg	Neg
Spray tower	NA	NA	0.0372	0.0186	Neg	Neg
Dry electrostatic precipitator	1.13	0.57	0.93	0.47	Neg	Neg
Self-induced spray	0.06	0.03	0.0372	0.0186	Neg	Neg
Prebaked reduction cell[h]						
Uncontrolled	81.3 (11.9 to 177.0)	40.65 (5.95 to 88.5)	24.7 (13.8 to 34.8)	12.35 (6.9 to 17.4)	20.4 (9.8 to 35.5)	10.2 (4.9 to 17.8)
Multiple cyclone	17.9	8.95	24.7	12.35	4.49	2.25
Fluid-bed dry scrubber system	2.02	1.01	0.247	0.124	0.507	0.253
Coated filter dry scrubber	1.62	0.81	1.98 to 5.93	0.99 to 2.97	0.408	0.204
Dry electrostatic precipitator	1.62 to 8.94	0.81 to 4.47	24.7	12.35	0.408 to 2.24	0.204 to 1.12
Spray tower	16.2	8.1	0.494 to 2.72	0.247 to 1.36	4.08	2.04
Floating-bed scrubber	16.2	8.1	0.494	0.247	4.08	2.04
Chamber scrubber	12.2	6.1	2.96	1.48	3.06	1.53
Vertical flow packed bed	12.2	6.1	8.4	4.2	3.06	1.53
Dry alumina adsorption	1.62	0.81	0.494	0.247	0.408	0.204
Horizontal-stud Soderberg cell[i]						
Uncontrolled	98.4 (93.6 to 104.0)	49.2 (46.8 to 52.0)	26.6 (25.2 to 28.8)	13.3 (12.6 to 14.4)	15.6 (14.4 to 16.2)	7.8 (7.2 to 8.1)
Spray tower	19.6 to 36.4	9.8 to 18.2	1.86 to 2.39	0.93 to 1.195	3.12 to 5.77	1.56 to 2.885
Floating-bed scrubber	21.6	10.8	0.532	0.266	0.343	0.1715
Wet electrostatic precipitator	7.10	3.55	26.6	13.3	1.13	0.563
Vertical-stud Soderberg cell[j]						
Uncontrolled	78.4	39.2	30.4 (20.0 to 35.0)	15.2 (10.0 to 17.5)	10.6 (5.6 to 55.3)	5.3 (2.8 to 27.7)
Spray tower	19.6	9.8	0.304	0.152	2.65	1.325
Self-induced spray	NA	NA	0.304	0.152	NA	NA
Venturi scrubber	3.14	1.57	0.304	0.152	0.424	0.212
Wet electrostatic precipitator	0.784 to 7.84	0.392 to 3.92	30.4	15.2	0.106 to 1.06	0.053 to 0.53

(continued)

TABLE 122: (continued)

Type of operation	Total particulates[b]		Gaseous fluorides (HF)		Particulate fluorides (F)	
	lb/ton	kg/MT	lb/ton	kg/MT	lb/ton	kg/MT
Multiple cyclones	3.92 to 4.7	1.96 to 2.35	30.4	15.2	5.30 to 6.36	2.65 to 3.18
Dry alumina adsorption	1.57	0.784	0.608	0.304	0.212	0.106
Materials handling[c]						
Uncontrolled	10.0	5.0	Neg	Neg	NA	NA
Spray tower	3.0	1.5	Neg	Neg	NA	NA
Floating-bed scrubber	2.8	1.4	Neg	Neg	NA	NA
Quench tower and spray screen	1.7	0.85	Neg	Neg	NA	NA
Electrostatic precipitator	0.20	0.10	Neg	Neg	NA	NA

[a]Emission factors for bauxite grinding expressed as pounds per ton (kg/MT) of bauxite processed. Factors for calcining of aluminum hydroxide expressed as pounds per ton (kg/MT) of alumina produced. All other factors in terms of tons (MT) of molten aluminum produced.
[b]Includes particulate fluorides.
[c]References 1 and 3.
[d]No information available.
[e]Controlled emission factors are based on average uncontrolled factors and on average observed collection efficiencies.
[f]References 1,2, and 4 through 6.
[g]Numbers in parentheses are ranges of uncontrolled values observed.
[h]References 2 and 4 through 6.
[i]Reference 1.
[j]References 2 and 6.

Source: EPA Publication AP-42, 2nd Edition

Water Pollution

A composite flow diagram of water use, treatment and disposal for primary aluminum plants using wet scrubbing methods for air pollution control is shown in Figure 17. In any specific plant, the data will vary and a stream or unit illustrated may not exist. Treatment of water at the source 1 depends upon the quality required and varies from simple chlorination at wellheads for control of algae and bacteria to full clarification and treatment of river intake water. Stream 2 is made potable and the effluent is discharged through a sewage disposal unit. Stream 3 is makeup water to a closed-loop cryolite recovery stream 8 which may or may not include a potroom secondary air scrubber.

The combination of materials added during cryolite recovery varies from plant to plant as some cryolite recovery systems are highly sophisticated and proprietary chemical manufacturing facilities, while others are operated as by-product recovery or water-treatment units with either disposal or recycling of the solids. The bleed stream 9 is required to limit the buildup of sulfates in the recovery loop. Some plants do not practice cryolite recovery, in which case stream 9 represents the once-through discharge.

Stream 4 through a casthouse furnace air scrubber is common but plans exist to eliminate the stream in several plants by changes in degassing techniques to minimize noxious fumes or by the installation of a dry system for collection of alumina and occluded hydrogen chloride. Streams 5 and 6 are not common since dry processes prevail; however, where there is a liquid effluent, the carbon particulates are usually settled in ponds. Segments of stream 7 are treated to promote wetting and to inhibit corrosion and algae growth.

The chief effluent constituent is oil from rod mills or other fabrication units, when present; thus the use of commercial oil separators is common. From this generalized picture a number of potential sources of wastewater can be identified, including wet scrubbers (primary potline, secondary potroom, anode bake plant, cast house); cooling water (casting, rectifiers, fabrication); and boiler blow-down. The constituents of the wastewater from several commercial plants are identified in Table 123.

FIGURE 17: SCHEMATIC COMPOSITE FLOW DIAGRAM FOR PLANTS USING WET SCRUBBING

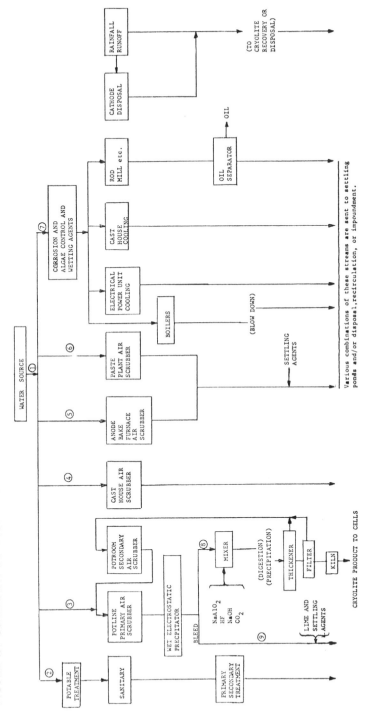

In a specific plant any particular stream or unit may not exist as alternate technology is applied.

Various combinations of these streams are sent to settling ponds and/or disposal, recirculation, or impoundment.

Source: Report EPA 440/1-73/019-2

TABLE 123: QUANTITIES OF SELECTED CONSTITUENTS IN WATER EFFLUENT FROM SELECTED PRIMARY ALUMINUM PLANTS IN THE UNITED STATES

[kg/metric ton of Al produced (lb/ton Al produced)]

	A, N	B, G(a)	C, N	D, G	F, N(b)
Alkalinity	0.01 (0.02)	---	---	---	---
Chemical oxygen demand	0.6 (1.2)	3.5 (7.1)	0.3 (0.6)	1.1 (2.1)	12.7 (25.5)
Total solids	0.5 (1.0)	---	8.9 (17.8)	0.8 (1.7)	23.8 (47.6)
Dissolved solids	0.45 (0.9)	---	4.4 (8.7)	10.8 (21.7)	---
Suspended solids	0.55 (1.1)	1.1 (2.2)	3.8 (7.5)	10.5 (21.0)	16.5 (33.1)
Sulfate	---	4.2 (8.4)	---	0.38 (0.8)	---
Chloride	---	50.3 (100.6)	0.078 (0.2)	4.6 (9.2)	---
Cyanide	---	---	---	1.5 (2.9)	---
Fluoride	---	0.6 (1.2)	5.0 (10.0)	0.34 (0.7)	15.3 (30.6)
Aluminum	---	---	0.17 (0.34)	0.03 (0.07)	2.7 (5.5)
Calcium	---	---	---	---	---
Copper	---	---	0.001 (0.002)	---	0.005 (0.01)
Magnesium	---	---	0.31 (0.6)	---	---
Nickel	---	---	---	---	0.024 (0.04)
Sodium	---	---	0.47 (0.9)	2.1 (4.3)	---
Zinc	---	---	0.015 (0.03)	0.0017 (0.003)	---
Oil and grease	0.1 (0.2)	0.5 (1.0)	0.45 (0.9)	0.04 (0.08)	1.2 (2.4)
Phenol	---	---	---	---	---

N = net values. Concentration of each constituent in intake water subtracted from concentration in effluent and the difference used to calculate values given.

G = gross values. Data for intake water not available.

(a) Data reported as "not to exceed" so quantities are limits, not actually present.

(b) Does not include effluent from separate cryolite manufacturing facility operated on primary plant site.

Source: Report EPA 440/1-73/019-2

References

(1) "Engineering and Cost Effectiveness Study of Fluoride Emissions Control," Volume 1, TRW Systems and Resources Research Corp., Reston, Va. Prepared for Environmental Protection Agency, Office of Air Programs, Research Triangle Park, N.C., under Contract Number EHSD-71-14, January 1972.

(2) "Air Pollution Control in the Primary Aluminum Industry," Volume I, Singmaster and Breyer, New York, N.Y. Prepared for Environmental Protection Agency, Office of Air Programs, Research Triangle Park, N.C., under Contract Number CPA-70-21, March 1972.

(3) "Particulate Pollutant System Study," Volume I, Midwest Research Institute, Kansas City, Mo. Prepared for Environmental Protection Agency, Office of Air Programs, Research Triangle Park, N.C., May 1971.

(4) "Source Testing Report: Emissions from Wet Scrubbing System," York Research Corp., Stamford, Conn. Prepared for Environmental Protection Agency, Office of Air Programs, Research Triangle Park, N.C., Report Number Y-7730-E.

(5) "Source Testing Report: Emissions from Primary Aluminum Smelting Plant," York Research Corp., Stamford, Conn. Prepared for Environmental Protection Agency, Office of Air Programs, Research Triangle Park, N.C., Report Number Y-7730-B, June 1972.

(6) "Source Testing Report: Emissions from the Wet Scrubber System," York Research Corp., Stamford, Conn. Prepared for Environmental Protection Agency, Office of Air Programs, Research Triangle Park, N.C., Report Number Y-7730-F, June 1972.

(7) U.S. Environmental Protection Agency, "Compilation of Air Pollutant Emission Factors," 2nd Edition, *Publication Number AP-42*, Research Triangle Park, N.C., Office of Air & Water Programs, April 1973.

(8) U.S. Environmental Protection Agency, "Development Document for Proposed Effluent Limitations Guidelines & New Source Performance Standards for the Primary Aluminum Smelting Subcategory of the Aluminum Segment of the Nonferrous Metals Manufacturing Point Source Category," *Report EPA 440/1-73/019-2*, Washington, D.C., October 1973.

(9) Hallowell, J.B., Shea, J.F., Smithson, G.R., Jr., Tripler, A.B. and Gonser, B.W., "Water Pollution Control in the Primary Nonferrous Metals Industry, II - Aluminum, Mercury, Gold, Silver, Molybdenum & Tungsten," *Report EPA-R2-73-2476*, Washington, D.C., U.S. Government Printing Office, September 1973.

ALUMINUM INDUSTRY, SECONDARY

Secondary aluminum operations involve making lightweight metal alloys for industrial castings and ingots. Copper, magnesium, and silicon are the most common alloying constituents. Aluminum alloys for castings are melted in small crucible furnaces charged by hand with pigs and foundry returns. Larger melting operations use open-hearth reverberatory furnaces charged with the same type of materials but by mechanical means. Small operations sometimes use sweating furnaces to treat dirty scrap in preparation for smelting(1).

To produce a high-quality aluminum product, fluxing is practiced to some extent in all secondary aluminum melting. Aluminum fluxes are expected to remove dissolved gases and oxide particles from the molten bath. Sodium and various mixtures of potassium or sodium chloride with cryolite and chlorides of aluminum and zinc are used as fluxes. Chlorine gas is usually lanced into the molten bath to reduce the magnesium content by reacting to form magnesium and aluminum chlorides(2)(3).

Air Pollution

Emissions from secondary aluminum operations include fine particulate matter and gaseous chlorine. A large part of the material charged to a reverberatory furnace is low-grade scrap and chips. Paint, dirt, oil, grease, and other contaminants from this scrap cause large quantities of smoke and fumes to be discharged. Even if the scrap is clean, large surface-to-volume ratios require the use of more fluxes, which can cause serious air pollution problems. Table 124 presents particulate emission factors for secondary aluminum operations(5).

TABLE 124: PARTICULATE EMISSION FACTORS FOR SECONDARY ALUMINUM OPERATIONS[a], EMISSION FACTOR RATING: B

Type of operation	Uncontrolled		Baghouse		Electrostatic precipitator	
	lb/ton	kg/MT	lb/ton	kg/MT	lb/ton	kg/MT
Sweating furnace	14.5	7.25	3.3	1.65	—	—
Smelting						
Crucible furnace	1.9	0.95	—	—	—	—
Reverberatory furnace	4.3	2.15	1.3	0.65	1.3	0.65
Chlorination station[b]	1000	500	50	25	—	—

[a]Reference(4). Emission factors expressed as units per unit weight of metal processed.
[b]Pounds per ton (kg/MT) of chlorine used.

Source: EPA Publication AP-42, 2nd Edition

Water Pollution

The secondary aluminum industry generates wastewaters in the following processes: ingot cooling and shot quenching; scrubbing of furnace fumes during demagging; and wet milling of residues or residue fractions. Data on the quantity of water used for metal cooling in the secondary industry are very sparse and of questionable quality. Only a small number of plants had even approximate gallonage figures. Data gathered were converted to liters used per metric tons of metal cooled and are given in Table 125.

TABLE 125: COOLING WATER USAGE BY SECONDARY SMELTERS

Plant	Water Use of Metal Cooled liters/metric ton (gallons/short ton)	
	Ingot Cooling	Shot Quenching
C-7	680 (160)	
C-26	250 (60)	
C-20	2,300 (550)	60,000 (14,400)
D-6	570 (140)	
B-11	11,500 (2,760)	

Source: Report EPA 440/1-73/019-b

As is evident, the values vary widely. It is not certain whether these great differences are real or whether they are due to grossly inaccurate estimates of water flow. Each of the plants listed in Table 125 is discharging the cooling wastewater after one passage through the circuit. Plants recycling their cooling water had very limited information on the amount of water used per ton of product cooled.

Of the 69 secondary smelters surveyed(b), one plant, designated B-11, had analytical data on cooling wastewater (for a Corps of Engineers' permit). To better characterize the nature of cooling wastewater, sampling teams were sent to other plants for water samples. Samples obtained were analyzed for appropriate constituents and related to pollutant loadings per metric ton of alloy cooled. Data on Plant B-11 are given in Table 126. The table shows that pollutant levels in the cooling wastewaters, with the exception of oils and grease, (which were from 50 to 5,000 milligrams per liter in the other plant sampled) are relatively low.

TABLE 126: CHARACTER OF COOLING WATER OF PLANT B-11[a]

	Intake Water Municipal mg/l	Discharge mg/l[a] Avg.	Net loading in Waste-water gram/mton[b] Average
Alkalinity	95	95	---
COD	NA[c]	15	172
Total solids	192	198	69
Total dissolved solids	190	180	---
Total suspended solids	2	18	182
Ammonia	0.01	1.1	12.5
Nitrate	0.06	0.07	0.11
Chloride	25	29	46
Fluoride	1.01	0.9	---
Aluminum, µg	--	0.7	0.008
Oil and grease, lb/day	--	5 (?) (7.5 mg/l)	86
pH	4.5-6.5	4.5-6.5	---
Temperature, F	NA	97-112	---
Temperature, C		36-44	---

Volume: 80,000 gal/day = 302,800 l/day.   Product: 25-33 tons/day = 23-30 mton/day.

(a) Corps of Engineers data.
(b) $\dfrac{\text{[Conc effluent - conc intake (mg/l)] x liters/day}}{\text{Avg. amount of metal cooled, mtons/day}}$ x 10^{-3} gram/mg = loading, gram/mton
(c) NA = Not applicable.

Source: Report EPA 440/1-73/019-b

A great deal of variability in waste loading is noted in some of the parameters. For instance, total dissolved solid loadings range between 0 and 1.34 kilograms per metric ton of alloy cooled.

Recirculation of cooling water produces sludge and accumulates oil and grease contamination. The sources of sludge include collection of airborne solids from ambient air during spray cooling of the water, buildup of hydrated alumina from chemical reaction with the molten aluminum and debris and dust from the plant floor. Flux salt buildup (NaCl) occurs in recirculated water used for shot cooling. Water used once and discharged will contain oil and grease contaminants. There are operations in which the rate of water flow for cooling is controlled to assure total evaporation.

Aluminum scrap normally charged into the furnace contains a higher percentage of magnesium than is desired for the alloy produced. It is, therefore, necessary to remove a portion of this element from the melt. Magnesium removal, or "demagging", is normally accomplished by either passing chlorine through the melt (chlorination), with the formation of magnesium chloride ($MgCl_2$), or by mixing aluminum fluoride (AlF_3) with the melt, with the removal of magnesium as MgF_2. Heavy fuming results from the demagging of a melt and these fumes are often controlled by passing them through a wet scrubbing system. Water used in the scrubbing thus gains resulting pollutants and is the source of a wastewater stream.

Wastewater from AlF_3 demagging gas scrubbers can normally be recirculated because of the relative insolubility of fluorides (which can be settled out). Wastewater from the scrubbing of chlorine demagging fumes, however, can be recycled only to a very limited degree. This is because the chloride salts are highly soluble and would soon build up to make water unusable. Thus, the discharge of this effluent is the source of wastewater from fume scrubbing. Table 127 gives data on present smelter practices in regard to scrubbing wastewater. Of 69 plants surveyed, 46 are demagging their melts. No demagging wastewater discharges are reported from those plants using AlF_3. All plants using chlorine are discharging demagging scrubber wastewater, whether to navigable waters, public sewage, or holding ponds.

TABLE 127: FUME SCRUBBING WASTEWATER, GENERATION AND DISPOSAL

Practice		Number of Plants
● Use AlF_3 for demagging		14
No air pollution control	5	
Dry air pollution control	7	
Wet air pollution control	2	
- Water recycled continuously	2	
● Use Cl_2 for demagging		28
No air pollution control	3	
Dry air pollution control	1	
Wet air pollution control	24	
Wastewater discharged:		
- with no recycling	12	
- with some recycling	6	
- no discharge-continuously recycled	0	
- to evaporation pond	7	
- with neutralization	17	
- with solids removal	12	
● Use both AlF_3 and Cl_2 for demagging		4
No air pollution control	1	
Dry air pollution control	1	
Wet air pollution control	2	
Wastewater discharged:		
- with no recycling	1	
- to evaporation pond	1	
- with neutralization	2	
- with settling	2	
Total Number of Plants Demagging		46

Source: Report EPA 440/1-73/019-b

Very few smelters in the secondary industry have reliable water use data for their fume scrubbing systems. In one plant, D-6, water usage measured by the project sampling team was one-third of the usage estimated by company personnel. In general, data given out by the plants should be used with caution.

Data on the quantities of water used in scrubbing, which were most consistent in terms of their content, are given in Table 128. Water usage is given in liters per kilogram of magnesium removed during the demagging operation. Basing the water use on magnesium removal provides a common unit for all smelters. The values in Table 128 are fairly consistent, with the average water use being 150 liters per kilogram of magnesium removed.

TABLE 128: QUANTITIES OF WASTEWATER GENERATED IN THE WET SCRUBBING OF CHLORINATION FUMES

Company (code)	Wastewater Generated		
	1/kg	of Mg Removed	(gal/lb)
C-7		95.2	(11)
D-6		182	(22)
D-8		190	(23)
C-26		133[1]	(16)

(1) Estimated from data provided by plant on water usage and rate of Mg removal.

Source: Report EPA 440/1-73/019-b

The character of the raw wastewater generated during the scrubbing of chlorination fumes is given in Table 129. No similarly detailed data on this wastewater were available in the secondary aluminum industry. The data on plants C-7 and D-6 were obtained by sending project water sampling teams to the plant sites for representative samples. The wastewater samples were then analyzed for appropriate constituents.

At Plant C-7 fumes were scrubbed in a tower followed by neutralization and settling of the raw wastewater in separate unit operations. This arrangement permitted sampling the acidic effluent from the scrubber before it was treated and is one example of raw fume scrubber wastewater collected by a tower.

At Plant D-6, the fumes were trapped under a proprietary bell-shaped device in contact with the molten metal and were scrubbed with water. This arrangement also permitted sampling of raw untreated wastewater from a different method of fume scrubbing.

Residues used by the secondary aluminum industry are generally composed of 10 to 30% aluminum, with attached aluminum oxide, fluxing salts (mostly NaCl and KCl), dirt, and various other chlorides, fluorides, and oxides. Separation of the metal from the nonmetals is done by milling and screening and is done wet or dry. When done dry, dust collection is necessary to reduce air emissions. Milling of dross and skimmings will produce a dust that when scrubbed wet will contain in suspension insoluble solids such as aluminum oxide, hydrated alumina, and soluble salts from the flux cover residues such as a sodium chloride and potassium chloride. Drosses also contain aluminum nitride which hydrolyzes in water to yield ammonia. When slags are milled, the wastewater from dust control contains more dissolved sodium and potassium chloride and fluoride salts from the cryolite, than from drosses or skimmings. Some of the oxides of heavy metals are solubilized in the slag and leachable from the dust.

TABLE 129: CHARACTER OF WASTEWATER FROM CHLORINATION FUME SCRUBBING (NO TREATMENT)

Parameter	C-7 [a]		D-6 [b]	
	Conc., mg/ℓ	Loading, grams/KgMg [c]	Conc., mg/ℓ	Loading, grams/KgMg [c]
COD	123	12.1	536	95.8
Total solids	2910	301		
Total dissolved solids	1885	194	10,500	1856
Total suspended solids	225	22.3	480	83.0
Sulfate	11	0.51	481	84.4
Chloride	4420	446	8,671	1560
Cyanide	<0.02	0	--	--
Fluoride	0.24	-0.08 [d]	0.7	-0.324
Aluminum	472	50.9	6.12	0.615
Calcium	0.12	-0.215	990	176
Copper	0.25	0.02	1.31	0.236
Magnesium	41.2	3.86	55.8	9.81
Nickel	0.050	0.003	0.74	0.106
Sodium	3.11	-0.007	770	32.7
Potassium	--	--	206	37.1
Zinc	0.952	0.091	3.58	0.64
Cadmium	0.066	0.006	0.30	0.054
Lead	0.061	0.004	0.24	0.025
Manganese	0.449	0.049	2.34	0.349
Chlorine residue	0.257	0.027		
Oils and grease	13.9	0.590	6.24	0.403
Phenols (ppb)	20.7	-0.002	--	
pH	2.1	--	1.0	--

(a) Average of three composite samples.

(b) Average of five composite samples.

(c) Loading calculated as: [conc. effluent (mg/ℓ) - conc. intake (mg/ℓ)] x

$$\frac{\text{quantity of water used } (ℓ)}{\text{quantity of Mg removed } (kg)}$$

(d) Negative numbers indicate that the process apparently reduced the concentration of this parameter, and are derived from the reports of analytical results as shown above.

Analytical methods from Standard Methods for the Examination of Water and Wastewater, 13th Edition (1971).

Source: Report EPA 440/1-73/019-b

With wet milling, the dust problem is minimized but the operation produces a wastewater stream that is similar to the scrubber waters in makeup but more concentrated in dissolved solids contaminants. The aluminum and alumina fines are settled rapidly and are used to assist the settling of more-difficult-to-settle components obtained as sludges from related wastewater discharges.

Of the 23 plants recovering aluminum values from residues, 8 use wet techniques which lead to the generation of highly saline wastewaters. Table 130 lists the general character of these 8 coded plants. Wastewater is generated by wet dust removal systems (dust generated by dry milling of residue), the washing of residue fractions (sized), and by wet milling the residue to liberate metallic aluminum. In every case the wastewater is passed into a settling pond before discharge. Water use for the wet milling of residues has been based on the tonnage of aluminum recovery rather than the tonnage of residues processed. This is because the former quantity is generally known more accurately by the smelters than the latter.

Table 131 gives available data on the quantity of wastewater generated in the wet milling of residues in liters per metric ton of aluminum recovered. Values for Plants D-3 and D-8 are fairly close, while the value for Plant D-4 is roughly an order of magnitude higher.

TABLE 130: RESIDUE WASTEWATER GENERATION AND DISPOSAL PRACTICE

Practice	D-1	D-2	D-3	D-4	D-5	D-6	D-7	D-8
Wastewater generated by:								
Wet dust removal system	X		X					
Washing of residue fractions						X		
Wet milling of residues		X	X	X	X		X	X
Disposal of wastewater:								
Discharge with some recycling							X	X
Discharge to settling pond	X	X	X	X	X	X	X	X
Chemically treat wastewater to aid settling			X	X			X	X
Discharge to navigable waters via settling pond			X	X				X
No direct discharge streams from settling ponds	X	X			X	X	X	

Source: Report EPA 440/1-73/019-b

TABLE 131: QUANTITIES OF WASTEWATER GENERATED IN THE WET MILLING OF RESIDUES PER TON OF ALUMINUM RECOVERED

Company (code)	Wastewater Generation ℓ/mton of Al recovered (gal/ton)
D-3	16,690[1]
D-4	218,000
D-8	28,838

(1) Corps of Engineers' data.

Source: Report EPA 440/1-73/019-b

The character of wastewater generated during wet milling of residues or residue fractions is given in Table 132. Two plants, D-4 and D-3, had some analytical data on their wastewater from Corps of Engineers' permits. To provide better characterization of the wastewater, sampling teams were sent to Plants D-6, D-8, and D-4 to gather water samples for analysis.

TABLE 132: CHARACTER OF SETTLED WASTEWATER FROM RESIDUE PROCESSING

			Plants			
	D-3 [a]	D-6 [b]	D-4 [c]		D-8 [d]	
Parameter	Loading (Kg/mton Al)	Conc. (mg/ℓ)	Conc. (mg/ℓ)	Loading (Kg/mton) [e]	Conc. (mg/ℓ)	Loading (Kg/mton) [e]
Alkalinity	6.47	314	586	102	500	-7.5 [f]
COD	0.97	2,045			29	0.17
Total solids			24,264	5,144	17,800	326
Total dissolved solids	13.51	12,920			17,400	324
Total suspended solids	0.121	4,961	15	1.5	159	-5.6
Sulfate		1,100	47	1.5	151	1.8
Chloride	0.319	6,492	15,465	3,264	8,903	150
Cyanide		0.04			0.05	0
Fluoride	0.129	2.9	8.7	1.81	16.5	0.38
Ammonia	0.33	0.75	350	73	0.30	-0.03
Aluminum	0.002	0.3	16.4	3.5	28	-1.49
Calcium		58.8	23	-7.4	48	0.17
Copper	<0.001	0.174	0.070	0.008	0.137	0.003
Magnesium		32.5	6	3.9	76	1.39
Nickel		1.2	0.240	0.009	0.20	0
Sodium		2,560	11,600	2,528	3,103	46.2
Potassium		1,087	6,470	1,407	4,802	102
Zinc	0	0.015	0.10	0	0.198	-9.1
Cadmium		0.05	0.002	0	0.005	-0.001
Lead		0.20	0.020	0.004	0.028	-0.001
Manganese		0.16	0.045	0.002	0.060	0
Chlorine residue		.	--	--	--	--
Oils and grease	0.053	55.4	0	0	0.5	0
Phenols (ppb)		--	--	--	0.03	0
pH	8.68	8.3	9.09		9.2	
Nitrates	0.032					

(a) Calculated from U. S. Corps of Engineers, concentration data not given.

(b) From residue milling solid waste washing, tonnage values of residue waste processed not available - loading cannot be calculated. Water flow is 151 ℓpm.

(c) Data from 7 month and 9 month average and verification data from state: metals verified, composite of 18 samples collected over a period of 6 days.

(d) Represents composite of 9 samples collected over 3 days. Milling waste stream is blended with scrubber waste stream.

(e) Loading calculated as: [conc. effluent (mg/ℓ) - conc. intake (mg/ℓ)] x

$$\frac{\text{quantity of water used } (\ell)}{\text{quantity of Al recovered from residue (mton)}}$$

(f) Negative values indicate that the process reduced the concentration of this parameter. and are derived from reported analytical values.

Source: Report EPA 440/1-73/019-b

References

(1) Hammond, W.F. and Simon, H., "Secondary Aluminum-Melting Processes," In: *Air Pollution Engineering Manual,* Danielson, J.A., ed. U.S. DHEW, PHS, National Center for Air Pollution Control, Cincinnati, Ohio, *Publication Number 999-AP-40,* 1967, pages 284 through 290.

(2) "Technical Progress Report: Control of Stationary Sources," Los Angeles County Air Pollution Control District, 1: April 1960.

(3) Allen, G.L. et al, "Control of Metallurgical and Mineral Dusts and Fumes in Los Angeles County," Bureau of Mines, Washington, D.C., *Information Circular Number 7627,* April 1952.

(4) Hammond, W.F. and Weiss, S.M., unpublished report on air contaminant emissions from metallurgical operations in Los Angeles County, Los Angeles County Air Pollution Control District, (Presented at Air Pollution Control Institute, July 1964).

(5) U.S. Environmental Protection Agency, "Compilation of Air Pollutant Emission Factors," 2nd Edition, *Publication Number AP-42*, Research Triangle Park, N.C., Office of Air & Water Programs, April 1973.

(6) U.S. Environmental Protection Agency, "Development Document for Proposed Effluent Limitations Guidelines and New Source Performance Standards for the Secondary Aluminum Smelting Subcategory of the Aluminum Segment of the Nonferrous Metals Manufacturing Point Source Category," *Report EPA 440/1-73/019-b*, Washington, D.C., October 1973.

ALUMINUM SULFATE MANUFACTURE

Water Pollution

Aluminum sulfate is prepared by reaction of bauxite ore with sulfuric acid. The ore and sulfuric acid are reacted in a digester and the resulting aluminum sulfate solution, containing muds and other insolubles from the ore, is then fed to a settling tank, wherein the insolubles are removed by settling and filtration. The filtered product liquor is either shipped as liquid aluminum sulfate solution or evaporated to recover a solid product.

Raw wastes from the process include muds (insolubles) from the digester, settling tank and filtration unit as well as washwaters from vessel cleanouts. At one facility these wastes are treated in a settling basin to remove the muds and the waters and then recycled for reuse. A similar recycling system is used in the other facility. Raw wastes from aluminum sulfate manufacture are listed below.

Waste Products	Process Source	kg/kkg of Product (lb/ton)
Spent aluminum sulfate muds*	Mud washing	170 (340) (two different
		100 (200) facilities)
Low aluminum sulfate water	Mud washing	800 (1,600)

*The raw material bauxite contains 54 to 56% of soluble Al_2O_3, about 3.5% TiO_2, about 5.5% SiO_2, about 1.5% Fe_2O_3 and the rest water of hydration. The muds have approximately the following compositions: 40% SiO_2, 40% TiO_2, 20% Al_2O_3, 0.5% $Al_2(SO_4)_3$.

At these plants all waters are fed to a settling basin where muds are removed and impounded, and the clear effluent is then used back in the process. A breakdown of water use at both facilities is shown below.

		- - - - - - - - - - - - - Quantity - - - - - - - - - - - - - -		
Input Type	Plant	Cubic Meters per Day	Liters per kkg	Comments
Well	049	47 (12,400 GPD)	1,650 (396 gallons/ton)	No pretreatment
Well	063	76 (20,000 GPD)	2,090 (500 gallons/ton)	required for
				either

- - Process Water - -		- - - - - - - - - - - - - Quantity - - - - - - - - - - - - - -		% of Process
Type	Plant	Cubic Meters per Day	Liters per kkg	Stream Recycled
Process	049	77 (20,400 GPD)	2,720 (652 gallons/ton)	30*
Process	063	87 (23,000 GPD)	2,400 (575 gallons/ton)	All excess pro-
				cess water*

*Remaining water shipped with product. Aluminum sulfate solutions are made at both plants.

References

(1) U.S. Environmental Protection Agency, "Development Document for Proposed Effluent Limitations Guidelines & New Source Performance Standards for the Major Inorganic

Products Segment of the Inorganic Chemicals Manufacturing Point Source Category," *Report EPA 440/1-73/007*, Washington, D.C., August 1973.

AMMONIA MANUFACTURE

The manufacture of ammonia (NH_3) is accomplished primarily by the catalytic reaction of hydrogen and nitrogen at high temperatures and pressures. In a typical plant a hydrocarbon feed stream (usually natural gas) is desulfurized, mixed with steam, and catalytically reformed to carbon monoxide and hydrogen. Air is introduced into the secondary reformer to supply oxygen and provide a nitrogen to hydrogen ratio of 1 to 3. The gases then enter a two-stage shift converter that allows the carbon monoxide to react with water vapor to form carbon dioxide and hydrogen.

The gas stream is next scrubbed to yield a gas containing less than 1% CO_2. A methanator may be used to convert quantities of unreacted CO to inert CH_4 before the gases, now largely nitrogen and hydrogen in a ratio of 1 to 3, are compressed and passed to the converter. Alternatively, the gases leaving the CO_2 scrubber may pass through a CO scrubber and then to the converter. The synthesis gases finally react in the converter to form ammonia.

Air Pollution

When a carbon monoxide scrubber is used before sending the gas to the converter, the regenerator off-gases contain significant amounts of carbon monoxide (73%) and ammonia (4%). This gas may be scrubbed to recover ammonia and then burned to utilize the CO fuel value(1).

The converted ammonia gases are partially recycled, and the balance is cooled and compressed to liquefy the ammonia. The noncondensable portion of the gas stream, consisting of unreacted nitrogen, hydrogen, and traces of inerts such as methane, carbon monoxide, and argon, is largely recycled to the converter. To prevent the accumulation of these inerts, however, some of the noncondensable gases must be purged from the system.

The purge or bleed-off gas stream contains about 15% ammonia(1). Another source of ammonia is the gases from the loading and storage operations. These gases may be scrubbed with water to reduce the atmospheric emissions. In addition, emissions of CO and ammonia can occur from plants equipped with CO-scrubbing systems. Emission factors are presented in Table 133.

TABLE 133: EMISSION FACTORS FOR AMMONIA MANUFACTURING WITHOUT CONTROL EQUIPMENT[a], EMISSION FACTOR RATING: B

Type of source	Carbon monoxide		Hydrocarbons[b]		Ammonia	
	lb/ton	kg/MT	lb/ton	kg/MT	lb/ton	kg/MT
Plants with methanator						
Purge gas[c]	Neg	Neg	90	45	3	1.5
Storage and loading[c]	—	—	—	—	200	100
Plants with CO absorber and regeneration system						
Regenerator exit[d]	200	100	—	—	7	3.5
Purge gas[c]	Neg	Neg	90	45	3	1.5
Storage and loading[c]	—	—	—	—	200	100

[a] References(1)(2)
[b] Expressed as methane.
[c] Ammonia emissions can be reduced by 99 percent by passing through three stages of a packed-tower water scrubber. Hydrocarbons are not reduced.
[d] A two-stage water scrubber and incineration system can reduce these emissions to a negligible amount.

Source: EPA Publication AP-42, 2nd Edition

Water Pollution

In the production of ammonia the principal contaminated waste streams are process condensates, blowdown from cooling water systems and wash solutions. Boiler water blowdown may also contribute to the wasteload.

When synthesis gas is prepared by steam reforming of natural gas at high pressures, various steps in the process require cooling or quenching of the gas stream. When the gas is cooled, water condenses and the condensate must be drained from the system. Whenever possible, the condensate is returned to the process. However, it is impossible to return all of the condensate to the process and that which is not recycled must be disposed of. For example, when the synthesis gas is cooled and compressed before entering the synthesis circuit, a knockout pot is used to collect condensed water and oil from the gas stream; and this stream must be disposed of.

The condensate streams will contain CO_2 and some monoethanolamine (an aqueous monoethanolamine (MEA) solution is used to strip the CO_2 from the synthesis gas and some MEA enters the gas stream). Any process leaks and spills and plant cleanup are normally washed to a sump system and collected. They and the process condensates are usually combined to give a single waste stream. Specific information on the individual waste stream volumes and compositions is limited. Most plants do not monitor the individual streams, but only the combined stream. Typical composition and volume ranges for the combined waste stream are presented in Table 134.

TABLE 134: TYPICAL AMMONIA PLANT WASTEWATER VOLUME AND COMPOSITION*

Contaminant	mg/ℓ	Contaminant	mg/ℓ
Ammonia	20-100	BOD	50-150
CO_2	150-750	COD	60-200
MEA	50-100	Oil	100-10,000

Wastewater Volume   100-1000 gal/ton NH_3

*Does not include cooling water or boiler water blowdowns.

Source: EPA Report 12020 FPD 09/71

Ammonia production requires extremely large volumes of cooling water. Volume requirements are reported to range from 20,000 to 185,000 gallons per ton of ammonia product. For a 600 ton per day ammonia plant this corresponds to a cooling water flow of 8,000 to 74,000 gpm. When the cooling water is used on a once-through basis contamination of the cooling water effluent is normally not a problem, except for thermal pollution. The water is normally discharged to the receiving water without treatment except for temperature reduction.

The large volumes of cooling water required make once-through operation impractical in most cases. Normally the cooling water is recycled using cooling towers or ponds. The number of cycles of concentration that can be used depends primarily on the purity and availability of the fresh makeup water. In most cases the number of cycles of concentration will be in the range of 3 to 7. The volume of blowdown required to control impurities is usually in the range of 1 to 2% of the cooling water flow. Makeup water volume will usually be several times the blowdown volume, because of water lost by evaporation and windage in the cooling tower or pond.

The volume of blowdown water from an ammonia plant cooling water system can vary over a wide range and values from 200 to 3,500 gallons per ton of product have been reported. Normally, however, the blowdown volume will run from 400 to 1,000 gallons per ton NH_3. The composition of the blowdown water will depend on the feed water composition, cycles of concentration, process leaks, and materials added to the cooling water for corrosion, scaling and plant growth control. Typical ranges of concentration for various impurities are given in Table 135.

TABLE 135: COMPOSITION OF AMMONIA PLANT COOLING WATER BLOWDOWN

Contaminant	mg/ℓ	Contaminant	mg/ℓ
Chromate	0-300	NH_3	10-100
Phosphate	0-50	MEA	0-10
Zinc	0-30	Sulfate	500-5000
Heavy metals	0-60	TDS	500-10,000
Fluoride	0-10	BOD	10-300
Biocides	0-200	COD	15-400
Misc. organics	0-100	Oil	10-1000
Blowdown volume	400-1000 gal/ton NH_3		

Source: EPA Report 12020 FPD 09/71

Ammonia production requires substantial quantities of steam and process waste heat boilers are used to generate this steam whenever possible. In some plants additional steam generation may be required. Condensate recycle to the steam boiler can lead to generation of a boiler water blowdown waste which is high in dissolved solids and corrosion and scale inhibitors.

References

(1) Burns, W.E. and McMullan, R.R., "No Noxious Ammonia Odor Here," *Oil and Gas Journal,* pages 129 to 131, February 25, 1967.
(2) Axelrod, L.C. and O'Hare, T.E., "Production of Synthetic Ammonia," New York, M.W. Kellogg Company, 1964.
(3) U.S. Environmental Protection Agency, "Compilation of Air Pollutant Emission Factors," 2nd Edition, *Publication Number AP-42,* Research Triangle Park, N.C., Office of Air and Water Programs, April 1973.
(4) Fullam, H.T. and Faulkner, B.P., "Inorganic Fertilizer and Phosphate Mining Industries - Water Pollution and Control," *Report Number 12020 FPD 09/71,* Washington, D.C., U.S. Government Printing Office, September 1971.

AMMONIUM NITRATE MANUFACTURE

Water Pollution

In the production of prilled ammonium nitrate there are various operations which can result in the generation of gas streams which can contain ammonia or nitrate ion or both. These operations include the neutralizer, where both ammonia and nitric acid may be vaporized, the concentrator, and the prilling tower. Normally scrubbing is required to remove the nitrogen species from the gas streams. Water scrubbing is usually used and the scrub water may or may not be recycled. The ammonia and nitrate recovered in the scrub system is normally not returned to the process. The volume of nitrogen containing wastewater generated is difficult to define because of the many operating variables involved. However, ammonia losses can run as high as 1% or about 4 pounds per ton of NH_4NO_3 product. Nitrate losses are usually much less.

Significant quantities of cooling water may be required in ammonium nitrate production (up to 35,000 gallons per ton of product). When large volumes are required cooling water recycle is usually used. This means cooling water blowdown is required. The volume of blowdown required will depend primarily on makeup water purity. The composition of the blowdown stream will be quite similar to that generated in ammonia production, except that there will be a significant nitrate concentration in the blowdown from ammonium nitrate production.

Process spills and resultant cleanup can generate a significant waste stream. Solid spills are usually recovered if possible. Liquid spills, however, are usually washed to a sump or drainage system. Normally the wash solutions are not returned to the process because of the possibility of oil or grease contamination and its resultant hazard.

Reference

(1)   U.S. Environmental Protection Agency, "Inorganic Fertilizer and Phosphate Mining Industries - Water Pollution and Control," *Report 12020 FPD 09/71,* Washington, D.C., U.S. Government Printing Office, September 1971.

AMMONIUM SULFATE PRODUCTION

Water Pollution

In the direct production of ammonium sulfate, ammonia and sulfuric acid are reacted in a crystallizer to form the sulfate. Heat of reaction causes the evaporation of water from the crystallizer solution and a loss of ammonia and sulfate. The ammonia normally must be removed from the gas stream prior to discharge. This can be done using a water scrubber, barometric condenser, or noncontact condenser.

Where a water scrubber or barometric condenser is used, a water flow of 1 to 5 gallons per minute per daily ton of product is required. For a 300 ton per day plant this will correspond to a water flow of 300 to 1,500 gpm. If the water is used on a once-through basis the ammonia concentration will vary between 10 and 100 milligrams per liter depending on the crystallizer design. (The sulfate concentration will be less.) If the water is recycled, the ammonia concentration will be much higher. Where indirect condensation is used, the water and ammonia condense to give a stream which can be returned to the process.

The off-gas from the ammonium sulfate dryer contains ammonium sulfate particles and small amounts of ammonia. The particulates are collected by dry means such as cyclones and returned to the process. Depending on local regulations, recovery of the ammonia from the gas stream may or may not be necessary. When recovery is required, water scrubbing is used. The ammonia concentration in the scrub solution is quite low. If water is used on a once-through basis, scrub volumes vary from 60 to 600 gallons per minute for a 300 ton per day plant. The ammonia concentration is usually less than 10 grams per liter. Process spills are usually washed to a retention sump with a minimum of water. Frequently the liquor from the sump is used as the wash solution. The sump liquor is normally returned to the process.

Reference

(1)   Fullam, H.T. and Faulkner, B.P., "Inorganic Fertilizer and Phosphate Mining Industries - Water Pollution and Control," *Report Number 12020 FPD 09/71,* Washington, D.C., U.S. Government Printing Office, September 1971

ASBESTOS PRODUCTS INDUSTRY

Air Pollution

The reader is referred to the section on "Asbestos" in the first part of the volume for information on this topic.

Water Pollution

Water is commonly used in asbestos manufacturing as an ingredient, a carrying medium, for cooling, and for various auxiliary purposes such as in pump seals, wet saws, pressure testing of pipe, and others. Water is used only for cooling in the manufacture of asbestos roofing and floor tile products. In most asbestos manufacturing plants the wastewaters from all sources are combined and discharged in a single sewer.

Asbestos manufacturing, in almost all cases, involves forming the product from a dilute water slurry of the mixed raw ingredients. The product is brought to the desired size, thickness, or shape by accumulating the solid materials and removing most of the carriage water. The water is removed at several places in the machine and it, together with any excess slurry, is piped to the save-all system.

The mixing operations are carried out on a batch, or semicontinuous basis. Water and materials are returned from the save-alls as needed during mixing. Excess water and, in some cases, materials are discharged from the save-all system. Fresh water and additional raw materials are added during mixing. The fresh water is often used first as vacuum pump seal water before going into the mixing operations.

The major source of process wastewater in asbestos manufacturing is the machine that converts the slurry into the formed wet product. It is not practical to isolate individual sources of wastewater within the machine system. The water is commonly transported from the machine to the save-all system and back to the machine in a closed system. To measure the quantity of water flowing in the machine-save-all recycle system involves a rather elaborate monitoring program that was beyond the scope of this study. Only one manufacturing plant provided data on in-plant water flows that were more than rough estimates.

An important factor influencing both the volume and strength of the raw wastewaters is the save-all capacity in the plant. Save-alls are basically settling tanks in which solid-liquid separation is accomplished by gravity. Their purpose is first to recover raw materials (solids) and, second, water. The efficiency of separation is primarily dependent upon the hydraulic loading on the save-all. Plants with greater save-all capacity have greater flexibility in operation, more water storage volume, and a cleaner raw wastewater leaving the manufacturing process. In many asbestos manufacturing plants, the solids in the save-alls are dumped when the product is to be changed or when it is necessary to remove the accumulated waste solids at the bottom. It may also be necessary to dump the save-alls when the manufacturing process is shut down.

The manufacture of asbestos-cement pipe in a typical plant increases the levels of the major constituents in the water by the amounts (approximate) shown on the following page.

	mg/l	kg/kkg	lb/ton
Total solids	1,500	9	18
Suspended solids	500	3.1	6.3
BOD (5-day)	2	0.01	0.02
Alkalinity	700	4.4	8.8

The dissolved salts are reported to be primarily calcium and potassium sulfates with lesser amounts of sodium chloride. The magnesium levels are not known to be high. The alkalinity is primarily caused by hydroxide with a small carbonate contribution. The pH ranges as high as 12.9, but is generally close to 12.0, or slightly lower.

The manufacture of asbestos-cement sheet products in a typical plant increases the level of constituents in the water by the following approximate amounts.

	mg/l	kg/kkg	lb/ton
Total solids	2,000	15	30
Suspended solids	850	6.5	13
BOD (5-day)	2	0.015	0.03
Alkalinity	1,000	7.5	15

Little information is available on the dissolved salts in sheet wastewaters, but they should be similar to those from asbestos-cement pipe manufacture. The alkalinity is caused primarily by hydroxide with a pH averaging 11.7 and ranging from 11.4 to 12.4 in all reporting plants.

The manufacture of asbestos paper in a typical plant increases the levels of the constituents in the water by the following approximate amounts.

	mg/l	kg/kkg	lb/ton
Total solids	1,900	26	52
Suspended solids	680	9.5	19
BOD (5-day)	110	1.5	3
COD	160	2.2	4.4

The pH of raw wastewaters from asbestos paper manufacturing is 8.0 or lower.

At an asbestos millboard plant that discharges its wastewaters to the lagoon system, the constituents added to the water were measured as follows.

	mg/l	kg/kkg	lb/ton
Suspended solids	35	1.8	3.5
BOD (5-day)	5	0.25	0.5

The total solids and COD levels in the water leaving the millboard save-alls were the same as those of the makeup water. The pH of the raw wastewater ranged from 8.3 to 9.2. Some millboard is manufactured with portland cement and the pH would be higher in such cases.

The effluent from the save-all system at the millboard plant that operates with a completely closed water system had the characteristics listed below.

	Average	Range
Total solids, milligrams per liter	6,100	3,950 - 7,800
Suspended solids	5,100	3,060 - 6,270
BOD (5-day)	2	- - - - - - - -
COD	60	10 - 145

In such a plant, the waste constituents accumulate until a steady-state level is reached. The contribution of each manufacturing cycle cannot be determined directly and, consequently, raw waste loadings expressed in terms of production units are meaningless. The pH ranged from 11.8 to 12.1 and the alkalinity from 2,000 to 2,700 milligrams per liter, mostly in the hydroxide form.

The characteristics of spent cooling water from roofing manufacture are developed from sampling data taken at one plant. This plant employs surface sprays and discharges the contact and noncontact cooling water into a common sewer. The combined wastewater was sampled. At the time of sampling, the roofing was being made from organic (non-asbestos) paper. Since the water spray contacts only the outer bituminous surface and not the base paper, it is believed that the samples are representative of wastes from contact cooling of asbestos-based roofing. The added quantities of the major constituents were as follows.

	mg/l	kg/kkg	lb/ton
Suspended solids	150	0.06	0.13
BOD (5-day)	6	0.003	0.005
COD	20	0.008	0.016

The pH of the wastewater averaged 8.2.

The added waste constituents in a typical asbestos floor tile plant are as follows.

	mg/l	kg/1,000 pc*	lb/1,000 pc*
Suspended solids	150	0.18	0.40
BOD (5-day)	15	0.02	0.04
COD	300	0.36	0.80

*pc = pieces of tile, 12" by 12" by $\frac{3}{32}$".

The reported pH of tile plant wastewaters ranges from 6.9 to 8.3, averaging 7.3.

Reference

(1) U.S. Environmental Protection Agency, "Development Document for Proposed Effluent Limitations Guidelines and New Source Performance Standards for the Building, Construction and Paper Segment of the Asbestos Manufacturing Point Source Category," Report EPA 440/1-73/017, Washington, D.C., October 1973.

ASPHALT ROOFING MANUFACTURE

The manufacture of asphalt roofing felts and shingles involves saturating fiber media with asphalt by means of dipping and/or spraying. Although it is not always done at the same site, preparation of the asphalt saturant is an integral part of the operation. This preparation, called "blowing," consists of oxidizing the asphalt by bubbling air through the liquid asphalt for 8 to 16 hours. The saturant is then transported to the saturation tank or spray area. The saturation of the felts is accomplished by dipping, high-pressure sprays, or both. The final felts are made in various weights: 15, 30, and 55 pounds per 100 square feet (0.72, 1.5, and 2.7 kilograms per square meter). Regardless of the weight of the final product, the makeup is approximately 40% dry felt and 60% asphalt saturant.

Air Pollution

The major sources of particulate emissions from asphalt roofing plants are the asphalt blowing operations and the felt saturation. Another minor source of particulates is the covering of the roofing material with roofing granules. Gaseous emissions from the saturation process have not been measured but are thought to be slight because of the initial driving off of contaminants during the blowing process.

A common method of control at asphalt saturating plants is the complete enclosure of the spray area and saturator with good ventilation through one or more collection devices, which include combinations of wet scrubbers and two-stage low-voltage electrical precipitators, or cyclones and fabric filters. Emission factors for asphalt roofing are presented in Table 136.

TABLE 136: EMISSION FACTORS FOR ASPHALT ROOFING MANUFACTURING WITHOUT CONTROLS[a], EMISSION FACTOR RATING: D

Operation	Particulates[b]		Carbon monoxide		Hydrocarbons (CH_4)	
	lb/ton	kg/MT	lb/ton	kg/MT	lb/ton	kg/MT
Asphalt blowing[c]	2.5	1.25	0.9	0.45	1.5	0.75
Felt saturation[d]						
Dipping only	1	0.5	–	–	–	–
Spraying only	3	1.5	–	–	–	–
Dipping and spraying	2	1	–	–	–	–

[a]Approximately 0.65 unit of asphalt input is required to produce 1 unit of saturated felt. Emission factors expressed as units per unit weight of saturated felt produced.

[b]Low-voltage precipitators can reduce emissions by about 60 percent; when they are used in combination with a scrubber, overall efficiency is about 85 percent.

[c]Reference(1).

[d]References(2)(3).

Source: EPA Publication AP-42, 2nd Edition

References

(1) Von Lehmden, D.J., Hangebrauck, R.P. and Meeker, J.E., "Polynuclear Hydrocarbon Emissions from Selected Industrial Processes," *J. Air Pol. Control Assoc.,* 15:306-312, July 1965.
(2) Weiss, S.M., "Asphalt Roofing Felt Saturators," In: *Air Pollution Engineering Manual,* Danielson, J.A., ed. U.S. DHEW, PHS, National Center for Air Pollution Control, Cincinnati, Ohio, *Publication Number 999-AP-40,* 1967, pages 378 to 383.
(3) Goldfield, J. and McAnlis, R.G., "Low-Voltage Electrostatic Precipitators to Collect Oil Mists from Roofing-Felt Asphalt Saturators and Stills," *J. Industrial Hygiene Assoc.,* July-August 1963.
(4) U.S. Environmental Protection Agency, "Compilation of Air Pollutant Emission Factors," 2nd Edition, *Publication Number AP-42,* Research Triangle Park, N.C., Office of Air and Water Programs, April 1973.

ASPHALTIC CONCRETE PLANTS (ASPHALT BATCHING)

Selecting and handling the raw material is the first step in the production of asphaltic concrete, a paving substance composed of a combination of aggregates uniformly mixed and coated with asphalt cement. Different applications of asphaltic concrete require different aggregate size distributions, so that the raw aggregates are crushed and screened at the quarries. The coarse aggregate usually consists of crushed stone and gravel, but waste materials, such as slag from steel mills or crushed glass, can be used as raw material.

Plants produce finished asphaltic concrete through either batch or continuous aggregate mixing operations. The raw aggregate is normally stock-piled near the plant at a location where the moisture content will stabilize between 3 and 5% by weight.

As processing for either type of operation begins, the aggregate is hauled from the storage piles and placed in the appropriate hoppers of the cold-feed unit. The material is metered from the hoppers onto a conveyor belt and is transported into a gas- or oil-fired rotary dryer. Because a substantial portion of the heat is transferred by radiation, dryers are equipped with flights that are designed to tumble the aggregate and promote drying. As it leaves the dryer, the hot material drops into a bucket elevator and is transferred to a set of vibrating screens where it is classified by size into as many as four different grades. At this point it enters the mixing operation.

In a batch plant, the classified aggregate drops into one of four large bins. The operator controls the aggregate size distribution by opening individual bins and allowing the classified aggregate to drop into a weigh hopper until the desired weight is obtained. After all the material is weighed out, the sized aggregates are dropped into a mixer and mixed dry for about 30 seconds. The asphalt, which is a solid at ambient temperatures, is pumped from heated storage tanks, weighed, and then injected into the mixer. The hot, mixed batch is then dropped into a truck and hauled to the job site.

In a continuous plant, the classified aggregate drops into a set of small bins, which collect and meter the classified aggregate to the mixer. From the hot bins, the aggregate is metered through a set of feeder conveyors to another bucket elevator and into the mixer. Asphalt is metered into the inlet end of the mixer, and retention time is controlled by an adjustable dam at the end of the mixer. The mix flows out of the mixer into a hopper from which the trucks are loaded.

Air Pollution

Dust sources are the rotary dryer; the hot aggregate elevators; the vibrating screens; and the hot-aggregate storage bins, weigh hoppers, mixers, and transfer points. The largest dust emission source is the rotary dryer. In some plants, the dust from the dryer is handled separately from emissions from the other sources. More commonly, however, the dryer, its vent lines, and other fugitive sources are treated in combination by a single collector and fan system.

The choice of applicable control equipment ranges from dry, mechanical collectors to scrubbers and fabric collectors; attempts to apply electrostatic precipitators have met with little success. Practically all plants use primary dust collection equipment, such as large diameter cyclone, skimmer, or settling chambers. These chambers are often used as classifiers with the collected materials being returned to the hot aggregate elevator to combine with the dryer aggregate load. The air discharge from the primary collector is seldom vented to the atmosphere because high emission levels would result. The primary collector effluent is therefore ducted to a secondary or even to a tertiary collection device.

Emission factors for asphaltic concrete plants are presented in Table 137. Particle size information has not been included because the particle size distribution varies with the aggregate being used, the mix being made, and the type of plant operation.

TABLE 137: PARTICULATE EMISSION FACTORS FOR ASPHALTIC CONCRETE PLANTS, EMISSION FACTOR RATING: A

Type of Control	Emissions	
	lb/ton	kg/MT
Uncontrolled[a]	45.0	22.5
Precleaner	15.0	7.5
High-efficiency cyclone	1.7	0.85 (continued)

TABLE 137: (continued)

	----- Emissions -----	
	lb/ton	kg/MT
Spray tower	0.4	0.20
Multiple centrifugal scrubber	0.3	0.15
Baffle spray tower	0.3	0.15
Orifice-type scrubber	0.04	0.02
Baghouse[b]	0.1	0.05

[a] Almost all plants have at least a precleaner following the rotary dryer.
[b] Emissions from a properly designed, installed, operated, and maintained
   collector can be as low as 0.005 to 0.020 lb/ton (0.0025 to 0.010 kg/MT).

Source: EPA Publication AP-42, 2nd Edition

Reference

(1) U.S. Environmental Protection Agency, "Compilation of Air Pollutant Emission Factors," 2nd Edition, *Publication Number AP-42*, Research Triangle Park, N.C., Office of Air and Water Programs, April 1973.

AUTOMOBILE BODY INCINERATION

Auto incinerators consist of a single primary combustion chamber in which one or several partially stripped cars are burned(1). (Tires are removed.) Approximately 30 to 40 minutes are required to burn two bodies simultaneously(2). As many as 50 cars per day can be burned in this batch-type operation, depending on the capacity of the incinerator. Continuous operations in which cars are placed on a conveyor belt and passed through a tunnel-type incinerator have capacities of more than 50 cars per 8-hour day.

Air Pollution

Both the degree of combustion as determined by the incinerator design and the amount of combustible material left on the car greatly affect emissions. Temperatures on the order of 1200°F (650°C) are reached during auto body incineration(2). This relatively low combustion temperature is a result of the large incinerator volume needed to contain the bodies as compared with the small quantity of combustible material. The use of overfire air jets in the primary combustion chamber increases combustion efficiency by providing air and increased turbulence.

In an attempt to reduce the various air pollutants produced by this method of burning, some auto incinerators are equipped with emission control devices. Afterburners and low-voltage electrostatic precipitators have been used to reduce particulate emissions; the former also reduces some of the gaseous emissions(3)(4). When afterburners are used to control emissions, the temperature in the secondary combustion chamber should be at least 1500°F (815°C). Lower temperatures result in higher emissions. Emission factors for auto body incinerators are presented in Table 137.

TABLE 137: AUTO BODY INCINERATION[a], EMISSION FACTOR RATING: B

Pollutants	Uncontrolled		With afterburner	
	lb/car	kg/car	lb/car	kg/car
Particulates[b]	2	0.9	1.5	0.68
Carbon monoxide[c]	2.5	1.1	Neg	Neg

(continued)

TABLE 137: (continued)

Pollutants	Uncontrolled		With afterburner	
	lb/car	kg/car	lb/car	kg/car
Hydrocarbons (CH$_4$)^c	0.5	0.23	Neg	Neg
Nitrogen oxides (NO$_2$)^d	0.1	0.05	0.02	0.01
Aldehydes (HCOH)^d	0.2	0.09	0.06	0.03
Organic acids (acetic)^d	0.21	0.10	0.07	0.03

^{a}Based on 250 lb (113 kg) of combustible material on stripped car body.
^{b}References 2 and 4.
^{c}Based on data for open burning and References 2 and 5.
^{d}Reference 3.

Source: EPA Publication AP-42, 2nd Edition

References

(1) U.S. Environmental Protection Agency, "Compilation of Air Pollutant Emission Factors," 2nd Edition, *Publication Number AP-42,* Research Triangle Park, N.C., Office of Air and Water Programs, April 1973.
(2) Kaiser, E.R. and Tolcias, J., "Smokeless Burning of Automobile Bodies," *J. Air Pol. Control Assoc.,* 12, 64-73, February 1962.
(3) Alpiser, F.M., "Air Pollution from Disposal of Junked Autos," *Air Engineering,* 10, 18-22, November 1968.
(4) Private communication with D.F. Walters, U.S. DHEW, PHS, Division of Air Pollution, Cincinnati, Ohio, July 19, 1963.
(5) Gerstle, R.W. and Kenmitz, D.A., "Atmospheric Emissions from Open Burning," *J. Air Pol. Control Assoc.,* 17, 324-327, May 1967.

AUTOMOTIVE VEHICLE OPERATION

Passenger cars and light trucks, heavy-duty trucks, and motorcycles comprise the three main categories of highway vehicles. Within each of these categories, power-plant and fuel variations result in significantly different emission characteristics. For example, passenger cars may be powered by gasoline or diesel fuel or operate on a gaseous fuel such as compressed natural gas (CNG). Similarly, a motorcycle may have either a four-stroke or a two-stroke engine.

Air Pollution

Highway vehicle emission factors are presented here in the form of average emission factors based on statistical information for all major types of highway vehicles combined, (i.e., light- and heavy-duty, gasoline-powered vehicles and heavy-duty, diesel-powered vehicles). These values are presented in grams of pollutant per mile traveled (and in grams of pollutant per kilometer).

It is important to note that highway vehicle emission factors change with time and, therefore, must be calculated for a specific time period, normally one calendar year. The major reason for this time dependence is the gradual replacement of vehicles without emission control equipment by vehicles with control equipment. Emission factors are also available in detail(1) for individual classes of highway vehicles: [1] light-duty, gasoline-powered vehicles, [2] light-duty, diesel-powered vehicles, [3] heavy-duty, gasoline-powered vehicles, [4] heavy-duty, diesel-powered vehicles, [5] gaseous-fueled vehicles using compressed natural gas (CNG) or liquefied natural gas (LNG), [6] motorcycles, and [7] off-highway vehicles such as garden tractors.

Average Emission Factors for Highway Vehicles: Emission factors in this section update emission factors for gasoline-powered motor vehicles presented in the February 1972 "Compilation of Air Pollutant Emission Factors"(2). These new factors are based on nation-wide statistical data for light-duty, gasoline-powered vehicles; heavy-duty, gasoline-powered vehicles; and heavy-duty, diesel-powered vehicles. Average emission factors are intended to assist those individuals interested in compiling approximate emission estimates for large areas, such as an individual state or the nation. The emission factor calculation techniques presented for specific vehicle types(1) are strongly recommended for the formulation of localized emission estimates required for air quality modeling or for the evaluation of air pollutant control strategies.

Average emission factors by calendar year based on statistical data for the United States are presented in Table 138. These factors were calculated using the techniques described(1). Because the majority of highway vehicle emissions are produced (on a nationwide basis) by gasoline-powered, light-duty vehicles and heavy-duty, gasoline and diesel-powered vehicles, these are the only vehicles considered in Table 138. The emission contribution from diesel-powered, light-duty vehicles, from gaseous-fuel-powered vehicles, and from motorcycles is assumed to be insignificant for the purpose of developing these approximate factors.

TABLE 138: AVERAGE EMISSION FACTORS FOR HIGHWAY VEHICLES BASED ON NATIONWIDE STATISTICS*

| | Carbon monoxide | | Hydrocarbons | | | | Nitrogen oxides (NO_x as NO_2) | | Particulates | | | | Sulfur oxides (SO_2) | |
| | | | Exhaust | | Crankcase and evaporation | | | | Exhaust | | Tire wear | | | |
Year	g/mi	g/km	g/mi	g/km	g/mi	g/km	g/mi	g/km	g/mi	g/km	g/mi	g/km	g/mi	g/km
1965	89	55	9.2	5.7	5.8	3.6	4.8	3.0	0.38	0.24	0.20	0.12	0.20	0.12
1970	78	48	7.8	4.8	3.9	2.4	5.3	3.3	0.38	0.24	0.20	0.12	0.20	0.12
1971	74	46	7.2	4.5	3.5	2.2	5.4	3.4	0.38	0.24	0.20	0.12	0.20	0.12
1972	68	42	6.6	4.1	2.9	1.8	5.4	3.4	0.38	0.24	0.20	0.12	0.20	0.12
1973	62	39	6.1	3.8	2.4	1.5	5.4	3.4	0.38	0.24	0.20	0.12	0.20	0.12
1974	56	35	5.5	3.4	2.0	1.2	5.2	3.2	0.38	0.24	0.20	0.12	0.20	0.12
1975	50	31	4.9	3.0	1.5	0.93	4.9	3.0	0.38	0.24	0.20	0.12	0.20	0.12
1976	42	26	4.2	2.6	1.3	0.81	4.7	2.9	0.38	0.24	0.20	0.12	0.20	0.12
1977	36	22	3.6	2.2	1.0	0.62	4.2	2.6	0.38	0.24	0.20	0.12	0.20	0.12
1978	31	19	3.1	1.9	0.83	0.52	3.7	2.3	0.38	0.24	0.20	0.12	0.20	0.12
1979	26	16	2.7	1.7	0.67	0.42	3.4	2.1	0.38	0.24	0.20	0.12	0.20	0.12
1980	22	14	2.4	1.5	0.53	0.33	3.1	1.9	0.38	0.24	0.20	0.12	0.20	0.12
1990	14	8.7	1.6	0.99	0.38	0.24	2.2	1.4	0.38	0.24	0.20	0.12	0.20	0.12

*Based on References (1) through (5).
Note: This table does not reflect interim standards promulgated by the EPA Administrator on April 11, 1973.

Source: EPA Publication AP-42, 2nd Edition

The exhaust emission values presented in Table 138 for carbon monoxide, hydrocarbons, and and nitrogen oxides are for an average speed of approximately 19.6 miles per hour (31.5 kilometers per hour). These values can be modified to make them representative of the area for which emission estimates are being prepared, by using the average speed adjustment factors contained in Figure 18(6)(7)(8). For example, if carbon monoxide emissions in 1970 are to be estimated for a state where the average speed is 35 miles per hour, the appropriate emission factor would be 0.6 times 78 or 47 grams per mile. This value would then be multiplied by the total vehicle miles of travel (VMT) to arrive at a carbon monoxide emission estimate.

Crankcase and evaporative hydrocarbons, particulate, and sulfur oxide emission factors are average values that can be considered independent of speed. Emission estimates for these pollutants are calculated by simply multiplying the VMT by the emission factor.

FIGURE 18: AVERAGE SPEED CORRECTION FACTORS FOR ALL MODEL YEARS

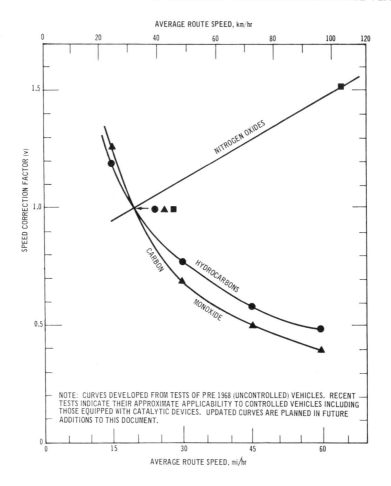

Source: EPA Publication AP-42, 2nd Edition

The emission factor data presented for highway vehicles in this chapter are based on a generalized test cycle that involves operation typical of every-day driving patterns. Because this driving cycle is intended to represent typical driving, it cannot apply in specific instances, i.e., to a particular segment of a particular roadway at a particular time. In order to estimate vehicular emissions under a specific set of conditions, modal emission factor data are required. Driving modes include: idle, constant speed, acceleration, and deceleration. Because all driving patterns can be divided into one of these four modes, emissions can be determined by summing the modal emissions for a particular driving pattern. The Environmental Protection Agency is currently evaluating the use of modal emission data. Emission data for idle, various constant speeds, and various initial and final speeds (accelerations and decelerations) are being collected and analyzed.

References

(1) U.S. Environmental Protection Agency, "Compilation of Air Pollutant Emission

Factors," 2nd Edition, *Publication Number AP-42,* Research Triangle Park, N.C., Office of Air and Water Programs, April 1973.

(2) U.S. Environmental Protection Agency, "Compilation of Air Pollutant Emission Factors," *Publication Number AP-42,* Research Triangle Park, N.C., Office of Air and Water Programs, February 1972.

(3) "Highway Statistics 1970," U.S. Department of Transportation, Federal Highway Administration, Washington, D.C., 1971.

(4) "Census of Transportation - Truck Inventory and Use Survey," U.S. Department of Commerce, Bureau of the Census, Washington, D.C., July 1970.

(5) "Automotive Facts and Figures," Automobile Manufacturers Association, Washington, D.C., July 1970.

(6) McMichael, W.F. and Rose, A.H., Jr., *A Comparison of Emissions from Automobiles in Cities at Two Different Altitudes,* U.S. DHEW, PHS, Cincinnati, Ohio, July 1965.

(7) "Study of Emissions from Light-Duty Vehicles in Six Cities," Automotive Environmental Systems Inc., San Bernardino, Calif., Prepared for the Environmental Protection Agency, Research Triangle Park, N.C., under Contract Number 68-04-0042, June 1972.

(8) Walsh, M.P., Unpublished data on emissions from a catalyst-equipped light-duty vehicle, The City of New York Department of Air Resources, Bureau of Motor Vehicle Pollution Control, New York, N.Y., November 1972.

BAUXITE REFINING

Air Pollution

The reader is referred to the preceding section on "Aluminum Production, Primary" for data on air pollution from bauxite grinding and alumina calcining.

Water Pollution

The dominant waste from a bauxite refinery is the gangue material from the ore, known as red or brown mud, which is produced on a very large scale (500 to nearly 4,000 kkg per day). The most common solution to this red mud waste problem is total impoundment, but the tonnages to be disposed of can make the problem difficult.

Depending upon the type of bauxite used, from one-third ton to approximately one ton of red mud will be produced per ton of alumina. In the case of brown mud from Arkansas bauxite, this increases to two to two and one-half tons per ton. The red mud is the major waste stream from a bauxite refinery. It will generally issue from the washing thickeners at approximately 17 to 20% solids, and be pumped to a disposal lake. Iron impurities impart the red color to the mud. If derived from Jamaican or Arkansas bauxite, the red mud may contain as much as 50% iron.

Table 139 shows a typical chemical analysis of the insoluble solids in Jamaican red mud. Although some 23 elements have been indicated as analyzed, 98.5% of the material consists of the oxides of only 8 elements plus water and carbon dioxide as indicated by ignition losses. The remaining 1.5% consists of the oxides of metallic elements, such as MgO, K_2O, Cr_2O_3, ZnO, ZrO_2, NiO, V_2O_5, SrO, and others. Thus, red mud has a complex chemical makeup, dependent upon its parent bauxite ore, aluminum extraction techniques, and impurity control.

The principal soluble constituents found in a typical Jamaican red mud liquor at 17% solids are shown in Table 139. The concentrations of metallic elements, including those listed in Table 140, are small. The hydroxides of these nonamphoteric elements are quite insoluble. Also, the alkaline leaching process is almost totally specific for aluminum and results in a highly pure alumina product.

TABLE 139: RED MUD INSOLUBLE SOLIDS*

	Percent
LOI (Loss on Ignition)	11.0
SiO_2	5.5
Al_2O_3	12.0
Fe_2O_3	49.5
P_2O_5	2.0
CaO	8.0
Na_2O	3.5
TiO_2	5.0
MnO_2	1.0
Miscellaneous	1.5

*Specific gravity = 3.6.

Source: Report EPA 440/1-73/019

TABLE 140: RED MUD SLURRY SOLUBLE SOLIDS

Al_2O_3	2.5 g/kg liq.
NaOH	3.7 g/kg
Na_2CO_3	1.6 g/kg
Na_2SO_4	0.4 g/kg
NaCl	0.7 g/kg
$Na_2C_2O_4$	0.1 g/kg
Specific gravity	1.008
pH	12.5
BOD	6 ppm
COD	148 ppm

Source: Report EPA 440/1-73/019

One of the characteristics of Jamaican red mud is its fine size. It has been reported that wet screening showed a particle size distribution as shown in Table 141.

TABLE 141: SCREEN ANALYSIS OF RED MUD

Screen Mesh		Percent Dry Solids
	+10	0.0
-10	+20	0.2
-20	+50	0.8
-50	+100	0.8
-100	+200	0.8
-200	+325	1.9
-325		95.5

Source: Report EPA 440/1-73/019

Red mud wastes contain significant amounts of suspended solids and alkalinity. Depending upon the number of mud washing stages, the water associated with the mud may contain 3 to 10 grams per liter alkalinity (expressed as Na_2CO_3), and one-third gram per liter of sodium aluminate. (Convention in the U.S. alumina industry is to express total alkalinity, including both NaOH and Na_2CO_3.)

The ideal solution to the red mud problem would be to develop a use for it. An obvious possible application utilizes its high iron content (Table 142). A process based on sintering the red mud with carbon and limestone and melting the sinter in an electric furnace has been developed to produce a low-purity iron which could be further processed into steel. Although the basic process has been further developed, it has not yet found commercial application. It may have some economic value in countries which have bauxite refineries but produce little or no steel.

TABLE 142: RANGE OF CHEMICAL ANALYSES OF RED MUDS

	Weight Percent		
Component	Alcoa Mobile, Ala. (Surinam)	Reynolds Bauxite, Ark. (Arkansas)	Reynolds Corpus Christi, Texas (Jamaica)
Fe_2O_3	30-40	55-60	50-54
Al_2O_3	16-20	12-15	11-13
SiO_2	11-14	4-5	2.5-6
TiO_2	10-11	4-5	trace
CaO	5-6	5-10	6.5-8.5
Na_2O	6-8	2	1.5-5.0
Loss on ignition	10.7-11.4	5-10	10-13

Source: Report EPA 440/1-73/019

Other investigators have examined its applicability to the manufacture of portland cement, bricks, road construction, but no domestic markets have yet been developed which will economically justify processing the waste red mud. Total impoundment, however, affords the opportunity for reclamation of the red mud when an economic recovery process and adequate markets are developed.

Thermal efficiency of a bauxite refinery, economically important, depends on the efficiency of the heat exchanger used to transfer heat from hot- to cold-process streams. Many streams contain substantial dissolved solids, and scaling of exchanger surfaces is a recurring problem. Acid cleaning is universally used, usually with sulfuric, but also with small quantities of inhibited hydrochloric or acetic acid. Similar cleaning is used for the problem scaling of filtration equipment and filter cloths.

Normally, the resulting spent acid is primarily a solution high in sulfates, but with only low to moderate free acid concentrations. These sulfate solutions are disposed of by most plants to active or, preferably, to abandoned mud lakes where the neutralization is completed. In a few instances they are neutralized and discharged to waterways but this procedure is being replaced by impoundment. The quantities of sulfuric acid used are not large averaging about 1,000 to 2,000 pounds of acid per day.

Possible pollutants from the operation of barometric condensers are heat and alkali. As described earlier, sizable barometric condensers are found at nearly all bauxite refineries, where they are used on the evaporative coolers and the spent liquor evaporators. Illustrative of the heat duty are the data from one plant with five spent liquor evaporators. Here the barometric condensers averaged 126 liters per second (2,000 gallons per minute) and the temperature rise was approximately 14°C (25°F) for a total heat duty of about 31 million kilogram-calories per hour (125 million Btu/hr). Entrainment of significant amounts of alkali to the barometric condenser effluent should be negligible except during periods of upset or abnormal operation. However, barometric condenser effluents will tend to have a pH over 7.

Air compressor aftercoolers may also contribute heat to process streams. Compressed air is generally used in sizable amounts for agitation in the numerous precipitators in a bauxite refinery. Air from compressors is frequently passed through water-cooled aftercoolers to remove the heat of compression and cool the air. This service is a noncontact cooling application and the only pollutant the cooling water can acquire is heat.

Similarly, there may be other noncontact water-cooling applications, such as, seal rings on rotary calcining kilns, from which thermal discharges may result. Heat discharges from these ancillary services are nominal.

Several bauxite refineries operate rotary lime kilns to produce the lime needed to compensate for the carbonate accumulated in the process liquor, or for use in the combination process. Some plants use wet scrubbers on these lime kilns, from which a potential waste stream results, but the resultant hydrated lime slurry is invariably fed back to the process in order to utilize the contained lime, and is never discharged.

One miscellaneous waste stream from a bauxite refinery difficult to characterize is the "housekeeping" or "hose-down" stream. This results from minor spills and leaks and wastes resulting from cleanups. In most plants the in-plant drains are connected to the storm sewer, which may be discharged to the storm water lake or to the red mud lake.

In most plants all process areas where aqueous spills are possible are floored with concrete, and curb about 6 inches high surrounds the entire area. Any spill is thus contained for recovery and controlled disposal. Several other waste streams may also be associated with the operation of a bauxite refinery. Examples of these are: [1] sludge from treatment and softening of the raw intake water, [2] spent regenerant from ion-exchange treatment of intake water, [3] boiler blowdown, [4] cooling tower blowdown, and [5] treated sanitary waste effluent.

None of these streams is unique to bauxite refining. However, it should be noted that all of them fit very well into the total impoundment philosophy of disposal of process wastes from a bauxite refinery and would represent a one increment percent, to the normal red mud load. Effluent limitations for these streams have not been established inasmuch as they are not considered process waste streams. The characterization of process waste streams from the refining of bauxite is summarized in Table 143.

TABLE 143: CHARACTERIZATION OF PRINCIPAL WASTE STREAMS FROM U.S. BAUXITE REFINERIES

Waste	Quantity	Characterization
Red mud	500-3600 T/D (dry basis)	15-20 % solids
	1,000-7,200 T/D (wet, settled)	5-12 g/l soda
	3,000-20,000 T/D (slurry at 18% solids)	2-5 g/l aluminum
		pH - 12.5

(continued)

TABLE 143: (continued)

Waste	Quantity	Characterization
Spent Cleaning Acid	Variable, 5-10 T/week intermittently discharged	Na_2SO_4, plus some free H_2SO_4 HCl or HAc may also be used pH - 0
Salts from salting-out evaporator	Variable-estimated up to several thousand kg/day	Na_2SO_4 - alkaline pH - 12.5
Barometric condenser, C.W.*	Millions of liters/hr	Temp. rise of up to $15°C(25°F)$ may contain traces entrained alkali
Boiler and cooling tower blowdown	Variable - thousands of liters/day	Dilute alkaline solutions pH - 12.5
Water softener sludge	one to few T/D	Lime and suspended solids from intake water
Sanitary waste	375 liters/D/capita	B.O.D. 70 g(0.15 lb)/day/capita

*C.W.= Cooling Water

Source: Report EPA 440/1-73/019

Reference

(1) U.S. Environmental Protection Agency, "Development Document for Proposed Effluent Limitations Guidelines and New Source Performance Standards for the Bauxite Refining Subcategory of the Aluminum Segment of the Nonferrous Metals Manufacturing Point Source Category," *Report EPA 440/1-73/019,* Washington, D.C., October 1973.

BEET SUGAR MANUFACTURE

Water Pollution

Water is commonly used in a beet sugar processing plant for six principal reasons: [1] transporting (fluming) of beets to the processing operation, [2] washing beets, [3] processing (extraction of sugar from the beets), [4] transporting beet pulp and lime mud cake waste, [5] condensing vapors from evaporators and pans and [6] cooling.

The quantity of fresh water intake to plants ranges between 1,250 and 25,000 l/kkg (300 to 6,000 gal/ton) of beets sliced. Fresh water use is highly contingent upon in-plant water conservation practices and reuse techniques. Average water use in the industry approximates 9,200 l/kkg (2,200 gal/ton) of beets processed. Total water used, including reused water, varies much less and totals approximately 20,900 l/kkg (5,000 gal/ton) of beets sliced. Most of the water used in beet sugar processing plants is employed for condensing vapors from evaporators and for the conveying and washing of beets (see Table 144).

Since many process uses do not require water of high purity, considerable recirculation is possible without extensive treatment. The nature and amounts of these water reuses as influenced by in-plant controls and operational practices have a substantial effect on resulting wastewater quantities and characteristics. Reduction in water use with minimum wastewater volumes promises less difficulties in waste handling and disposal, and greater economy of treatment. Water uses for various operations in a beet sugar processing plant are further described below.

The most widely recognized and representative data of waste characterization for the beet sugar processing segment of the sugar processing industry is included in "An Industrial

Waste Guide to the Beet Sugar Industry" published by the U.S. Public Health Service. This waste data, circa 1950, is included in Table 144. The waste loads are representative of once-through water use without recycling or treatment. The data given in Table 144 serves as a reliable base for determining the total waste load potential of a beet sugar processing plant.

TABLE 144: REPRESENTATIVE WASTEWATER CHARACTERISTICS AND TOTAL FLOW DATA FOR A TYPICAL BEET SUGAR PROCESSING PLANT(*)

Waste Source	Flow l/kkg of beets sliced (gal/ton)	BOD_5(mg/l)	BOD_5 kg/kkg of beets sliced (lb/ton)	Suspended Solids(mg/l)	Suspended Solids kg/kkg of beets sliced (lb/ton)
Flume Water	10,842 (2600)	210	2.25 (4.5)	800-4,300	8.5-41.5 (17-93)
Process Water					
Screen (Pulp Transport) Water	1668 (400)	910	1.50 (3.0)	1,020	1.7 (3.4)
Press Water	751 (180)	1,700	1.30 (2.6)	420	0.3 (0.6)
Silo Water	876 (210)	7,000	6.15 (12.3)	270	0.25 (0.5)
Lime Mud Slurry**	375 (90)	8,600	3.25 (6.5)	120,000	45 (90)
Condenser Water	8340 (2000)	40	0.35 (0.7)	-	
Steffen Filtrate	500 (120)	10,500	5.20 (10.4)	100-700	0.05-0.35 (0.1-0.7)
Totals	23,352 (5600)		20.0 (40.0)		55.8-94.1 (111.6-188.2)

(*) All values are based upon no recirculation or treatment of waste waters

(**) Relates to non-Steffen or straight house process.

Source: Report EPA 440/1-73/002

Because of the wide diversity of in-plant control, recycling, and treatment practices at present beet sugar processing plants, the data in Table 144 does not reflect the combination of conditions existing at any single plant within the industry today. The data does reflect total waste load and wastewater flow values associated with the individual waste source components, which may be predicably amended by various methods of controlling and handling these individual wastewater sources within the industry. The total potential waste load and water requirement attributed to each of the waste producing production processes has particular significance and constancy throughout the industry. In addition to providing a baseline of total pollutional load attributed to individual waste components, the data also serve to provide a basis for comparison between former and current waste-handling techniques.

The former practice of beet sugar processing plants of discharging wastes containing between 15 and 20 kilograms BOD_5/kkg (30 to 40 lb/ton) of beets sliced had been reduced to an average of less than 2.5 kilograms (5 lb) by 1968. A further reduction in BOD_5 load has taken place in most recent years with all plants soon to accomplish a discharge from 0 to less than 1.0 kilogram BOD_5/kkg (2.0 lb/ton) of beets sliced to surface streams. The total waste discharge to streams from the entire beet sugar processing industry in the U.S. in 1968 was estimated at about 215 billion liters (57 billion gallons) which contained a total of about 37 million kilograms (82 million pounds) of BOD_5. However, the 24 million kilokilograms (26 million-ton) crop in 1968 was unusually large. A more normal crop would have been about 20 million kilokilograms (22 million tons) of beets processed. A number of plants currently recycle much of the flume and condenser waters, and some plants do not discharge any wastewater to navigable waters at all.

Every beet sugar processing plant today employs some degree of wastewater recycling or reuse. Under present practices, process waters (pulp screen water, pulp press water, and pulp silo drainage), Steffen waste, and lime mud slurry have essentially been eliminated as polluting waste sources in terms of discharge to navigable waters. Process waters are universally recycled within the plant, Steffen waste is disposed of with by-product use or land disposal, and lime mud slurry receives land disposal. Flume water and barometric condenser water are presently two primary polluting sources.

Reference

(1) U.S. Environmental Protection Agency, "Development Document for Proposed Effluent Limitations Guidelines and New Source Performance Standards: Beet Sugar Segment of the Sugar Processing Point Source Category," *Report EPA 440/1-73/002,* Washington, D.C., August 1973.

BOAT AND SHIP OPERATION

Inboard Powered Vessels

Fuel oil is the primary fuel used in vessels powered by inboard engines, power steamships, motor ships, and gas turbine powered ships(1). Gas turbines presently are not in wide-spread use and thus are not included in this section. Within the next few years, however, their use may become increasingly common(2)(3). Steamships are any ships that have steam turbines driven by an external combustion engine. Motor ships, on the other hand, have internal combustion engines operated on the diesel cycle.

Outboard Powered Vessels

Most of the approximately 7 million outboard motors in use in the United Stares are 2-stroke engines with an average available horsepower of about 25. Because of the predominately leisure time use of outboard motors, emissions related to their operation occur primarily during nonworking hours, in rural areas, and during the three summer months. Nearly 40% of the outboards are operated in the states of New York, Texas, Florida, Michigan, California, and Minnesota. This distribution results in the concentration of a large portion of total nationwide outboard emissions in these states(1).

Air Pollution

The air pollutant emissions resulting from inboard powered vessel operations may be divided into two categories: emissions that occur as the ship is under way and emissions that occur when the ship is dockside or in-berth. Underway emissions may vary considerably for vessels that are maneuvering or docking because of the varying fuel consumption. During such a time a vessel is operated under a wide range of power demands for a period of 15 minutes to 1 hour. The high demand may be 15 times the low demand; however, once the vessel has reached and sustained a normal operation speed, the fuel consumed is reasonably constant. Table 145 shows that 29 to 65 gallons of fuel oil is consumed per nautical mile (60 to 133 liters per kilometer) for steamships and 7 to 30 gallons of oil per nautical mile (14 to 62 liters per kilometer) for motorships.

Unless a ship receives auxiliary steam provided by the port, goes immediately into drydock, or is out of operation after arrival in port, she continues her emissions at dockside. Power must be generated for the ship's light, heat, pumps, refrigeration, ventilation, etc. A few steamships use auxiliary engines to supply power, but they generally operate one or two main boilers under reduced draft and lowered fuel rates, a much less efficient process. Motor ships generally use diesel-powered generators to furnish auxiliary power.

As shown in Table 145, fuel oil consumption at dockside varies appreciably. Based on the data presented in this table and the emission factors for residual fuel oil combustion and diesel oil combustion, emission factors have been determined for vessels and are presented in Table 146.

TABLE 145: FUEL CONSUMPTION RATES FOR STEAMSHIPS AND MOTOR SHIPS

Fuel consumption	Steamships		Motor ships	
	Range	Average	Range	Average
Underway				
lb/hphr	0.51 to 0.65	0.57	0.28 to 0.44	0.34
kg/hphr	0.23 to 0.29	0.26	0.13 to 0.20	0.15
gal/naut. mi	29 to 65	44	7 to 30	19
liters/km	59.4 to 133.0	90	14 to 62	38.8
In-berth				
gal/day	840 to 3,800	1,900	240 to 1,260	660
liters/day	3,192 to 14,400	7,200	910 to 4,800	2,500

Source: EPA Publication AP-42, 2nd Edition

TABLE 146: EMISSION FACTORS FOR INBOARD VESSELS, EMISSION FACTOR RATING: D

Pollutant	Steamships[a]				Motor ships[b]			
	Underway		In-berth		Underway		In-berth	
	lb/mi	kg/km	lb/day	kg/day	lb/mi	kg/km	lb/day	kg/day
Particulate	0.4	0.098	15	6.8	2	0.49	16.5	7.5
Sulfur dioxide[c]	7S	1.71S	300S	136S	(SO_x) 1.5	0.37	43	19.5
Sulfur trioxide[c]	0.1S	0.02S	4S	1.8S	—	—	—	—
Carbon monoxide	0.002	0.0005	0.08	0.036	1.2	0.29	46	20.8
Hydrocarbons	0.2	0.05	9	4.1	0.9	0.22	33	14.9
Nitrogen oxides (NO_2)	4.6	1.13	200	90.7	1.4	0.34	50	22.7
Aldehydes (HCHO)	0.04	0.01	2	0.9	0.07	0.017	2.6	1.2

[a]Based on data in Table 145 and emission factors for fuel oil.
[b]Based on data in Table 145 and emission factors for diesel fuel.
[c]S = weight percent sulfur in fuel, assumed to be 0.5% for diesel.

Source: EPA Publication AP-42, 2nd Edition

Because the vast majority of outboards have underwater exhaust, emission measurement is very difficult. The values presented in Table 147 are the approximate atmospheric emissions from outboards. These data are based on tests of four outboard motors ranging from 4 to 65 horsepower(4). The emission results from these motors are a composite based on the nationwide breakdown of outboards by horsepower.

Emission factors are presented two ways in this section: in terms of fuel use and in terms of work output (horsepower hour). The selection of the factor used depends on the source inventory data available. Work output factors are used when the number of outboards in use is available. Fuel-specific emission factors are used when fuel consumption data are obtainable.

In the following table the term hydrocarbons includes exhaust hydrocarbons only. No crankcase emissions occur because the majority of outboards are 2-stroke engines that use crankcase induction. Evaporative emissions are limited by the widespread use of unvented tanks.

TABLE 147: AVERAGE EMISSION FACTORS FOR OUTBOARD MOTORS[a],
EMISSION FACTOR RATING: B

Pollutant[b]	Based on fuel consumption		Based on work output[c]	
	lb/10^3 gal	kg/10^3 liter	g/hphr	g/metric hphr
Sulfur oxides[d] (SO_x as SO_2)	6.4	0.77	0.49	0.49
Carbon monoxide	3300	400	250	250
Hydrocarbons	1100	130	85	85
Nitrogen oxides (NO_x as NO_2)	6.6	0.79	0.50	0.50

[a] Reference(4). Data in this table are emissions to the atmosphere. A portion of the exhaust remains behind in the water.

[b] Particulate emission factors are not available because of the problems involved with measurement from an underwater exhaust system but are considered negligible.

[c] Horsepower hours are calculated by multiplying the average power produced during the hours of usage by the population of outboards in a given area. In the absence of data specific to a given geographic area, the hphr value can be estimated using average nationwide values from Reference(4). Reference(4) reports the average power produced (not the available power) as 9.1 hp and the average annual usage per engine as 50 hours. Thus, hphr = (number of outboards) (9.1 hp) (50 hours/outboard-year). Metric hphr = 0.9863 hphr.

[d] Based on fuel sulfur content of 0.043 percent from Reference 2 and on a density of 6.17 lb/gal.

Source: EPA Publication AP-42, 2nd Edition

Water Pollution

Specific data were not available on water pollution from boat and ship operation. Such pollution obviously occurs and includes both sanitary wastes and oily wastes resulting from tank cleaning, oil spills, etc.

References

(1) Pearson, J.R., Ships as Sources of Emissions, Presented at the Annual Meeting of the Pacific Northwest International Section of the Air Pollution Control Association, Portland, Oregon, November 1969.

(2) Standard Distillate Fuel for Ship Propulsion, Report of a Committee to the Secretary of the Navy, U.S. Department of the Navy, Washington, D.C., October 1968.

(3) "GTS Admiral William M. Callahan Performance Results," *Diesel and Gas Turbine Progress,* 35 (9):78, September 1969.

(4) Hare, C.T. and Springer, K.J., Exhaust Emissions from Uncontrolled Vehicles and Related Equipment Using Internal Combustion Engines, Part II, Outboard Motors, Final Report, Southwest Research Institute, San Antonio, Texas. Prepared for the Environmental Protection Agency, Research Triangle Park, N.C., under Contract Number EHS 70-108, January 1973.

(5) Hare, C.T. and Springer, K.J., Study of Exhaust Emissions from Uncontrolled Vehicles and Related Equipment Using Internal Combustion Engines, Emission Factors and Impact Estimates for Light-Duty Air-Cooled Utility Engines and Motorcycles, Southwest Research Institute, San Antonio, Texas. Prepared for the Environmental Protection Agency, Research Triangle Park, N.C., under Contract Number EHS 70-108, January 1972.

BRASS AND BRONZE INGOT PRODUCTION (SECONDARY COPPER INDUSTRY)

Obsolete domestic and industrial copper-bearing scrap is the basic raw material of the brass and bronze ingot industry. The scrap frequently contains any number of metallic and non-metallic impurities, which can be removed by such methods as hand sorting, magnetizing, heat methods such as sweating or burning, and gravity separation in a water medium. Brass and bronze ingots are produced from a number of different furnaces through a combination of melting, smelting, refining, and alloying of the processed scrap material. Reverberatory, rotary, and crucible furnaces are the ones most widely used, and the choice depends on the size of the melt and the alloy desired. Both the reverberatory and the rotary furnaces are normally heated by direct firing, in which the flame and gases come into direct contact with the melt. Processing is essentially the same in any furnace except for the differences in the types of alloy being handled. Crucible furnaces are usually much smaller and are used principally for special-purpose alloys.

Air Pollution

The principal source of emissions in the brass and bronze ingot industry is the refining furnace. The exit gas from the furnace may contain the normal combustion products such as fly ash, soot, and smoke. Appreciable amounts of zinc oxide are also present in this exit gas. Other sources of particulate emissions include the preparation of raw materials and the pouring of ingots.

The only air pollution control equipment that is generally accepted in the brass and bronze ingot industry is the baghouse filter, which can reduce emissions by as much as 99.9%. Table 148 summarizes uncontrolled emissions from various brass and bronze melting furnaces(1).

TABLE 148: PARTICULATE EMISSION FACTORS FOR BRASS AND BRONZE MELTING FURNACES WITHOUT CONTROLS[a], EMISSION FACTOR RATING: A

Type of furnace	Uncontrolled emissions[b]	
	lb/ton	kg/MT
Blast[c]	18	9
Crucible	12	6
Cupola	73	36.5
Electric induction	2	1
Reverberatory	70	35
Rotary	60	30

[a]Reference(2). Emission factors expressed as units per unit weight of metal charged.
[b]The use of a baghouse can reduce emissions by 95 to 99.6 percent.
[c]Represents emissions following precleaner.

Source: EPA Publication AP-42, 2nd Edition

References

(1) U.S. Environmental Protection Agency, "Compilation of Air Pollutant Emission Factors," 2nd Edition, *Publication Number AP-42,* Research Triangle Park, N.C., Office of Air and Water Programs, April 1973.
(2) "Air Pollution Aspects of Brass and Bronze Smelting and Refining Industry," U.S. DHEW, PHS, EHS, National Air Pollution Control Administration, Raleigh, N.C., *Publication Number AP-58,* November 1969.

BRICK AND RELATED CLAY PRODUCTS MANUFACTURE

The manufacture of brick and related products such as clay pipe, pottery, and some types of refractory brick involves the mining, grinding, screening, and blending of the raw materials, and the forming, cutting or shaping, drying or curing, and firing of the final product. Surface clays and shales are mined in open pits; most fine clays are found underground. After mining, the material is crushed to remove stones and stirred before it passes onto screens that are used to segregate the particles by size.

At the start of the forming process, clay is mixed with water, usually in a pug mill. The three principal processes for forming brick are: stiff-mud, soft-mud, and dry-process. In the stiff-mud process, sufficient water is added to give the clay plasticity; bricks are then formed by forcing the clay through a die and using cutter wire to separate the bricks. All structural tile and most brick are formed by this process. The soft-mud process is usually used when the clay contains too much water for the stiff-mud process. The clay is mixed with water until the moisture content reaches 20 to 30%, and the bricks are formed in molds. In the dry-press process, the clay is mixed with a small amount of water and formed in steel molds by applying a pressure of 500 to 1,500 psi.

Before firing, the wet clay units that have been formed are almost completely dried in driers that are usually heated by waste heat from the kilns. Many types of kilns are used for firing brick; however, the most common are the tunnel kiln and the periodic kiln. The downdraft periodic kiln is a permanent brick structure that has a number of fireholes where fuel is fired into the furnace. The hot gases from the fuel are drawn up over the bricks, down through them by underground flues, and out of the oven to the chimney. Although fuel efficiency is not as high as that of a tunnel kiln because of lower heat recovery, the uniform temperature distribution through the kiln leads to a good quality product. In most tunnel kilns, cars carrying about 1,200 bricks each travel on rails through the kiln at the rate of one 6-foot car per hour. The fire zone is located near the middle of the kiln and remains stationary.

In all kilns, firing takes place in six steps: evaporation of free water, dehydration, oxidation, vitrification, flashing, and cooling. Normally, gas or residual oil is used for heating, but coal may be used. Total heating time varies with the type of product; for example, 9-inch refractory bricks usually require 50 to 100 hours of firing. Maximum temperatures of about 2000°F (1090°C) are used in firing common brick.

Air Pollution

Particulate matter is the primary emission in the manufacture of bricks. The main source of dust is the material handling procedure, which includes drying, grinding, screening, and storing the raw material. Combustion products are emitted from the fuel consumed in the curing, drying, and firing portion of the process.

Fluorides, largely in gaseous form, are also emitted from brick manufacturing operations. Sulfur dioxide may be emitted from the bricks when temperatures reach 2500°F (1370°C) or greater; however, no data on such emissions are available.

A variety of control systems may be used to reduce both particulate and gaseous emissions. Almost any type of particulate control system will reduce emissions from the material handling process, but good plant design and hooding are also required to keep emissions to a minimum.

The emissions of fluorides can be reduced by operating the kiln at temperatures below 2000°F (1090°C) and by choosing clays with low fluoride content. Satisfactory control can be achieved by scrubbing kiln gases with water; wet cyclonic scrubbers are available that can remove fluorides with an efficiency of 95%, or higher.

Emission factors for brick manufacturing are presented in Table 149 which appears on the following page. Insufficient data are available to present particle size information.

TABLE 149: EMISSION FACTORS FOR BRICK MANUFACTURING WITHOUT CONTROLS[a], EMISSION FACTOR RATING: C

Type of process	Particulates		Sulfur oxides (SO$_x$)		Carbon monoxide (CO)		Hydrocarbons (HC)		Nitrogen oxides (NO$_x$)		Fluorides[b] (HF)	
	lb/ton	kg/MT	lb/ton	kg/MT	lb/ton	kg/MT	lb/ton	kg/MT	lb/ton	kg/MT	lb/ton	kg/MT
Raw material handling[c]												
Dryers, grinders, etc.	96	48	–	–	–	–	–	–	–	–	–	–
Storage	34	17	–	–	–	–	–	–	–	–	–	–
Curing and firing[d]												
Tunnel kilns												
Gas-fired	0.04	0.02	Neg[e]	Neg	0.04	0.02	0.02	0.01	0.15	0.08	1.0	0.5
Oil-fired	0.6	0.3	4.0S[f]	2.0S	Neg	Neg	0.1	0.05	1.1	0.55	1.0	0.5
Coal-fired	1.0A	0.5A[g]	7.2S	3.6S	1.9	0.95	0.6	0.3	0.9	0.45	1.0	0.5
Periodic kilns												
Gas-fired	0.11	0.05	Neg	Neg	0.11	0.05	0.04	0.02	0.42	0.21	1.0	0.5
Oil-fired	0.9	0.45	5.9S	2.95S	Neg	Neg	0.1	0.05	1.7	0.85	1.0	0.5
Coal-fired	1.6A	0.8A	12.0S	6.0S	3.2	1.6	0.9	0.45	1.4	0.70	1.0	0.5

[a]One brick weighs about 6.5 pounds (2.95 kg). Emission factors expressed as units per unit weight of brick produced.
[b]Based on data from Reference(1).
[c]Based on data from sections on ceramic clays and cement manufacturing in this publication. Because of process variation, some steps may be omitted. Storage losses apply only to that quantity of material stored.
[d]Based on data from Reference(1) and emission factors for fuel combustion.
[e]Negligible.
[f]S is the percent sulfur in the fuel.
[g]A is the percent ash in the coal.

Source: EPA Publication AP-42, 2nd Edition

Reference

(1) U.S. Environmental Protection Agency, "Compilation of Air Pollutant Emission Factors," 2nd Edition, *Publication Number AP-42,* Research Triangle Park, N.C., Office of Air and Water Programs, April 1973.

BUILDING PAPER MANUFACTURE

Water Pollution

A building paper and/or roofing felt mill utilizes water in its process, exclusive of steam generation, for the following purposes.

[1] To act as an agent for separating the raw materials into discrete fibers which is essential for the formation of the end product, the removal of contaminants and undesirable fibers from the stock system, and the control and metering of stock to the paper machine. This water, which is generally recycled, acts as a vehicle for transporting the fiber to the process.

[2] To clean those areas, particularly on the wet end of the machine, which tend to develop fiber buildup. These areas are the paper-forming section of the machine and the felts used to carry the formed sheet through the machine and press sections. This water enters the system via shower nozzles and represents the largest contribution to the volume of raw wastewater generated since it is nearly all excess water in terms of process water needs.

[3] To keep production equipment throughout the mill operational or permit the equipment to perform its design function. Typical applications are the seal and cooling waters used on pumps, agitators, drives, bearings, vacuum pumps, and process controls. Also cooling water is required by those mills that include the asphalt saturating process for the production of roofing felts and shingles. This water represents the second largest contributor to the volume of wastewater generated by the process.

[4] To supply emergency makeup water, under automatic control, to various storage tanks to avoid operational problems resulting in reduced production or complete mill shutdown.

[5]   To provide power boiler condenser, heat exchange condensate, and noncontact
      cooling water that can be segregated and discharged separately without treat-
      ment. However, there are many mills that still permit all or part of this water
      to enter the wastewater sewer system which increases the volume of water
      requiring treatment.

The manufacture of building paper involves three relatively discrete process systems in
terms of quantity and quality of water utilization: stock preparation and the wet end
and dry end of the machine.

The stock preparation area uses water for purposes described in Items [1], [3], [4], and
[5] above. Water in the form of steam may also be used directly to maintain stock temperature which contributes to the volume of wastewater generated since it represents excess
water in terms of the process water balance.

Process water is mixed with baled waste paper in the pulper or beater and the resulting
slurry is then carried through the stock cleaning system where additional process water is
introduced. The stock is then thickened to increase consistency for refining or jordaning
(fiber control). The process water removed by the thickener or decker is recirculated back
to the pulper and cleaning system. A mill utilizing wood flour instead of wood pulp from
an attrition mill adds the flour in the above waste paper stock system ahead of the jordans
or refiners. However, those that use wood chips and/or rags and/or inorganic materials such
as asbestos require a preparation process for each type of furnish used. These are generally
low-volume water users although each system contributes to the waste load generated. The
various stock components are blended and passed through the refiners and discharged to
a machine stock chest.

In the wet end areas, the stock is pumped to a head box which meters the quantity of
stock of the paper machine. At this point process water is added to reduce the stock
consistency to 0.25 to 0.5% in the vat which is the forming section of the machine. The
stock deposits on a cylinder wire and the excess machine white water passes through the
wire. A large portion of this white water is recycled back through the machine stock loop
and the excess is pumped to a white water collection chest for reuse in the stock preparation area. It is on the wet end that excess water is created by the use of fresh water
showers as described in Item [2]. The sheet is carried by felts to the press sections where
additional quantities of water are removed. Felt cleaning showers add more excess water,
but are necessary for the maintenance of the drainability of the felt.

The sheet passes through the drier section to the dry end where water use is generally low
in volume consisting principally of cooling water and sheet moisture control. The product
at this point may be the finished product or it may be subject to additional processes in
the mill. For some products, the saturating process is the next waste-generating step after
the papermaking process. However, the production of deadening or flooring felts from the
paper produced does not require processing which generates a wastewater load.

The paper is carried through one or two stations for asphalt saturation and application of
a coat of talc on one side of the sheet. This requires the utilization of cooling water applied
by spray nozzles after each saturation which represents the waste load sewered from the
area. This process has the capability of making roofing shingles as well as roofing felts;
therefore, a section for coating the saturated felt with a granular stone and/or mica is part
of the operation. These particles fall to the floor and are washed to the sewer and represent
the principal source of inert suspended solids in the wastewater generated in the area.

Definitive data on individual waste loads from each of the above process sources have
been difficult to develop. First, many, if not the majority, of the mills in this subcategory change raw materials and products manufactured in response to short-term pricing,
availability, and demand. A great variety of process options may be used in even a single
mill in response to these factors. Second, the pronounced tendency in these mills toward
increased recycle could erroneously attribute a waste load to one unit process which actually

originated in another. Such recycle, as explained below, reduces pollutant levels in the raw waste and in the final discharge.

Definition of "total raw waste load" from mills in this subcategory is subject to interpretation dependent upon the particular scheme of recycle used. Three principal schemes have been identified, each being effective insofar as reduction of final discharge pollutants is concerned, and each dependent upon product quality, mill layout, and other factors:

[1] An internal device such as a save-all or DSM screen is used to remove suspended solids. Both the solids and the clarified process water may then be recycled, at least in part, resulting in a low raw waste level of suspended solids.

[2] An external device such as a mechanical clarifier is used to serve the same functions. The influent to the clarifier may technically be called raw waste but any effluent not reused would be the definition comparable to scheme [1].

[3] The third scheme relies principally upon internal recycle, with internal or external storage facilities to hold surge flows due to grade changes and other process upsets. Most of these surge flows are then returned to the process as production equilibrium is again approached, with only a small and sometimes intermittent final waste flow occurring.

Thus, raw waste loads from mills in this subcategory vary widely, depending upon the definition used. Data developed in 1971 illustrate this point. Of 13 mills in this subcategory, raw waste suspended solids varied typically from 2.5 kilograms per metric ton (5 pounds per short ton) to 30 kilograms per metric ton (60 pounds per short ton).

Raw waste suspended solids for the two exemplary mills ranged from 4 kg/kkg (8 lb/ton) to 42 kg/kkg (84 lb/ton). Raw waste BOD_5 for the 2 exemplary mills ranged from 7 kg/kkg (14 lb/ton) to 15 kg/kkg (30 lb/ton). The above raw waste characteristics are shown in Table 150.

TABLE 150: RAW WASTE CHARACTERISTICS

Mill	BOD_5 kg/kkg (lb/ton)	TSS kg/kkg (lb/ton)
a*	15 (30)	41 (82)
a**	9.5 (19)	42 (84)
b**	7.2 (14.3)	4.1 (8.3)

*Mill records
**Short-term survey data (3 to 7 days)

Source: Report EPA 440/1-74/026

Although no definition of total raw waste load fits all cases, the "primary effluent not recycled" probably meets most field conditions as the best definition. Final effluent flow is a measure of the degree of reuse employed by a given mill. The first surveyed mill employed extensive recycle and used only 4,200 liters per metric ton (1,000 gallons per short ton) during the 4 days of the survey. The second mill, which did not employ extensive recycle, used 54,000 liters per metric ton (13,000 gallons per short ton) during the survey.

Longer term data from the 13 mills mentioned above show a wide variation in water usage, primarily as a function of recycle. The typical range among these mills was from 8,400 liters per metric ton (2,000 gallons per short ton) to 42,000 liters per metric ton (10,000 gallons per short ton).

Reference

(1)   U.S. Environmental Protection Agency, "Development Document for Proposed
      Effluent Limitations Guidelines and New Source Performance Standards for the
      Builders Paper and Roofing Felt Segment of the Builders Paper and Board Mills Point
      Source Category," *Report EPA 440/1-74/026,* Washington, D.C., January 1974.

CALCIUM CARBIDE MANUFACTURE

Calcium carbide is manufactured by heating a mixture of quicklime (CaO) and carbon in
an electric-arc furnace, where the lime is reduced by the coke to calcium carbide and carbon monoxide(1). Metallurgical coke, petroleum coke, or anthracite coal is used as the
source of carbon. About 1,900 pounds (860 kg) of lime and 1,300 pounds (600 kg) of
coke yield 1 ton (1 MT) of calcium carbide. There are two basic types of carbide furnaces:
(1) the open furnace, in which the carbon monoxide burns to carbon dioxide when it comes
in contact with air above the charge; and (2) the closed furnace, in which the gas is collected
from the furnace. The molten calcium carbide from the furnace is poured into chill cars
or bucket conveyors and allowed to solidify. The finished calcium carbide is dumped into
a jaw crusher and then into a cone crusher to form a product of the desired size.

Air Pollution

Particulates, acetylene, sulfur compounds, and some carbon monoxide are emitted from
the calcium carbide plants. Table 151(2) contains emission factors based on one plant in
which some particulate matter escapes from the hoods over each furnace and the remainder
passes through wet-impingement-type scrubbers before being vented to the atmosphere
through a stack. The coke dryers and the furnace room vents are also sources of emissions.

TABLE 151: EMISSION FACTORS FOR CALCIUM CARBIDE PLANTS[a], EMISSION
FACTOR RATING: C

Type of source	Particulates		Sulfur oxides		Acetylene	
	lb/ton	kg/MT	lb/ton	kg/MT	lb/ton	kg/MT
Electric furnace						
Hoods	18	9	—	—	—	—
Main stack	20	10	3	1.5	—	—
Coke dryer	2	1	3	1.5	—	—
Furnace room vents	26	13	—	—	18	9

[a]Reference 3. Emission factors expressed as units per unit weight of calcium carbide produced.

Source: EPA Publication AP-42, 2nd Edition

References

(1)   "Carbide," In: *Kirk-Othmer Encyclopedia of Chemical Technology,* New York, John
      Wiley and Sons, Inc., 1964
(2)   U.S. Environmental Protection Agency, "Compilation of Air Pollutant Emission
      Factors," 2nd Edition, *Publication Number AP-42,* Research Triangle Park, N.C.,
      Office of Air and Water Programs, April 1973.
(3)   The Louisville Air Pollution Study, U.S. DHEW, PHS, R.A. Taft, Sanitary Engineer-
      ing Center, Cincinnati, Ohio, 1961.

CALCIUM CHLORIDE MANUFACTURE

Calcium chloride is produced by extraction from natural brines. Some material is also recovered as a by-product of soda ash manufacture by the Solvay process.

Water Pollution

In the manufacture of calcium chloride from brines, the salts are solution-mined and the resulting brines are first partly evaporated to remove sodium chloride by precipitation. The brine is further purified by addition of other materials to remove sodium, potassium and magnesium salts by precipitation and further evaporation, and is then evaporated to dryness to recover calcium chloride which is packaged and sold. Bromides and iodides are first separated from the brines before sodium chloride recovery is performed. There is a large degree of brine recycling to remove most sodium chloride values. The composition of the brine is:

$CaCl_2$	19.3%
$MgCl_2$	3.1%
NaCl	4.9%
KCl	1.4%
Bromides	0.25%
Other minerals	0.5%
Water	Balance

The raw wastes expected from calcium chloride manufacture arise from blowdowns as well as from the several partial evaporation steps used. Most of the wastes are weak brine solutions:

Waste Products	Process Source	Average kg/kkg of Product (lb/ton)
NH_3	Evaporators	0.55 (1.1)
$CaCl_2$	Evaporators	29 (58)
NaCl	Evaporators	0.5 (1.0)
$CaCl_2$	Packaging	0.7 (1.4)
NaCl and KCl*	Brine separation	45.5 (91)
NaCl*	Secondary brine separation	110 (220)

*Recycled or used elsewhere.

The waste brine streams are passed through an activated sludge treatment to remove organics and are then passed to a settling basin to remove suspended matter, adjusted to neutral pH, fed into a second pond to further settle suspended matter, and finally discharged. Future plans call for changes in the evaporators to reduce calcium chloride discharges and eliminate ammonia from the discharges. More recycling of spent brines is also planned. The effluent consists mostly of weak brine solutions (neutral pH). These discharges are expected to be reduced in the near future.

In the calcium chloride recovery process, the waste stream is first cycled through a number of partial evaporation and filtration steps to concentrate the waste solutions. After this, further partial evaporation is used to selectively remove the sodium chloride from solution and then total evaporation is used to recover calcium chloride from the remaining solution. Table 152 shows the raw wastes produced in this recovery operation and some other data. The principal waste is a contaminated sodium chloride coproduct which is discarded, as well as some calcium chloride from condensates and spills.

The present recovery unit reduces the effluent calcium chloride by about 21%. This is because of the limited market for calcium chloride. According to the manufacturer, if more of the material could be marketed, more would be recovered. Thus, the major problem with soda ash wastes lies in finding a use or disposal for the by-product calcium chloride.

An evaporation process for its recovery, as can be seen from this discussion, is already operative. This recovery step, as it is now practiced, also reduces the sodium chloride effluent of the Solvay process by 4%.

TABLE 152: CALCIUM CHLORIDE RECOVERY PROCESS

A. Raw Materials for Product

1. Soda ash distiller waste
2. Chlorine
3. Carbon dioxide 40% CO_2
4. Captive steam and power

B. Raw Waste Loads

Waste Products	Process Source	kg/kkg (1b/ton) of Product*
1. Ash and cinders	Steam and power	42.5(85)
2. Water purification sludge	Steam	0.75(1.5)
3. NaCl co-product	Evaporation	235(470)
4. CaCl2	Condensates and spills	35-50(70-100)

C. Comments

Ratio of $CaCl_2$ to NaCl available in distiller waste is approximately 1.4. Market demand at this location is at a ratio of 10.6 to 1.

---

*Product is 100% calcium chloride.

Source: Report EPA 440/1-73/007

Reference

(1) U.S. Environmental Protection Agency, "Development Document for Proposed Effluent Limitations Guidelines and New Source Performance Standards for the Major Inorganic Products Segment of the Inorganic Chemicals Manufacturing Point Source Category," Report EPA 440/1-73/007, Washington, D.C., August 1973.

CARBON BLACK MANUFACTURE

Carbon black is produced by the reaction of hydrocarbon fuel such as oil or gas, or both, with a limited supply of air at temperatures of 2500° to 3000°F (1370° to 1650°C). Part of the fuel is burned to CO_2, CO, and water, thus generating heat for the combustion of fresh feed. The unburned carbon is collected as a black fluffy particle. The three basic processes for producing this compound are the furnace process, accounting for about 83% of production; the older channel process, which accounts for about 6% of production; and the thermal process.

Channel Black Process

In the channel black process, natural gas is burned with a limited air supply in long, low buildings. The flame from this burning impinges on long steel channel sections that swing continuously over the flame. Carbon black is deposited on the channels, is scraped off, and falls into collecting hoppers. The combustion gases containing the solid carbon that is not collected on the channels, in addition to carbon monoxide and other combustion products, are then vented directly from the building.

Approximately 1 to 1.5 pounds of carbon black is produced from the 32 pounds of carbon available in 1,000 cubic feet of natural gas (16 to 24 kilograms carbon black from the 513 kilograms in 1,000 cubic meters) (2)(3)(4). The balance is lost as CO, CO_2, hydrocarbons and particulates.

Furnace Process

The furnace process is subdivided into either the gas or oil process depending on the primary fuel used to produce the carbon black. In either case, the fuel—gas in the gas process or gas and oil in the oil process—is injected into a reactor with a limited supply of combustion air. The combustion gases containing the hot carbon are then rapidly cooled to a temperature of about 500°F (260°C) by water sprays and by radiant cooling.

The largest and most important portion of the furnace process consists of the particulate or carbon black removal equipment. While many combinations of control equipment exist, an electrostatic precipitator, a cyclone, and a fabric filter system in series are most commonly used to collect the carbon black. Gaseous emissions of carbon monoxide and hydrocarbons are not controlled in the United States.

Thermal Black Process

In thermal black plants, natural gas is decomposed by heat in the absence of air or flame. In this cyclic operation, methane is pyrolyzed or decomposed by passing it over a heated brick checkerwork at a temperature of about 3000°F (1650°C). The decomposed gas is then cooled and the carbon black removed by a series of cyclones and fabric filters.

The exit gas, consisting largely of hydrogen (85%), methane (5%), and nitrogen, is then either recycled to the process burners or used to generate steam in a boiler. Because of the recycling of the effluent gases, there are essentially no atmospheric emissions from this process, other than from product handling.

Air Pollution

Table 153 presents the emission factors from the various carbon black processes. Nitrogen oxide emissions are not included but are believed to be low because of the lack of available oxygen in the reaction.

TABLE 153: EMISSION FACTORS FOR CARBON BLACK MANUFACTURING[a]
Emission Factor Rating: C

Type of process	Particulate		Carbon monoxide		Hydrogen sulfide		Hydrocarbons[b]	
	lb/ton	kg/MT	lb/ton	kg/MT	lb/ton	kg/MT	lb/ton	kg/MT
Channel	2,300	1,150	33,500	16,750	—	—	11,500	5,750
Thermal	Neg	Neg	Neg	Neg	Neg	Neg	Neg	Neg
Furnace								
Gas	c	c	5,300	2,650	—	—	1,800	900
Oil	c	c	4,500	2,250	38S[d]	19S[d]	400	200
Gas or oil	220[e]	110[e]						
	60[f]	30[f]						
	10[g]	5[g]						

[a]Based on data in References 2, 3, 5, and 6.

[b]As methane.

[c]Particulate emissions cannot be separated by type of furnace and are listed for either gas or oil furnaces.

[d]S is the weight percent sulfur in feed.

[e]Overall collection efficiency was 90 percent with no collection after cyclone.

[f]Overall collection efficiency was 97 percent with cyclones followed by scrubber.

[g]Overall collection efficiency was 99.5 percent with fabric filter system.

Source: EPA Publication AP-42, 2nd Edition

References

(1) U.S. Environmental Protection Agency, "Compilation of Air Pollutant Emission Factors," 2nd Edition, *Publication No. AP-42,* Research Triangle Park, N.C., Office of Air and Water Programs (April 1973).
(2) Drogin, I., "Carbon Black," *J. Air Pol. Control Assoc.* 18, 216–228, (April 1968).
(3) Cox, J.T., "High Quality, High Yield Carbon Black," *Chem. Eng.* 57, 116–117 (June 1950).
(4) Shreve, R.N., *Chemical Process Industries,* 3rd Ed. New York, McGraw-Hill Book Company, 1967, p 124–130.
(5) Reinke, R.A. and Ruble, T.A., "Oil Black," *Ind. Eng. Chem.* 44, 685–694 (April 1952).
(6) Allan, D.L., "The Prevention of Atmospheric Pollution in the Carbon Black Industry," *Chem. Ind.,* p 1320–1324 (October 15, 1955).

CASTABLE REFRACTORY MANUFACTURE

Castable or fused-cast refractories are manufactured by carefully blending such components as alumina, zirconia, silica, chrome, and magnesia; melting the mixture in an electric-arc furnace at temperatures of 3200° to 4500°F (1760° to 2480°C); pouring it into molds; and slowly cooling it to the solid state. Fused refractories are less porous and more dense than kiln-fired refractories.

Air Pollution

Particulate emissions occur during the drying, crushing, handling, and blending of the components; during the actual melting process; and in the molding phase. Fluorides, largely in the gaseous form, may also be emitted during the melting operations. The general types of particulate controls may be used on the materials handling aspects of refractory manufacturing. Emissions from the electric-arc furnace, however, are largely condensed fumes and consist of very fine particles. Fluoride emissions can be effectively controlled with a scrubber. Emission factors for castable refractories manufacturing are presented in Table 154.

TABLE 154: PARTICULATE EMISSION FACTORS FOR CASTABLE REFRACTORIES MANUFACTURING[a]

Emission Factor Rating: C

Type of process	Type of control	Uncontrolled		Controlled	
		lb/ton	kg/MT	lb/ton	kg/MT
Raw material dryer	Baghouse	30	15	0.3	0.15
Raw material crushing	Scrubber			7	3.5
and processing	Cyclone	120	60	45	22.5
Electric-arc melting	Baghouse	50	25	0.8	0.4
	Scrubber			10	5
Curing oven	–	0.2	0.1	–	–
Molding and shakeout	Baghouse	25	12.5	0.3	0.15

[a]Fluoride emissions from the melt average about 1.3 pounds of HF per ton of melt (0.65 kg HF/MT melt). Emission factors expressed as units per unit weight of feed material.

Source: EPA Publication AP-42, 2nd Edition

Reference

(1) U.S. Environmental Protection Agency, "Compilation of Air Pollutant Emission Factors," 2nd Edition, *Publ. No. AP-42*, Research Triangle Park, N.C., Office of Air and Water Programs (April 1973).

CEMENT MANUFACTURE

Portland cement manufacture accounts for about 98% of the cement production in the U.S. The more than 30 raw materials used to make cement may be divided into four basic components: lime (calcareous), silica (siliceous), alumina (argillaceous), and iron (ferriferous). Approximately 3,200 pounds of dry raw materials are required to produce 1 ton of cement. Approximately 35% of the raw material weight is removed as carbon dioxide and water vapor. The raw materials undergo separate crushing after the quarrying operation, and, when needed for processing, are proportioned, ground, and blended using either the wet or dry process.

In the dry process, the moisture content of the raw material is reduced to less than 1% either before or during the grinding operation. The dried materials are then pulverized into a powder and fed directly into a rotary kiln. Usually, the kiln is a long, horizontal, steel cylinder with a refractory brick lining. The kilns are slightly inclined and rotate about the longitudinal axis. The pulverized raw materials are fed into the upper end and travel slowly to the lower end.

The kilns are fired from the lower end so that the hot gases pass upward and through the raw material. Drying, decarbonating, and calcining are accomplished as the material travels through the heated kiln, finally burning to incipient fusion and forming the clinker. The clinker is cooled, mixed with about 5% gypsum by weight, and ground to the final product fineness. The cement is then stored for later packaging and shipment. With the wet process, a slurry is made by adding water to the initial grinding operation. Proportioning may take place before or after the grinding step. After the materials are mixed, the excess water is removed and final adjustments are made to obtain a desired composition. This final homogeneous mixture is fed to the kilns as a slurry of 30 to 40% moisture or as a wet filtrate of about 20% moisture. The burning, cooling, addition of gypsum, and storage are carried out as in the dry process.

Air Pollution

Particulate matter is the primary emission in the manufacture of Portland cement. Emissions also include the normal combustion products of the fuel used to supply heat for the kiln and drying operations, including oxides of nitrogen and small amounts of oxides of sulfur. Sources of dust at cement plants include: quarrying and crushing, raw material storage, grinding and blending (dry process only), clinker production, finish grinding, and packaging. The largest source of emissions within cement plants is the kiln operation, which may be considered to have three units: the feed system, the fuel-firing system, and the clinker-cooling and handling system.

The most desirable method of disposing of the collected dust is injection into the burning zone of the kiln and production of clinkers from the dust. If the alkali content of the raw materials is too high, however, some of the dust is discarded or leached before returning to the kiln. In many instances, the maximum allowable alkali content of 0.6% (calculated as sodium oxide) restricts the amount of dust that can be recycled. Additional sources of dust emissions are raw material storage piles, conveyors, storage silos, and loading/unloading facilities.

The complications of kiln burning and the large volumes of materials handled have led to the adoption of many control systems for dust collection. Depending upon the emission, the temperature of the effluents in the plant in question, and the particulate emission standards in the community, the cement industry generally uses mechanical collectors, electrical precipitators, fabric filter (baghouse) collectors, or combinations of these devices to control emissions.

Table 155 summarizes emission factors for cement manufacturing and also includes typical control efficiencies of particulate emissions. Table 156 indicates the particle size distribution for particulate emissions from kilns and cement plants before control systems are applied. Sulfur dioxide may be generated from the sulfur compounds in the ores as well as from combustion of fuel. The sulfur content of both ores and fuels will vary from plant to plant and with geographic location. The alkaline nature of the cement, however, provides for direct absorption of SO_2 into the product.

The overall control inherent in the process is approximately 75% or greater of the available sulfur in ore and fuel if a baghouse that allows the SO_2 to come in contact with the cement dust is used. Control, of course, will vary according to the alkali and sulfur content of the raw materials and fuel (2).

TABLE 155: EMISSION FACTORS FOR CEMENT MANUFACTURING WITHOUT CONTROLS[a, b, c]

Emission Factor Rating: B

Pollutant	Dry Process		Wet process	
	Kilns	Dryers, grinders, etc.	Kilns	Dryers, grinders, etc.
Particulate[d]				
lb/ton	245.0	96.0	228.0	32.0
kg/MT	122.0	48.0	114.0	16.0
Sulfur dioxide[e]				
Mineral source[f]				
lb/ton	10.2	–	10.2	–
kg/MT	5.1	–	5.1	–

(continued)

TABLE 155: (continued)

Pollutant	Dry Process		Wet process	
	Kilns	Dryers, grinders, etc.	Kilns	Dryers, grinders, etc.
Gas combustion				
lb/ton	Neg[g]	–	Neg	–
kg/MT	Neg	–	Neg	–
Oil combustion				
lb/ton	4.2S[h]	–	4.2S	–
kg/MT	2.1S	–	2.1S	–
Coal combustion				
lb/ton	6.8S	–	6.8S	–
kg/MT	3.4S	–	3.4S	–
Nitrogen oxides				
lb/ton	2.6	–	2.6	–
kg/MT	1.3	–	1.3	–

[a]One barrel of cement weighs 376 pounds (171 kg).

[b]These emission factors include emissions from fuel combustion, which should not be calculated separately.

[c]References (3) and (4).

[d]Typical collection efficiencies for kilns, dryers, grinders, etc., are: multicyclones, 80 percent; electrostatic precipitators, 95 percent; electrostatic precipitators with multicyclones, 97.5 percent; and fabric filter units, 99.8 percent.

[e]The sulfur dioxide factors presented take into account the reactions with the alkaline dusts when no baghouses are used. With baghouses, approximately 50 percent more SO_2 is removed because of reactions with the alkaline particulate filter cake. Also note that the total SO_2 from the kiln is determined by summing emission contributions from the mineral source and the appropriate fuel.

[f]These emissions are the result of sulfur being present in the raw materials and are thus dependent upon source of the raw materials used. The 10.2 lb/ton (5.1 kg/MT) factors account for part of the available sulfur remaining behind in the product because of its alkaline nature and affinity for SO_2.

[g]Negligible.

[h]S is the percent sulfur in fuel.

Source: EPA Publication AP-42, 2nd Edition

TABLE 156: SIZE DISTRIBUTION OF DUST EMITTED FROM KILN OPERATIONS WITHOUT CONTROLS (3)(5)

Particle size, μm	Kiln dust finer than corresponding particle size, %
60	93
50	90
40	84
30	74
20	58
10	38
5	23
1	3

Source: EPA Publication AP-42, 2nd Edition

Water Pollution

The operations where the largest volumes of water are used in cement plants are essentially nonpolluting. Process water in wet plants is evaporated; most cooling water is not contaminated; the change usually noted is an increase in temperature. Any contaminated discharges contain constituents that are present in the raw materials, collected kiln dust, or cement dust. These constituents, which include aluminum, iron, calcium, magnesium, sodium, potassium, sulfate, and chloride, may occur in any water that has contact with these materials. The presence of these constituents will be reflected as total dissolved solids, total suspended solids, and high pH and alkalinity.

Other constituents, reported as BOD, COD, Kjeldahl nitrogen, total volatile solids, and phenols, have been noted in the effluents of some plants. However, these are related to the presence of organic materials not directly related to the process of cement manufacture, but arising from sanitary effluents, spills of fuel oil, runoff from coal piles, and drainage from quarries, settling ponds and cooling ponds, which may contain decayed vegetation. Plants in the leaching subcategory have a higher pollutant loading than other plants. This is illustrated by the average loading for 35 parameters reported in Table 157 for plants in both subcategories. Water usage for the cement industry is summarized in Table 158. These uses and the characteristics of the associated discharges are discussed below.

Cooling Water: The major use of water at most cement plants is for cooling. This water is used to cool bearings on the kiln and grinding equipment, air compressors, burner pipes and the finished cement prior to storage or shipment. A summary of average volumes of cooling water used for specific purposes is given in Table 159. While cooling water is mostly noncontact, it can become contaminated to some extent through poor water management practices. This contamination may include oil and grease, suspended solids, and even some dissolved solids. If cooling towers are used, blow down discharges may contain residual algicides.

TABLE 157: COMPARISON OF WASTE LOADINGS FOR LEACHING AND NONLEACHING SUBCATEGORIES AS REPORTED

Parameter	Units	Mean Value for Leaching Subcategory	Number of Plants	Mean Value for Non-leaching Subcategory	Number of Plants
Alkalinity	kg/kkg (1b/ton)	1.381 (2.76)	10	0.087 (0.17)	61
BOD, 5 day	kg/kkg (1b/ton)	0 (0)	9	0 (0)	57
COD	kg/kkg (1b/ton)	0.032 (0.06)	9	0 (0)	53
Total Solids	kg/kkg (1b/ton)	7.495 (14.99)	10	0.314 (0.63)	61
Total Dissolved Solids	kg/kkg (1b/ton)	6.622 (13.24)	10	0.272 (0.54)	60
Total Suspended Solids	kg/kkg (1b/ton)	0.906 (1.81)	10	0 (0)	58
Total Volatile Solids	kg/kkg (1b/ton)	0.825 (1.65)	8	0 (0)	57
Ammonia	kg/kkg (1b/ton)	0 (0)	8	0 (0)	53
Kjeldahl Nitrogen	kg/kkg (1b/ton)	0 (0)	8	0 (0)	52
Nitrate Nitrogen	kg/kkg (1b/ton)	0 (0)	8	0 (0)	53
Phosphorus	kg/kkg (1b/ton)	0 (0)	8	0 (0)	55
Oil and Grease	kg/kkg (1b/ton)	0 (0)	4	0 (0)	47
Chloride	kg/kkg (1b/ton)	1.202 (2.40)	6	0 (0)	56
Sulfate	kg kkg (1b/ton)	3.667 (7.33)	6	0 (0)	56
Sulfide	kg kkg (1b/ton)	0 (0)	4	0 (0)	41
Sulfite	kg/kkg (1b/ton)	– –	0	0 (0)	5
Phenols	g/kkg (.001 1b/ton)	0 (0)	4	0 (0)	47
Chromium	g/kkg (.001 1b/ton)	0.080 (0.16)	6	0 (0)	51
Acidity	kg/kkg (1b/ton)	– –	0	0 (0)	6
Total Organic Carbon	kg/kkg (1b/ton)	– –	0	0 (0)	4
Total Hardness	kg/kkg (1b/ton)	2.207 (4.41)	4	0.864 (1.73)	21
Flouride	kg/kkg (1b/ton)	0 (0)	1	0 (0)	5
Aluminum	g/kkg (.001 1b/ton)	0.638 (1.28)	3	0.009 (0.02)	10
Calcium	kg/kkg (1b/ton)	0.965 (1.93)	4	0.094 (0.19)	18
Copper	g/kkg (.001 1b/ton)	– –	0	0 (0)	5
Iron	g/kkg (.001 1b/ton)	4.765 (9.53)	3	0.156 (0.31)	15
Lead	g/kkg (.001 1b/ton)	0.990 (1.98)	2	0 (0)	3
Magnesium	kg/kkg (1b/ton)	0.014 (0.03)	4	0.156 (0.31)	15
Mercury	g/kkg (.001 1b/ton)	– –	0	0 (0)	3
Nickel	g/kkg (.001 1b/ton)	– –	0	0 (0)	4
Potassium	kg/kkg (1b/ton)	3.298 (6.60)	4	0.077 (0.15)	11
Sodium	kg/kkg (1b/ton)	0.371 (0.74)	4	0.238 (0.48)	12
Zinc	g/kkg (.001 1b/ton)	0 (0)	2	0 (0)	9

Source: Report EPA 440/1-73/005

TABLE 158: SUMMARY OF WATER USAGE FOR THE CEMENT INDUSTRY

Use	Number of Plants	Average	Reported Flow Minimum	Maximum	Units
Cooling	117	1,550 (450)	17 (5)	72,000 (21,000)	liters/kkg of product
Raw material washing and beneficiation	4	100	2.1	405	liters/kkg of raw material
Process	78	860 (29)	246 (0.7)	1,740 (108)	liters/kkg of product
Dust control	13	264,000 (250)	1,890 (72)	600,000 (510)	liters/day
Dust leaching	11	2,410	2,150	2,650	liters/kkg of dust
Dust disposal	5	190 (55)	7.9 (2.3)	490 (140)	liters/kkg of product
Wet scrubber	3	28,000 (8,100)	4,150 (1,200)	42,500 (12,300)	liters/kkg of product

Note: Values in parentheses are gallons/ton.
      Data derived from 88 RAPP applications and 29 questionnaires.

TABLE 159: REPORTED COOLING WATER USAGE IN CEMENT PLANTS

Use	Average Flow, l/kkg Product	Number of Plants	Range Minimum	Maximum
Bearing cooling	1,080 (284)	39	3.8 (1.0)	5,800 (1,530)
Cement cooling	760 (200)	22	1.9 (0.5)	3,750 (985)
Clinker cooling	60 (23)	12	2.1 (0.6)	242 (64)
Kiln-gas cooling	322 (85)	4	92 (24)	770 (203)
Burner-pipe cooling	265 (70)	2	258 (68)	272 (72)

Note: Values in parentheses are gallons/ton.
      Data derived from 39 questionnaires.

Source: Report EPA 440/1-73/005

References

(1)   U.S. Environmental Protection Agency, "Compilation of Air Pollutant Emission Factors," 2nd Ed., *Publ. No. AP-42*, Research Triangle Park, N.C., Office of Air and Water Programs (April 1973).

(2)   Restriction of Emissions from Portland Cement Works, VDI Richtlinien, Dusseldorf, Germany (February 1967).

(3)   Kreichelt, T.E., Kemnitz, D.A., and Cuffe, S.T., "Atmospheric Emissions from the Manufacture of Portland Cement," U.S. DHEW, Public Health Service, Cincinnati, Ohio, *PHS Publication Number 999-AP-17,* 1967.

(4)   Unpublished standards of performance for new and substantially modified portland cement plants, Environmental Protection Agency, Bureau of Stationary Source Pollution Control, Research Triangle Park, N.C. (August 1971).

(5)   Particulate Pollutant System Study, Midwest Research Institute, Kansas City, Mo. Prepared for Environmental Protection Agency, Air Pollution Control Office, Research Triangle Park, N.C., under Contract Number CPA-22-69-104, (May 1971).

(6)   U.S. Environmental Protection Agency, "Development Document for Proposed Effluent Limitations Guidelines and New Source Performance Standards for the Cement Manufacturing Point Source Category," *Report EPA 440/1-73/005,* Washington, D.C. (August 1973).

CERAMIC CLAY MANUFACTURE

The manufacture of ceramic clay involves the conditioning of the basic ores by several methods. These include the separation and concentration of the minerals by screening, floating, wet and dry grinding, and blending of the desired ore varieties. The basic raw materials in ceramic clay manufacture are kaolinite ($Al_2O_3 \cdot 2SiO_2 \cdot 2H_2O$) and montmorillonite [(Mg, Ca) $O \cdot Al_2O_3 \cdot 5SiO_2 \cdot nH_2O$] clays. These clays are refined by separation and bleaching, blended, kiln-dried, and formed into such items as whiteware, heavy clay products (brick, etc.), various stoneware, and other products such as diatomaceous earth, which is used as a filter aid.

Air Pollution

Emissions consist primarily of particulates, but some fluorides and acid gases are also emitted in the drying process. The high temperatures of the firing kilns are also conducive to the fixation of atmospheric nitrogen and the subsequent release of NO, but no published information has been found for gaseous emissions. Particulates are also emitted from the grinding process and from storage of the ground product.

Factors affecting emissions include the amount of material processed, the type of grinding (wet or dry), the temperature of the drying kilns, the gas velocities and flow direction in the kilns, and the amount of fluorine in the ores. Common control techniques include settling chambers, cyclones, wet scrubbers, electrostatic precipitators, and bag filters. The most effective control is provided by cyclones for the coarser material, followed by wet scrubbers, bag filters, or electrostatic precipitators for dry dust. Emission factors for ceramic clay manufacturing are presented in Table 160.

TABLE 160: PARTICULATE EMISSION FACTORS FOR CERAMIC CLAY MANUFACTURING[a]

Emission Factor Rating: A

Type of process	Uncontrolled		Cyclone[b]		Multiple-unit cyclone and scrubber[c]	
	lb/ton	kg/MT	lb/ton	kg/MT	lb/ton	kg/MT
Drying	70	35	18	9	7	3.5
Grinding	76	38	19	9.5	–	–
Storage	34	17	8	4	–	–

[a]Emission factors expressed as units per unit weight of input to process.

[b]Approximate collection efficiency: 75 percent.

[c]Approximate collection efficiency: 90 percent.

Source: EPA Publication AP-42, 2nd Edition

Reference

(1) U.S. Environmental Protection Agency, "Compilation of Air Pollutant Emission Factors," 2nd Edition, *Publ. No. AP-42,* Research Triangle Park, N.C., Office of Air and Water Programs (April 1973).

CHARCOAL MANUFACTURE

Charcoal is generally manufactured by means of pyrolysis, or destructive distillation, of wood waste from members of the deciduous hardwood species. In this process, the wood is placed in a retort where it is externally heated for about 20 hours at 500° to 700°F (260° to 370°C). Although the retort has air intakes at the bottom, these are only used during start-up and thereafter are closed. The entire distillation cycle takes approximately 24 hours, the last 4 hours being an exothermic reaction. Four units of hardwood are required to produce one unit of charcoal.

Air Pollution

In the pyrolysis of wood, all the gases, tars, oils, acids, and water are driven off, leaving virtually pure carbon. All of these except the gas, which contains methane, carbon monoxide, carbon dioxide, nitrogen oxides, and aldehydes, are useful by-products if recovered. Unfortunately, economics has rendered the recovery of the distillate by-products unprofitable, and they are generally permitted to be discharged to the atmosphere.

If a recovery plant is utilized, the gas is passed through water-cooled condensers. The condensate is then refined while the remaining cool, noncondensable gas is discharged to the atmosphere. Gaseous emissions can be controlled by means of an afterburner because the unrecovered by-products are combustible. If the afterburner operates efficiently, no organic pollutants should escape into the atmosphere. Emission factors for the manufacture of charcoal are shown in Table 161.

TABLE 161: EMISSION FACTORS FOR CHARCOAL MANUFACTURING[a]
Emission Factor Rating: C

Pollutant	Type of operation			
	With chemical recovery plant		Without chemical recovery plant	
	lb/ton	kg/MT	lb/ton	kg/MT
Particulate (tar, oil)	—	—	400	200
Carbon monoxide	320[b]	160[b]	320[b]	160[b]
Hydrocarbons[c]	100[b]	50[b]	100[b]	50[b]
Crude methanol	—	—	152	76
Acetic acid	—	—	232	116
Other gases (HCHO, N_2 NO)	60	30	60[b]	30[b]

[a]Calculated values based on data in Reference 2.
[b]Emissions are negligible if afterburner is used.
[c]Expressed as methane.

Source: EPA Publication AP-42, 2nd Edition

References

(1) U.S. Environmental Protection Agency, "Compilation of Air Pollutant Emission Factors," 2nd Edition, *Publ. No. AP-42,* Research Triangle Park, N.C., Office of Air and Water Programs (April 1973).
(2) Shreve, R.N., *Chemical Process Industries,* 3rd Edition, New York, McGraw-Hill Book Company, p 619 (1967).

CHLOR-ALKALI MANUFACTURE

Chlorine and caustic are produced concurrently by the electrolysis of brine in either the diaphragm or mercury cell. In the diaphragm cell, hydrogen is liberated at the cathode and a diaphragm is used to prevent contact of the chlorine produced at the anode with either the alkali hydroxide formed or the hydrogen. In the mercury cell, liquid mercury is used as the cathode and forms an amalgam with the alkali metal. The amalgam is removed from the cell and is allowed to react with water in a separate chamber, called a denuder, to form the alkali hydroxide and hydrogen.

Chlorine gas leaving the cells is saturated with water vapor and then cooled to condense some of the water. The gas is further dried by direct contact with strong sulfuric acid. The dry chlorine gas is then compressed for in-plant use or is cooled further by refrigeration to liquefy the chlorine. Caustic as produced in a diaphragm-cell plant leaves the cell as a dilute solution along with unreacted brine. The solution is evaporated to increase the concentration to a range of 50 to 73%; evaporation also precipitates most of the residual salt, which is then removed by filtration. In mercury-cell plants, high-purity caustic can be produced in any desired strength and needs no concentration.

Air Pollution

Emissions from diaphragm- and mercury-cell chlorine plants include chlorine gas, carbon dioxide, carbon monoxide, and hydrogen. Gaseous chlorine is present in the blow gas from liquefaction, from vents in tank cars and tank containers during loading and unloading, and from storage tanks and process transfer tanks. Other emissions include mercury vapor from mercury cathode cells and chlorine from compressor seals, header seals, and the air blowing of depleted brine in mercury-cell plants.

Chlorine emissions from chlor-alkali plants may be controlled by one of three general methods: (1) use of the gas in other plant processes, (2) neutralization in alkaline scrubbers, and (3) recovery of chlorine from effluent gas streams. The effect of specific control practices is shown to some extent in the table on emission factors (Table 162).

TABLE 162: EMISSION FACTORS FOR CHLOR-ALKALI PLANTS[a]

Emission Factor Rating: B

Type of source	Chlorine gas	
	lb/100 tons	kg/100 MT
Liquefaction blow gases		
Diaphragm cell, uncontrolled	2,000 to 10,000	1,000 to 5,000
Mercury cell[b], uncontrolled	4,000 to 16,000	2,000 to 8,000
Water absorber	25 to 1,000	12.5 to 500
Caustic or lime scrubber	1	0.5
Loading of chlorine		
Tank car vents	450	225
Storage tank vents	1,200	600
Air-blowing of mercury-cell brine	500	250

[a]References 1 and 2.
[b]Mercury cells lose about 1.5 pounds mercury per 100 tons (0.75 kg/100 MT) of chlorine liquefied.

Source: EPA Publication AP-42, 2nd Edition

Water Pollution

Raw waste loads for mercury-cell plants are presented in Table 163, which gives overall figures based on twenty-one facilities, plus partial data as furnished from Plants 098 and 130. The chief raw wastes include purification muds [$CaCO_3$, $Mg(OH)_2$ and $BaSO_4$] from brine purification, some spent brine materials from caustic recovery, and condensates from chlorine and hydrogen compressions. The sulfuric acid used to dry the chlorine is not a waste in Plant 130 as it is recovered for sale.

In the caustic potash plant, Plant 130, the brine muds and potassium chloride make up the bulk of the primary waste. A small amount of copper sulfate catalyst is also wasted. This catalyst is used in treatment of waste chlorine. Specifically, the chlorine is reacted with excess sodium hydroxide in the presence of copper sulfate to produce sodium chloride, water and oxygen. The sodium chloride so produced is sent to the waste treatment facilities.

TABLE 163: RAW WASTE LOADS FROM MERCURY CELL PROCESS*

| | Based on 21 Facilities | | Plant 098 | - - - - Plant 130 - - - - | |
	Mean	Range		Mean	Range
Purification muds,					
$CaCO_3$ & $Mg(OH)_2$	16.5	0.5 – 35	7.25	7.5	6.8 – 7.9
NaOH	13.5	0.5 – 32	--	--	--
NaCl	211	15 – 500	--	40	35 – 45
KCl	0	--	0	50	45 – 54
H_2SO_4	16	0 – 50	11.3	0	--
Chlorinated					
hydrocarbons**	0.7	0 – 1.5	--	--	--
Na_2SO_4	15.5	0 – 63	--	--	--
Cl_2 (as $CaOCl_2$)	11	0 – 75	--	--	--
Filter aids	0.85	0 – 5	1.83	--	--
Mercury	0.15	0.02 – 0.28	0.0018	--	--
Carbon, graphite	20.3	0.35 – 340	--	--	--
$CuSO_4$	0	--	0	0.004	--

　　*All amounts in kg/kkg of chlorine which can be converted to lb/ton of product by
　　　multiplication by 2.0.
　　**Depends markedly on grade of chlorine produced.

Source:　Report EPA 440/1-73/007

At one plant, the wastes emerging from chlor-alkali manufacture are sent to a series of two settling ponds, with the exception of those from the cell building, which are sent to a mercury treatment unit first. The wastes from chlorine drying, brine preparation, salt saturation and caustic loading are sent directly to the two settling ponds described above, where suspended solids are removed and the pH adjusted prior to discharge. Two emergency ponds are in parallel with these two ponds and wastes can be diverted to them for special treatment if needed. Mercury-containing wastes from the cell building are first treated prior to being sent to the central waste treatment system. The effectiveness of treatment based on six months of data (129 days of measurements) is, in summary:

Mercury Concentration to Secondary Treatment (mg/l)		Mercury Concentration after Treatment (mg/l)	Average Removal Efficiency (%)
Average	44.3	0.43	99.0
Maximum values	1920.0	15.0	--
Minimum values	0.48	0.01	--

Approximately 99% removal of mercury is achieved with the mercury losses from the facility being kept to about 0.0045 – 0.0237 kg/day (0.01 – 0.05 lb/day) for the most part. The mean value of this discharge parameter is 0.0178 kg/day (0.03882 lb/day) or 0.000070 kg/kkg of chlorine (0.000140 lb/ton of chlorine). 91% of the measurements fell below twice these mean values.

At Plant 098 several of the streams are completely recycled to minimize brine wastes. Treatment of mercury-containing streams makes use of sodium sulfide to precipitate mercury and mercury sulfides. These materials are filtered from the streams, recovered as solids and treated with sodium hypochlorite to recover mercury (as chloride). The leached solids can then be safely discarded and the mercury chloride-containing solutions can be used for brine makeup and returned to the cells where the mercury chloride is decomposed to elemental mercury for reuse. The mercury effluent and chlorine treatment effectiveness at Plant 098 are as follows:

Method	Qualitative Rating	Waste Reduction Accomplished
Mercury recovery	Excellent	97% recovery of mercury
Chlorine neutralization system	Excellent	100% removal of chlorine from waste gas stream
Hydrogen peroxide treatment of liquid effluent	Good	100% removal of available chlorine

Now, with regard to diaphragm cell plants, one facility studied (1) is part of an integrated complex, making use of a considerable amount of recycling and reuse technology. The study of this facility demonstrates that this facility comes fairly close to the "zero discharge goal." In this overall facility, there is an older 1,810 kkg/day (2,000 ton) diaphragm-cell chlorine-caustic soda plant. A new 2,080 kkg (2,300 ton) per day chlorine-caustic soda plant also exists in this facility. The sodium hydroxide product from these two plants is concentrated in another portion of the facility. There are no brine wastes from this facility and several of the other waste streams are diverted for other uses in the complex. This stream diversion and maximal raw material utilization has served to minimize the wastes to be treated.

References

(1) "Atmospheric Emissions from Chlor-Alkali Manufacture," U.S. EPA, Air Pollution Control Office, Research Triangle Park, N.C., *Publication Number AP-80* (January 1971).

(2) Duprey, R.L., "Compilation of Air Pollutant Emission Factors," U.S. DHEW, PHS, National Center for Air Pollution Control, Durham, N.C., *PHS Publication Number 999-AP-42*, p 49 (1968).

(3) U.S. Environmental Protection Agency, "Compilation of Air Pollutant Emission Factors," 2nd Edition, *Publ. No. AP-42*, Research Triangle Park, N.C., Office of Air and Water Programs (April 1973).

(4) U.S. Environmental Protection Agency, "Development Document for Proposed Effluent Limitations Guidelines and New Source Performance Standards for the Major Inorganic Products Segment of the Inorganic Chemicals Manufacturing Point Source Category," *Report EPA 440/1-73/007* (August 1973).

CLAY AND FLY ASH SINTERING

Although the processes for sintering fly ash and clay are similar, there are some distinctions that justify a separate discussion of each process. Fly-ash sintering plants are generally located near the source, with the fly ash delivered to a storage silo at the plant. The dry fly ash is moistened with a water solution of lignin and agglomerated into pellets or balls. This material goes to a traveling-grate sintering machine where direct contact with hot combustion gases sinters the individual particles of the pellet and completely burns off the residual carbon in the fly ash. The product is then crushed, screened, graded, and stored in yard piles. Clay sintering involves the driving off of entrained volatile matter. It is desirable that the clay contain a sufficient amount of volatile matter so that the resultant aggregate will not be too heavy. It is thus sometimes necessary to mix the clay with finely pulverized coke (up to 10% coke by weight). In the sintering process the clay is first mixed with pulverized coke, if necessary, and then pelletized. The clay is next sintered in a rotating kiln or on a traveling grate. The sintered pellets are then crushed, screened, and stored, in a procedure similar to that for fly ash pellets.

Air Pollution

In fly-ash sintering, improper handling of the fly ash creates a dust problem. Adequate design features, including fly-ash wetting systems and particulate collection systems on all transfer points and on crushing and screening operations, would greatly reduce emissions. Normally, fabric filters are used to control emissions from the storage silo, and emissions are low. The absence of this dust collection system, however, would create a major emission problem. Moisture is added at the point of discharge from the silo to the agglomerator, and very few emissions occur there. Normally, there are few emissions from the sintering machine, but if the grate is not properly maintained, a dust problem is created. The consequent crushing, screening, handling, and storage of the sintered product also create dust problems.

In clay sintering, the addition of pulverized coke presents an emission problem because the sintering of coke-impregnated dry pellets produces more particulate emissions than the sintering of natural clay. The crushing, screening, handling, and storage of the sintered clay pellets creates dust problems similar to those encountered in fly-ash sintering. Emission factors for both clay and fly-ash sintering are shown in Table 164 (6).

TABLE 164: PARTICULATE EMISSION FACTORS FOR SINTERING OPERATIONS[a]
Emission Factor Rating: C

Type of material	Sintering operation[b]		Crushing, screening, and yard storage[b,c]	
	lb/ton	kg/MT	lb/ton	kg/MT
Fly ash[d]	110	55	e	e
Clay mixed with coke[f,g]	40	20	15	7.5
Natural clay[h,i]	12	6	12	6

[a] Emission factors expressed as units per unit weight of finished product.
[b] Cyclones would reduce this emission by about 80 percent.
 Scrubbers would reduce this emission by about 90 percent.
[c] Based on data in section on stone quarrying and processing.
[d] Reference 1.
[e] Included in sintering losses.
[f] 90 percent clay, 10 percent pulverized coke; traveling-grate, single-pass, up-draft sintering machine.
[g] References 3 through 5.
[h] Rotary dryer sinterer.
[i] Reference 2.

Source: EPA Publication AP-42, 2nd Edition

References

(1) "Air Pollutant Emission Factors," Final Report, Resources Research, Inc., Reston, Va., prepared for National Air Pollution Control Administration, Durham, N.C., under *Contract Number CPA-22-69-119* (April 1970).
(2) Communication between Resources Research, Incorporated, Reston, Va. and a clay sintering firm (October 2, 1969).
(3) Communication between Resources Research, Incorporated, Reston, Va. to a State Air Pollution Control Agency (October 16, 1969).

(4) Henn, J.J. et al, "Methods for Producing Alumina from Clay: An Evaluation of Two Lime Sinter Processes," Department of the Interior, Bureau of Mines, Washington, D.C., Report of Investigation Number 7299 (September 1969).

(5) Peters, F.A. et al, "Methods for Producing Alumina from Clay: An Evaluation of the Lime-Soda Sinter Process," Department of the Interior, Bureau of Mines, Washington, D.C., Report of Investigation Number 6927 (1967).

(6) U.S. Environmental Protection Agency, "Compilation of Air Pollutant Emission Factors," 2nd Edition, *Publ. No. AP-42*, Research Triangle Park, N.C., Office of Air and Water Programs (April 1973).

COAL CLEANING

Coal cleaning is the process by which undesirable materials are removed from bituminous and anthracite coal and lignite. The coal is screened, classified, washed, and dried at coal preparation plants. The major sources of air pollution from these plants are the thermal dryers. Seven types of thermal dryers are presently used: rotary, screen, cascade, continuous carrier, flash or suspension, multilouver, and fluidized bed. The three major types, however, are the flash, multilouver, and fluidized bed (1).

In the flash dryer, coal is fed into a stream of hot gases where instantaneous drying occurs. The dried coal and wet gases are drawn up a drying column and into the cyclone for separation. In the multilouver dryer, hot gases are passed through falling curtains of coal. The coal is raised by flights of a specially designed conveyor. In the fluidized bed the coal is suspended and dried above a perforated plate by rising hot gases.

Air Pollution

Particulates in the form of coal dust constitute the major air pollution problem from coal cleaning plants. The crushing, screening, or sizing of coal are minor sources of dust emissions; the major sources are the thermal dryers. The range of concentration, quantity, and particle size of emissions depends upon the type of collection equipment used to reduce particulate emissions from the dryer stack. Emission factors for coal-cleaning plants are shown in Table 165 (2). Footnote b of the table lists various types of control equipment and their possible efficiencies.

TABLE 165: PARTICULATE EMISSION FACTORS FOR THERMAL COAL DRYER[a]

Emission Factor Rating: B

| Type of dryer | Uncontrolled emissions[b] | |
	lb/ton	kg/MT
Fluidized bed	20	10
Flash	16	8
Multilouvered	25	12.5

[a] Emission factors expressed as units per unit weight of coal dried.
[b] Typical collection efficiencies are: cyclone collectors (product recovery), 70 percent; multiple cyclones (product recovery), 85 percent; water sprays following cyclones, 95 percent; and wet scrubber following cyclones, 99 to 99.9 percent.

Source: EPA Publication AP-42, 2nd Edition

References

(1) Midwest Research Institute, "Particulate Pollutant System Study," *Report PB, 203,522,* Springfield, Va., National Tech. Information Service (May 1, 1971).
(2) U.S. Environmental Protection Agency, "Compilation of Air Pollutant Emission Factors," 2nd Edition, *Publ. No. AP-42,* Research Triangle Park, N.C., Office of Air and Water Programs (April 1973).

COAL COMBUSTION, ANTHRACITE

Because of its low volatile content and the nonclinkering characteristics of its ash, anthracite coal is used in medium-sized industrial and institutional boilers with stationary or traveling grates. Although it is not used in spreader stokers because of its low volatile content and relatively high ignition temperature, anthracite coal may be burned in pulverized-coal-fired units, but this practice is limited to only a few plants in Eastern Pennsylvania because of ignition difficulties. Anthracite coal has also been widely used in hand-fired furnaces.

Air Pollution

Particulate emissions from anthracite coal combustion are greatly affected by the rate of firing and by the ash content of the fuel. Smoke emissions from anthracite coal combustion are rarely a problem. High grate loadings result in excessive emissions because of the underfire air required to burn the fuel. Large units equipped with forced-draft fans may also produce high rates of particulate emmissions. Hand-fired furnaces and some small natural-draft units have fewer particulate emissions because underfire air is not usually supplied by mechanical means.

The quantity of sulfur dioxide emissions from coal combustion, as from other fuels, is directly related to the sulfur content of the coal. Nitrogen oxide and carbon monoxide emissions are similar to those found in bituminous-coal-fired units because excess air rates and combustion temperatures are similar. Because the volatile matter content of anthracite is lower than that of bituminous, hydrocarbon emissions from combustion of anthracite are somewhat lower than those from bituminous coal combustion. The factors for uncontrolled emissions from anthracite coal combustion are presented in Table 166.

References

(1) "Air Pollutant Emission Factors," Final Report, Resources Research, Inc., Reston, Va., prepared for National Air Pollution Control Administration, Durham, N.C., under *Contract Number CPA-22-69-119* (April 1970).
(2) Hovey, H., Risman, A. and Cunnan, J., "The Development of Air Contaminant Emission Tables for Nonprocess Emissions," *J. Air Pol. Con. Assoc.,* 16:362–366, (July 1966).
(3) Unpublished stack test data on emissions from anthracite coal combustion, Pennsylvania Air Pollution Commission, Harrisburg, Pa. (1969)
(4) Unpublished stack test data on emissions from anthracite coal combustion, New Jersey Air Pollution Control Program, Trenton, N.J. (1969)
(5) Anderson, D.M., Lieben, J. and Susman, V.H., "Pure Air for Pennsylvania," Pennsylvania Department of Health, Harrisburg, Pa., p 15 (November 1961).
(6) Blackie, A., "Atmospheric Pollution from Domestic Appliances," The Report of the Joint Conference of the Institute of Fuel and the National Smoke Abatement Society, London (February 23, 1945).
(7) Smith, W.S., "Atmospheric Emissions from Coal Combustion, U.S. DHEW, PHS, National Center for Air Pollution Control, Cincinnati, Ohio, *PHS Publication Number 999-AP-24,* p 76 (April 1966).
(8) Crumley, P.H. and Fletcher, A.W., "The Formation of Sulfur Trioxide in Flue Gases," *J. Inst. of Fuel Combustion,* 30:608–612 (August 1957).

TABLE 166: EMISSIONS FROM ANTHRACITE COAL COMBUSTION WITH CONTROL EQUIPMENT

EMISSION FACTOR RATING: B

Type of furnace	Particulate[a,b]		Sulfur dioxide[c]		Sulfur trioxide[c,d]		Hydrocarbons[e,f]		Carbon monoxide[g]		Nitrogen oxides[d,h]	
	lb/ton	kg/MT	lb/ton	kg/MT	lb/ton	kg/MT	lb/ton	kg/MT	lb/ton	kg/MT	lb/ton	kg/MT
Pulverized (dry bottom), no fly-ash reinjection	17A	8.5A	38S	19S	0.5S	0.25S	0.03	0.015	1	0.5	18	9
Overfeed stokers, no fly-ash reinjection[i]	2A	1A	38S	19S	0.5S	0.25S	0.2	0.1	(2 to 10)[j]	(1 to 5)	(6 to 15)[k]	(3 to 7.5)
Hand-fired units	10	5	36S	18S	0.8S	0.4S	2.5	1.25	90	45	3	1.5

[a] References 2 through 7.
[b] A is the ash content expressed as weight percent.
[c] S is the sulfur content expressed as weight percent.
[d] References 5, 7, and 8.
[e] Based on Reference 2 and bituminous coal combustion.
[f] Expressed as methane.
[g] Based on bituminous coal combustion.
[h] Emitted as NO, but calculated as NO_2.
[i] Based on data obtained from traveling-grate stokers in the 12 to 180 Btu/hr (3 to 45 kcal/hr) heat input range. Anthracite is not burned in spreader stokers.
[j] Use high side of range for smaller-sized units [less than 10 x 10⁶ Btu/hr (2.5 x 10⁶ kcal/hr) heat input].
[k] Use low side of range for smaller-sized units [less than 10 x 10⁶ Btu/hr (2.5 x 10⁶ kcal/hr) heat input].

NOTE: Approximate efficiencies of control devices used for anthracite are: cyclone, 75 to 85 percent, and electrostatic precipitator, 85 percent.

Source: EPA Publication AP-42, 2nd Edition

COAL COMBUSTION, BITUMINOUS

Coal, the most abundant fossil fuel in the United States, is burned in a wide variety of furnaces to produce heat and steam. Coal-fired furnaces range in size from small hand-fired units with capacities of 10 to 20 pounds (4.5 to 9 kilograms) of coal per hour to large pulverized-coal-fired units, which may burn 300 to 400 tons (275 to 300 MT) of coal per hour. Although predominantly carbon, coal contains many compounds in varying amounts. The exact nature and quantity of these compounds are determined by the location of the mine producing the coal and will usually affect the final use of the coal.

Air Pollution

Particulates emitted from coal combustion consist primarily of carbon, silica, alumina, and iron oxide in the fly-ash. The quantity of atmospheric particulate emissions is dependent upon the type of combustion unit in which the coal is burned, the ash content of the coal, and the type of control equipment used. Table 167 gives the range of collection efficiencies for common types of fly-ash control equipment.

Factors for uncontrolled emissions are shown in Table 168. The emission factor for sulfur oxides indicates a conversion of 95% of the available sulfur to sulfur oxide. The balance of the sulfur is emitted in the fly-ash or combines with the slag or ash in the furnace and is removed with them. Increased attention has been given to the control of sulfur oxide emissions from the combustion of coal. The use of low-sulfur coal has been recommended in many areas; where low-sulfur coal is not available, other methods in which the focus is on the removal of sulfur oxide from the flue gas before it enters the atmosphere must be given consideration.

A number of flue-gas desulfurization processes have been evaluated; effective methods are undergoing full-scale operation. Processes included in this category are: limestone-dolomite injection, limestone wet scrubbing, catalytic oxidation, magnesium oxide scrubbing, and the Wellman-Lord process. Emissions of oxides of nitrogen result not only from the high temperature reaction of atmospheric nitrogen and oxygen in the combustion zone, but also from the partial combustion of nitrogenous compounds contained in the fuel. The important factors that affect NO_x production are: flame and furnace temperature, residence time of combustion gases at the flame temperature, rate of cooling of the gases, and amount of excess air present in the flame.

TABLE 167: RANGE OF COLLECTION EFFICIENCIES FOR COMMON TYPES OF OF FLY-ASH CONTROL EQUIPMENT[a]

Type of furnace	Range of collection efficiencies, %			
	Electrostatic precipitator	High-efficiency cyclone	Low-resistance cyclone	Settling chamber expanded chimney bases
Cyclone furnace	65 to 99.5[b]	30 to 40	20 to 30	10[b]
Pulverized unit	80 to 99.5[b]	65 to 75	40 to 60	20[b]
Spreader stoker	99.5[b]	85 to 90	70 to 80	20 to 30
Other stokers	99.5[b]	90 to 95	75 to 85	25 to 50

[a]Reference (1).
[b]The maximum efficiency to be expected for this collection device applied to this type source.

Source: EPA Publication AP-42, 2nd Edition

TABLE 168: EMISSION FACTORS FOR BITUMINOUS COAL COMBUSTION WITHOUT
CONTROL EQUIPMENT

Emission Factor Rating: A

Furnace size, 10^6 Btu/hr heat input[a]	Particulates[b]		Sulfur oxides[c]		Carbon monoxide		Hydro-carbons[d]		Nitrogen oxides		Aldehydes	
	lb/ton coal burned	kg/MT coal burned	lb/ton coal burned	kg/MT coal burned	lb/ton coal burned	kg/MT coal burned	lb/ton cqal burned	kg/MT coal burned	lb/ton coal burned	kg/MT coal burned	lb/ton coal burned	kg/MT coal burned
Greater than 100[e] (Utility and large industrial boilers)												
Pulverized												
General	16A	8A	38S	19S	1	0.5	0.3	0.15	18	9	0.005	0.0025
Wet bottom	13A[f]	6.5A	38S	19S	1	0.5	0.3	0.15	30	15	0.005	0.0025
Dry bottom	17A	8.5A	38S	19S	1	0.5	0.3	0.15	18	9	0.005	0.0025
Cyclone	2A	1A	38S	19S	1	0.5	0.3	0.15	55	27.5	0.005	0.0025
10 to 100[g] (large commercial and general industrial boilers)[g]												
Spreader stoker[h]	13A[i]	6.5A	38S	19S	2	1	1	0.5	15	7.5	0.005	0.0025
Less than 10[j] (commercial and domestic furnaces)												
Spreader stoker	2A	1A	38S	19S	10	5	3	1.5	6	3	0.005	0.0025
Hand-fired units	20	10	38S	19S	90	45	20	10	3	1.5	0.005	0.0025

[a] 1 Btu/hr = 0.252 kcal/hr.
[b] The letter A on all units other than hand-fired equipment indicates that the weight percentage of ash in the coal should be multiplied by the value given.
    Example: If the factor is 16 and the ash content is 10 percent, the particulate emissions before the control equipment would be 10 times 16, or 160 pounds of particulate per ton of coal (10 times 8, or 80 kg of particulates per MT of coal).
[c] S equals the sulfur content (see footnote b above).
[d] Expressed as methane.
[e] Reference (1).
[f] Without fly-ash reinjection.
[g] Reference (1).
[h] For all other stokers use 5A for particulate emission factor.
[i] Without fly-ash reinjection. With fly-ash reinjection use 20. This value is not an emission factor but represents loading reaching the control equipment.[i]
[i] Reference (1).

Source: EPA Publication AP-42, 2nd Edition

The efficiency of combustion primarily determines the carbon monoxide and hydrocarbon content of the gases emitted from bituminous coal combustion. Successful combustion that results in a low level of carbon monoxide and organic emissions requires a high degree of turbulence, a high temperature, and sufficient time for the combustion reaction to take place. Thus, careful control of excess air rates, the use of high combustion temperature, and provision for intimate fuel-air contact will minimize these emissions. The size range in Btu per hour for the various types of furnaces as shown in Table 168 is only provided as a guide in selecting the proper factor and is not meant to distinguish clearly between furnace applications.

Reference

(1) U.S. Environmental Protection Agency, "Compilation of Air Pollutant Emission Factors," 2nd Edition, *Publ. No. AP-42,* Research Triangle Park, N.C., Office of Air and Water Programs (April 1973).

COFFEE ROASTING

Coffee, which is imported in the form of green beans, must be cleaned, blended, roasted, and packaged before being sold. In a typical coffee roasting operation, the green coffee beans are freed of dust and chaff by dropping the beans into a current of air. The cleaned beans are then sent to a batch or continuous roaster. During the roasting, moisture is driven off, the beans swell, and chemical changes take place that give the roasted beans their typical color and aroma. When the beans have reached a certain color, they are quenched, cooled, and stoned.

Air Pollution

Dust, chaff, coffee bean oils (as mists), smoke, and odors are the principal air contaminants emitted from coffee processing. The major source of particulate emissions and practically the only source of aldehydes, nitrogen oxides, and organic acids is the roasting process. In a direct-fired roaster, gases are vented without recirculation through the flame. In the indirect-fired roaster, however, a portion of the roaster gases are recirculated and particulate emissions are reduced. Emissions of both smoke and odors from the roasters can be almost completely removed by a properly designed afterburner (1)(2).

Particulate emissions also occur from the stoner and cooler. In the stoner, contaminating materials heavier than the roasted beans are separated from the beans by an air stream. In the cooler, quenching the hot roasted beans with water causes emissions of large quantities of steam and some particulate matter (3). Table 169 summarizes emissions from the various operations involved in coffee processing.

TABLE 169: EMISSION FACTORS FOR ROASTING PROCESSES WITHOUT CONTROLS

Emission Factor Rating: B

Type of process	Pollutant							
	Particulates[a]		NO_x[b]		Aldehydes[b]		Organic acids[b]	
	lb/ton	kg/MT	lb/ton	kg/MT	lb/ton	kg/MT	lb/ton	kg/MT
Roaster								
Direct-fired	7.6	3.8	0.1	0.05	0.2	0.1	0.9	0.45
Indirect-fired	4.2	2.1	0.1	0.05	0.2	0.1	0.9	0.45
Stoner and cooler[c]	1.4	0.7	—	—	—	—	—	—
Instant coffee spray dryer	1.4[d]	0.7[d]	—	—	—	—	—	—

[a]Reference 3.
[b]Reference 1.
[c]If cyclone is used, emissions can be reduced by 70 percent.
[d]Cyclone plus wet scrubber always used, representing a controlled factor.

Source: EPA Publication AP-42, 2nd Edition

References

(1) Polglase, W.L., Dey, H.F. and Walsh, R.T., "Coffee Processing," *Air Pollution Engineering Manual,* Danielson, J.A. (ed.), U.S. DHEW, PHS, National Center for Air Pollution Control, Cincinnati, Ohio, *Publication Number 999-AP-40,* p 746–749 (1967).

(2) Duprey, R.L., "Compilation of Air Pollutant Emission Factors," U.S. DHEW, PHS, National Center for Air Pollution Control, Durham, N.C., *PHS Publication Number 999-AP-42,* p 19–20 (1968).

(3) Partee, F., "Air Pollution in the Coffee Roasting Industry," Revised Ed., U.S. DHEW, PHS, Division of Air Pollution, Cincinnati, Ohio, *Publication Number 999-AP-9* (1966).

(4) U.S. Environmental Protection Agency, "Compilation of Air Pollutant Emission Factors," 2nd Edition, *Publ. No. AP-42,* Research Triangle Park, N.C., Office of Air and Water Programs (April 1973).

COKE (METALLURGICAL) MANUFACTURE

Coking is the process of heating coal in an atmosphere of low oxygen content, i.e., destructive distillation. During this process, organic compounds in the coal break down to yield gases and a residue of relatively nonvolatile nature. Two processes are used for the manufacture of metallurgical coke, the beehive process and the by-product process; the by-product process accounts for more than 98% of the coke produced.

Beehive oven: The beehive is a refractory-lined enclosure with a dome-shaped roof. The coal charge is deposited onto the floor of the beehive and leveled to give a uniform depth of material. Openings to the beehive oven are then restricted to control the amount of air reaching the coal. The carbonization process begins in the coal at the top of the pile and works down through it. The volatile matter being distilled escapes to the atmosphere through a hole in the roof. At the completion of the coking time, the coke is watered out or quenched.

By-product process: The by-product process is oriented toward the recovery of the gases produced during the coking cycle. The rectangular coking ovens are grouped together in a series, alternately interspersed with heating flues, called a coke battery. Coal is charged to the ovens through ports in the top, which are then sealed. Heat is supplied to the ovens by burning some of the coke gas produced. Coking is largely accomplished at temperatures of 2000° to 2100°F (1100° to 1150°C) for a period of about 16 to 20 hours. At the end of the coking period, the coke is pushed from the oven by a ram and quenched with water.

Air Pollution

Visible smoke, hydrocarbons, carbon monoxide, and other emissions originate from the following by-product coking operations: (1) charging of the coal into the incandescent ovens, (2) oven leakage during the coking period, (3) pushing the coke out of the ovens, and (4) quenching the hot coke. Virtually no attempts have been made to prevent gaseous emissions from beehive ovens. Gaseous emissions from the by-product ovens are drawn off to a collecting main and are subjected to various operations for separating ammonia, coke-oven gas, tar, phenol, light oil (benzene, toluene, xylene), and pyridine. These unit operations are potential sources of hydrocarbon emissions.

TABLE 170: EMISSION FACTORS FOR METALLURGICAL COKE MANUFACTURE WITHOUT CONTROLS[a]

Emission Factor Rating: C

Type of operation	Particulates lb/ton	kg/MT	Sulfur dioxide lb/ton	kg/MT	Carbon monoxide lb/ton	kg/MT	Hydrocarbons[b] lb/ton	kg/MT	Nitrogen oxides (NO₂) lb/ton	kg/MT	Ammonia lb/ton	kg/MT
By-product coking[c]												
Unloading	0.4	0.2	—	—	—	—	—	—	—	—	—	—
Charging	1.5	0.75	0.02	0.01	0.6	0.3	2.5	1.25	0.03	0.015	0.02	0.01
Coking cycle	0.1	0.05	—	—	0.6	0.3	1.5	0.75	0.01	0.005	0.06	0.03
Discharging	0.6	0.3	—	—	0.07	0.035	0.2	0.1	—	—	0.1	0.05
Quenching	0.9	0.45	—	—	—	—	—	—	—	—	—	—
Underfiring[d]	—	—	4	2	—	—	—	—	—	—	—	—
Beehive ovens[e]	200	100	—	—	1	0.5	8	4	—	—	2	1

[a]Emission factors expressed as units per unit weight of coal charged.
[b]Expressed as methane.
[c]References 2 and 3.
[d]Reference 5. The sulfur dioxide factor is based on the following representative conditions: (1) sulfur content of coal charged to oven is 0.8 percent by weight; (2) about 33 percent by weight of total sulfur in the coal charged to oven is transferred to the coke-oven gas; (3) about 40 percent of coke-oven gas is burned during the underfiring operation and the remainder is used in other parts of the steel operation where the rest of the sulfur dioxide is discharged—about 6 lb/ton (3 kg/MT) of coal charged; and (4) gas used in underfiring has not been desulfurized.
[e]Reference 4.

Source: EPA Publication AP-42, 2nd Edition

Oven-charging operations and leakage around poorly sealed coke-oven doors and lids are major sources of gaseous emissions from by-product ovens. Sulfur is present in the coke-oven gas in the form of hydrogen sulfide and carbon disulfide. If the gas is not desulfurized, the combustion process will emit sulfur dioxide. Associated with both coking processes are the material-handling operations of unloading coal, storing coal, grinding and sizing of coal, screening and crushing coke, and storing and loading coke. All of these operations are potential particulate emission sources. In addition, the operations of oven charging; coke pushing and quenching produce particulate emissions. The emission factors for coking operations are summarized in Table 170.

Water Pollution

Raw waste loads for by-product coke plants may vary due to the nature of the process, water use systems, moisture and volatility of the coal, and the carbonizing temperature of the ovens. Minimum and maximum values for plant effluents in an EPA study (6) ranged from 167 to 18,800 liters/kkg (40 to 4,150 gal/ton) coke produced. The most significant liquid wastes produced from the coke plant process are excess ammonia liquor, final cooling water overflow, light oil recovery wastes, and indirect cooling water. In addition, small volumes of water may result from coke wharf drainage, quench water overflow and coal pile runoff.

The volume of ammonia liquor produced varies from 100 to 170 liter/kkg (24 to 41 gal/ton) of coke produced at plants using the semidirect ammonia recovery process to 350 to 500 liters/kkg (84 to 127 gal/ton) for the indirect process. This excess flushing liquor is the major single source of contaminated water from coke making. Indirect (non-contact) cooling water is not normally considered waste but leaks in coils or tubes may contribute a significant source of pollution.

Direct contact of the gas in the final cooler with sprays of water dissolve any remaining soluble gas components and physically flush out crystals of condensed naphthalene, which is then recovered by skimming or filtration. This final cooler water becomes so highly contaminated that most plants must cool and recirculate this water. When a closed recycle system is not used, this wastewater may exceed the raw ammonia liquor as the source of high contaminant loads.

Condensed steam from the stripping operations and cooling water constitute the bulk of liquid wastes discharged to the sewer. Light oil recovery wastes will vary with the plant process. Flows may vary from 2,100 to 6,300 liters/kkg (500 to 1,500 gal/ton) of coke at plants which discharge cooling once-through water to 125 to 625 liters/kkg (30 to 150 gal/ton) where cooling water is recycled. Effluent from the light oil recovery plant contains primarily phenol, cyanide, ammonia, and oil. The quenching of coke requires about 2,100 liters of water per kkg of coke (500 gal/ton). Approximately 35% of this water is evaporated by the hot coke and discharges from the quench tower as steam.

Table 171 summarizes the net plant raw waste loads for the plants studied. Raw waste loads are presented only for the critical parameters which include ammonia, BOD_5, cyanide, oil, phenol, sulfide, and suspended solids.

TABLE 171: CHARACTERISTICS OF BY-PRODUCT COKE PLANT WASTES – NET PLANT RAW WASTE LOAD

Characteristics	Plants			
	A	B	C	D*
Flow, l/kkg	580	530	154	19,200
Ammonia, mg/l	1,900	1,380	7,330	39
BOD_5, mg/l	1,550	1,280	1,120	12
Cyanide, mg/l	102	110	91	7.7
Oil and grease, mg/l	--	240	101	2.1

(continued)

TABLE 171: (continued)

| Characteristics | Plants | | | |
	A	B	C	D*
Phenol, mg/l	450	350	910	6.1
Sulfide, mg/l	--	629	197	4.2
Suspended solids, mg/l	--	36	421	23

*Concentrations are low due to the addition of the final once-through cooler stream which contained significant cyanide.

Source: Report EPA 440/1-73/024

TABLE 172: CHARACTERISTICS OF BEEHIVE COKE PLANT WASTES – NET
PLANT RAW WASTE LOAD

| Characteristics | Plants | | |
	E	F*	G*
Flow, l/kkg	2,040	2,040	513
Ammonia, mg/l	0.33	0	0
BOD$_5$, mg/l	3.00	0	0
Cyanide, mg/l	0.002	0	0
Phenol, mg/l	0.011	0	0
Suspended solids, mg/l	--	29	722

*Unless a significant pick-up is found in a given constituent in recycle systems, it is not possible to determine a meaningful net raw waste load.

Source: Report EPA 440/1-73/024

Beehive Coke Subcategory: Raw waste loads for the beehive will vary due to coking time, water use systems, moisture and volatility of the coal, and carbonizing temperature of the ovens. However, the raw waste is affected most by the type of water use systems, that is once-through or recycle. Test data indicated that with a recycle system, the net plant raw waste loads after quenching are less than the recycled water that is used for quenching. Minimum and maximum values for plant effluents in the study ranged from 0 to 2,040 liters/kkg (0 to 490 gal/ton) coke produced. Table 172 summarizes the net plant raw waste loads for the plants studied. Raw waste loads are presented only for the critical parameters which include ammonia, BOD$_5$, cyanide, phenol, and suspended solids.

References

(1) U.S. Environmental Protection Agency, "Compilation of Air Pollutant Emission Factors," 2nd Edition, *Publ. No. AP-42*, Research Triangle Park, N.C., Office of Air and Water Programs (April 1973).

(2) "Air Pollution by Coking Plants," United Nations Report: Economic Commission for Europe, *ST/ECE/Coal/26*, p 3-27 (1968).

(3) Fullerton, R.W., "Empingement Baffles to Reduce Emissions from Coke Quenching," *J. Air Pol. Control Assoc.*, 17:807-809 (December 1967).

(4) Sallee, G., "Particulate Pollutant Study," Midwest Research Institute, Kansas City, Mo., prepared for National Air Pollution Control Administration, Durham, N.C., under *Contract Number 22-69-104* (June 1970).

(5) Varga, J. and Lownie, H.W., Jr., Final Technological Report on "A Systems Analysis Study of the Integrated Iron and Steel Industry," Battelle Memorial Institute, Columbus, Ohio, prepared for U.S. DHEW, National Air Pollution Control Administration, Durham, N.C., under *Contract Number PH 22-68-65* (May 1969).

(6) U.S. Environmental Protection Agency, "Development Document for Proposed Effluent Limitations Guidelines and New Source Performance Standards for the Steel Making Segment of the Iron and Steel Manufacturing Point Source Category," *Report EPA 440/1-73/024/*, Washington, D.C. (January 1974).

COMBUSTION SOURCES

External combustion sources include steam-electric generating plants, industrial boilers, commercial and institutional boilers, and commercial and domestic combustion units. Coal, fuel oil, and natural gas are the major fossil fuels used by these sources. Other fuels used in relatively small quantities are liquefied petroleum gas, wood, coke, refinery gas, blast furnace gas, and other waste- or by-product fuels. Coal, oil, and natural gas currently supply about 95% of the total thermal energy consumed in the United States. In 1970 over 500 million tons (454×10^6 MT) of coal, 623 million barrels (99×10^9 liters) of distillate fuel oil, 715 million barrels (114×10^9 liters) of residual fuel oil, and 22 trillion cubic feet (623×10^{12} liters) of natural gas were consumed in the United States (1).

Air Pollution

Power generation, process heating, and space heating are some of the largest fuel-combination sources of sulfur oxides, nitrogen oxides, and particulate emissions. Individual sections of this volume present emission factor data for the major fossil fuels — coal, fuel oil, and natural gas — as well as for liquefied petroleum gas and wood waste combustion in boilers.

Reference

(1)   Ackerson, D.H. "Nationwide Inventory of Air Pollutant Emissions," unpublished report, Office of Air and Water Programs, EPA, Research Triangle Park, N.C. (May 1971).

CONCRETE BATCHING

Concrete batching involves the proportioning of sand, gravel, and cement by means of weigh hoppers and conveyors into a mixing receiver such as a transit mix truck. The required amount of water is also discharged into the receiver along with the dry materials. In some cases, the concrete is prepared for on-site building construction work or for the manufacture of concrete products such as pipes and prefabricated construction parts.

Air Pollution

Particulate emissions consist primarily of cement dust, but some sand and aggregate gravel dust emissions do occur during batching operations. There is also a potential for dust emissions during the unloading and conveying of concrete and aggregates at these plants and during the loading of dry-batched concrete mix. Another source is the traffic of heavy equipment over unpaved or dusty surfaces in and around the plant. Control techniques include the enclosure of dumping and loading areas and of conveyors and elevators, filters on storage bin vents, and water sprays. Table 173 presents emission factors for concrete batch plants.

TABLE 173: PARTICULATE EMISSION FACTORS FOR CONCRETE BATCHING*

Emission Factor Rating: C

Concrete Batching	Emission	
	lb/yd^3 of Concrete	kg/m^3 of Concrete
Uncontrolled	0.2	0.12
Good control	0.02	0.012

*One cubic yard of concrete weighs 4,000 pounds (1 m^3 = 2,400 kg). The cement content varies with the type of concrete mixed, but 735 lb of cement per yard (436 kg/m^3) may be used as a typical value.

Source: EPA Publication AP-42, 2nd Edition

Reference

(1)   EPA, "Compilation of Air Pollutant Emission Factors," 2nd Ed., *Publ. No. AP-42*, Research Triangle Park, N.C., Office of Air and Water Programs (April 1973).

CONICAL BURNERS

Conical burners are generally a truncated metal cone with a screened top vent. The charge is placed on a raised grate by either conveyor or bulldozer; however, the use of a conveyor results in more efficient burning. No supplemental fuel is used, but combustion air is often supplemented by underfire air blown into the chamber below the grate and by overfire air introduced through peripheral openings in the shell.

Air Pollution

The quantities and types of pollutants released from conical burners are dependent on the composition and moisture content of the charged material, control of combustion air, type of charging system used, and the condition in which the incinerator is maintained. The most critical of these factors seems to be the level of maintenance on the incinerators. It is not uncommon for conical burners to have missing doors and numerous holes in the shell, resulting in excessive combustion air, low temperatures, and, therefore, high emission rates of combustible pollutants (1). Particulate control systems have been adapted to conical burners with some success. These control systems include water curtains (wet caps) and water scrubbers. Emission factors for conical burners are shown in Table 174.

References

(1)   Kreichelt, T.E., "Air Pollution Aspects of Teepee Burners," U.S. DHEW, PHS, Division of Air Pollution, Cincinnati, Ohio, *PHS Publication Number 999-AP-28* (September 1966).
(2)   U.S. Environmental Protection Agency, "Compilation of Air Pollutant Emission Factors," 2nd Edition, *Publ. No. AP-42,* Research Triangle Park, N.C. Office of Air and Water Programs (April 1973).
(3)   Magill, P.L. and Benoliel, R.W., "Air Pollution in Los Angeles County: Contribution of Industrial Products," *Ind. Eng. Chem.* 44, 1347-1352 (June 1952).
(4)   Private communication with Public Health Service, Bureau of Solid Waste Management, Cincinnati, Ohio (October 31, 1969).
(5)   Anderson, D.M., Lieben, J. and Sussman, V.H., "Pure Air for Pennsylvania," Pennsylvania State Department of Health, Harrisburg (November 1961), p 98.
(6)   Boubel, R.W. et al, "Wood Waste Disposal and Utilization," Engineering Experiment Station, Oregon State University, Corvallis, *Bulletin Number 39,* p 57 (June 1958).
(7)   Netzley, A.B. and Williamson, J.E., "Multiple Chamber Incinerators for Burning Wood Waste," *Air Pollution Engineering Manual,* Danielson, J.A. (ed.), U.S. DHEW, PHS, National Center for Air Pollution Control, Cincinnati, Ohio, *PHS Publication Number 999-AP-40,* p 436-445 (1967).
(8)   Droege, H. and Lee, G., "The Use of Gas Sampling and Analysis for the Evaluation of Teepee Burners," Bureau of Air Sanitation, California Department of Public Health, (presented at the 7th Conference on Methods in Air Pollution Studies, Los Angeles, January 1965).
(9)   Boubel, R.W., "Particulate Emissions from Sawmill Waste Burners," Engineering Experiment Station, Oregon State University, Corvallis, *Bulletin Number 42,* p 7, 8 (August 1968).

COPPER (PRIMARY) SMELTING

Copper is produced primarily from low-grade sulfide ores, which are concentrated by gravity and flotation methods. Copper is recovered from the concentrate by four steps: roasting, smelting, converting, and refining. Copper sulfide concentrates are normally roasted in either multiple-hearth or fluidized-bed roasters to remove the sulfur and then calcined in preparation for smelting in a reverberatory furnace. For about half the smelters the roasting step is eliminated. Smelting removes other impurities as a slag with the aid of

TABLE 174: EMISSION FACTORS FOR WASTE INCINERATION IN CONICAL BURNERS WITHOUT CONTROLS[a]

EMISSION FACTOR RATING: B

Type of waste	Particulates		Sulfur oxides		Carbon monoxide		Hydrocarbons		Nitrogen oxides	
	lb/ton	kg/MT	lb/ton	kg/MT	lb/ton	kg/MT	lb/ton	kg/MT	lb/ton	kg/MT
Municipal refuse[b]	20(10 to 60)[c,d]	10	2	1	60	30	20	10	5	2.5
Wood refuse[e]	1[f]	0.5	0.1	0.05	130	65	11	5.5	1	0.5
	7[g]	3.5								
	20[h]	10								

[a] Moisture content as fired is approximately 50 percent for wood waste.
[b] Except for particulates, factors are based on comparison with other waste disposal practices.
[c] Use high side of range for intermittent operations charged with a bulldozer.
[d] Based on Reference 3.
[e] References 4 through 9.
[f] Satisfactory operation: properly maintained burner with adjustable underfire air supply and adjustable, tangential overfire air inlets, approximately 500 percent excess air and 700°F (370°C) exit gas temperature.
[g] Unsatisfactory operation: properly maintained burner with radial overfire air supply near bottom of shell, approximately 1200 percent excess air and 400°F (204°C) exit gas temperature.
[h] Very unsatisfactory operation: improperly maintained burner with radial overfire air supply near bottom of shell and many gaping holes in shell, approximately 1500 percent excess air and 400°F (204°C) exit gas temperature.

Source: EPA Publication AP-42, 2nd Edition

fluxes. The matter that results from smelting is blown with air to remove the sulfur as sulfur dioxide, and the end product is a crude metallic copper. A refining process further purifies the metal by insertion of green logs or natural gas. This is often followed by electrolytic refining.

Air Pollution

The high temperatures attained in roasting, smelting, and converting cause volatilization of a number of the trace elements present in copper ores and concentrates. The raw waste gases from these processes contain not only these fumes but also dust and sulfur oxide. Carbon monoxide and nitrogen oxides may also be emitted, but no quantitative data have been reported in the literature. The value of the volatilized elements dictates efficient collection of fumes and dusts. A combination of cyclones and electrostatic precipitators seems to be most often used. Table 175 summarizes the uncontrolled emissions of particulates and sulfur oxides from copper smelters.

TABLE 175: EMISSION FACTORS FOR PRIMARY COPPER SMELTERS WITHOUT CONTROLS[a]

Emission Factor Rating: C

Type of operation	Particulates[b,c]		Sulfur oxides[d]	
	lb/ton	kg/MT	lb/ton	kg/MT
Roasting	45	22.5	60	30
Smelting (reverberatory furnace)	20	10	320	160
Converting	60	30	870	435
Refining	10	5	—	—
Total uncontrolled	135	67.5	1250	625

[a]Approximately 4 unit weights of concentrate are required to produce 1 unit weight of copper metal. Emission factors expressed as units per unit weight of concentrated ore produced.
[b]References 2 through 4.
[c]Electrostatic precipitators have been reported to reduce emissions by 99.7 percent.
[d]Sulfur oxides can be reduced by about 90 percent by using a combination of sulfuric acid plants and lime slurry scrubbing.

Source: EPA Publication AP-42, 2nd Edition

Water Pollution

Specific sources of wastewater in the primary copper industry are as follows:

(1) Water-treatment wastes, including filter backwash, sludges from primary settling and ion-exchange regeneration solutions.

(2) Sanitary wastes.

(3) Indirect and direct cooling water from the cooling of furnaces, machinery, casting operations, etc. Generally, such water is recirculated to cooling towers of spray ponds for reuse, and is eventually discharged as blowdown. Some plants, notably smelters and refineries located near large bodies of water such as the sea, may use once-through cooling water.

(4) Process wastes including mine drainage, flotation-plant discharges, discarded leaching solutions, scrubber water, smelter wastewater, discarded electrolyte from refineries, sanitary waste, clean-up water, etc. Often these may be combined in tailings ponds for treatment and recirculation.

(5) Boiler and power plant wastes including boiler blowdown and ashpit overflow.

Data on the quantity of these individual waste streams were not available to this study (5). Some data on the characteristics of such wastes were obtained however. Table 176 shows the composition of raw waste to tailings ponds and of tailings-ponds effluents for several mine-concentrator combinations. Table 177 gives similar data for several operations involving mines, concentrators, and smelters. Table 178 shows compositions of waste streams from copper smelters and refineries.

TABLE 176: COMPOSITIONS OF WASTE WATERS FROM MINE AND CONCENTRATOR OPERATIONS*

Components	Raw Waste to Tailings Ponds			Tailings Pond Effluent
	Discharge from Concentrator	Plant Drain A	Plant Drain B	
As	0.01	0.09	0.03	0.006
BOD	5.8	164	50	--
Cd	0.01	--	--	nil
CN	0.06	--	--	
Cu	0.10	0.33	0.28	0.01 --
F	2.9	2.1	1.3	
Fe	0.12	--	--	0.05
Mn	0.08	--	--	0.8 nil
Pb	0.005	--	--	0.18 0.02
pH	7.3	7.0	6.7	7.64 --
TDS	3412	2310	3950	--
TSS	48	549	53	--
Zn	0.03	--	--	2.0 --
COD	--	343	106	--
Coliform	--	211,000	20,000	--
Metal Sulfides	--	0.23	0.18	--

*All analyses in mg/l.

Source: Report EPA R2 73 247a

References

(1) U.S. Environmental Protection Agency, "Compilation of Air Pollutant Emission Factors," 2nd Edition, *Publ. No. AP-42*, Research Triangle Park, N.C. Office of Air and Water Programs (April 1973).

(2) Stern, A. (ed.), "Sources of Air Pollution and Their Control," *Air Pollution*, Vol III, 2nd Ed., New York, Academic Press, p 173–179 (1968).

(3) Sallee, G., "Particulate Pollutant Study," Midwest Research Institute, Kansas City, Mo., prepared for National Air Pollution Control Administration under *Contract Number 22-69-104* (June 1970).

(4) *Systems Study for Control of Emissions in the Primary Nonferrous Smelting Industry*, 3 Volumes, San Francisco, California, Arthur G. McKee and Company (June 1969).

(5) Hallowell, J.B., Shea, J.F., Smithson G.R., Jr., Tripler, A.B. and Gonser, B.W., "Water Pollution Control in the Primary Nonferrous Metal Industry, I-Copper, Zinc and Lead Industries," *Report No. EPA R2-73-247a*, Washington, D.C., U.S. Government Printing Office (September 1973).

TABLE 177: WATER ANALYSES ASSOCIATED WITH COPPER SMELTING OPERATIONS

Units:	mg/l		mg/l		ppm			
Description:	Intake	Tailings Pond Overflow	Intake	Tailings Pond Overflow	Intakes Min	Intakes Max	Receiver of Tailings Pond Outlet Min	Receiver of Tailings Pond Outlet Max
Components								
As	<0.01	<0.01	--	--	--	--	--	--
BOD	1.6	1.0	--	--	--	--	--	--
Ca	14.	60.	172.	--	24.	205.	136.	144.
Cd	<0.005	<0.005	--	--	--	--	--	--
Cl	3.0	120.	195.	200.	1.	17.	--	--
CN	--	--	--	--	--	--	22.	--
COD	--	--	--	--	--	--	--	--
Cu	<0.010	<0.010	--	--	0.04	0.15	0.05	--
F	0.16	0.35	--	--	--	--	--	--
Fe	0.03	0.1	--	--	--	--	--	--
Hardness	35.	151.	540.	--	94.	600.	700.	736.
HCl	--	--	--	--	--	--	--	--
Hg	<0.001	<0.001	--	--	--	--	--	--
H2SO4	--	--	--	--	--	--	--	--
K	0.5	11.0	--	--	--	--	--	--
Mg	3.7	0.3	26.4	--	8.	22.	92.	110.
Mn	--	--	--	--	--	--	--	--
(M)S2	--	--	--	--	--	--	--	--
Na	1.5	65.0	--	--	2.	19.	25.	27.
Ni	<0.010	<0.010	--	--	--	--	--	--
NO3-N	0.25	0.75	--	--	--	--	--	--
Oxygen (dissolved)	8.7-13.1	8.7-12.0	--	--	--	--	--	--
P (total)	0.01	0.12	--	--	--	--	--	--
Pb	<0.005	<0.005	--	--	--	--	--	--
pH	8.2	10.0	8.6	7.9	7.2	8.5	7.9	7.9
Se	--	--	--	--	--	--	--	--
SiO2	--	--	--	--	7.	22.	20.	--
SO4	3.0	1.5	750.	90.5	5.	740.	309.	475.
TDS	60.	400.	--	102.8	140.	1100.	1050.	1300.
TSS	1-35	10-20	--	14.0	--	--	--	--
Total Acidity	--	--	--	--	--	--	--	--
Zn	<0.005	<0.005	--	--	--	--	--	--

Source: Report EPA R2-73-247a

TABLE 178: COMPOSITION OF WASTE STREAMS FROM COPPER SMELTERS AND REFINERIES*

Components	Smelter Acid Plant Water	Smelter Combined Plant Waste	Electrolytic Refineries I Intake	Electrolytic Refineries I Discharge	Electrolytic Refineries II Combined	III	IV
As	<5	22.1	--	--	--	--	0.55
BOD	--	211	10	70	--	--	--
Cd	--	0.6	--	--	<0.1	--	--
COD	--	457	--	--	--	--	--
Cu	<9	7.4	0.07	0.15	50-100	0.35	2.8
Metal Sulfides	--	11.7	--	--	--	--	--
Pb	<5	7.2	--	--	1	0.05	--
pH	2.5-3.0	1.9	7.4	7.7	1-2	6.1	7.6
Se	--	0.7	--	--	26,000	--	0.7
SO_4	--	1716	--	--	26,000	--	--
TSS	--	167	--	--	--	48	--
TDS	--	--	--	--	2-30,000	1024	--
Zn	<6	7.3	0.3	0.4	--	--	--
Fe	--	--	0.2	1.00	50	--	--
Ni	--	--	<.1	<0.1	10-20	--	--
Cr	--	--	--	--	--	<0.03	--
Hg	--	--	--	--	--	<0.1	--
Oil	--	--	--	--	--	10.5	--
F	--	--	--	--	--	--	0.6

*Analyses in mg/l.

Source: Report EPA-R2-73-247a

COTTON GINNING

The primary function of a cotton gin is to take raw seed cotton and separate the seed and the lint. A large amount of trash is found in the seed cotton, and it must also be removed. The problem of collecting and disposing of gin trash is two-fold. The first problem consists of collecting the coarse, heavier trash such as burrs, sticks, stems, leaves, sand, and dirt. The second problem consists of collecting the finer dust, small leaf particles, and fly lint that are discharged from the lint after the fibers are removed from the seed. From 1 ton (0.907 MT) of seed cotton, approximately one 500-pound (226-kilogram) bale of cotton can be made.

Air Pollution

The major sources of particulates from cotton ginning include the unloading fan, the cleaner, and the stick and burr machine. From the cleaner and stick and burr machine, a large percentage of the particles settle out in the plant, and an attempt has been made in Table 179 (1) to present emission factors that take this into consideration. Where cyclone collectors are used, emissions have been reported to be about 90% less (2).

TABLE 179: EMISSION FACTORS FOR COTTON GINNING OPERATIONS WITHOUT CONTROLS[a, b]

Emission Factor Rating: C

Process	Estimated total particulates		Particles > 100 μm settled out, %	Estimated emission factor (released to atmosphere)	
	lb/bale	kg/bale		lb/bale	kg/bale
Unloading fan	5	2.27	0	5.0	2.27
Cleaner	1	0.45	70	0.30	0.14
Stick and burr machine	3	1.36	95	0.20	0.09
Miscellaneous	3	1.36	50	1.5	0.68
Total	12	5.44	—	7.0	3.2

[a]References (2) and (3).
[b]One bale weighs 500 pounds (226 kilograms).

Source: EPA Publication AP-42, 2nd Edition

References

(1) U.S. Environmental Protection Agency, "Compilation of Air Pollutant Emission Factors," 2nd Edition, *Publ. No. AP-42,* Research Triangle Park, N.C. Office of Air and Water Programs (April 1973).

(2) "Air-Borne Particulate Emissions from Cotton Ginning Operations," U.S. DHEW, PHS, Taft Sanitary Engineering Center, Cincinnati, Ohio (1960).

(3) "Control and Disposal of Cotton Ginning Wastes," a Symposium sponsored by National Center for Air Pollution Control and Agricultural Research Service, Dallas, Texas (May 1966).

DAIRY INDUSTRY

Water Pollution

The main sources of waste in dairy plants are the following:

(1) The washing and cleaning out of product remaining in tank trucks, cans, piping, tanks, and other equipment performed routinely after every processing cycle.

(2) Spillage produced by leaks, overflow, freezing-on, boiling-over, equipment malfunction, or careless handling.

(3) Processing losses, including:
   (a) Sludge discharges from CIP clarifiers;

 (b) Product wasted during HTST pasteurizer start-up,
      shut-down, and product change-over;
 (c) Evaporator entrainment;
 (d) Discharges from bottle and case washers;
 (e) Splashing and container breakage in automatic packag-
      ing equipment, and;
 (f) Product change-over in filling machines.

(4) Wastage of spoiled products, returned products, or by-products such
     as whey.
(5) Detergents and other compounds used in the washing and sanitizing
     solutions that are discharged as waste.
(6) Entrainment of lubricants from conveyors, stackers and other equip-
     ment in the wastewater from cleaning operations.
(7) Routine operation of toilets, washrooms, and restaurant facilities at
     the plant.
(8) Waste constituents that may be contained in the raw water which ul-
     timately goes to waste.

The first five sources listed relate to the product handled and contribute the greatest
amount of waste (1). The general topic of pollution control in the dairy industry has
been reviewed by Jones (2).

Materials that are discharged to the waste streams in practically all dairy plants include:

(1) Milk and milk products received as raw materials.
(2) Milk products handled in the process and end products manufactured.
(3) Lubricants (primarily soap and silicone based) used in certain handling
     equipment.
(4) Sanitary and domestic sewage from toilets, washrooms and kitchens.

Other products that may be wasted include:

(1) Nondairy ingredients (such as sugar, fruits, flavors, nuts, and fruit juices)
     utilized in certain manufactured products (including ice cream, flavored
     milk, frozen desserts, yogurt, and others).
(2) Milk by-products that are deliberately waste, significantly whey, and
     sometimes, buttermilk.
(3) Returned products that are wasted.

Uncontaminated water from coolers, refrigeration systems, evaporators and other equipment which does not come in contact with the product is not considered waste. Such
water is recycled in many plants. If wasted, it increases the volume of the effluent and
has an effect on the size of the piping and treatment system needed for disposal. Roof
drainage will have the same effect unless discharged through separate drains. Sanitary
sewage from plant employees and domestic sewage from washrooms and kitchens is usually
disposed of separately from the process wastes, and represents a very minor part of the load.

The effect on the waste load of the raw water used by the plant has often been over-
looked. Raw water can be drawn from wells or a municipal system and may be contributing substantially to the waste load unless periodic control of its quality indicates otherwise. The principle organic constituents in the milk products are the natural milk solids,
namely fat, lactose and protein. Sugar is added in significant quantities to ice cream and
has an important effect in the waste loads of plants producing that product. The average
composition of selected milk, milk products and other selected materials is shown in
Table 180.

Cleaning products used in dairy plants include alkalis (caustic soda, soda ash) and acids
(muriatic, sulfuric, phosphoric, acetic, and others) in combination with surfactants, phosphates, and calcium sequestering compounds. BOD_5 is contributed by acids and surfactants
in the cleaning product. However, the amounts of cleaning products used ard relatively
small and highly diluted.

TABLE 180: COMPOSITION OF COMMON DAIRY PRODUCTS PROCESSING MATERIALS

Material	% Protein	% Fat	% Carbohydrate	BOD5 Kg/100Kg (lb/100lb)
Almonds (dried)	18.6	54.2	19.5	80.89
Blackberries (canned, light syrup)	0.8	0.6	17.3	13.30
Buttermilk				
Fluid(cultured skim milk)	3.6	0.1	5.1	7.22
Dried	34.6	5.3	50.0	74.63
Chocolate (semisweet)	4.2	35.7	57.0	65.49
Cheese				
Brick	22.2	30.5	1.9	51.35
Cheddar	25.0	32.2	2.1	55.89
Cottage (uncreamed)	17.0	0.3	2.7	19.66
Cherries (sweet, light syrup)	0.9	0.2	16.5	12.51
Cocoa (dry powder, low-medium fat)	19.2	12.7	53.8	68.17
Cream (fluid)				
Half-and-half	3.2	11.7	4.6	16.89
Light (coffee or table)	3.0	20.6	4.3	24.39
Light whipping	2.5	31.3	3.6	32.93
Heavy whipping	2.2	37.6	3.1	37.87
Milk (fluid)				
Whole, 3.7 % Fat	3.5	3.7	4.9	10.39
Whole, 3.5 % Fat	3.5	3.5	4.9	10.23
Skim	3.6	0.1	5.1	7.44
Milk (canned)				
Evaporated (unsweetened)	7.0	7.9	9.7	21.74
Condensed (sweetened)	8.1	8.7	54.3	53.76
Milk (dried)				
Whole	26.4	27.5	38.2	78.85
Skim	35.9	0.8	52.3	75.01
Orange Juice				
All commercial varieties	0.7	0.2	10.4	7.85
Peaches, canned				
Water pack	0.4	0.1	8.1	6.11
Juice pack	0.6	0.1	11.6	8.75
Pecans	9.2	71.2	14.6	83.17
Strawberries				
Canned, water pack	0.4	0.1	5.6	4.40
Frozen, sweetened	0.4	0.2	23.5	17.06
Sugar	0.0	0.0	99.5	68.75
Walnuts, black	20.5	59.3	14.8	85.15
Whey				
Fluid	0.9	0.3	5.1	4.72
Dried	12.9	1.1	73.5	65.07

Source: Report EPA 440/1-73/021

Sanitizers utilized in dairy facilities include chlorine compounds, iodine compounds, quaternary ammonium compounds, and in some cases acids. Their significance in relation to dairy wastes has not been fully evaluated, but it is believed that their contribution to the BOD_5 load is quite small.

Most lubricants used in the dairy industry are soap or silicones. They are employed principally in casers, stackers and conveyors. Soap lubricants contain BOD_5 and are more widely used than silicone based lubricants. The organic substances in dairy wastewaters are contributed primarily by the milk and milk products wasted, and to a much lesser degree, by cleaning products, sanitizing compounds, lubricants, and domestic sewage that are discharged to the waste stream. The importance of each source of organic matter in dairy wastewaters is illustrated in Table 181.

The inorganic constituents of dairy wastewaters have been given much less attention as sources of pollution than the organic wastes simply because the products manufactured are edible materials which do not contain hazardous quantities of inorganic substances. However, the nonedible materials used in the process, do contain inorganic substances which by themselves, or added to those of milk products and raw water, potentially pose a pollution problem. Such inorganic constituents include phosphates (used as deflocculants and emulsifiers in cleaning compounds), chlorine (used in detergents and sanitizing products) and nitrogen (contained in wetting agents and sanitizers). Wastewater volume data are shown in Table 182 (in metric units).

Wastewater flow for identified plants covers a very broad range from a mean of 542 liters per kkg milk equivalent (65 gal per 1,000 lb, ME) for receiving stations to a mean of over 9,000 liters per kkg milk equivalent (over 1,000 gal per 1,000 lb ME) for certain multiproduct plants. It should be noted that wastewater flow does not necessarily represent total water consumed, because many plants recycle condenser and cooling water and/or use water as a necessary ingredient in the product.

TABLE 181: ESTIMATED CONTRIBUTION OF WASTED MATERIALS TO THE BOD_5 LOAD OF DAIRY WASTEWATER (FLUID MILK PLANT)

	kg BOD_5/kkg (lb/1,000 lb) Milk Equivalent Processed	Percent
Milk, milk products, and other edible materials	3.0	94
Cleaning products	0.1	3
Sanitizers	Undetermined, but probably very small	--
Lubricants	Undetermined, but probably small	--
Employee wastes (sanitary and domestic)	0.1	3

Source: Report EPA 440/1-73/021

References

(1) U.S. Environmental Protection Agency, "Development Document for Proposed Effluent Limitations Guidelines and New Source Performance Standards for the Dairy Product Processing Point Source Category," *Report EPA 440/1-73/021*, Washington, D.C. (January 1974).

(2) Jones, H.R., *Pollution Control in the Dairy Industry*, Park Ridge, N.J., Noyes Data Corp. (1974).

TABLE 182: SUMMARY OF LITERATURE REPORTED AND IDENTIFIED PLANT SOURCE RAW WASTEWATER VOLUME DATA

Type of Plant	Literature Reported Plant Sources			Identified Plant Sources				
	Number of Plants Reporting	Liters Waste Water per 1,000 kg Milk Equivalent Received Range	Mean	Number of Plants Reporting	Liters Waste Water per 1,000 kg Milk Equivalent Received Range	Mean	Liters Waste Water per 100 kg BOD₅ Received Range	Mean
A. Single Product								
Receiving Station (Cans)	6	525-1,251	676	5	317-1,868	826	317-1,868	826
Receiving Station (Bulk)	16	108-9,091	3,077	11	434-8,507	3,870	434-8,507	3,886
Fluid Products	–	–	–	–	–	–	–	–
Cultured Products	10	1,334-6,547	2,602	1	–	542	–	542
Butter	5	834-12,543	2,740	–	–	–	–	2,093
Cottage Cheese	20	200-5,846	2,135	1	–	801	–	–
Natural Cheese	7	776-5,563	2,977	5	275-959	567	275-1,384	676
Ice Cream	4	1,000-3,336	1,985	12	525-5,039	1,251	767-13,144	7,427
Ice Cream Mix	8	984-12,835	4,720	2	801-7,289	4,053	–	1,968
Condensed Milk	3	909-1,026	967	3	751-3,836	1,810	801-7,289	4,045
Dry Milk	3	5,079-7,081	5,396	7	917-1,151	992	917-5,529	2,502
Condensed Whey	–	–	–	5	509-2,152	1,076	2,285-2,852	2,444
Dry Whey	–	–	–	–	–	–	1,259-5,534	2,669
B. Multi-Products								
Fluid-Cottage	10	575-2,135	1,193	6	234-4,645	2,177	234-4,645	2,177
Fluid-Cultured	8	751-3,336	1,676	7	459-7,948	3,453	709-7,948	3,536
Fluid-Butter	1	–	7,106	–	–	–	–	–
Fluid-Natural Cheese	–	–	–	–	–	–	–	–
Fluid-Ice Cream Mix-Cottage-Cultured	–	–	–	–	–	–	–	–
Fluid-Ice Cream Mix-Cond. Milk-Cultured	–	–	–	–	–	–	–	–
Fluid-Cultured-Juice	–	–	–	1	–	3,678	–	3,678
Fluid-Cottage-Cultured	12	801-11,518	3,545	6	617-2,819	5,980	617-2,819	13,861
Fluid-Cottage-Ice Cream	9	500-4,253	2,002	1	–	2,002	–	2,002
Fluid-Butter-Natural Cheese	1	–	1,618	1	–	2,319	–	2,319
Fluid-Cottage-Dry Milk	–	–	–	–	–	–	–	–
Fluid-Cottage-Cultured-Dry Whey	–	–	–	1	–	2,210	–	2,210
Fluid-Cottage-Cultured-Ice Cream	–	–	–	3	1,134-3,753	2,783	1,518-3,886	2,955
Fluid-Cottage-Cultured-Cond. Milk	–	–	–	1	–	5,921	–	5,921
Fluid-Cottage-Butter-Ice Cream-Dry Milk	–	–	–	–	–	–	–	–
Butter-Dry Milk	6	834-2,519	1,735	4	542-1,126	2,619	709-1,126	2,769
Butter-Cond. Milk	–	–	–	1	–	851	–	984
Butter-Dry Milk-Dry Whey	–	–	–	1	–	2,685	–	3,286
Butter-Natural Cheese	19	417-6,505	2,777	1	–	2,802	–	4,287
Butter-Dry Milk-Ice Cream	1	–	1,526	–	–	–	–	–
Cottage-Cond. Milk	–	–	–	1	–	1,084	–	1,084
Cottage-Cultured-Dry Milk-Dry Whey-Fluid	–	–	–	1	–	1,368	–	1,535
Cottage-Natural Cheese	–	–	–	1	–	6,297	–	6,297
Natural Cheese-Dry Whey	1	–	2,085	3	1,401-20,333	9,207	1,401-20,333	9,207
Natural Cheese-Cultured-Rec. Sta.	1	–	–	3	–	6,572	–	6,572
Natural Cheese-Cond. Whey	–	–	–	3	3,786-8,040	5,271	3,987-8,040	5,880

Source: Report EPA 440/1-73/021

DRY CLEANING

Clothing and other textiles may be cleaned by treating them with organic solvents. This treatment process involves agitating the clothing in a solvent bath, rinsing with clean solvent, and drying with warm air. There are basically two types of dry-cleaning installations: those using petroleum solvents [Stoddard at 140°F (60°C] and those using chlorinated synthetic solvents (perchloroethylene).

The trend in dry-cleaning operations today is toward smaller package operations located in shopping centers and suburban business districts that handle approximately 1,500 pounds (675 kg) of clothes per week on the average. These plants almost exclusively use perchloroethylene, whereas the older, larger dry-cleaning plants use petroleum solvents. It has been estimated that perchloroethylene is used on 50% of the weight of clothes dry-cleaned in the United States today and that 70% of the dry-cleaning plants use perchloroethylene.

Air Pollution

The major source of hydrocarbon emissions in dry cleaning is the tumbler through which hot air is circulated to dry the clothes. Drying leads to vaporization of the solvent and consequent emissions to the atmosphere, unless control equipment is used. The primary control element in use in synthetic solvent plants is a water-cooled condenser that is an integral part of the closed cycle in a tumbler or drying system. Up to 95% of the solvent that is evaporated from the clothing is recovered here. About half of the remaining solvent is then recovered in an activated-carbon adsorber, giving an overall control efficiency of 97 to 98%. There are no commercially available control units for solvent recovery in petroleum-based plants because it is not economical to recover the vapors. Emission factors for dry-cleaning operations are shown in Table 183 (1).

It has been estimated that about 18 pounds (8.2 kilograms) per capita per year of clothes are cleaned in moderate climates and about 25 pounds (11.3 kilograms) per capita per year in colder areas. Based on this information and the facts that 50% of all solvents used are petroleum-based and 25% of the synthetic solvent plants are controlled, emission factors can be determined on a pounds- (kilograms-) per-capita basis. Thus approximately 2 pounds (0.9 kilogram) per capita per year are emitted from dry-cleaning plants in moderate climates and 2.7 pounds (1.23 kilograms) per capita per year in colder areas.

TABLE 183: HYDROCARBON EMISSION FACTORS FOR DRY-CLEANING OPERATIONS
Emission Factor Rating: C

Control	Petroleum solvents		Synthetic solvents	
	lb/ton	kg/MT	lb/ton	kg/MT
Uncontrolled	305	152.5	210	105
Average control	—	—	95	47.5
Good control	—	—	35	17.5

Source: EPA Publication AP-42, 2nd Edition

Reference

(1) U.S. Environmental Protection Agency, "Compilation of Air Pollutant Emission Factors," 2nd Ed., *Publ. No. AP-42,* Research Triangle Park, N.C., Office of Air and Water Programs (April 1973).

ELECTROPLATING

Air Pollution

The purpose and type of plating determine the details of the process employed and, indirectly, the air pollution potential, which is a function of the type and rate of gassing, or release of gas bubbles from plating solutions with entrainment of droplets of solution as a mist. The degree of severity of air pollution from these processes may vary from being an insignificant problem to a nuisance. A potential air pollution problem can also occur in the preparation of articles for plating, evident in the process of degreasing, for example. These procedures, primarily cleaning processes, are as important as the plating operation itself for the production of high-quality finishes of impervious, adherent metal coatings. The cleaning of metals before electroplating generally requires a multistage procedure as follows:

(1) Precleaning by vapor degreasing or by soaking in a solvent, an emulsifiable solvent, or an emulsion (used for heavily soiled items);

(2) Intermediate cleaning with an alkaline bath soak treatment;

(3) Electrocleaning with an alkaline anodic or cathodic bath treatment, or both [the chemical and mechanical (gassing) action created by passing a current through the bath between the immersed article and an electrode produces the cleaning] ;

(4) Pickling with an acid bath soak treatment, with or without electricity.

The electrolytic processes do not operate with 100% efficiency, and some of the current decomposes water in the bath, evolving hydrogen and oxygen gases. In fact, the chief advantage of electrocleaning is the mechanical action produced by the vigorous evolution of hydrogen at the cathode, which tends to lift off films of oil, grease, paint, and dirt. The rate of gassing varies widely with the individual process. If the gassing rate is high, entrained mists of acids, alkaline materials, or other bath constituents are discharged to the atmosphere.

Most of the electrolytic plating and cleaning processes are of little interest from a standpoint of air pollution because the emissions are inoffensive and of negligible volume, owing to low gassing rates. Generally, air pollution control equipment is not required for any of these processes except the chromium-plating process. In this process, large volumes of hydrogen and oxygen gases are evolved. The bubbles rise and break the surface with considerable energy, entraining chromic acid mist, which is discharged to the atmosphere.

Chromic acid mist is very toxic and corrosive and its discharge to the atmosphere should be prevented. Chromic acid emissions have caused numerous nuisance complaints and frequently cause property damage. Particularly vulnerable are automobiles parked downwind of chrome-plating installations. The acid mist spots car finishes severely. The amounts of acid involved are relatively small but are sufficient to cause damage. In a typical decorative chromium-plating installation with an exhaust system but without mist control equipment, a stack test disclosed that 0.45 pound of chromic acid per hour was being discharged from a 1,300 gallon tank.

Chromium-plating processes can be divided into two general classes, one of which offers a considerably greater air pollution problem than the other. Hard chrome plating, which causes the more severe problem, produces a thick, hard, smooth, corrosion-resistant coating. This plating process requires a current density of about 250 amperes per square foot, which results in a high rate of gassing and a heavy evolution of acid mist. The less severe problem is presented by the process called decorative chrome plating, which requires a current density of only about 100 amperes per square foot and results in a definitely lower gassing rate.

Water Pollution

Water is used in electroplating operations in the following ways:

(1) Rinsing to remove films of processing solution from the surface of work
     pieces at the site of each operation;
(2) Washing away spills in the areas of the operations;
(3) Washing the air that passes through ventilation ducts so as to remove
     spray from the air before it is exhausted;
(4) Dumps of operating solutions, mostly pretreatment and posttreatment
     solutions;
(5) Rinse water and dumps of solutions from auxiliary operations such as
     rack stripping;
(6) Washing of equipment (e.g., pumps, filters, tanks); and
(7) Cooling water used in heat exchangers to cool solutions in electroplating
     processes.

Rinsing: A large proportion (perhaps 90%) of the water usage is in the rinsing operations. That used as cooling water usually does second duty in rinsing steps. The water is used to rinse away the films of processing solutions from the surface of the work pieces. In performing this task, the water acquires the constituents of the operating solutions and is not directly reusable. Thus, the cost of water is an operating expense to which is added the cost of treating the water to clean it up for reuse or for discard. Dilute water solutions result from the raw waste from each operation. Therefore, the location of rinse steps is important relative to the operations performed in the electroplating process.

At least 95% of the products being electroplated (or electroformed) to provide resistance to corrosion, wear, and other destructive forces are processed in medium sized or large plants (4,000 to 5,000 in number), each deploying at least 11 kg/day (25 pounds/day) of raw waste into rinse water. The potentially toxic waste in the form of heavy metal salts and cyanide salts from these sources is approximately 340,000 kg/day or 750,000 pounds per day (data is from a survey of chemicals consumed by electroplating conducted by Battelle's Columbus Laboratories in 1965, adjusted to reflect trends in process modifications since 1965), equivalent to about 110,000 kg/day or 250,000 pounds/day of heavy metal and cyanide ions. Of the total salts, about two-thirds or 228,000 kg/day (505,000 pounds/day) is contributed by copper, nickel, chromium, and zinc plating operations, as shown in Table 184.

TABLE 184: ESTIMATED DAILY RAW WASTE LOAD OF PRINCIPAL SALTS USED IN COPPER, NICKEL, CHROMIUM, ZINC PLATING AND RELATED PROCESSES

Operation	Principal Salts Identity	kg/day	pounds/day	Percent of Total Salts Consumed by Plating
Copper plating	Copper cyanide, sodium cyanide, and copper sulfate	54,000	120,000	13
Nickel plating	Nickel chloride and nickel sulfate	54,000	120,000	17
Chromium plating	Chromic acid	45,000	100,000	13
Zinc plating	Zinc oxide, zinc cyanide, sodium cyanide, and zinc sulfate	68,000	150,000	23
Zinc chromating	Sodium chromate and sodium dichromate	6,800	15,000	2
		227,800	505,000	68

Source: Report EPA 440/1-73/003

Supplementing the chemicals listed in Table 184, at least 225,000 kg/day (500,000 lb per day) of alkalies and 450,000 kg/day (1,000,000 pounds/day) of acids are contributed to the total waste by cleaning and pickling operations that precede copper, nickel, chromium, and zinc plating. The proportion of phosphates in alkaline cleaning chemicals is unknown, but is believed to be 25% of the total alkalies. Some of the alkaline solution waste and nearly all of the acid solution waste contain heavy metals resulting from the dissolution of metal products to be plated. Hence, the total amount of wastewater constituents generated by copper, nickel, chromium, and zinc electroplating probably exceeds 900,000 kg/day (2,000,000 pounds/day). From the estimated plating salts in Table 184, the total metal and cyanide load was estimated as follows:

	kg/day	lb/day
Copper	11,000	24,000
Nickel	12,000	27,000
Chromium	25,000	55,000
Zinc	19,000	42,000
Cyanide	46,000	102,000
Total:	113,000	250,000

The estimated alkali load of 230,000 kg/day (500,000 pounds/day) and acid load of 450,000 kg/day (1,000,000 pounds/day) are usually in about the same ratio in most plants (i.e., combined acid/alkali wastewaters are mostly acid). Assuming the alkalinity as sodium hydroxide (NaOH) and acidity as sulfuric acid (H_2SO_4), combination/neutralization (about 0.9 kg NaOH/kg H_2SO_4) would indicate a total net acid load of 350,000 kg/day (750,000 pounds/day).

In electroplating facilities the wastes are derived from the basis materials receiving electroplates and the contents of operating solutions used for electroplating processes. The processes used and the steps involved are shown in Tables 185, 187, 189, 191 and 193. The principal ionic constituents of wastewater from typical processes for plating on five basis materials are listed in Tables 186, 188, 190, 192 and 194. Wastes associated with preparation for plating, electroplating, and postplating are combined in these tables.

TABLE 185: PROCESSES FOR PLATING ON STEEL

	Decorative Chromium Plating 1	Decorative Chromium Plating 2	Hard Chromium Plating 3	Copper Cladding 4	Protective Zinc Plating 5	Protective Zinc Plating 6	Carburizing Resist 7	Protective Nickel Plating 8
Alkaline clean/ rinse	X	X	X	X	X	X	X	X
Acid dip/rinse	X	X	X	X	X	X	X	X
Copper strike/ rinse	X			X				
Acid dip/rinse	X			X				
Copper/rinse	X			X			X	
Semibright nickel/ rinse	X	X						X
Bright nickel/ rinse	X	X						
Anodic treat/ rinse			X					
Chromium/rinse	X	X	X					
Zinc rinse					X	X		
Chromate/rinse						X		

Source: Report EPA 440/1-73/003

TABLE 186: PRINCIPAL WASTEWATER CONSTITUENTS IN WASTES FROM PROCESSES FOR PLATING ON STEELS*

Constituent	1	2	3	4	5	6	7	8
Iron, ferrous, Fe^{+2}	X	X	X	X	X	X	X	X
Copper, cuprous, Cu^{+1}							X	
Copper, cupric, Cu^{+2}	X	X		X			X	
Nickel, Ni^{+2}	X	X		X				X
Chromium, chromate, Cr^{+6}	X	X	X			X		
Chromium, chromic, Cr^{+3}	X	X	X			X		
Zinc, Zn^{+2}					X	X		
Cyanide, CN^{-1}	X				X		X	
Sulfate, SO_4^{-2}	X	X	X	X	X	X	X	X
Chloride, Cl^{-1}	X	X	X	X	X	X	X	X
Carbonate, CO_3^{-2}		X	X	X	X	X	X	
Silicate, SiO_3^{-2}	X	X	X	X	X	X	X	X
Phosphate, PO_4^{-3}	X	X	X	X	X	X	X	X
Fluoborate, BF_6^{-1}								
Sulfamate, $NH_2SO_3^{-1}$	X	X		X				X
Nitrate, NO_3^{-1}								
Ammonium, NH_4^{+1}								
Organics	X	X		X			X	X

*Process numbers correspond to those in Table 185.

Source: Report EPA 440/1-73/003

TABLE 187: PROCESSES FOR PLATING ON ZINC DIE CASTINGS

Operation	Decorative Chromium Plate 1	Decorative Chromium Plate 2	Protective Finish 3	Protective Finish 4
Alkaline clean/ rinse	X	X	X	X
Acid dip/rinse	X	X	X	X
Copper strike/ rinse	X	X		
Acid dip/rinse		X		
Copper/rinse	X	X		
Nickel/rinse	X	X		
Nickel/rinse	X	X		
Anodic treat/ rinse			X	
Chromium/rinse	X	X		
Chromate/rinse				X

Source: Report EPA 440/1-73/003

TABLE 188: PRINCIPAL WASTEWATER CONSTITUENTS IN WASTE FROM PROCESSES FOR PLATING ON ZINC DIE CASTINGS*

Constituent	1	2	3	4
Fe^{+2}				
Cu^{+1}				
Cu^{+2}	X			
Ni^{+2}	X			
Cr^{+6}	X		X	
Cr^{+3}	X		X	
Zn^{+2}	X	X		
CN^{-1}	X			
SO_4^{-2}	X	X	X	
Cl^{-1}				
CO_3^{-2}	X	X	X	
SiO_3^{-2}	X	X	X	
PO_4^{-3}	X	X	X	
BF_6^{-1}				
$NH_2SO_3^{-1}$				
NO_3^{-1}				
NH_4^{+1}				
Organics	X	X		

*Processes correspond to those in Table 187.

Source: Report EPA 440/1-73/003

TABLE 189: PROCESSES FOR PLATING ON BRASS

Operation	Chromium Plate 1	Decorative Chromium Plate 2	Decorative Chromium Plate 3	Protective Nickel Plate 4
Alkaline clean/ rinse	X	X	X	X
Acid dip/rinse	X	X	X	X
Copper strike/ rinse	X	X	X	X
Acid dip/rinse				X
Copper/rinse				
Nickel/rinse	X	X	X	X
Nickel/rinse			X	
Anodic treat/ rinse				
Chromium/rinse	X	X	X	

Source: Report EPA 440/1-73/003

TABLE 190: PRINCIPAL WASTEWATER CONSTITUENTS IN WASTE FROM PROCESSES FOR PLATING ON BRASS*

Constituent	1	2	3	4
Fe^{+3}				
Cu^{+1}				
Cu^{+2}	X	X	X	X
Ni^{+2}	X	X	X	X
Cr^{+6}	X	X	X	
Cr^{+3}	X	X	X	
Zn^{+2}	X	X	X	
CN^{-1}	X	X	X	X
SO_4^{-2}				
Cl^{-1}	X		X	X
CO_3^{-2}	X	X	X	X
SiO_3^{-2}	X	X	X	X
PO_4^{-3}	X	X	X	X
BF_6^{-1}				
$NH_2SO_3^{-1}$				
NO_3^{-1}	X	X	X	X
NH_4^{+1}				
Organics	X	X	X	X

*Processes correspond to those in Table 189.

Source: Report EPA 440/1-73/003

TABLE 191: PROCESSES FOR PLATING ON ALUMINUM

Operation	Decorative Chromium Plate 1	Decorative Chromium Plate 2	Decorative Chromium Plate 3	Protective Zinc Plate 4
Alkaline Clean/rinse	X	X	X	X
Acid dip/rinse	X	X	X	X
Activate/rinse	X	X	X	X
Zinc strike/ rinse	X	X		X
Copper strike/ rinse	X	X		
Copper/rinse	X	X		
Nickel/rinse	X	X		
Nickel/rinse		X		
Chromium/rinse	X	X	X	
Zinc/rinse				X
Chromate/rinse				X

Source: Report EPA 440/1-73/003

TABLE 192: PRINCIPAL WASTEWATER CONSTITUENTS IN WASTE FROM PROCESSES FOR PLATING ON ALUMINUM*

Constituent		1	2	3	4
	Fe^{+3}				
	Cu^{+1}				
	Cu^{+2}				
	Ni^{+2}	X			
	Cr^{+6}	X	X	X	
	Cr^{+3}	X	X	X	
	Zn^{+2}	X	X		
Aluminum,	Al^{+3}	X	X	X	
	CN^{-1}				
	SO_4^{-2}				
	Cl^{-1}	X	X	X	
	CO_3^{-2}	X	X	X	
	SiO_3^{-2}	X	X	X	
	PO_4^{-3}				
	BF_6^{-1}	X	X	X	
	$NH_2SO_3^{-1}$				
	NO_3^{-1}				
	NH_3^{-1}				
Organics		X			

*Processes correspond to those in Table 191.

Source: Report EPA 440/1-73/003

TABLE 193: PROCESSES FOR PLATING ON PLASTICS

Operation	Decorative Chromium Plate 1	Decorative Chromium Plate 2	Basis for Coating 3	Basis for Magnetic Coating 4
Alkaline Clean/rinse	X	X	X	X
Acid dip rinse	X	X	X	X
Activate rinse	X	X	X	X
Catalyze rinse	X	X	X	X
Electroless Deposit/rinse	X	X	X	X
Copper strike/ rinse	X			
Copper/rinse	X	X	X	
Nickel/rinse	X	X	X	X
Nickel/rinse		X		
Chromium/rinse	X	X		

Source: Report EPA 440/1-73/003

TABLE 194: PRINCIPAL WASTEWATER CONSTITUENTS IN WASTE FROM PROCESSES
FOR PLATING ON PLASTICS*

Constituent		1	2	3	4
	Fe^{+3}				
	Cu^{+1}				
	Cu^{+2}	X	X	X	
	Ni^{+2}	X	X	X	X
	Cr^{+6}	X	X		
	Cr^{+3}	X	X		
	Zn^{+2}				
Tin,	Sn^{+2}	X	X	X	X
Palladium,	Pd^{+2}	X	X	X	X
	CN^{-1}				
	SO_4^{-1}	X	X	X	X
	Cl^{-1}				
	CO_3^{-2}	X	X	X	X
	SiO_3^{-2}	X	X	X	X
	PO_4^{-3}				
	BF_6^{-1}				
Organics					
	NO_3^{-1}				
	NH_3^{-1}				

*Processes correspond to those in Table 193.

Source: Report EPA 440/1-73/003

References

(1) Watson, M.R., "Pollution Control in Metal Finishing," Park Ridge, N.J., Noyes Data
     Corp. (1973).
(2) U.S. Environmental Protection Agency, "Development Document for Proposed Efflu-
     ent Limitations Guidelines and New Source Performance Standards for the Copper,
     Nickel, Chromium and Zinc Segment of the Electroplating Point Source Category,"
     Report EPA 440/1-73/003, Washington, D.C. (August 1973).

EXPLOSIVES MANUFACTURE

An explosive is a material that, under the influence of thermal or mechanical shock, decomposes rapidly and spontaneously with the evolution of large amounts of heat and gas.
Explosives fall into two major categories: high explosives and low explosives. Although
a multitude of different types of explosives exists, this section will deal only with an example of each major category: TNT as the high explosive and nitrocellulose as the low explosive.

TNT Production

TNT is usually prepared by a batch three-stage nitration process using toluene, nitric acid,
and sulfuric acid as raw materials. A combination of nitric acid and fuming sulfuric acid
(oleum) is used as the nitrating agent. Spent acid from the nitration vessels is fortified
with make-up nitric acid before entering the next nitrator. The spent acid from the primary nitrator and the fumes from all the nitrators are sent to the acid-fume recovery sys-

tem. This system supplies the make-up nitric acid needed in the process. After nitration, the undesired by-products are removed from the TNT by agitation with a solution of sodium sulfite and sodium hydrogen sulfite (Sellite process). The wash waste (commonly called red water) from this purification process is either discharged directly into a stream or is concentrated to a slurry and incinerated. The TNT is then solidified, granulated, and moved to the packing house for shipment or storage.

Nitrocellulose Production

Nitrocellulose is prepared in the United States by the mechanical dipper process. This batch process involves dripping the cellulose into a reactor (niter pot) containing a mixture of concentrated nitric acid and a dehydrating agent such as sulfuric acid, phosphoric acid, or magnesium nitrate. When nitration is complete, the reaction mixtures are centrifuged to remove most of the spent acid. The centrifuged nitrocellulose is then drowned in water and pumped as a water slurry to the final purification area.

Air Pollution

Emissions of sulfur oxides and nitrogen oxides from processes that produce some of the raw materials for explosives production, such as nitric acid and sulfuric acid, can be considerable. Because all of the raw materials are not manufactured at the explosives plant, it is imperative to obtain detailed process information for each plant in order to estimate emissions. The emissions from the manufacture of nitric acid and sulfuric acid are not included in this section as they are discussed in other sections of this publication.

The major emissions from the manufacturing of explosives are nitrogen oxides. The nitration reactors for TNT production and the reactor pots and centrifuges for nitrocellulose represent the largest nitrogen oxide sources. Sulfuric acid regenerators or concentrators, considered an integral part of the process, are the major sources of sulfur oxide emissions. Emission factors for explosives manufacturing are presented in Table 195.

TABLE 195: EMISSION FACTORS FOR EXPLOSIVES MANUFACTURING WITHOUT CONTROL EQUIPMENT

Emission Factor Rating: C

Type of process	Particulate		Sulfur oxides (SO_2)		Nitrogen oxides (NO_2)	
	lb/ton	kg/MT	lb/ton	kg/MT	lb/ton	kg/MT
High explosives						
TNT						
Nitration reactors[a]	—	—	—	—	160	80
Nitric acid concentrators	—	—	—	—	1	0.5
Sulfuric acid regenerators	0.4	0.2	18	9	—	—
Red water incincerator[b]	36	18	13	6.5	6	3
Nitric acid manufacture	(See section on nitric acid)					
Low explosives						
Nitrocellulose						
Reactor pots	—	—	—	—	12	6
Sulfuric acid concentrators	—	—	65	32.5	29	14.5

[a]With bubbles cap absorption, system is 90 to 95% efficient.
[b]Not employed in manufacture of TNT for commercial use.

Source: EPA Publication AP-42, 2nd Edition

Reference

(1)   U.S. Environmental Protection Agency, "Compilation of Air Pollutant Emission Factors," 2nd Edition, *Publ. No. AP-42,* Research Triangle Park, N.C., Office of Air and Water Programs (April 1973).

FEEDLOTS

Water Pollution

In accordance with the Federal Water Pollution Control Amendments of 1972, animal feedlots are defined as point sources of pollution. It is necessary, therefore, to distinguish between animals grown in feedlots and those grown in nonfeedlot situations. For the purposes of this document, the term feedlot is defined by the following three conditions:

(1)   A high concentration of animals held in a small area for periods of time in conjunction with one of the following purposes:

       (a)   Production of meat
       (b)   Production of milk
       (c)   Production of eggs
       (d)   Production of breeding stock
       (e)   Stabling of horses

(2)   The transportation of feeds to the animals for consumption.
(3)   By virtue of the confinement of animals or poultry, the land or area will neither sustain vegetation nor be available for crop production.

These criteria must be met by a facility in order to be classified as a feedlot. Facilities which meet the first condition invariably meet all conditions also. However, pasture and range operations do not meet the first condition but on occasion do meet the second condition. In pasture and range situations, the animals are at such a low density in terms of numbers of animals per acre that the growth of grasses and other plants is not inhibited.

In these cases the animals receive the major portion of their sustenance from these plants and in turn return nutrients to the soil in the form of wastes. These wastes are then assimilated by the plants in a natural recycle system. Under pure pasture or range conditions no pollutional source is ever reliably identifiable. In some instances supplementary feed may be brought to the animals (usually in the winter) but this level of feeding does not introduce a situation wherein the ability of the natural ecosystem to absorb the animal wastes is exceeded.

Animal feedlot wastes generally includes the following components:

(1)   Bedding or litter (if used) and animal hair or feathers
(2)   Water and milking center wastes
(3)   Spilled feed
(4)   Undigested or partially digested food or feed additives
(5)   Digestive juices
(6)   Biological products of metabolism
(7)   Microorganisms from the digestive tract
(8)   Cells and cell debris from the digestive tract wall
(9)   Residual soil and sand

The greatest influences on waste characteristics are animal type, type of facility used and diet. The latter two considerations are usually the only factors which lead to any substantial variation in waste characteristics for any given animal type. For example, bedding materials used in certain facilities, or amount of roughage used in various feeds will affect the character of waste loads. In the case of diet, variations encountered in manure constituents for one particular type of animal are not usually significant although solids content can be increased due to high roughage feeds. Some of the trace elements and all of the pharmaceuticals present in the wastes are a result of additives in the diet obtained from other than natural sources (i.e., other than crops used for feed).

Figures 19 to 27 show the material balances for some typical animal feeding operation and the waste quantities evolved. For details of open feedlot operation and housed feedlot operation, the reader is referred to a lengthy EPA report (1).

FIGURE 19: TYPICAL BEEF FEEDLOT FLOW DIAGRAM

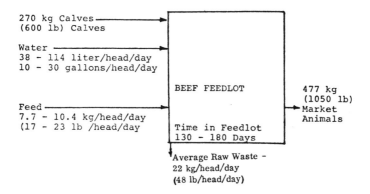

FIGURE 20: TYPICAL DAIRY FARM FLOW DIAGRAM

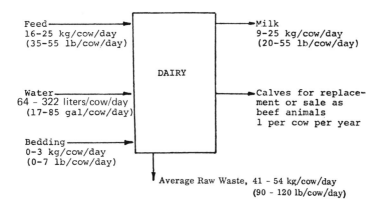

Source: Report EPA 440/1-73/004

FIGURE 21: TYPICAL SWINE FEEDLOT FLOW DIAGRAM

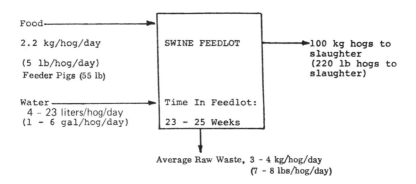

FIGURE 22: TYPICAL BROILER FEEDLOT FLOW DIAGRAM

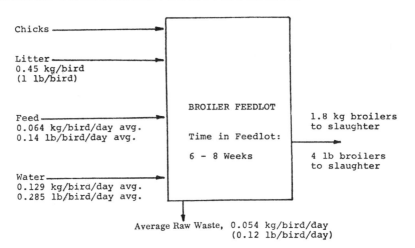

FIGURE 23: TYPICAL LAYING OPERATION FLOW DIAGRAM

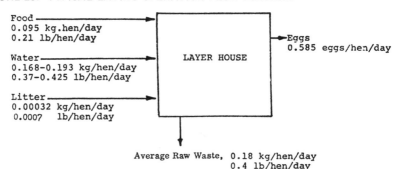

Source: Report EPA 440/1-73/004

FIGURE 24: TYPICAL LAMB FEEDLOT FLOW DIAGRAM

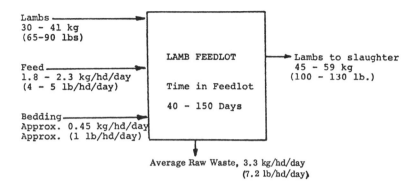

Average Raw Waste, 3.3 kg/hd/day
(7.2 lb/hd/day)

FIGURE 25: TYPICAL TURKEY FEEDLOT FLOW DIAGRAM

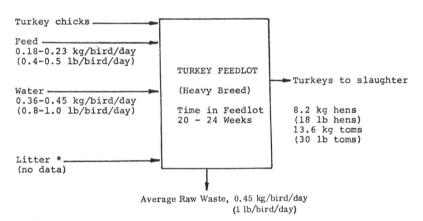

Average Raw Waste, 0.45 kg/bird/day
(1 lb/bird/day)

* For Housed Feedlots Only

FIGURE 26: TYPICAL DUCK FEEDLOT FLOW DIAGRAM

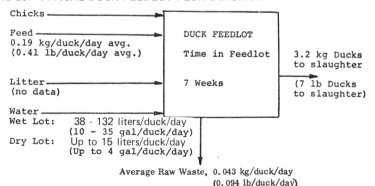

Average Raw Waste, 0.043 kg/duck/day
(0.094 lb/duck/day)

Source: Report EPA 440/1-73/004

FIGURE 27: FLOW DIAGRAM OF TYPICAL RACETRACK

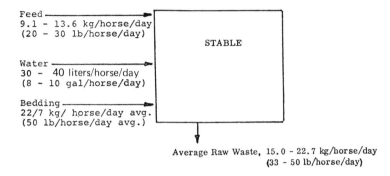

Average Raw Waste, 15.0 - 22.7 kg/horse/day
(33 - 50 lb/horse/day)

Source: Report EPA 440/1-73/004

Reference

(1) U.S. Environmental Protection Agency, "Development Document for Proposed
Effluent Limitations Guidelines and New Source Performance Standards for the
Feedlots Point Source Category," *Report EPA 440/1-73/004,* Washington, D.C.
(April 1973).

FERMENTATION INDUSTRY

For the purpose of this report only the fermentation industries associated with food will
be considered. This includes the production of beer, whiskey, and wine. The manufacturing process for each of these is similar. The four main brewing production stages and
their respective substages are:

(1) Brewhouse operations, which include
    (a) malting of the barley,
    (b) addition of adjuncts (corn, grits, and rice) to barley mash,
    (c) conversion of starch in barley and adjuncts to maltose sugar
          by enzymatic processes,
    (d) separation of wort from grain by straining, and
    (e) hopping and boiling of the wort;
(2) Fermentation, which includes
    (a) cooling of the wort,
    (b) additional yeast cultures,
    (c) fermentation for 7 to 10 days,
    (d) removal of settled yeast, and
    (e) filtration and carbonation;
(3) Aging, which lasts from 1 to 2 months under refrigeration; and
(4) Packaging, which includes
    (a) bottling-pasteurization, and
    (b) racking draft beer.

The major differences between beer production and whiskey production are the purification and distillation necessary to obtain distilled liquors and the longer period of aging.
The primary difference between wine making and beer making is that grapes are used as
the initial raw material in wine rather than grains.

Air Pollution

Emissions from fermentation processes are nearly all gases and primarily consist of carbon dioxide, hydrogen, oxygen, and water vapor, none of which present an air pollution problem. Emissions of particulates, however, can occur in the handling of the grain for the manufacture of beer and whiskey. Gaseous hydrocarbons are also emitted from the drying of spent grains and yeast in beer and from the whiskey-aging warehouses. No significant emissions have been reported for the production of wine. Emission factors for the various operations associated with beer, wine, and whiskey production are shown in Table 196.

TABLE 196: EMISSION FACTORS FOR FERMENTATION PROCESSES

Emission Factor Rating: E

Type of product	Particulates		Hydrocarbons	
	lb/ton	kg/MT	lb/ton	kg/MT
Beer				
Grain handling[a]	3	1.5	—	—
Drying spent grains, etc.[a]	5	2.5	NA[b]	NA
Whiskey				
Grain handling[a]	3	1.5	—	—
Drying spent grains, etc.[a]	5	2.5	NA	NA
Aging	—	—	10[c]	0.024[d]
Wine	Neg[e]	Neg	Neg[e]	Neg

[a]Based on section on grain processing.
[b]No emission factor available, but emissions do occur.
[c]Pounds per year per barrel of whiskey stored.[2]
[d]Kilograms per year per liter of whiskey stored.
[e]No significant emissions.

Source: EPA Publication AP-42, 2nd Edition

Water Pollution

Some data on water pollution problems presented by the fermentation industry have been presented (2) as in Table 197.

TABLE 197: WASTE FROM A GRAIN DISTILLERY

Total Effluent: Pound of BOD per Bushel 0.51

Operation	% of Total BOD
Cooking and fermenting	12.3
Distilling	1.8
Power plant	2.5
Feed recovery	83.4
	100.0

Source: Gurnham, C.F., *Industrial Wastewater Control,* 1965

References

(1) U.S. Environmental Protection Agency, "Compilation of Air Pollutant Emission Factors," 2nd Edition, *Publ. No. AP-42,* Research Triangle Park, N.C., Office of Air and Water Programs (April 1973).

(2) Gurnham, C. Fred, Ed., *Industrial Wastewater Control,* New York, Academic Press (1965).

FERROALLOY PRODUCTION

Ferroalloy is the generic term for alloys consisting of iron and one or more other metals. Ferroalloys are used in steel production as alloying elements and deoxidants. There are three basic types of ferroalloys: silicon-based alloys, including ferrosilicon and calcium-silicon; manganese-based alloys, including ferromanganese and silicomanganese; and chromium-based alloys, including ferrochromium and ferrosilicochrome.

The four major procedures used to produce ferroalloy and high-purity metallic additives for steelmaking are: blast furnace, electrolytic deposition, alumina silico-thermic process, and electric smelting furnace. Because over 75% of the ferroalloys are produced in electric smelting furnaces, this section deals only with that type of furnace.

The oldest, simplest, and most widely used electric furnaces are the submerged-arc open type, although semicovered furnaces are also used. The alloys are made in the electric furnaces by reduction of suitable oxides. For example, in making ferrochromium the charge may consist of chrome ore, limestone, quartz (silica), coal and wood chips, along with scrap iron.

Air Pollution

The production of ferroalloys has many dust- or fume-producing steps. The dust resulting from raw material handling, mix delivery, and crushing and sizing of the solidified product can be handled by conventional techniques and is ordinarily not a pollution problem. By far the major pollution problem arises from the ferroalloy furnaces themselves. The conventional submerger-arc furnace utilizes carbon reduction of metallic oxides and continuously produces large quantities of carbon monoxide. This escaping gas carries large quantities of particulates of submicron size, making control difficult.

In an open furnace, essentially all of the carbon monoxide burns with induced air at the top of the charge, and CO emissions are small. Particulate emissions from the open furnace, however, can be quite large. In the semiclosed furnace, most or all of the CO is withdrawn from the furnace and burns with dilution air introduced into the system. The unburned CO goes through particulate control devices and can be used as boiler fuel or can be flared directly. Particulate emission factors for electric smelting furnaces are presented in Table 198. No carbon monoxide emission data have been reported in the literature.

TABLE 198: EMISSION FACTORS FOR FERROALLOY PRODUCTION IN ELECTRIC SMELTING FURNACES*

Emission Factor Rating: C

Type of Furnace and Product	Particulates	
	lb/ton	kg/MT
Open furnace		
50% FeSi	200	100
75% FeSi	315	157.5
90% FeSi	565	282.5
Silicon metal	625	312.5
Silicomanganese	195	97.5
Semicovered furnace		
Ferromanganese	45	22.5

*Emission factors expressed as units per unit weight of specified product produced.

Source: EPA Publication AP-42, 2nd Edition

Water Pollution

The waste characteristics to be determined may be considered on the basis of the industry categories and the various water uses as follows:

(1) Cooling water for electric furnace smelting
(2) Water for wet air pollution control devices namely:
    (a) Electric furnace smelting (disintegrator-type scrubbers, high energy scrubbers, electrostatic precipitator spray towers, or steam/hot water scrubbers)
    (b) Exothermic smelting processes
(3) Sanitary uses, boiler feed, air conditioning, etc.
(4) Slag processing uses.

The data from Plant D provide raw waste loads for open submerged arc furnaces in which the off-gases are scrubbed with steam/hot water scrubbers as shown in Table 199

TABLE 199: RAW WASTE LOADS — OPEN CHROMIUM ALLOY AND FERROSILICON FURNACES WITH STEAM/HOT WATER SCRUBBERS

Constituent	kg/mwhr	lb/mwhr
Suspended solids	8.2	18.1
Phosphate	0.010	0.023
Phenol	0.0004	0.001
Oil	0.002	0.004
CN, total	0.0001	0.0002
CN, free	0.00005	0.0001
Iron	0.069	0.152
Manganese	0.005	0.010
Zinc	0.068	0.149
Cr, total	0.003	0.007
Cr(VI)	0.002	0.004
Lead	0.008	0.018
Aluminum	0.158	0.348
	l/mwhr	gal/mwhr
Flow	2,691	711

Source: Report EPA 440/1-73/008

The data from Plant E provide an additional raw waste load for an open electric furnace using a venturi scrubber, as shown in Table 200.

TABLE 200: RAW WASTE LOAD — HIGH ENERGY SCRUBBER ON OPEN ELECTRIC FURNACE

Constituent	Venturi Scrubber	
	kg/mwhr	lb/mwhr
Suspended solids	23.74	52.29
Phosphate	0	0
Phenol	0.0005	0.001
Oil	0.006	0.014
Cyanide, total	0	0
Iron, total	0.041	0.089
Manganese	10.06	22.15
Zinc	0.33	0.72
Chromium, total	0.002	0.005
Lead	0.07	0.16
Aluminum	1.13	2.49
	l/mwhr	gal/mwhr
Flow	6,382	1,686

Source: Report EPA 440/1-73/008

The data from Plant G providing raw waste loads for open submerged arc furnaces in which the off-gases are conditioned in a spray tower preceding an electrostatic precipitator are shown in Table 201. The data from Plant B provide information on the wastewater from disintegrator scrubbers operating on covered furnaces producing silicon alloys. Raw waste loads of suspended solids and cyanides are given in Table 202 on the basis of the furnace power during the 16-hour sampling periods. The data from Plant C provide raw waste load data for a sealed silicomanganese furnace where the off-gases are scrubbed in a spray tower and a disintegrator scrubber. These data are shown in Table 203.

TABLE 201: RAW WASTE LOADS – OPEN CHROMIUM ALLOY FURNACES WITH ELECTROSTATIC PRECIPITATORS

Constituent	kg/mwhr	lb/mwhr
Suspended solids	0.289	0.636
Phosphate	0.00001	0.00002
Oil	0.0001	0.0002
Iron	0.001	0.002
Manganese	0.0012	0.0026
Zinc	0.001	0.003
Chromium, total	0.0016	0.0036
Aluminum	0.0070	0.0155
	l/mwhr	gal/mwhr
Flow	84.0	22.2

TABLE 202: RAW WASTE LOADS FOR COVERED FURNACES WITH DISINTEGRATOR SCRUBBERS

Product	Suspended Solids kg/mwhr	lb/mwhr	Cyanides kg/mwhr	lb/mwhr	Flow l/mwhr	gal/mwhr
SiMnZr	20.1	44.3	0.0338	0.0745	8,270	2,185
75% FeSi	39.2	86.3	--	--	8,967	2,369
50% FeSi	5.1	11.3	0.0001	0.0002	8,823	2,331
75% FeSi	6.8	15.0	0.0139	0.0307	7,562	1,998

TABLE 203: RAW WASTE LOADS – SEALED SILICOMANGANESE FURNACE WITH DISINTEGRATOR SCRUBBER

Constituent	kg/mwhr	lb/mwhr
Suspended solids	16.6	36.6
Phosphate	0.052	0.114
Phenol	0.009	0.019
Oil	0.038	0.084
Cyanide, total	0.044	0.098
Cyanide, free	0.011	0.024
Iron	0.056	0.123
Zinc	0.224	0.493
Chromium, total	0.0004	0.001
Lead	0.033	0.072
Aluminum	0.413	0.91
Manganese	4.858	10.70
	l/mwhr	gal/mwhr
Flow	10,863	2,870

Source: EPA 440/1-73/008

The data from Plant E also provide data on scrubber raw wastewater loads from covered furnaces equipped with high energy and disintegrator scrubbers (see Table 204).

TABLE 204: RAW WASTE LOAD – COVERED FURNACES WITH SCRUBBERS

Constituent	kg/mwhr	lb/mwhr
Suspended solids	4.01	8.83
Phosphate	0.004	0.008
Phenol	0.002	0.004
Oil	0.022	0.048
Cyanide, total	0.007	0.015
Iron, total	0.08	0.17
Manganese	0.016	0.034
Zinc	0.21	0.45
Chromium, total	0.002	0.004
Lead	0.023	0.051
Aluminum	0.07	0.16
	l/mwhr	gal/mwhr
Flow	9,746	2,575

Source: EPA 440/1-73/008

The data from Plant H provide data on the raw waste loads from aluminothermic production of chromium alloys in which the off-gases are treated in a combination wet scrubber and baghouse and are given in Table 205.

TABLE 205: RAW WASTE LOADS – ALUMINOTHERMIC SMELTING WITH
                    COMBINATION WET SCRUBBERS AND BAGHOUSE

Constituent	kg/kkg	lb/ton
Suspended solids	3.6	7.1
Phosphate	0	0
Phenol	0	0
Oil	0.048	0.095
Cyanide, total	0	0
Cyanide, free	0	0
Iron, total	0	0
Manganese	0.0005	0.001
Zinc	0	0
Chromium, total	2.98	5.95
Cr(VI)	0.95	1.90
Lead	0	0
Aluminum	0	0
	l/kkg	gal/ton
Flow	26,332	6,310

Source: EPA 440/1-73/008

The data from Plant E provide information on the raw waste loads from slag processing operations. That from slag concentrating is shown in Table 206.

TABLE 206: RAW WASTE LOADS – SLAG CONCENTRATION PROCESS

Constituent	kg/kkg	lb/ton
Suspended solids	46.0	91.9
Phosphate	0	0
Phenol	0	0
Oil	0.064	0.128
Cyanide, total	0.0003	0.007
Iron, total	0.543	1.085
Manganese	0.245	0.489

(continued)

TABLE 206: (continued)

Constituent	kg/kkg	lb/ton
Zinc	0.012	0.023
Aluminum	0.569	1.138
Lead	0	0
Chromium, total	0.109	0.217
	l/kg	gal/ton
Flow	48,259	12,750

Source: EPA 440/1-73/008

Plant A provides information on the wastewater resulting from the recirculation and reuse of cooling water on electric smelting furnaces producing silicon alloys and utilizing no wet air pollution control devices. A softener is used on one cooling water circuit handling about 50% of the total cooling water flow. Treatment chemicals are used with a phosphate reagent used rather than a chromate. The raw waste loads are given in Table 207 on the basis of the total furnace power during the sampling period.

TABLE 207: RAW WASTE LOADS – NONCONTACT COOLING WATER FOR
SUBMERGED ARC FURNACES

Parameter	kg/mwhr	lb/mwhr
Suspended solids	0.09	0.19
Phosphate	0.001	0.003
Phenol	0.00005	0.0001
Oil	0.002	0.004
Iron, total	0.001	0.002
Manganese	0.0005	0.001
Zinc	0.00001	0.00003
Chromium, total	0.000005	0.00001
Aluminum	0.002	0.004
	l/mwhr	gal/mwhr
Flow	502.6	132.8

Source: EPA 440/1-73/008

The data from Plant F provide raw waste loads resulting from the recirculation and reuse of cooling water on electric smelting furnaces utilizing chromate treatment for corrosion inhibition, as shown in Table 208.

TABLE 208: RAW WASTE LOADS OF NONCONTACT COOLING WATER FOR
SUBMERGED ARC FURNACES

Parameter	kg/mwhr	lb/mwhr
Suspended solids	2.71	5.98
Total iron	0.011	0.025
Manganese	0.007	0.016
Zinc	0.14	0.30
Oil	0.08	0.18
Total chromium	0.26	0.58
Cr(VI)	0.0001	0.0003
Aluminum	0.013	0.03
	l/mwhr	gal/mwhr
Flow	734	194

Source: EPA 440/1-73/008

References

(1)   U.S. Environmental Protection Agency, "Compilation of Air Pollutant Emission
      Factors," 2nd Edition, *Publ. No. AP-42,* Research Triangle Park, N.C., Office of Air
      and Water Programs (April 1973).
(2)   U.S. Environmental Protection Agency, "Development Document for Proposed
      Effluent Limitations Guidelines and New Source Performance Standards for the
      Smelting and Slag Processing Segment of the Ferroalloy Manufacturing Point
      Source Category," *Report EPA 440/1-73/008,* Washington, D.C. (August 1973).

FERTILIZER INDUSTRY

The fertilizer industry is quite complex and a variety of types and grades of fertilizer
materials are manufactured.

Air Pollution

The reader is referred for additional detail to other sections of this volume on Ammonia
Manufacture, Nitrogen Fertilizer Manufacture and Phosphate Fertilizer Manufacture.

The points of atmospheric emission by various fertilizer manufacturing processes have been
reviewed by Jones (1), drawing to a considerable extent on the volume by Stern (2).

Water Pollution

In analyzing the aqueous wastes generated by the fertilizer industry one fact quickly becomes evident. The principal pollutants found in the aqueous waste from most fertilizer
plants are inorganic materials. Except in a few instances, organic contaminants are relatively unimportant. Therefore, pollutional criteria such as BOD and COD are less important in evaluating fertilizer plant effluents than is the case with most industrial waste.
There are in general four types of aqueous effluent streams which can be generated in a
fertilizer plant: cooling water, steam condensate, process effluents, and sanitary waste.

On an overall industry basis, aqueous process effluents represent the major source of
highly contaminated waste streams. Process water accounts for only about 20 to 25% of
the water used by the fertilizer industry but contains the bulk of the contaminants generated by the various processes. Aqueous waste streams can normally be divided into five
general classes.

    (1)   By-product streams
    (2)   Scrubber solutions (from gas scrubbing equipment)
    (3)   Process spills
    (4)   Wash solutions (from equipment cleanup)
    (5)   Barometric condenser water

All fertilizer processes will generate one or more of the above streams. In the production
of the various major fertilizer materials, each process will generate one or more of the several waste streams described above. Table 209 lists the potential aqueous waste streams.
In every process, spills will occur which can lead to an additional waste stream. However,
it is impossible to define the magnitude and concentrations of such streams. In a phosphate fertilizer complex, waste streams generated by spills are usually discharged to the
acid plant gypsum pond where they become a part of the overall wastewater system for
the complex. In ammonia plant complexes the spill waste is usually combined with the
other highly contaminated waste streams.

The reader is referred to the discussion elsewhere in this volume of the individual fertilizer
production processes for more detail on pollutant sources and emissions. See particularly

Ammonia Manufacture, Nitrate Fertilizer Manufacture, and Phosphate Fertilizer Manufacture.

TABLE 209: PROBABLE AQUEOUS WASTE STREAMS FROM FERTILIZER PROCESSING

Product	Potential Waste Streams*
Ammonia	Process condensate Cooling water Steam condensate
Ammonium nitrate	Cooling water Steam condensate Off-gas scrubber solutions Condenser water
Ammonium sulfate	Condenser water Off-gas scrubber solutions
Urea	Cooling water Steam condensate Off-gas scrubber solutions
Calcined phosphate rock	Off-gas scrubber solution
Wet process phosphoric acid	Cooling water Gypsum slurry Off-gas scrubber solution Acid sludge Steam condensate
Furnace phosphoric acid	Cooling water Sludge Phossy water Off-gas scrubber solution
Normal superphosphate	Off-gas scrubber solutions
Triple superphosphate	Off-gas scrubber solutions
Ammonium phosphate	Off-gas scrubber solutions
Ammoniated superphosphate	Off-gas scrubber solutions
Potassium chloride (sylvinite)	Recycle brine bleed Salt-slimes tailings
Potassium chloride (brine)	Recycle brine bleed Salt tailings
NPK fertilizers	Off-gas scrubber solutions
Liquid blends	Equipment washdown
Bulk blends	Equipment washdown

*Waste streams generated by spill cleanup not included

Source: EPA Report 12020 FPD 09/71

References

(1) Jones, H.R., "Environmental Control in the Inorganic Chemical Industry," Park Ridge, New Jersey, Noyes Data Corp. (1972).
(2) Heller, A.N., Cuffe, S.T. and Goodwin, D.R. in *Air Pollution, III,* A.P. Stern, Ed., New York, Academic Press (1968).
(3) Fullam, H.T. and Faulkner, B.P., "Inorganic Fertilizer and Phosphate Mining Industries — Water Pollution and Control," *Report 12020 FPD 09/71,* Washington, D.C., U.S. Government Printing Office (September 1971).

FIBER GLASS MANUFACTURE

Glass fiber products are manufactured by melting various raw materials to form glass (predominantly borosilicate), drawing the molten glass into fibers, and coating the fibers with an organic material. The two basic types of fiber glass products, textile and wool, are manufactured by different processes.

In the manufacture of textiles, the glass is normally produced in the form of marbles after refining at about 2800°F (1540°C) in a regenerative, recuperative, or electric furnace. The marble-forming stage can be omitted with the molten glass passing directly to orifices to be formed or drawn into fiber filaments. The fiber filaments are collected as spools as continuous fibers and staple yarns, or in the form of a fiber glass mat on a flat, moving surface. An integral part of the textile process is treatment with organic binder materials followed by a curing step.

In the manufacture of wool products, which are generally used in the construction industry as insulation, ceiling panels, etc., the molten glass is most frequently fed directly into the forming line without going through a marble stage. Fiber formation is accomplished by air blowing, steam blowing, flame blowing, or centrifuge forming. The organic binder is sprayed onto the hot fibers as they fall from the forming device. The fibers are collected on a moving, flat surface and transported through a curing oven at a temperature of 400° to 600°F (200° to 315°C) where the binder sets. Depending upon the product, the wool may also be compressed as a part of this operation.

Air Pollution

The major emissions from the fiber glass manufacturing processes are particulates from the glass-melting furnace, the forming line, the curing oven, and the product cooling line. In addition, gaseous organic emissions occur from the forming line and curing oven. Particulate emissions from the glass-melting furnace are affected by basic furnace design, type of fuel (oil, gas, or electricity), raw material size and composition, and type and volume of the furnace heat-recovery system.

TABLE 210: EMISSION FACTORS FOR FIBER GLASS MANUFACTURING WITHOUT CONTROLS[a]

Emission Factor Rating: A

Type of process	Particulate		Sulfur oxides (SO_2)		Carbon monoxide		Nitrogen oxides (NO_2)		Fluorides	
	lb/ton	kg/MT	lb/ton	kg/MT	lb/ton	kg/MT	lb/ton	kg/MT	lb/ton	kg/MT
Textile products Glass furnace[b]										
Regenerative	16.4	8.2	29.6	14.8	1.1	0.6	9.2	4.6	3.8	1.9
Recuperative	27.8	13.9	2.7	1.4	0.9	0.5	29.2	14.6	12.5	6.3
Electric	ND[c]	–	–	–	–	–	–	–	–	–
Forming	1.6	0.8	–	–	–	–	–	–	–	–
Curing oven	1.2	0.6	–	–	1.5	0.8	2.6	1.3	–	–
Wool products[d] Glass furnace[b]										
Regenerative	21.5	10.8	10.0	5.0	0.25	0.13	5.0	2.5	0.12	0.06
Recuperative	28.3	14.2	9.5	4.8	0.25	0.13	1.70	0.9	0.11	0.06
Electric	0.6	0.3	0.04	0.02	0.05	0.03	0.27	0.14	0.02	0.01
Forming	57.6	28.8	–	–	–	–	–	–	–	–
Curing oven	3.5	1.8	ND	–	1.7	0.9	1.1	0.6	–	–
Cooling	1.3	0.7	–	–	0.2	0.1	0.2	0.1	–	–

[a]Emission factors expressed as units per unit weight of material processed.
[b]Only one process is generally used at any one plant.
[c]No data available.
[d]In addition, 0.09 lb/ton (0.05 kg/MT) phenol and 3.3 lbs/ton (1.7 mg/MT) aldehyde are released from the wool curing and cooling operations.

Source: EPA Publication AP-42

Organic and particulate emissions from the forming line are most affected by the composition and quality of the binder and by the spraying techniques used to coat the fibers; very fine spray and volatile binders increase emissions. Emissions from the curing ovens are affected by oven temperature and binder composition, but direct-fired afterburners with heat exchangers may be used to control these emissions. Emission factors for fiber glass manufacturing are summarized in Table 210.

Water Pollution

A general water flow diagram for an insulation fiber glass plant is pictured in Figure 28. Nonprocess waters identified in this diagram include boiler blowdown and water treatment backwashes. Those parameters that are likely to be found in significant quantities in each of the waste streams are listed in Table 211. A more detailed analysis of each waste flow (i.e., concentration ranges) is not possible since the combined waste stream only has been of interest to the industry from whom most of the data was obtained. The principal process waste streams within the process are the chain cleaning water and water sprays used on the exiting forming air.

FIGURE 28: GENERAL WATER FLOW DIAGRAM FOR AN INSULATION FIBER GLASS PLANT

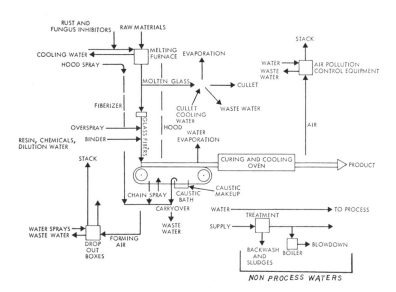

Source: EPA, July 13, 1973 (2)

The principal uses for steam are for building heating and steam attenuation. In the latter case the industry has been converting to compressed air attenuation. The accompanying boiler blowdown in this case is replaced by noncontact cooling water for air compressors. Water usages vary significantly between plants. Factors such as design of furnace, method of chain cleaning and method of air emissions control will affect quantities of water. For example, plants at which marbles are remelted require very little furnace cooling water, since the remelt furnaces are small melting pots. Large continuous drawing furnaces, however, need large quantities of water to control oven temperatures and to protect the furnace bricks.

TABLE 211: CONSTITUENTS OF INSULATION – FIBER GLASS PLANT WASTE STREAMS

Waste Stream	Phenols	BOD5	COD	Dissolved Solids	Suspended Solids	Oil & Grease	Ammonia	pH	Color	Turbidity	Temperature	Specific Conductance
Air Scrubbing	x	x	x	x	x			x		x	x	x
Boiler Blowdown				x	x					x	x	x
Caustic Blowdown				x	x			x		x		
Chain Spray	x	x	x	x	x	x	x	x	x	x		x
Cullet Cooling				x	x						x	
Fresh Water Treatment				x	x			x		x	x	x
Hood Spray	x	x	x	x	x	x	x	x	x	x		x
Noncontact Cooling Water			x	x							x	x

Source: EPA, July 13, 1973 (2)

Table 212 summarizes the raw waste concentrations for several plants. Although the numbers are not completely comparable because of treatment differences and different blowdown percentages, the table nevertheless shows a wide variance in wastewater composition. Other factors affecting the raw waste load include binder composition, chain temperature, and other thermal and time factors affecting the rate of resin polymerization. Annual raw waste loads in metric tons are computed in Table 213. The values are based upon an average of five parameters at four plants.

TABLE 212: RAW WASTE LOADS FOR INSULATION FIBER GLASS PLANTS

Plant	Phenol mg/l	BOD5 mg/l	COD mg/l	TSS mg/l	TDS mg/l	Turbidity	pH	Percent Blowdown[3]
H	363	156	2500-4000	116-561				
F[1]	2564	7800	43,603	360	3000-5000			8.3
G	4.11			76	822		7.7-8.9	13.0
A	212	991	6532	769	10,000-20,000[2]			1.5
B[1]	240	6200	23,000	200	16,000	200	8.0	1.0
D[1]					40,000[2]			2.3
I	11-98	900	3,290	690	2,080		6.1-12.2	

[1]Sample taken from water recirculation system.

[2]Given by company with no backup data.

[3]Defined as percent total process water used as overspray or binder dilution.

Source: EPA, July 13, 1973 (2)

Water usage and raw waste loads are not relatable in a practical manner to production levels or techniques. Of the nineteen existing plants, there may be as many different formulas for relating these factors. There are significant differences between plants even within the same company. A compensating factor, however, is the fact that all such wastes are amenable to the same general type of chemical and/or physical treatment.

TABLE 213: ANNUAL RAW WASTE LOADS

| Plant | Estimated Size (1000 metric tons per yr.) | Kilograms Pollutant Per Metric Ton Product | | | | |
		Phenol	Suspended Solids	BOD_5	COD	Dissolved Solids
A	120	0.36	1.29	1.67	11.0	
E	18	0.06	4.45	8.90	31.5	18.0
H	16	0.90	0.40	8.1		
I	131	0.33	5.60	6.65	24.2	14.1
Average		0.41	2.90	4.40	18.7	16.0
Annual Raw Waste Load[1] (Metric tons per yr.)		316	2240	3390	14,400	12,300

[1] Derived by multiplying kg/metric ton by 771,000 metric tons product per year by 1/1000 metric ton per kg.

Source: EPA, July 13, 1973 (2)

References

(1) U.S. Environmental Protection Agency, "Compilation of Air Pollutant Emission Factors," 2nd Edition, *Publication No. AP-42,* Research Triangle Park, N.C., Office of Air and Water Programs (April 1973).
(2) U.S. Environmental Protection Agency, "Development Document for Proposed Effluent Limitations Guidelines and New Source Performance Standards: Insulation Fiber Glass Manufacturing Segment of the Glass Manufacturing Point Source Category," Washington, D.C. (July 13, 1973).

FISH PROCESSING

The reader is also referred to the section of this book on Seafood Processing. The canning, dehydration, and smoking of fish, and the manufacture of fish meal and fish oil are the important segments of fish processing. There are two types of fish-canning operations: the wet-fish method, in which the trimmed fish are cooked directly in the can, and the precooked process, in which the whole fish is cooked and then hand-sorted before canning.

A large fraction of the fish received in a cannery is processed into by-products, the most important of which is fish meal. In the manufacture of fish meal, fish scrap from the canning lines is charged to continuous live-steam cookers. After the material leaves the cooker, it is pressed to remove oil and water. The pressed cake is then broken up, usually in a hammer mill, and dried in a direct-fired rotary drier or in a steam-tube rotary drier.

Air Pollution

The biggest problem from fish processing is odorous emissions. The principal odorous gases generated during the cooking portion of fish-meal manufacturing are hydrogen sulfide and trimethylamine. Some of the methods used to control odors include adsorption by activated carbon, scrubbing with oxidizing solution, and incineration. The only significant sources of dust emissions in fish processing are the driers and grinders used to handle dried fish meal. Emission factors for fish meal manufacturing are shown in Table 214.

TABLE 214: EMISSION FACTORS FOR FISH MEAL PROCESSING
Emission Factor Rating: C

Emission source	Particulates		Trimethylamine $(CH_3)_3N$		Hydrogen sulfide (H_2S)	
	lb/ton	kg/MT	lb/ton	kg/MT	lb/ton	kg/MT
Cookers, lb/ton (kg/MT) of fish meal produced[a]						
Fresh fish	—	—	0.3	0.15	0.01	0.005
Stale fish	—	—	3.5	1.75	0.2	0.10
Driers, lb/ton (kg/MT) of fish scrap[b]	0.1	0.05	—	—	—	—

[a]Reference 2.
[b]Reference 1.

Source: EPA Publication AP-42, 2nd Edition

References

(1) Walsh, R.T., Luedtke, K.D. and Smith, L.K., "Fish Canneries and Fish Reduction Plants," *Air Pollution Engineering Manual,* Danielson, J.A. (ed), US. DHEW, PHS, National Center for Air Pollution Control, Cincinnati, Ohio, *Publication Number 999-AP-40,* p 760-770 (1967).
(2) Summer, W., *Methods of Air Deodorization,* New York, Elsevier Publishing Company, p 284-286.

FRIT MANUFACTURE

Frit is used in enameling iron and steel and in glazing porcelain and pottery. In a typical plant, the raw materials consist of a combination of materials such as borax, feldspar, sodium fluoride or fluorspar, soda ash, zinc oxide, litharge, silica, boric acid, and zircon. Frit is prepared by fusing these various minerals in a smelter, and the molten material is then quenched with air or water. This quenching operation causes the melt to solidify rapidly and shatter into numerous small glass particles, called frit. After a drying process, the frit is finely ground in a ball mill where other materials are added.

Air Pollution

Significant dust and fume emissions are created by the frit-smelting operation. These emissions consist primarily of condensed metallic oxide fumes that have volatilized from the molten charge. They also contain mineral dust carryover and sometimes hydrogen fluoride.

Emissions can be reduced by not rotating the smelter too rapidly (to prevent excessive dust carryover) and by not heating the batch too rapidly or too long (to prevent volatilizing the more fusible elements). The two most feasible control devices for frit smelters are baghouses and venturi water scrubbers. Emission factors for frit smelters are shown in Table 215. Collection efficiencies obtainable for venturi scrubbers are also shown in the table.

TABLE 215:　EMISSION FACTORS FOR FRIT SMELTERS WITHOUT CONTROLS[a]

Emission Factor Rating:　C

Type of furnace	Particulates[b]		Fluorides[b]	
	lb/ton	kg/MT	lb/ton	kg/MT
Rotary	16	8	5	2.5

[a]Reference 2. Emission factors expressed as units per unit weight of charge.

[b]A venturi scrubber with a 21-inch (535-mm) water-gauge pressure drop can reduce particulate emissions by 67 percent and fluorides by 94 percent.

Source:　EPA Publication AP-42

References

(1)　Duprey, R.L. "Compilation of Air Pollutant Emission Factors," U.S. DHEW, PHS, National Center for Air Pollution Control, Durham, N.C., *PHS Publication Number 999-AP-42,* p 37–38 (1968).

(2)　Spinks, J.L., "Frit Smelters," *Air Pollution Engineering Manual,* Danielson, J.A. (ed), U.S. DHEW, PHS, National Center for Air Pollution Control, Cincinnati, Ohio, *PHS Publication Number 999-AP-40,* - 738–744 (1967).

FRUIT AND VEGETABLE INDUSTRY

Water Pollution

Water is extensively used in all phases of the food processing industry. For example, it is used as one of the following: a cleaning agent to remove dirt and foreign material, a heat transfer medium for heating and cooling, a solvent for removal of undesirable ingredients from the product, a carrier for the incorporation of additives into the product, and a method of transporting and handling the product.

Many of the steps used in the process of canning and freezing fruits and vegetables are common to the industry as a whole, and the character of the wastewaters are similar in that they contain biodegradable organic matter. Typically, the fruit or vegetable is received, washed and sorted to prepare it for subsequent processing. Commodities such as apples, citrus and potatoes are then usually peeled when the end product style is in a solid form (slices, cubes, or powder).

If the final product is a juice or liquid product, the peel is not removed from either the citrus or the apples. Subsequent process steps following the peel removal in which water may be used are trimming, slicing, blanching, cooling, concentrating and can washing/cooling. Water transport may be used, and cleanup is common to all processes. Although the steps used in processing the various commodities display a general similarity, there are variations in the equipment used and in the amount and character of the wastewaters produced. The parameters used to characterize the raw effluent were the flow, BOD, and SS; BOD_5 and SS are generally considered to be the best available measure of the waste load.

The following raw waste characteristics have been tabulated from the best available in-plant unit process waste characteristics. Total raw waste effluent values can be calculated but caution must accompany such a tubulation. The tabulations should not be used to develop effluent limitation guidelines. The waste characteristics of primary concern are BOD_5 and suspended solids. The following tabulations summarize BOD_5, SS and water usage values by process steps for apples, citrus, and potatoes. They have been synthesized from available data acquired through in-plant sampling with some supplemental in-plant data acquired

from processors (1). In only a few cases was complete in-plant data available. Information from 10 apple plants, 20 citrus plants, and 15 potato plants was used to develop these
tabulations. The tabulations are not used to develop effluent guidelines. The purpose of
this presentation is to show where substantial water savings can be realized and where substantial waste reductions can be accomplished. They should not be used to develop effluent limitations.

Washing, as listed in Table 216, includes receiving and sorting as well as fruit cleaning.
The apples are dumped into a water filled tank and are washed with water sprays after
leaving the tank or the associated water transport system. Mechanical peeling, slicing and
deaeration are treated as separate process steps. Cooking and cooling waters can be kept
separate and are shown as individual values. Cleanup (floor and equipment) normally occurs in a separate work shift, following one or two processing shifts.

TABLE 216: WATER USAGE AND WASTE CHARACTERIZATION IN APPLE PROCESSING

Process Step	Water Usage l/kkg	Water Usage G/T	BOD5 kg/kkg	BOD5 lb/T	Suspended Solids kg/kkg	Suspended Solids lb/T
Washing	142	34	0.09	0.18	0.03	.06
Peeling						
Mechanical	104	25	0.16	0.31	0.015	0.03
Slicing	638	158	2.49	4.97	0.182	0.36
Deaeration	71	17	2.21	4.42	0.12	0.24
Cooking	267	64	0.14	0.27	0.05	0.10
Cooling (1)	58	14	0.02	0.03	0.005	0.01
Transport	58	14	0.02	0.03	0.005	0.01
Clean-up	1,558	372	1.90	3.80	0.30	0.60

(1) 95% recirculated

Source: Report EPA 440/1-73/027

Although it is not yet general practice in the industry, some plants recycle the water used
for can cooling through a cooling tower or spray pond. When recycling, the small amount
of spray water used to clean the cans following cooking is kept separate from the cooling
water in order to keep organic material out of the cooling water. In this and subsequent
tabulations, the can wash water is included in the water used for cooking. It is estimated
that the cooling water requirements can be reduced to about 5% of the once-through requirement of 1,182 l/kkg (283 G/T) or to a level of 58 l/kkg (14 G/T) as used in Table 216.

The latter figure is used for the cooling step in apple processing. Seven of ten plants contributing data listed in Table 216 are primarily sampled plants processing stored fruit near
the end of the canning season. The ten plants make different apple products and product
mixes and range in size from less than 3.6 kkg/hr (4T/hr) to more than 28 kkg/hr (31 T/hr).
The production of each product style (sauce, slices and juice) employs a different set of
operations. Water usage and characterization can be determined for the production of
slices, sauce and juice. The process steps employed in the manufacture of sauce are washing, peeling, slicing, cooking, cooling, transport and cleanup. The process steps for slices
are washing, peeling, slicing, deaerating, cooking, cooling, transport and cleanup. The production of juice involves only washing (including receiving and sorting), transport, cooling
and cleanup. Data from about 20 citrus plants processing different citrus products and
coproducts contributed to the tabulation given in Table 217.

Fruit cleaning, as used in Table 217, includes washing, as well as receiving and sorting.
The citrus is sometimes stored in bins and upon leaving the bins is washed with water
sprays and/or roller brushes with sprays and sometimes detergent. Juice extraction may
be accomplished by slicing the citrus in half and reaming each half simultaneously. After
extraction, the peel and the majority of the pulp are separated from the juice and may,
or may not, be processed for citrus oil and other by-products.

Depending on the extractor, oil may or may not be liberated from the peel at this point. The juice is next passed through a finisher and may then be either processed into single strength (ss), which involves juice pasteurization/homogenization and can cooling, or concentrated which involves evaporation of the juice. The majority of the cleanup normally occurs in a separate work shift, following one, two, or two and one-half processing shifts.

TABLE 217: WATER USAGE AND WASTE CHARACTERIZATION IN CITRUS PROCESSING

Process Steps	Water Usage 1/kkg	(G/T)	BOD5 kg/kkg	(lb/T)	Suspended Solids kg/kkg	(lb/T)
Fruit Cleaning	303	(73)	0.08	(0.16)	0.04	(0.07)
Extracting	389	(93)	0.40	(0.79)	0.27	(0.54)
Pasteurizing/Homogenizing	62	(15)	0	(0)	0	(0)
Cooling (1)						
Juice Products	221	(53)	0.03	(0.05)	0.02	(0.03)
Segments			0.01	(0.02)	0.01	(0.02)
Juice Condensing	400	(96)	0.06	(0.12)	0.02	(0.03)
Barometric Condensing (2)						
Juice Products	50	(12)	0.07	(0.13)	0.09	(0.17)
Waste Heat Evaporator	71	(17)	0.15	(0.29)	0.09	(0.18)
Peeled Fruit Washing	129	(31)	0.04	(0.07)	0.01	(0.01)
Caustic Treatment	1	(0.3)	0.01	(0.02)	0.01	(0.01)
Centrifuging	144	(35)	3.07	(6.14)	0.51	(1.02)
Container Washing	75	(18)	0		0	(0)
Waste Heat Evaporator Condensate	334	(80)	0.33	(0.66)	0.11	(0.22)
Waste Heat Evaporator Scrubber Effl.	351	(84)	0.22	(0.43)	0.08	(0.15)
Oil Lean Residue From Separator	126	(30)	0.16	(0.32)	0.25	(0.49)
Boiler Blowdown	60	(14)	0.01	(0.02)	0.01	(0.02)
Regeneration Brine	13	(3)	0	(0)	0	(0)
Cleanup						
Juice Products	705	(169)	0.16	(0.32)	0.16	(0.31)
Segments	371	(89)	0.36	(0.72)	0.07	(0.13)
Peel Products	484	(116)	0.07	(0.14)	0.11	(0.22)

(1) 90% recirculated
(2) 2% cooling tower blowdown

TABLE 218: WATER USAGE AND WASTE CHARACTERIZATION IN POTATO PROCESSING

Process Steps	Water Usage 1/kkg	G/T	BOD5 kg/kkg	lb/T	Suspended Solids kg/kkg	lb/T
Washing	1,102	264	0.676	1.35	1.383	2.76
Peeling						
Dry Caustic	1,448	347	7.325	14.62	9.569	19.1
Wet Caustic	3,000	719	20.245	40.41	28.662	57.2
Steam	2,391	573	15.215	30.37	13.427	26.8
Trimming	793	190	0.777	1.55	0.26	0.52
Slicing						
Dehydrated	764	183	0.296	0.59	0.701	1.4
Frozen	1,519	364	2.630	5.25	1.303	2.6
Blanching						
Dehydrated	175	42	0.701	1.40	0.601	1.2
Frozen	1,043	250	5.461	10.9	2.104	4.2
Cooling	668	160	1.172	2.34	-	-
Cooking	448	117	1.192	2.38	-	-
Dewatering	513	123	0.471	0.94	0.351	0.70
Fryer Scrubber	417	100	-	-	-	-
Fryer Belt Spray	417	100	-	-	-	-
Refrigeration	1,602	384	-	-	-	-
Transport Water	292	70	0.261	0.52	-	-
Cleanup	951	228	2.725	5.44	-	-

Source: Report EPA 440/1-73/027

Oil/peel-pulp by-products are manufactured from plants that have some type of juice operation. Additional water flows involved include the waste heat evaporator condensate, the waste heat evaporator's barometric condensate, the waste heat evaporator's scrubber effluent, and the oil lean residue from the d-limonene residue separator. The production of segments involves wastewater from peeling, caustic treating, washing, cooking, cooling and cleanup. It is possible to develop water usage figures for plants making various product combinations.

Water use figures for juice and oil processing, segment processing and juice, oil and peel product processing can be determined. The figures are the summation of the water flows from each of the process steps required to produce juice, oil, segments and peel products with minimum water usage. There is a degree of variability for water usage and waste characterization among the products and product combinations. The majority of this variability is attributable to differences in plant operation and plant management and difference in availability of raw material, water, and waste treatment facilities.

Minor differences in size, age and location of plants also contribute to the total variability. Even without consideration of these sources of variability, there is sufficient similarity for water usage and waste character among the product combinations to support a single category for the citrus industry. Fifteen potato plants processing frozen and/or dehydrated potato products contributed to the waste characterization given in Table 218. In each of these potato processing subcategories, there are several processing steps using large quantities of water which are common to both categories.

For example, it is common practice to use water hoses to remove the potatoes from storage and direct them into a water transport system for delivery to the process area. In an exemplary water usage plant, the water which is used to receive and clean the potatoes is usually segregated from the process water. The receiving/cleaning water is recycled through a settling basin where there is sufficient retention time to allow the solids to settle out in the basin. The make-up water to this closed system is added by water sprays which are positioned to rinse the potatoes as they enter the process.

Three methods of peeling are in current industrial use within the frozen and dehydrated potato processing industry: dry caustic, conventional wet caustic and steam. With the conventional wet caustic and steam peeling systems, large quantities of water were used for removal of the treated peel. This results in large waste loads appearing in the plant waste effluent discharge as can be seen in Table 218.

During the slicing step, large quantities of water are used to remove any starch adhering to the surface of the pieces. This water is also used to convey the pieces to the blanching step. Water blanching is required for both frozen and dehydrated products since a large amount of the leachables must be removed from the potato pieces during the blanching step. In the case of frozen products, a three-step series blanching system is used. While for dehydrated products the water blanching step is followed by a water cooling step and then a cooking step. The frozen products are usually French fried while the majority of the dehydrated products are dried in a flake or granule form.

The production of dehydrated and frozen products employs different process steps. Water usage and waste characterization can be determined for the production of both products. As mentioned earlier, the tabulations should not be used to develop wastewater effluent limitation guidelines. The tabulations are presented only to show where substantial water savings can be realized and where substantial waste reductions can be accomplished.

References

(1) U.S. Environmental Protection Agency, "Development Document for Proposed Effluent Limitations Guidelines and New Source Performance Standards for the Citrus, Apple and Potato Segment of the Canned and Preserved Fruits and Vegetables Processing Point Source Category," *Report EPA 440/1-73/027,* Wash., D.C. (Nov. 1973).
(2) Jones, H.R., "Waste Disposal Control in the Fruit and Vegetable Industry," Park Ridge, N.J., Noyes Data Corp. (1973).

FUEL OIL COMBUSTION

Fuel oil is classified into two major types, residual and distillate. Distillate fuel oil is primarily a domestic fuel, but it is used in some commercial and industrial applications where a high-quality oil is required. Fuel oils are classified by grades: grades No. 1 and No. 2, distillate; No. 5 and No. 6, residual; and No. 3 and No. 4, blends. (Grade No. 3 has been practically discontinued.) The primary differences between residual oil and distillate oil are the higher ash and sulfur content of residual oil and the fact that it is much more viscous and therefore harder to burn properly. Residual fuel oils have a heating value of approximately 150,000 Btu/gallon (10,000 kcal/liter); the heating value for distillate oils is about 140,000 Btu/gallon (9,300 kcal/liter).

Air Pollution

Emissions from oil combustion are dependent on type and size of equipment, method of firing, and maintenance. Table 219 presents emission factors for fuel oil combustion. Note that the industrial and commercial category is split into residual and distillate because there is a significant difference in particulate emissions from the same equipment, depending on the fuel oil used. It should also be noted that power plants emit less particulate matter per quantity of oil consumed, reportedly because of better design and more precise operation of equipment.

In general, large sources produce more nitrogen oxides than small sources, primarily because of the higher flame and boiler temperatures characteristic of large sources. Large sources, however, emit fewer aldehydes than smaller sources as a result of more complete combustion and higher flame temperatures. Hydrocarbon and carbon monoxide emissions can be kept minimal if proper operating practices are employed; however, as the data suggest, this control is more often accomplished in larger equipment.

TABLE 219: EMISSION FACTORS FOR FUEL OIL COMBUSTION
EMISSION FACTOR RATING: A

| | | | | Type of unit Industrial and commercial | | | | | |
| | Power plant | | Residual | | Distillate | | Domestic | |
Pollutant	lb/10³ gal	kg/10³ liters	lb/10³ gal	kg/10³ liters	lb/10³ gal	kg/10³ liters	lb/10³ gal	kg/10³ liters
Particulate	8	1	23	2.75	15	1.8	10	1.2
Sulfur dioxide*	157S	19S	157S	19S	142S	17S	142S	17S
Sulfur trioxide*	2S	0.25S	2S	0.25S	2S	0.25S	2S	0.25S
Carbon monoxide	3	0.4	4	0.5	4	0.5	5	0.6
Hydrocarbons	2	0.25	3	0.35	3	0.35	3	0.35
Nitrogen oxides (NO₂)	105	12.6	(40 to 80)	(4.8 to 9.6)	(40 to 80)	(4.8 to 9.6)	12	1.5
Aldehydes (HCHO)	1	0.12	1	0.12	2	0.25	2	0.25

*S equals percent by weight of sulfur in the oil.

Source: EPA Publication AP-42, 2nd Edition.

Reference

(1) U.S. Environmental Protection Agency, "Compilation of Air Pollutant Emission Factors," 2nd Edition, *Publication Number AP-42,* Research Triangle Park, N.C., Office of Air and Water Programs (April 1973).

GLASS MANUFACTURE

Nearly all glass produced commercially is one of five basic types: soda-lime, lead, fused silica, borosilicate, and 96% silica. Of these, the modern soda-lime glass constitutes 90% of the total glass produced and will thus be the only type discussed in this section. Soda-lime glass is produced on a massive scale in large, direct-fired, continuous-melting furnaces in which the blended raw materials are melted at 2700°F (1480°C) to form glass (1)(2).

Air Pollution

Emissions from the glass-melting operation consist primarily of particulates and fluorides, if fluoride-containing fluxes are used in the process. Because the dust emissions contain particles that are only a few microns in diameter, cyclones and centrifugal scrubbers are not as effective as baghouses or filters in collecting particulate matter. Table 220 (5) summarizes the emission factors for glass melting.

TABLE 220: EMISSION FACTORS FOR GLASS MELTING
EMISSION FACTOR RATING: D

Type of glass	Particulates[a]		Fluorides[b]	
	lb/ton	kg/MT	lb/ton	kg/MT
Soda-lime	2	1	4F[c]	2F[c]

[a] Reference 3. Emission factors expressed as units per unit weight of glass produced.
[b] Reference 4.

[c] F equals weight percent of fluoride in input to furnace; e.g., if fluoride content is 5 percent, the emission factor would be 4F or 20 (2F or 10).

Source: EPA Publication AP-42, 2nd Edition.

Water Pollution

Water is used to some extent in all of the subcategories of the glass industry. Cooling and boiler water are required at all plants. Washwater is used in the plate, float, and automotive fabrication subcategories for washing, and water is the transfer medium for grinding sand and rouge in the plate process.

Noncontact cooling, boiler, and water treatment wastewaters are considered auxiliary wastes as distinguished from process wastewaters. Process wastewater is defined as water which has come into direct contact with the glass, and includes such sources as washing, quenching, and grinding and polishing. In sheet glass and rolled glass manufacture, only cooling water is used; no process water is used and therefore no process wastes are produced from the manufacture of these two types of glass.

Plate glass manufacturing is the production from raw materials of a high-quality thick glass sheet. This subcategory has historically been the greatest source of waste in the industry since large volumes of high-suspended-solids wastewater are produced. Owing to high production costs and related water pollution problems, plate glass is being replaced by float glass. Only three plate glass plants remain in the United States. The major process steps and points of water usage are shown in Figure 29.

Some typical characteristics for the combined wastewater stream are listed in Table 221. In all cases except for pH, the values listed are the quantities added to the water as a result of the plate glass process, and concentrations in the influent water have been subtracted.

FIGURE 29: PLATE GLASS MANUFACTURING

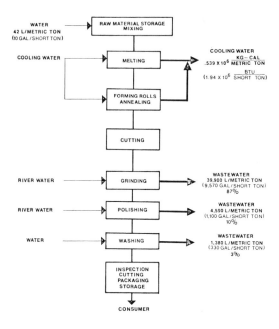

Source: Report EPA 440/1-73/00/2.

TABLE 221: RAW WASTEWATER[a]—PLATE GLASS MANUFACTURING PROCESS

Flow	45,900	l/metric ton		11,000 gal/short ton	
pH	9				
Temperature[b]	2.8°C		6°F		
Suspended Solids	690	kg/metric ton	1,375	lb/short ton	15,000 mg/l
COD[b]	4.6	kg/metric ton	9.2	lb/short ton	100 mg/l
Dissolved Solids[b]	8.0	kg/metric ton	16.1	lb/short ton	175 mg/l

(a) Represents typical plate glass process wastewater prior to treatment.
     Absolute value given for pH, increase over plant influent level given
     for other parameters.

(b) Indication of approximate level only; insufficient data are available
     to define actual level.

Source: Report EPA 440/1-73/00/2.

The float process may be considered the replacement for plate glass manufacturing. Float
glass production is substantially less expensive and process wastewater has all but been
eliminated. The major process steps and points of water usage are illustrated in Figure 30.

FIGURE 30: FLOAT GLASS MANUFACTURING

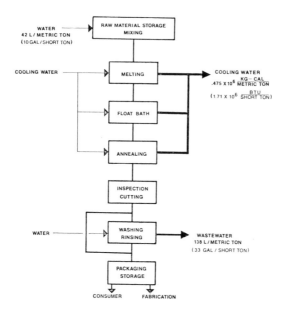

Source: Report EPA 440/1-73/00/2.

Some typical characteristics of float-glass washwater blowdown are listed in Table 222. In all cases except for pH, the values listed are the quantities added to the water as a result of the float glass process, and concentrations in the influent water have been subtracted.

TABLE 222: RAW WASTEWATER[a]—FLOAT GLASS MANUFACTURING PROCESS

Flow	138	l/metric ton	33	gal/ton		
pH	8					
Temperature[b]	37 C		98 F			
Suspended Solids	2	g/metric ton	.0041	lb/short ton	15	mg/l
Oil	.7	g/metric ton	.0014	lb/short ton	5	mg/l
COD	2	g/metric ton	.0041	lb/short ton	15	mg/l
BOD	.25	g/metric ton	.0005	lb/short ton	2	mg/l
Phosphorus	(c)					
Dissolved Solids	14	g/metric ton	.028	lb/short ton	100	mg/l

(a) Representative of typical float glass process wastewater. Absolute value given for pH and temperature, increase over plant influent level given for other parameters.

(b) Indication of approximate level only; insufficient data is available to define actual level.

(c) No information is available on wastewater containing phosphorus.

Source: Report EPA 440/1-73/00/2.

Solid tempered automotive glass fabrication is the fabrication from glass blanks of auto-mobile backlights (back windows) and sidelights (side windows). The major process steps and points of water usage are illustrated in Figure 31.

FIGURE 31: SOLID TEMPERED AUTOMOTIVE GLASS FABRICATION

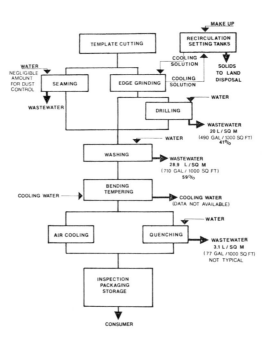

Source: Report EPA 440/1-73/00/2.

Some typical characteristics of the combined wastewater resulting from solid tempered automotive fabrication are listed in Table 223. In all cases except for pH, the values listed are the quantities added to the water as a result of solid tempered automotive fabrication. The background level in the influent water has been subtracted. The significant parameters are BOD, suspended solids, and oil.

TABLE 223: RAW WASTEWATER[a] SOLID TEMPERED AUTOMOTIVE GLASS FABRICATION

Flow	49 l/m²	1,200 gal/1,000 ft²	
pH	7		
Temperature[b]	8 °C	17 °F	
Suspended solids	4.9 g/m²	1 lb/1,000 ft²	100 mg/l
Oil	0.64 g/m²	0.13 lb/1,000 ft²	13 mg/l
COD[b]	1.22 g/m²	0.25 lb/1,000 ft²	25 mg/l

(continued)

TABLE 223: (continued)

BOD	0.73 g/m²	0.15 lb/1,000 ft²	15 mg/l
Dissolved solids	4.9 g/m²	1 lb/1,000 ft²	100 mg/l

[a]Representative of typical solid tempered automotive process wastewater. Absolute value given for pH, increase over plant influent level given for other parameters.

[b]Indication of approximate level only; insufficient data are available to define actual level.

Windshield fabrication is the manufacturing of laminated windshields from glass blanks and vinyl plastic. Oil resulting from oil autoclaving is the major constituent in this wastewater. The major process steps and points of water usage are illustrated in Figure 32.

FIGURE 32: WINDSHIELD FABRICATION

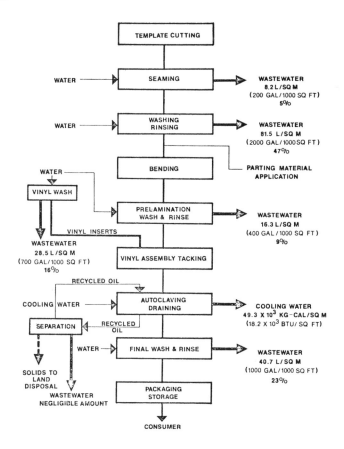

Source: Report EPA 440/1-73/00/2.

Some typical characteristics of the combined wastewater stream resulting from windshield fabrication are listed in Table 224. In all cases except for pH, the values listed are the quantities added to the water as a result of windshield fabrication. The influent water background levels have been subtracted. These data apply to a plant where an initial hot-water rinse is used for the post-lamination wash. No information is available for plants using an initial detergent wash.

TABLE 224:   RAW WASTEWATER[a] WINDSHIELD FABRICATION USING OIL
AUTOCLAVES

Flow	175 l/m²	4,300 gal/1,000 ft²	
pH	7–8		
Temperature	18.9 °C	40 °F	
Suspended solids	4.4 g/m²	0.9 lb/1,000 ft²	25 mg/l
Oil	298 g/m²	61 lb/1,000 ft²	1,700 mg/l
COD	298 g/m²	61 lb/1,000 ft²	1,700 mg/l
BOD	5.9 g/m²	1.2 lb/1,000 ft²	33 mg/l
Total phosphorus	0.98 g/m²	0.2 lb/1,000 ft²	5.6 mg/l

[a]Representative of typical process wastewater from the fabrication of windshields
using oil autoclaves. Absolute values are listed for pH; the increase over plant
influent level is given for other parameters.

Source: Report EPA 440/1-73/001-a

References

(1)   Netzley, A.B. and McGinnity, J.L., "Glass Manufacture." In: *Air Pollution Engineering Manual.* Danielson, J.A. (ed.). U.S. DHEW, PHS, National Center for Air Pollution Control. Cincinnati, Ohio. *PHS Publication Number 999-AP-40.* (1967) pages 720-730.

(2)   Duprey, R.L. Compilation of Air Pollutant Emission Factors. U.S. DHEW, PHS, National Center for Air Pollution Control, Durham, N.C. *PHS Publication Number 999-AP-42.* (1968) p. 38.

(3)   Technical Progress Report: Control of Stationary Sources. Los Angeles County Air Pollution Control District 1 (April 1960).

(4)   Semrau, K.T. "Emissions of Fluorides from Industrial Processes: A Review." *J. Air Pol. Control Assoc.* 7(2):92-108, (August 1957).

(5)   U.S. Environmental Protection Agency, "Compilation of Air Pollutant Emission Factors," 2nd Ed., *Publication No. AP-42,* Research Triangle Park, N.C., Office of Air and Water Programs (April 1973).

(6)   U.S. Environmental Protection Agency, "Development Document for Proposed Effluent Limitations Guidelines and New Source Performance Standards for the Flat Glass Segment of the Glass Manufacturing Point Source Category," *Report EPA 440/1-73/001-a, Washington, D.C (October 1973).*

GOLD AND SILVER MINING AND PRODUCTION

Water Pollution

Table 225 summarizes some data on gold and silver mining and production contribution to the water pollution problem.

TABLE 225: POLLUTION FROM GOLD AND SILVER MINING AND PRODUCTION

Metal	Overall Operation	Unit Operation	Use of Water	Wastewater Characteristics	Status of Pollution Control	Recommendations*
Gold	Placer mining	Dredging			No information	(1)
		Washing and screening			No information	(1)
		Gravity concentration			No information	(1)
	Lode mining	Underground mining	Mine drainage		No information	(1)
		Gravity concentration	Process		No information	(1)
	Extraction	Amalgamation	Process	Suspended solids (high slimes), high mercury	One case showed amalgamation discontinued; information incomplete.	(2)
	Extraction	Cyanide leaching	Process	Suspended solids (high slimes), high cyanide	One case deciding between oxidation in holding ponds or destruction with hypochlorite; information incomplete.	(3)
	Refining	Doré-furnace	Indirect cooling	As previously		
	Refining	Parting plant	Electrolytic chloride solutions	Unknown	Information lacking	(4)
Silver	Mining				There are 4 nominal silver mines in the U.S. (the balance of silver is mined at copper, lead, and zinc mines); information lacking on the 4 silver mines.	(5)
		Concentration and extraction			Information lacking for the four silver mines.	(5)

(continued)

TABLE 225: (continued)

Metal	Overall Operation	Unit Operation	Use of Water	Wastewater Characteristics	Status of Pollution Control	Recommendations*
	Refining	Doré-furnace, parting plant	Cooling electrolyte		Information lacking for the output of the four silver mines.	(5)

*Recommendations for Future Action.
    (1)   Need basic industry data and problem definition.
    (2)   Need basic information on number of industrial operations and feasibility of alternative extraction processes.
    (3)   Need better basic information on industry practice or the leading example could be used to set waste loads, costs, and effluent standards on the basis of holding pond oxidation of cyanides to cyanates.
    (4)   Need basic information on industrial practice for problem definition.
    (5)   Need basic information on four operations.

Source: Reports EPA-R-2-73-247a and b.

Reference

(1) Hallowell, J.B., Shea, J.F., Smithson, G.R., Jr., Tripler, A.B. and Gonser, B.W., "Water Pollution Control in the Primary Nonferrous Metals Industry" (2 volumes), *Reports No. EPA-R2-73-247a and b,* Washington, D.C., U.S. Government Printing Office (September 1973).

GRAIN AND FEED MILLS AND ELEVATORS

Grain elevators are primarily transfer and storage units and are classified as either the smaller, more numerous country elevators or the larger terminal elevators. At grain elevator locations the following operations can occur: receiving, transfer and storage, cleaning, drying, and milling or grinding. Many of the large terminal elevators also process grain at the same location. The grain processing may include wet and dry milling (cereals), flour milling, oil-seed crushing, and distilling. Feed manufacturing involves the receiving, conditioning (drying, sizing, cleaning), blending, and pelleting of the grains, and their subsequent bagging or bulk loading.

Air Pollution

Emissions from feed and grain operations may be separated into those occurring at elevators and those occurring at grain processing operations or feed manufacturing operations. Emission factors for these operations are presented in Table 226. Because dust collection systems are generally applied to most phases of these operations to reduce product and component losses, the selection of the final emission factor should take into consideration the overall efficiency of these control systems.

Emissions from grain elevator operations are dependent on the type of grain, the moisture content of the grain (usually 10 to 30%), the amount of foreign material in the grain (usually 5% or less), the degree of enclosure at loading and unloading areas, the type of cleaning and conveying, and the amount and type of control used.

Factors affecting emissions from grain processing operations include the type of processing (wet or dry), the amount of grain processed, the amount of cleaning, the degree of drying or heating, the amount of grinding, the temperature of the process, and the degree of control applied to the particulates generated.

TABLE 226: PARTICULATE EMISSION FACTORS FOR GRAIN HANDLING AND
PROCESSING, EMISSION FACTOR RATING: B

Type of source	Emissions	
	lb/ton	kg/MT
Terminal elevators[a]		
Shipping or receiving	1	0.5
Transferring, conveying, etc.	2	1
Screening and cleaning	5	2.5
Drying	6	3
Country elevators[b]		
Shipping or receiving	5	2.5
Transferring, conveying, etc.	3	1.5
Screening and cleaning	8	4
Drying	7	3.5
Grain processing		
Corn meal[c]	5	2.5
Soybean processing[b]	7	3.5
Barley or wheat cleaner[d]	0.2[e]	0.1[e]
Milo cleaner[f]	0.4[e]	0.2[e]
Barley flour milling[c]	3[e]	1.5[e]
Feed manufacturing		
Barley[f]	3[e]	1.5[e]

[a]References 2 and 3.
[b]Reference 3.
[c]References 3 and 4.
[d]References 5 and 6.
[e]At cyclone exit (only non-ether-soluble particulates).
[f]Reference 6.

Source: EPA Publication AP-42, 2nd Edition

Factors affecting emissions from feed manufacturing operations include the type and amount
of grain handled, the degree of drying, the amount of liquid blended into the feed, the type
of handling (conveyor or pneumatic), and the degree of control.

Water Pollution

Process water use and wastewater discharges vary markedly in the grain milling industry,
ranging from extremely high uses and discharges in the corn wet milling segment to virtually no process wastewaters in ordinary flour and rice milling. By far the largest water
users and hence, the greatest wastewater dischargers are the corn wet mills. The very
nature of corn wet milling processes is different from other segments of the grain milling
industry. In effect, these plants are large chemical complexes involving, as their name
implies, wet production methods.

Dry corn and normal wheat milling may employ water to clean the incoming grain, although
many plants, particularly the wheat mills, use mechanical methods for grain cleaning.
Bulgur and parboiled rice manufacturing techniques require water for steeping or cooking
and hence, generate modest quantities of process wastewaters.

Corn Milling: The reader is referred to the section on "Starch Manufacture" for details
of water pollution problems associated with that particular corn milling process.

Water use in corn dry milling is generally limited to corn washing, tempering, and cooling. Not all mills use water to clean the corn, probably because the resultant wastewaters constitute a pollution problem. It is believed that most of the larger mills, however, do wash the corn although data on the number of such installations are not available. Water use for this purpose ranges from about 0.45 to 1.2 m³/kkg (3 to 8 gal/SBu).

After washing, water is added to the corn to raise the moisture content to about 21 to 25% in order to make it more suitable for subsequent milling. Only enough water is added in this operation to reach the desired moisture content and no wastewater is generated.

Other than infrequent car washing, the only process wastewater in corn dry milling is that originating from the washing of corn. Data on raw waste characteristics from three plants are presented in Table 227. In one instance, the plant also processes soybeans and the wastewaters are combined. This same mill generates some wastewater from air pollution control equipment (wet scrubbers) on corn and soybean processing systems. These wastes are excluded from the data in Table 227 inasmuch as they originate from secondary processing of milled corn rather than the basic milling sequence covered in this document.

Average wastewater discharges from the three mills range from about 0.48 to 0.9 m³/kkg (3,200 to 6,000 gal/MSBu). The wastewaters are characterized by high BOD_5 and suspended solids concentrations. The raw wastewater BOD_5 values average 1.14 kg/kkg (64 lb/MSBu), and the suspended solids average 1.62 kg/kkg (91 lb/MSBu).

Insufficient data were available (7) to establish any relationships between wastewater characteristics and such factors as plant age, size, and operating procedures. Clearly the size of plant and the type and cleanliness of the corn will influence both the flow and waste characteristics, but in ways that cannot be defined at this time.

TABLE 227: WASTEWATER CHARACTERISTICS—CORN DRY MILLING

Plant	Flow cu m/day	Flow gals/day	BOD mg/l	COD mg/l	Suspended Solids mg/l	pH	Phosphorus as P mg/l	Total Nitrogen as N mg/l
1	227	60,000	900	–	1,522	–	–	–
2	900	238,000	603	1,795	1,038	7.8	–	–
3	818	216,000	2,748	4,901	3,485	3.7	43.0	13.2

Source: Report EPA 440/1-73/028

Wheat Milling: The normal milling of wheat into flour uses water only in tempering and cooling and no process wastewaters are discharged. A few normal flour mills do wash the wheat, but the vast majority use dry cleaning techniques. Accordingly, the remainder of this discussion will concentrate on bulgur production.

Water is added to the wheat in the soaking operation. Depending on the specific process employed, water may be added at as many as four locations, all essentially relating to the same soaking operation. Water usage for typical bulgur plants ranges from about 115 to 245 m³/day (30,000 to 65,000 gpd). Most of this water is used to raise the moisture content of the wheat from 12%, as received, to about 42%.

The only source of process wastewater, in the production of bulgur, is from steaming and cooking. As the grain is transferred from bin to bin, water is added on the conveyors and wastewater is discharged. The total quantities of wastewater from a given bulgur plant are quite small, ranging from 38 to 115 m³/day (10,000 to 30,000 gpd). Raw waste data from two plants are presented in Table 228. The BOD₅ values cited in the accompanying table correspond to an average of about 0.11 kg/kkg (5.9 lb/MSBu) of BOD₅ and 0.10 kg/kkg (5.5 lb/MSBu) of suspended solids.

TABLE 228: WASTEWATER CHARACTERISTICS—BULGUR PRODUCTION

	Concentration mg/l
BOD₅	238 - 521
COD	800
Suspended solids	294 - 414
Phosphorus as P	5.6
Total nitrogen as N	3.6
pH	5.8

Source: EPA Publication Number AP-42, 2nd Edition

Factors influencing wastewater characteristics undoubtedly include the particular production methods used, type of wheat, and operational procedures. Unfortunately, insufficient data are available to evaluate quantitatively the influence of these factors.

Rice Milling: The ordinary milling of rice to produce either brown or white rice utilizes no process waters and, hence, generates no wastewaters. Water is used in the production of parboiled rice and the remainder of this discussion will focus on this production method.

In the parboiled rice process, water is added in the steeping or cooking operation. Water use in the industry varies from about 1.4 to 2.1 m³/kkg (17 to 25 gal/cwt). Additional water is used in boilers for steam production for the parboiling process. At least one plant uses wet scrubbers for dust control thereby generating an additional source of wastewater.

Limited data are available on raw wastewater characteristics from rice parboiling. The information that is available is summarized in Table 229. The raw waste loads presented in the table correspond to 1.8 kg/kkg (0.18 lb/cwt) of BOD₅ and 0.07 kg/kkg (0.007 lb/cwt) of suspended solids. In general, the waste may be characterized as having a high soluble BOD₅ content and a low suspended solids level.

TABLE 229: WASTEWATER CHARACTERISTICS—PARBOILED RICE MILLING

	Concentration mg/l
BOD₅	1280 - 1305
COD	2810 - 3271
Suspended solids	33 - 77
Dissolved solids	1687
Phosphorus as P	98
Total nitrogen as N	7.0

Source: EPA Publication Number AP-42, 2nd Edition

Based on the very limited amount of data available, it appears that the waste characteristics from parboiled rice plants are quite similar. While there are some differences in flow volumes, the total waste loads per unit of production are similar.

References

(1) U.S. Environmental Protection Agency, "Compilation of Air Pollutant Emission Factors," 2nd Edition, *Publishing Number AP-42,* Research Triangle Park, N.C., Office of Air and Water Programs (April 1973).

(2) Thimsen, D.J. and Aften, P.W. "A Proposed Design for Grain Elevator Dust Collector." *J. Air Pol. Control Assoc.* 18(11):738-742 (November 1968).

(3) Communication between H.L. Kiser, Grain and Feed Dealers National Association, and Resources Research, Inc., Washington, D.C. (September 1969).

(4) Contribution of Power Plants and Other Sources to Suspended Particulate and Sulfur Dioxide Concentrations in Metropolis, Illinois. U.S. DHEW, PHS, National Air Pollution Control Administration (1966).

(5) Larson, G.P., Fischer, G.I. and Hamming, W.J., "Evaluating Sources of Air Pollution." *Ind. Eng. Chem.* 45:1070-1074 (May 1953).

(6) Donnelly, W.H., "Feed and Grain Mills." In: *Air Pollution Engineering Manual.* Danielson, J.A. (ed.). U.S. DHEW, PHS, National Center for Air Pollution Control. Cincinnati, Ohio. *Publication Number 999-AP-40* (1967) p. 359.

(7) U.S. Environmental Protection Agency, "Development Document for Proposed Effluent Limitations Guidelines and New Source Performance Standard for the Grain Processing Segment of the Grain Mills Point Source Category," *Report EPA 440/1-73/028,* Washington, D.C. (December 1973).

GYPSUM MANUFACTURE

Gypsum, or hydrated calcium sulfate, is a naturally occurring mineral that is an important building material. When heated gypsum loses its water of hydration, it becomes plaster of Paris, or when blended with fillers it serves as wall plaster. In both cases the material hardens as water reacts with it to form the solid crystalline hydrate.

The usual method of calcination of gypsum consists of grinding the mineral and placing it in large, externally heated calciners. Complete calcination of 1 ton (0.907 MT) of plaster takes about 3 hours and requires about 1.0 million Btu (0.25 million kcal).

Air Pollution

The process of calcining gypsum appears to be devoid of any air pollutants because it involves simply the relatively low-temperature removal of the water of hydration. However, the gases created by the release of the water of crystallization carry gypsum rock dust and partially calcined gypsum dust into the atmosphere. In addition, dust emissions occur from the grinding of the gypsum before calcining and from the mixing of the calcined gypsum with filler. Table 230 (1) presents emission factors for gypsum processing.

TABLE 230: PARTICULATE EMISSION FACTORS FOR GYPSUM PROCESSING[a]
EMISSION FACTOR RATING: C

Type of process	Uncontrolled emissions		With fabric filter		With cyclone and electrostatic precipitator	
	lb/ton	kg/MT	lb/ton	kg/MT	lb/ton	kg/MT
Raw-material dryer (if used)	40	20	0.2	0.1	0.4	0.2
Primary grinder	1	0.5	0.001	0.0005	–	–
Calciner	90	45	0.1	0.05	–	–
Conveying	0.7	0.35	0.001	0.0005	–	–

[a]Emission factors expressed as units per unit weight of process throughput.

Source: EPA Publication AP-42, 2nd Edition

The calcined material is known in the trade as "hot stucco." The finished stucco may be sold as such for use in various plasters. It may also be a major raw material for a board products plant where water, starch, chopped glass fiber etc. are added to form a mix which is fed to a board-forming machine. The only air pollutant from board forming is gypsum dust. As with gypsum calcination, however, there are of course SO_2 and NO_x emissions from the combustion operations which provide the process heat.

A rough estimate of nationwide particulate contamination from gypsum product plants has been made by using an average dust emission figure (2). From the data available this is 340 lb/hr for crushing, grinding, and calcining operations, and 10.3 lb/hr for board plants. An average board line will produce 400 square feet per minute, or

$$\frac{400\ (1440)\ 365}{1000} = 210{,}240 \text{ MSF per year.}$$

Since approximately 12,000,000 MSF per year are produced, the total emission for board lines is

$$\frac{12{,}000{,}000\ (10.3)}{210{,}240} = 588 \text{ pounds per hour nationwide.}$$

The total quantity of gypsum processed is approximately 17,500,000 tons per year. The average plant will process about 1000 tons of rock per day. Therefore, the approximate total emission for rock processing, is

$$\frac{17{,}500{,}000\ (340)}{365\ (1000)} = 16{,}300 \text{ pounds per hour nationwide}$$

This is based on continuous three-shift operation.

If the mill is operated on a one-shift basis, the dust emission rate would be three times this. Thus, the total dust emission rate could be as high as 49,500 pounds per hour on a nationwide basis.

References

(1) U.S. Environmental Protection Agency, "Compilation of Air Pollutant Emission Factors," 2nd Ed., *Publication Number AP-42,* Research Triangle Park, N.C., Office of Air and Water Programs (April 1973).
(2) Processes Research, Inc., "Screening Study for Background Information and Significant Emissions for Gypsum Product Manufacturing," *Report Number PB-222,736,* Springfield, Va., Nat. Tech. Info. Service (May 1973).

HYDROCHLORIC ACID MANUFACTURE

Hydrochloric acid is manufactured by a number of different chemical processes. Approximately 80% of the hydrochloric acid, however, is produced by the by-product hydrogen chloride process, which will be the only process discussed in this section. The synthesis process and the Mannheim process are of secondary importance.

By-product hydrogen chloride is produced when chlorine is added to an organic compound such as benzene, toluene, and vinyl chloride. Hydrochloric acid is produced as a by-product of this reaction. An example of a process that generates hydrochloric acid as a by-product is the direct chlorination of benzene. In this process benzene, chlorine, hydrogen, air, and some trace catalysts are the raw materials that produce chlorobenzene. The gases from the reaction of benzene and chlorine consist of hydrogen chloride, benzene, chlorobenzenes, and air. These gases are first scrubbed in a packed tower with a chilled mixture of monochlorobenzene and dichlorobenzene to condense and recover any benzene or chlorobenzene. The hydrogen chloride is then absorbed in a falling film absorption plant.

Air Pollution

The recovery of the hydrogen chloride from the chlorination of an organic compound is the major source of hydrogen chloride emissions. The exit gas from the absorption or scrubbing system is the actual source of the hydrogen chloride emitted. Emission factors for hydrochloric acid produced as by-product hydrogen chloride are presented in Table 231.

TABLE 231: EMISSION FACTORS FOR HYDROCHLORIC ACID MANUFACTURING[a]
EMISSION FACTOR RATING: B

Type of process	Hydrogen chloride emissions	
	lb/ton	kg/MT
By-product hydrogen chloride		
With final scrubber	0.2	0.1
Without final scrubber	3	1.5

[a]Reference 1.

Source: EPA Publication AP-42, 2nd Edition

Water Pollution

Raw waste loads from hydrochloric acid manufacture are presented below (2). Some of these are markedly dependent on conditions, with most of the wastes being produced during startups. There are no water-borne wastes during periods of normal operation.

Waste Products	Process Source	Amount of Product
1. Chlorine*	Burner Run – Chlorine-rich	Startup - 100 kg/kkg(200 lb/ ton) avg. 5-200 range(10-400) Operation - 5 kg/kkg(10 lb/ ton) avg. 0-10 range(0-20) Shutdown - no waste
2. HCl**	–	Startup - 4.5 kg/day(9 lb/ton) Operation - none Shutdown - none
3. NaOH*** reaction products (NaCl and NaOCl)	Neutralization	Startup - depends on HCl and Cl$_2$ to be neutralized Operation — none Shutdown - none

----
 *Emerges in vent gas during normal operation, neutralized
  during startup by NaOH.
 **All neutralized during startup.
***Caustic (NaOH) used has 12% NaCl present and is cell liquor
   from chlorine plant also in the complex.

References

(1)   Atmospheric Emissions from Hydrochloric Acid Manufacturing Processes. U.S. DHEW, PHS, CPEHS, National Air Pollution Control Administration, Durham, N.C. *Publication Number AP-54,* (September 1969).

(2)   U.S. Environmental Protection Agency, "Development Document for Proposed Effluent Limitations Guidelines and New Source Performance Standards for the Major Inorganic Products Segment of the Inorganic Chemicals Mfg. Point Source Category," *Report EPA 440/1-73/007* (August 1973).

HYDROFLUORIC ACID MANUFACTURE

All hydrofluoric acid in the United States is usually produced by the reaction of acid-grade fluorspar with sulfuric acid for 30 to 60 minutes in externally fired rotary kilns at a temperature of 400° to 500°F (204° to 260°C). The resulting gas is then cleaned, cooled, and absorbed in water and weak hydrofluoric acid to form a strong acid solution. Anhydrous hydrofluoric acid is formed by distilling 80% hydrofluoric acid and condensing the gaseous HF which is driven off.

Air Pollution

Air pollutant emissions are minimized by the scrubbing and absorption systems used to purify and recover the HF. The initial scrubber utilizes concentrated sulfuric acid as a scrubbing medium and is designed to remove dust, SO_2, SO_3, sulfuric acid mist, and water vapor present in the gas stream leaving the primary dust collector. The exit gases from the final absorber contain small amounts of HF, silicon tetrafluoride (SiF_4), CO_2, and SO_2 and may be scrubbed with a caustic solution to reduce emissions further. A final water ejector, sometimes used to draw the gases through the absorption system, will reduce fluoride emissions. Dust emissions may also result from raw fluorspar grinding and drying operations. Table 232 lists the emission factors for the various operations.

TABLE 232: EMISSION FACTORS FOR HYDROFLUORIC ACID MANUFACTURING
EMISSION FACTOR RATING: C

Type of operation	Fluorides		Particulates	
	lb/ton acid	kg/MT acid	lb/ton fluorspar	kg/MT fluorspar
Rotary kiln				
Uncontrolled	50	25	–	–
Water scrubber	0.2	0.1	–	–
Grinding and drying of fluorspar	–	–	20*	10*

*Factor given for well-controlled plant.

Source: EPA Publication AP-42, 2nd Edition

Water Pollution

At an exemplary plant, the calcium sulfate by-product from the reactor is slurried with water and sent to waste treatment. Also, all tail gases are scrubbed and the scrubbed water sent to the waste abatement system. The waste products from hydrofluoric acid manufacture are shown below. Wastes consist of materials from the furnaces, which include calcium sulfate, calcium fluoride and sulfuric acid, plus fluoride-containing scrubber wastes.

Waste Products	Process Source	Average kg/kkg of Product, (lb/ton)
$CaSO_4$	Kiln (reactor)	3,620 (7,240)
H_2SO_4	Kiln (reactor)	110 (220)
CaF_2	Kiln (reactor)	63 (126)
HF	Kiln (reactor)	1.5 (3)
H_2SiF_6	Scrubber	12.5 (25)
SiO_2	Kiln (reactor)	12.5 (25)
SO_2	Scrubber	5 (10)
HF	Scrubber	1 (2)

The water use within the sample is shown below.

Type	Total Quantity		Recycled
	cu m/day(GPD)	liters/kkg(gal/ton)	
Cooling (river water)	3,270(864,000)	90,140(21,600)	0
Slurry and Scrubber	3,270(864,000)	90,140(21,600)	100

All process and scrubber wastewaters are totally recycled in the exemplary plant. The waters used to slurry and remove the calcium sulfate from the furnaces and scrubber waters are fed to a pond system after being treated with caustic or soda ash and lime to precipitate fluorides and adjust the pH. In the pond system, the insolubles are settled out and the waters are then reused in the process.

Only cooling water is discharged from this facility. Low fluoride levels are easily maintained because of segregation of discharged cooling waters from the process water. Measurements verify that there is no fluoride discharge from this facility. The similarity of the intake and cooling water discharge verifies that there is no process water leakage into the cooling stream, and therefore there is no process water discharge from this exemplary hydrofluoric acid manufacturing plant.

References

(1) U.S. Environmental Protection Agency, "Compilation of Air Pollutant Emission Factors," 2nd Edition, *Publication Number AP-42,* Research Triangle Park, N.C., Office of Air and Water Programs (April 1973).
(2) U.S. Environmental Protection Agency, "Development Document for Proposed Effluent Limitations Guidelines and New Source Performance Standards for the Major Inorganic Products Segment of the Inorganic Chemicals Mfg. Point Source Category, *Report EPA 440/1-73/007* (August 1973).

HYDROGEN PEROXIDE MANUFACTURE

Hydrogen peroxide is manufactured by three different processes: (1) An electrolytic process; (2) An organic process involving the oxidation and reduction of anthraquinone; and (3) A by-product of acetone manufacture from isopropyl alcohol. In a recent study, the EPA considered only the first two processes.

In the organic process, anthraquinone (or an alkylanthraquinone) in an organic solvent is catalytically hydrogenated to yield a hydroanthraquinone. This material is then oxidized with oxygen or air back to anthraquinone, with hydrogen peroxide being produced as a by-product. The peroxide is water-extracted from the reaction medium, and the organic solvent and anthraquinone are recycled. The recovered peroxide is then purified and shipped.

In the electrolytic process, a solution of ammonium bisulfate is electrolyzed, hydrogen is liberated at the cathodes of the cells used, and ammonium persulfate is formed at the anode. The persulfate is then hydrolyzed to yield ammonium bisulfate and hydrogen peroxide which is separated by fractionation from the solution. The ammonium bisulfate solution is then recycled, and the peroxide is recovered for sale. The only waste is a stream of condensate from the fractionation condenser.

Water Pollution

The waste products from the organic process are as follows:

Waste Products	Process Source	Operation Avg. Range kg/kkg (lb/ton)
Sulfuric Acid	Ion Exchange Units	12.5-15 (25-30)
Trace Organics	Contact Cooling	0.17-0.35 (0.34-0.70)
Hydrogen Peroxide	Purification Washings	20-25 (40-50)

The process runs continuously except for shut-down approximately 10 days per year. Total discharge will normally be no higher during start-up and shut-down periods than under operation at capacity.

The data below describe the treatment of the waste stream emerging from the peroxide plant. Peroxide is decomposed by iron filings, and organic solvent losses are minimized by a skimming operation:

Waste Stream Source	m³/kkg (gal/ton)	Treatment	Final Disposal
Process Effluent	294 (70,200)	1. Peroxide reacted with iron filings	River
		2. Skimmers used to trap organics for recovery	
		3. Waste sulfuric acid is collected and discharged at a controlled rate	
		4. Solids (alumina & carbon) are hauled to landfill	

The effectiveness of the treatments in use is:

Method	Qualitative Rating	Waste Reduction Accomplished
Reduction	Generally satisfactory	80% reduction of peroxide to water and oxygen
Skimming	Generally satisfactory	60-70% of organics recovered

The wastes consist of unreacted peroxide and a small amount of organics and sulfates.

Table 233 lists the raw wastes from electrolytic peroxide manufacture. These consist of ammonium bisulfate losses, ion exchange losses, boiler blowdowns and some cyanide wastes from the special batteries used in electrolysis.

TABLE 233: RAW WASTE LOADS AT OPERATING PLANT

Waste Product	Process Source	kg/kkg of Peroxide (lb/ton) Operation	Startup	Shutdown
1. Blue prus-siate sludge	Purif.	0.18(0.36)	No significant dif-ference during start-	
2. Gray sludge	Battery rebuild	(5 times per year)	up & shutdown periods. Plant runs contin-	
3. Ion Exchange sludge	Deionizer regen.	---	uously; shuts down once per year.	
4. H₂SO₄	Plant solu-tion loss	0.0018(0.0036)		
5. (NH₄)₂SO₄	Plant solu-tion loss	0.012(0.024)		
6. Water flow	Cooling	2000-2900 (4000-58000)		
7. HCl	Deionizer regen.	1.3(2.6)		
8. NaOH	Deionizer regen.	0.33(0.66)		
9. Steam condensate	Boiler blowdown	581(1162)		

Source: EPA Report 440/1-73/007

In the table above H_2SO_4 and $(NH_4)_2SO_4$ are used to replenish plant solution. $Na_4Fe(CN)_6$ is converted to $(NH_4)_4Fe(CN)_6$ through ion exchange (yellow solution). NH_4SCN is oxidized in the batteries and is used for better current efficiency. HCl and NaOH are used for re-generation of demineralized water ion exchange resins.

Reference

(1) U.S. Environmental Protection Agency, "Development Document for Proposed Effluent Limitations Guidelines and New Source Performance Standards for the Major Inorganic Products Segment of the Inorganic Chemicals Manufacturing Point Source Category, *Report EPA 440/1-73/007,* Washington, D.C. (August 1973).

INCINERATION OF MUNICIPAL REFUSE

The most common types of incinerators consist of a refractory-lined chamber with a grate upon which refuse is burned. Some newer incinerators use water-walled furnaces. Combustion products are formed by heating and burning of refuse on the grate. In most cases, since insufficient underfire (undergrate) air is provided to enable complete combustion, additional overfire air is admitted above the burning waste to promote complete gas-phase combustion.

In multiple-chamber incinerators, gases from the primary chamber flow to a small secondary mixing chamber where more air is admitted, and more complete oxidation occurs. As much as 300% excess air may be supplied in order to promote oxidation of combustibles. Auxiliary burners are sometimes installed in the mixing chamber to increase the combustion tempera-ture. Many small-size incinerators are single-chamber units in which gases are vented from the primary combustion chamber directly into the exhaust stack. Single-chamber incinerators of this type do not meet modern air pollution codes. No exact definitions of incinerator size categories exist, but the following general categories and description have been selected for use here.

Municipal Incinerators — Multiple-chamber units often have capacities greater than 50 tons

(45.3 MT) per day and are usually equipped with automatic charging mechanisms, temperature controls, and movable grate systems. Municipal incinerators are also usually equipped with some type of particulate control device, such as a spray chamber or electrostatic precipitator.

Industrial/Commercial Incinerators — The capacities of these units cover a wide range, generally between 50 and 4,000 pounds (22.7 and 1,800 kg) per hour. Of either single- or multiple-chamber design, these units are often manually charged and intermittently operated. Some industrial incinerators are similar to municipal incinerators in size and design. Better designed emission control systems include gas-fired afterburners or scrubbing, or both.

Trench Incinerators — A trench incinerator is designed for the combustion of wastes having relatively high heat content and low ash content. The design of the unit is simple: a U-shaped combustion chamber is formed by the sides and bottom of the pit and air is supplied from nozzles along the top of the pit. The nozzles are directed at an angle below the horizontal to provide a curtain of air across the top of the pit and to provide air for combustion in the pit. The trench incinerator is not as efficient for burning wastes as the municipal multiple-chamber unit, except where careful precautions are taken to use it for disposal of low-ash, high-heat-content refuse, and where special attention is paid to proper operation.

Low construction and operating costs have resulted in the use of this incinerator to dispose of materials other than those for which it was originally designed. Emission factors for trench incinerators used to burn three such materials are included in Table 234.

Domestic Incinerators — This category includes incinerators marketed for residential use. Fairly simple in design, they may have single or multiple chambers and usually are equipped with an auxiliary burner to aid combustion.

Flue-fed Incinerators — These units, commonly found in large apartment houses, are characterized by the charging method of dropping refuse down the incinerator flue and into the combustion chamber. Modified flue-fed incinerators utilize afterburners and draft controls to improve combustion efficiency and reduce emissions.

Pathological Incinerators — These are incinerators used to dispose of animal remains and other organic material of high moisture content. Generally, these units are in a size range of 50 to 100 pounds (22.7 to 45.4 kg) per hour. Wastes are burned on a hearth in the combustion chamber. The units are equipped with combustion controls and afterburners to ensure good combustion and minimal emissions.

Controlled Air Incinerators — These units operate on a controlled combustion principle in which the waste is burned in the absence of sufficient oxygen for complete combustion in the main chamber. This process generates a highly combustible gas mixture that is then burned with excess air in a secondary chamber, resulting in efficient combustion. These units are usually equipped with automatic charging mechanisms and are characterized by the high effluent temperatures reached at the exit of the incinerators.

Air Pollution

Operating conditions, refuse composition, and basic incinerator design have a pronounced effect on emissions. The manner in which air is supplied to the combustion chamber or chambers has, among all the parameters, the greatest effect on the quantity of particulate emissions. Air may be introduced from beneath the chamber, from the side, or from the top of the combustion area. As underfire air is increased, an increase in fly-ash emissions occurs. Erratic refuse charging causes a disruption of the combustion bed and a subsequent release of large quantities of particulates. Large quantities of uncombusted particulate matter and carbon monoxide are also emitted for an extended period after charging of batch-fed units because of interruptions in the combustion process. In continuously fed units,

TABLE 234: EMISSION FACTORS FOR REFUSE INCINERATORS WITHOUT CONTROLS[a]—EMISSION FACTOR RATING: A

Incinerator type	Particulates		Sulfur oxides[b]		Carbon monoxide		Hydrocarbons[c]		Nitrogen oxides[d]	
	lb/ton	kg/MT	lb/ton	kg/MT	lb/ton	kg/MT	lb/ton	kg/MT	lb/ton	kg/MT
Municipal										
Multiple chamber, uncontrolled	30	15	2.5	1.25	35	17.5	1.5	0.75	3	1.5
With settling chamber and water spray system	14	7	2.5	1.25	35	17.5	1.5	0.75	3	1.5
Industrial/commercial										
Multiple chamber	7	3.5	2.5[e]	1.25	10	5	3	1.5	3	1.5
Single chamber	15	7.5	2.5[e]	1.25	20	10	15	7.5	2	1
Trench										
Wood	13	6.5	0.1[f]	0.05	NA	NA	NA	NA	4	2
Rubber tires	138	69	NA	NA	NA	NA	NA	NA	NA	NA
Municipal refuse	37	18.5	2.5[e]	1.25	NA	NA	NA	NA	NA	NA
Controlled air	1.4	0.7	1.5	0.75	Neg	Neg	Neg	Neg	10	5
Flue-fed single chamber	30	15	0.5	0.25	20	10	15	7.5	3	1.5
Flue-fed (modified)[g]	6	3	0.5	0.25	10	5	3	1.5	10	5
Domestic single chamber										
Without primary burner	35	17.5	0.5	0.25	300	150	100	50	1	0.5
With primary burner	7	3.5	0.5	0.25	Neg	Neg	2	1	2	1
Pathological	8	4	Neg	Neg	Neg	Neg	Neg	Neg	3	1.5

[a]Average factors given based on EPA procedures for incinerator stack testing.
[b]Expressed as sulfur dioxide.
[c]Expressed as methane.
[d]Expressed as nitrogen dioxide.
[e]Based on municipal incinerator data.
[f]Based on data for wood combustion in conical burners.
[g]With afterburners and draft controls.

Source: EPA Publication AP-42, 2nd Edition

furnace particulate emissions are strongly dependent upon grate type. The use of rotary kiln and reciprocating grates results in higher particulate emissions than the use of rocking or traveling grates. Emissions of oxides of sulfur are dependent on the sulfur content of the refuse. Carbon monoxide and unburned hydrocarbon emissions may be significant and are caused by poor combustion resulting from improper incinerator design or operating conditions. Nitrogen oxide emissions increase with an increase in the temperature of the combustion zone, an increase in the residence time in the combustion zone before quenching, and an increase in the excess air rates to the point where dilution cooling overcomes the effect of increased oxygen concentration.

Reference

(1)   U.S. Environmental Protection Agency, "Compilation of Air Pollutant Emission Factors," 2nd Edition, *Publication No. AP-42,* Research Triangle Park, N.C., Office of Air and Water Programs (April 1973).

INORGANIC CHEMICAL INDUSTRY

The reader is referred to individual chemical designations in this volume such as "Ammonia Manufacture" for details of air and water pollution problems. See also summary volumes by Jones (1) and by the EPA (2).

References

(1)   Jones, H.R., "Environmental Control in the Inorganic Chemical Industry," Park Ridge, N.J., Noyes Data Corp. (1972).
(2)   U.S. Environmental Protection Agency, "Development Document for Proposed Effluent Limitations Guidelines and New Source Performance Standard for the Major Inorganic Products Segment of the Inorganic Chemicals Manufacturing Point Source Category, *Report EPA 440/1-73/007,* Washington, D.C. (August 1973).

IRON FOUNDRIES

Three types of furnaces are used to produce gray iron castings: cupolas, reverberatory furnaces, and electric induction furnaces. The cupola is the major source of molten iron for the production of castings. In operation, a bed of coke is placed over the sand bottom in the cupola. After the bed of coke has begun to burn properly, alternate layers of coke, flux, and metal are charged into the cupola. Combustion air is forced into the cupola, causing the coke to burn and melt the iron. The molten iron flows out through a taphole.

Electric furnaces are commonly used where special alloys are to be made. Pig iron and scrap iron are charged to the furnace and melted, and alloying elements and fluxes are added at specific intervals. Induction furnaces are used where high-quality, clean metal is available for charging.

Air Pollution

Emissions from cupola furnaces include gases, dust, fumes, and smoke and oil vapors. Dust arises from dirt on the metal charge and from fines in the coke and limestone charge. Smoke and oil vapor arise primarily from the partial combustion and distillation of oil from greasy scrap charged to the furnace. Also, the effluent from the cupola furnace has a high carbon monoxide content that can be controlled by an afterburner. Emissions from reverberatory and electric induction furnaces consist primarily of metallurgical fumes and are relatively low. Table 235 presents emission factors for the manufacture of iron castings.

TABLE 235: EMISSION FACTORS FOR GRAY IRON FOUNDRIES [a,b]
EMISSION FACTOR RATING: B

Type of furnace	Particulates		Carbon monoxide	
	lb/ton	kg/MT	lb/ton	kg/MT
Cupola				
Uncontrolled	17	8.5	145[c]	72.5[c]
Wet cap	8	4	–	–
Impingement scrubber	5	2.5	–	–
High-energy scrubber	0.8	0.4	–	–
Electrostatic precipitator	0.6	0.3	–	–
Baghouse	0.2	0.1	–	–
Reverberatory	2	1	–	–
Electric induction	1.5	0.75	–	–

[a]Emission factors expressed as units per unit weight of metal
charged.
[b]Approximately 85% of the total charge is metal. For every
unit weight of coke in the charge, 7 unit weights of gray
iron are produced.
[c]A well-designed afterburner can reduce emissions to 9 pounds
per ton (4.5 kg/MT) of metal charged.

Source: EPA Publications AP-42, 2nd Edition

Reference

(1) U.S. Environmental Protection Agency, "Compilation of Air Pollutant Emission
Factors, 2nd Ed., *Publication Number AP-42,* Research Triangle Park, N.C.,
Office of Air and Water Programs (April 1973).

IRON AND STEEL MILLS

Iron and steel manufacturing processes may be grouped into five distinct sequential operations: [1] coke production; [2] pig iron manufacture in blast furnaces; [3] steel-making
processes using basic oxygen, electric arc, and open hearth furnaces; [4] rolling mill operations; and [5] finishing operations. The first three of these operations encompass nearly all
of the air pollution sources. Coke production is discussed in detail elsewhere in this publication.

Pig iron is produced in blast furnaces, which are large refractory-lined chambers into which
iron ore, coke, and limestone are charged and allowed to react with large amounts of hot
air to produce molten iron. Slag and blast furnace gases are by-products of this operation.
The production of one unit weight of pig iron requires an average charge of 1.55 unit
weights of iron-bearing charge, 0.55 unit weight of coke, 0.20 unit weight of limestone,
and 2.3 unit weights of air.

Blast furnace by-products consist of 0.2 weight of slag, 0.02 unit weight of flue dust,
and 2.5 unit weights of gas per unit of pig iron produced. Most of the coke used in the
process is produced in by-product coke ovens. The flue dust and other iron ore fines from
the process are converted into useful blast furnace charge via sintering operations.

In the open hearth process, a mixture of scrap iron, steel, and pig iron is melted in a shallow rectangular basin, or "hearth," for which various liquid gaseous fuels provide the heat. Impurities are removed in a slag.

The basic oxygen process, also called the Linz-Donawitz (LD) process, is employed to produce steel from a furnace charge composed of approximately 70% molten blast-furnace metal and 30% scrap metal by use of a stream of commercially pure oxygen to oxidize the impurities, principally carbon and silicon.

Electric furnaces are used primarily to produce special alloy steels or to melt large amounts of scrap for reuse. Heat is furnished by direct-arc electrodes extending through the roof of the furnace. In recent years, oxygen has been used to increase the rate of uniformity of scrap-melt-down and to decrease power consumption.

Scarfing is a method of surface preparation of semifinished steel. A scarfing machine removes surface defects from the steel billets and slabs, before they are shaped or rolled, by applying jets of oxygen to the surface of the steel, which is at orange heat, thus removing a thin upper layer of the metal by rapid oxidation.

Air Pollution

Blast furnace combustion gas and the gases that escape from bleeder openings constitute the major sources of particulate emissions in pig iron manufacture. The dust in the gas consists of 35 to 50% iron, 4 to 14% carbon, 8 to 13% silicon dioxide, and small amounts of aluminum oxide, manganese oxide, calcium oxide, and other materials. Because of its high carbon monoxide content, this gas has a low heating value (about 100 Btu/ft) and is utilized as a fuel within the steel plant. Before it can be efficiently oxidized, however, the gas must be cleaned of particulates.

Initially, the gases pass through a settling chamber or dry cyclone, where about 60% of the dust is removed. Next, the gases undergo a one- or two-stage cleaning operation. The primary cleaner is normally a wet scrubber, which removes about 90% of the remaining particulates. The secondary cleaner is a high-energy wet scrubber (usually a venturi) or an electrostatic precipitator, either of which can remove up to 90% of the particulates that have passed through the primary cleaner. Taken together, these control devices provide an overall dust removal efficiency of approximately 96%.

All of the carbon monoxide generated in the gas is normally used for fuel. Conditions such as "slips," however, can cause instantaneous emissions of carbon monoxide. Improvements in techniques for handling blast furnace burden have greatly reduced the occurrence of slips. In Table 236 particulate and carbon monoxide emission factors are presented for blast furnaces.

Emissions from open hearths consist of particulates and small amounts of fluorides when fluoride-bearing ore, fluorspar, is used in the charge. The particulates are composed primarily of iron oxides, with a large portion (45 to 50%) in the 0 to 5 micrometer size range. The quantity of dust in the off-gas increases considerably when oxygen lancing is used (see Table 236). The devices most commonly used to control the iron oxide and fluoride particulates are electrostatic precipitators and high-energy venturi scrubbers, both of which effectively remove about 98% of the particulates. The scrubbers also remove nearly 99% of the gaseous fluorides and 95% of the particulate fluorides.

The basic oxygen furnace that converts the molten iron into steel generates a considerable amount of particulate matter, largely in the form of iron oxide, although small amounts of fluorides may be present. Probably as the result of the tremendous agitation of the molten bath by the oxygen lancing, the dust loadings vary from 5 to 8 grains per standard cubic foot (11 to 18 grams/standard cubic meter) and high percentages of the particles are in the 0 to 5 micrometer size range. In addition, tremendous amounts of carbon monoxide (140 lb/ton of steel and more) are generated by the reaction. Combustion in the hood,

TABLE 236: EMISSION FACTORS FOR IRON AND STEEL MILLS[a,b]—EMISSION FACTOR RATINGS: A (PARTICULATES AND CARBON MONOXIDE) AND C (FLUORIDES)

| Type of operation | Total particulates | | Carbon monoxide | | Fluorides[c,d] | | | |
| | | | | | Gaseous (HF) | | Particulates (CaF$_2$) | |
	lb/ton	kg/MT	lb/ton	kg/MT	lb/ton	kg/MT	lb/ton	kg/MT
Pig iron production								
Blast furnaces[e]								
Ore charge, uncontrolled	110	55	—	—	—	—	—	—
Agglomerates charge, uncontrolled	40	20	—	—	—	—	—	—
Total, uncontrolled[f]	165 (130 to 200)	82.5 (65 to 100)	1750 (1400 to 2100)	875 (700 to 1050)	—	—	—	—
Settling chamber or dry cyclone	60	30	—	—	—	—	—	—
Plus wet scrubber	15	7.5	—	—	—	—	—	—
Plus venturi or electrostatic precipitator	1.5	0.75	—	—	—	—	—	—
Sintering[g]								
Windbox, uncontrolled[h]	20	10	—	—	—	—	—	—
Dry cyclone	2.0	1.0	—	—	—	—	—	—
Dry cyclone plus electrostatic precipitator	1.0	0.5	—	—	—	—	—	—
Dry cyclone plus wet scrubber	0.04	0.02	—	—	—	—	—	—
Discharge, uncontrolled	22	11	44	22	—	—	—	—
Dry cyclone	2.2	1.1	44	22	—	—	—	—
Dry cyclone plus electrostatic precipitator	0.11	0.055	44	22	—	—	—	—
Steel production								
Open hearth[i]								
No oxygen lance, uncontrolled	8.3 (5.8 to 12.0)	4.15 (2.9 to 6.0)	—	—	0.100	0.05	0.030	0.015
Venturi scrubber	0.17	0.085	—	—	0.011	0.0055	0.0015	0.0008
Electrostatic precipitator	0.17	0.085	—	—	0.100	0.050	0.0006	0.0003
Oxygen lance, uncontrolled	17.4 (9.3 to 22.0)	8.7 (4.65 to 11.0)	—	—	0.100	0.050	0.030	0.015

(continued)

TABLE 236: (continued)

Type of operation	Total particulates		Carbon monoxide		Fluorides[c,d]			
					Gaseous (HF)		Particulates (CaF$_2$)	
	lb/ton	kg/MT	lb/ton	kg/MT	lb/ton	kg/MT	lb/ton	kg/MT
Venturi scrubber	0.17	0.085	—	—	0.011	0.0055	0.0015	0.0008
Electrostatic precipitator	0.35	0.175	—	—	0.100	0.050	0.0006	0.0003
Basic oxygen, uncontrolled	51 (32 to 86)	25.5 (16 to 43)	139 (104 to 237)	69.5 (52.0 to 118.5)	Neg	Neg	0.200	0.100
Venturi scrubber	0.51	0.255	—	—	—	—	0.002	0.001
Electrostatic precipitator	0.51	0.255	—	—	—	—	0.002	0.001
Spray chamber	15.3	7.65	—	—	—	—	0.060	0.030
Electric arc[j]								
No oxygen lance[k], uncontrolled	9.2 (7.0 to 10.6)	4.6 (3.5 to 5.3)	18	9	0.012	0.006	0.238	0.119
Venturi scrubber	0.18	0.09	18	9	0.0018	0.0009	0.011	0.0055
Electrostatic precipitator	0.28 to 0.74	0.14 to 0.37	18	9	0.012	0.006	0.011	0.0055
Baghouse	0.09	0.045	18	9	0.012	0.006	0.0024	0.0012
Oxygen lance[l], uncontrolled	11	5.5	18	9	0.012	0.006	0.238	0.119
Venturi scrubber	0.22	0.11	18	9	0.0018	0.0009	0.011	0.0055
Electrostatic precipitator	0.33 to 0.88	0.165 to 0.44	18	9	0.012	0.006	0.011	0.0055
Baghouse	0.11	0.055	18	9	0.012	0.006	0.0024	0.0012
Scarfing[m], uncontrolled	1	0.5	—	—	—	—	—	—
Electrostatic precipitator	≤0.06	≤0.03	—	—	—	—	—	—
Venturi scrubber	≤0.02	≤0.01	—	—	—	—	—	—

a Emission factors expressed as units per unit weight of metal produced.
b Numbers in parentheses are ranges. Controlled factors are calculated using average uncontrolled factors and observed equipment efficiencies.
c Reference 4.
d Value included in "Total Particulates" figure.
e References 2, 3, and 5.
f These factors should be used to estimate particulate and CO emissions from the entire blast furnace operation. The total particulate factors for ore and agglomerates charging apply only to those operations.

g Reference 3.
h Approximately 3 lb SO$_2$/ton (1.5 kg/MT) of sinter is produced at windbox.
i References 2, 3, 5, and 6.
j Values are for carbon type electric arc furnaces. For alloy type furnaces multiply given values by 2.80.
k References 2 through 5.
l References 3 and 4.
m Factors are based on operating experiences and engineering judgement.

Source: EPA Publication AP-12, 2nd Edition.

direct flaring, or some other means of ignition is used in the stack to reduce the actual carbon monoxide emissions to less than 3 lb/ton (1.5 kg/MT). The particulate control devices used are venturi scrubbers and electrostatic precipitators, both of which have over-all efficiencies of 99%. Furthermore, the scrubbers are 99% efficient in removing gaseous fluorides (see Table 236).

The particulates, primarily oxides of iron, manganese, aluminum, and silicon, that evolve when steel is being processed in an electric furnace result from the exposure of molten steel to extremely high temperatures. The quantity of these emissions is a function of the cleanliness and composition of the scrap metal charge, the refining procedure used (with or without oxygen lancing), and the refining time. As with open hearths, many of the particulates (40 to 75%) are in the 0 to 5 micrometer range. Additionally, moderate amounts of carbon monoxide (15 to 20 lb/ton) are emitted.

Particulate control devices most widely used with electric furnaces are venturi scrubbers, which have a collection efficiency of approximately 98%, and bag filters, which have collection efficiencies of 99% or higher.

Emissions from scarfing operations consist of iron oxide fumes. The rate at which particulates are emitted is dependent on the condition of the billets or slabs and the amount of metal removal required (Table 236). Emission control techniques for the removal of fine particles vary among steel producers, but one of the most commonly used devices is the electrostatic precipitator, which is approximately 94% efficient.

Water Pollution

Raw wastes from the sintering process emanate from the material handling dust control equipment and the dust and volatilized oil in the process gases. Most plants built today have incorporated fabric-type dust collectors in this process. Therefore, newer plants generally have no aqueous discharge from the sintering operation. However, an attempt was made (7) to investigate several plants that utilized wet scrubbers and generated waste-water. Another problem that compounds the issue is that the sintering wastewaters are generally tied in with the blast furnace wastewaters for treatment.

The raw waste loads generated from the sintering operation are primarily dependent on the type of fume collection system installed. The fume collection systems are generally divided into two separate independent exhaust systems. One exhaust system serves the hot sinter bed, ignition furnace, sinter bed wind boxes, etc., while the other system serves as a dedusting system for sinter crushes, sinter fines conveyors, raw material, storage bins, feeders, etc.

The sinter bed fume collection and exhaust systems also furnish the necessary combustion air to maintain the coke burning which fuses the sinter mix bed on the moving sinter grates. The ignition furnace initially ignites the coke in the sinter bed and the combustion air maintains the burning of the moving bed. The ignition furnaces are fired by natural gas or fuel oils. The combustion air is drawn down through the sinter bed and hot gases and particulate are then exhausted. Any heavy sinter fines materials falling through the sinter grates are gravity settled in the wind box hoppers and discharged to the sinter fines return conveyor for reprocessing. The combustion exhaust systems require large quantities of air and generally dry electrostatic precipitators are installed at the charge end of sinter machine to clean the hot exhaust gas.

Table 237 summarizes the net plant raw waste loads for the plants studied. Raw waste loads are presented only for the critical parameters which include fluoride, oil, sulfide, and suspended solids.

The blast furnace has two basic water uses, cooling water and gas washer water. The blast furnace requires the continuous circulation of cooling water through hollow plates built into the walls of the bosh and stack. Without such cooling, a furnace wall would

TABLE 237: CHARACTERISTICS OF SINTERING PLANT WASTES—NET PLANT
RAW WASTE LOADS

	- - - - - -Plants - - - - - -	
Characteristics	A	B
Flow, l/kkg	434	1,420
Suspended solids, mg/l	4,340	19,500
Oil and grease, mg/l	504	457
Fluoride, mg/l	0.644	-14.9
Sulfide, mg/l	188	64.4

Source: Report EPA 440/1-73/024

quickly burn through. Furnace cooling water approximates 21,000 l/kkg (5,000 gal/ton).
The most significant parameter from this source is heat pick-up ranging from $2°$ to $8°C$.

The principal wastewaters result from the gas cleaning operation which is performed for
two basic reasons. The primary reason for cleaning the gas is to allow its use as a fuel.
The other reason is to prevent a considerable air pollution problem which would otherwise result. Gas washer water may range from 6,300 to 17,000 l/kkg (1,500 to 4,100 gal/ton)
depending upon the type of washer used. These wastewaters contain significant concentrations of cyanide, phenol, ammonia, sulfide, and suspended solids. The wastewaters from
ferromanganese furnaces have much higher concentrations of cyanides than do wash waters
from iron furnaces.

The suspended solids in blast furnace gas-washer water result from the fines in the burden
being carried out in the gas. The quantities depend upon the operation of the furnace
and the nature of the burden. Oils can be vaporized and carried into the gas when metal
turnings are part of the charge. Phenols, cyanides, and ammonia originate in the coke
and are particularly high if the coke has been quenched with wastewaters or if the coke
has not been completely coked. Cyanides are generated in the blast furnace in the reducing
atmosphere from carbon from the coke and nitrogen from the air; cyanide formation is
particularly high at the higher temperatures of a ferromanganese furnace.

Table 238 summarizes the net plant raw waste loads for the iron-making blast furnaces
studied. Table 239 presents comparable data for the ferromanganese furnace. Raw waste
loads are presented only for the critical parameters which include ammonia, cyanide, oil,
phenol, and sulfide with manganese added to the ferromanganese furnace.

TABLE 238: CHARACTERISTICS OF IRON-BLAST FURNACE PLANT WASTES—NET
PLANT RAW WASTE LOADS

	- - - - - - - - - - - - Plants - - - - - - - - - - - -			
Characteristics	C	D	E	F
Flow, l/kkg	22,500	8,050	14,000	13,000
Ammonia, mg/l	1.41	3.91	9.75	12.3
Cyanide, mg/l	1.44	0.858	-0.241	-0.231
Phenol, mg/l	0.578	-0.643	0.530	0.0853
Suspended solids, mg/l	1720	651	307	1170
Fluoride, mg/l	0.454	0.044	2.16	-2.59
Sulfide, mg/l	4.34	38.8	0.448	-1.14

Source: Report EPA 440/1-73/024

TABLE 239: CHARACTERISTICS OF FERROMANGANESE BLAST FURNACE
PLANT WASTES
NET PLANT RAW WASTE LOADS

Characteristics	Plant G
Flow, l/kkg	32,200
Ammonia, mg/l	114
Cyanide, mg/l	23.6
Phenol, mg/l	0.130
Suspended solids, mg/l	5,000
Sulfide, mg/l	-2.66
Manganese, mg/l	833

Source: Report EPA 440/1-73/024

The steelmaking process produces fume, smoke, and waste gases as the unwanted impurities which are burned off and the process vaporizes or entrains a portion of the molten steel into the off-gases. Wastewater results from the steelmaking processes when wet collection systems are used on the furnaces. Spray cooling, quenching, or the use of wet washers result in wastewaters containing particulates from the gas stream. Dry collection methods through the use of waste heat boilers, evaporation chambers, and spark boxes do not produce wastewater effluents. Tables 240 to 244 inclusive summarize the characteristics of net raw aqueous waste loads from the various steelmaking processes of:

[1] Basic oxygen furnace operation
[2] Open hearth furnace operation
[3] Electric arc furnace operation
[4] Vacuum degassing
[5] Continuous casting

In addition, wastewaters are produced in disposal of blast furnace slag. Hot blast furnace slag is usually dumped into a large pit, open at one end, to enable removal after quenching and quenched and cooled to a temperature at which it can be transported relatively safely to a final disposal site or a slag processing plant.

During quenching of the slag, there is little or no actual runoff from the site, the great majority of the water being evaporated. As the slag temperature is lowered, however, some excess quench water will remain unevaporated. The quench pits are normally graded so that this excess water will collect in the bottom of the pit rather than run off overland from the site. Once the cooled slag is removed for final disposal, the pooled water lying in the bottom of the quench pit will remain and be flashed off by the next hot slag charge.

However, during this period of slag cooling, some of the excess quench water may permeate into the ground, thus constituting a subsurface discharge. Samples of pooled quench water after contact with the slag, indicate that this is a highly alkaline (1,067 mg/l MO alkalinity) wastewater, low in suspended matter, but high in dissolved solids probably in the form of calcium and magnesium sulfates, sulfides, and sulfites (890 mg/l $SO_4^=$, 499 mg/l $S^=$, and 1,560 mg/l $SO_3^=$). The main source of the alkalinity is probably calcium carbonate leached out of the slag.

TABLE 240: BOF PLANT WASTES

Characteristics	Plants				
	H	I	J	K	L
Flow, l/kkg	542	4,270	2,570	3,040	1,080
Fluoride, mg/l	–	–	10.9	2.36	2.76
Suspended solids, mg/l	321	180	3730	396	5330

Source: Report EPA 440/1-73/024

TABLE 241: OPEN HEARTH PLANT WASTES

Characteristics	Plants	
	M	N
Flow, l/kkg	2,530	2,290
Suspended solids, mg/l	388	3,880
Fluoride, mg/l	21.4	16.2
Nitrate, mg/l	20.2	33.2
Zinc, mg/l	2.06	880

TABLE 242: ELECTRIC FURNACE PLANT WASTES

Characteristics	Plants			
	O	P	Q	R
Flow, l/kkg	406	1.01	1,250	751
Fluorine, mg/l	−28.7	–	14.8	11.3
Suspended solids, mg/l	863	77.4%	2,160	42,800
Zinc, mg/l	13	–	405	5637

TABLE 243: DEGASSING PLANT WASTES

Characteristics	Plants	
	S	T
Flow, l/kkg	3,750	813
Suspended solids, mg/l	23.2	70.7
Zinc, mg/l	2.01	7.76
Manganese, mg/l	5.72	13.3
Lead, mg/l	0.471	1.39
Nitrate, mg/l	25.3	3.03

TABLE 244: CONTINUOUS CASTING PLANT WASTES

Characteristics	Plants	
	U	V
Flow, l/kkg	17,100	6,172
Suspended solids, mg/l	7.87	74.0
Oil and grease, mg/l	20.5	22.0

Source: Report EPA 440/1-73/024

Reference

(1) Bramer, Henry C. "Pollution Control in the Steel Industry." *Environmental Science and Technology.* p. 1004-1008 (October 1971).
(2) Celenza, C.J. "Air Pollution Problems Faced by the Iron and Steel Industry." *Plant Engineering.* p. 60-63 (April 30, 1970).
(3) U.S. Environmental Protection Agency, "Compilation of Air Pollutant Emission Factors," 2nd Edition, *Publication Number AP-42,* Supplement Number 3, Research Triangle Park, N.C., Office of Air and Water Programs (July 1974).
(4) Communication between Kirkendall, E., American Iron and Steel Institute, and McGinnity, J., Environmental Protection Agency, Durham, N.C. (September 1970).
(5) *Particulate Pollutant Systems Study, Vol. 1.* Midwest Research Institute, Kansas City, Mo. Prepared for Environmental Protection Agency, Office of Air Programs, Research Triangle Park, N.C., under Contract Number CPA 22-69-104 (May 1971).

(6) Walker, A.B. and Brown, R.F. Statistics on Utilization, Performance, and Economics of Electrostatic Precipitation for Control of Particulate Air Pollution (Presented at 2nd International Clean Air Congress, International Union of Air Pollution Prevention Association), Washington, D.C. (December 1970).

(7) U.S. Environmental Protection Agency, "Development Document for Proposed Effluent Limitations Guidelines and New Source Performance Standards for the Steel Making Segment of the Iron and Steel Manufacturing Point Source Category," *Report EPA 440/1-73/024,* Washington, D.C. (January 1974).

LEAD SMELTING, PRIMARY

Lead is usually found in nature as a sulfide ore containing small amounts of copper, iron, zinc, and other trace elements. It is normally concentrated at the mine from an ore of 3 to 8% lead to an ore concentrate of 55 to 70% lead, containing from 13 to 19% free and uncombined sulfur by weight. Normal practice for the production of lead metal from this concentrate involves the following operations:

[1] Sintering, in which the concentrate lead and sulfur are oxidized to produce lead oxide and sulfur dioxide. (Simultaneously, the charge material, comprised of concentrates, recycle sinter, sand, and other inert materials, is agglomerated to form a dense, permeable material called sinter.)

[2] Reducing the lead oxide contained in the sinter to produce molten lead bullion.

[3] Refining the lead bullion to eliminate any impurities.

Air Pollution

Each of the three major lead smelting operations generates substantial quantities of particulates and/or sulfur dioxide. Nearly 85% of the sulfur present in the lead ore concentrate is eliminated in the sintering operation. In handling these process offgases, either a single weak stream is taken from the machine hood at less than 2% SO_2 or two streams are taken—one weak stream ($<0.5\%$ SO_2) from the discharge end of the machine and one strong stream (5 to 7% SO_2) taken from the feed end. Single stream operation is generally used when there is little or no market for the recovered sulfur, so that the uncontrolled weak SO_2 stream is emitted to the atmosphere. Where there is a potential sulfur market, however, the strong stream is sent to a sulfuric acid plant, and the weak stream is vented after particulate removal.

When dual gas stream operation is used with updraft sinter machines, the weak gas stream can be recirculated through the bed to mix with the strong gas stream, resulting in a single stream with an SO_2 concentration of about 6%. This technique has the overall effect of decreasing machine production capacity, but does permit a more convenient and economical recovery of the SO_2 via sulfuric acid plants and other control methods.

Without weak gas recirculation, the latter portion of the sinter machine acts as a cooling zone for the sinter and consequently assists in the reduction of dust formation during product discharge and screening. However, when recirculation is used, the sinter is usually discharged in a relatively hot state (400° to 500°C), with an attendant increase in particulate formation. Methods for reducing these dust quantities include recirculation of offgases through the sinter bed, relying upon the filtering effect of the latter, or ducting the gases from the discharge through a particulate collection device directly to the atmosphere. Because reaction activity has ceased in the discharge area in these cases, these latter gases contain little SO_2.

The particulate emissions from sinter machines consist of from 5 to 20% of the concentrated ore feed. When expressed in terms of product weight, these emissions are an estimated 106.5 kg/MT (213 lb/ton) of lead produced. This value, along with other particulate and SO_2 factors, appears in Table 245.

TABLE 245: EMISSION FACTORS FOR PRIMARY LEAD SMELTING PROCESSES WITHOUT CONTROLS[a], EMISSION FACTOR RATING: B

Process	Particulates		Sulfur dioxide	
	kg/MT	lb/ton	kg/MT	lb/ton
Ore crushing	1.0	2.0	—	—
Sintering (updraft)	106.5	213.0	275.0	550.0
Blast furnace	180.5	361.0	22.5	45.0
Dross reverberatory furnace	10.0	20.0	Neg	Neg
Materials handling	2.5	5.0	—	—

[a]Ore crushing emission factors expressed as kg/MT (lb/ton) of crushed ore; all other emission factors expressed as kg/MT (lb/ton) of lead product.

Source: EPA Publication AP-42, 2nd Edition

Typical material balances from domestic lead smelters indicate that about 10 to 20% of the sulfur in the ore concentrate fed to the sinter machine is eliminated in the blast furnace. However, only half of this amount (about 7% of the total) is emitted as SO_2; the remainder is captured by the slag. The concentration of this SO_2 stream can vary from 500 to 2,500 parts per million by volume, depending on the amount of dilution air injected to oxidize the carbon monoxide and cool the stream before baghouse treatment for particulate removal.

Particulate emissions from blast furnaces contain many different kinds of material, including a range of lead oxides, quartz, limestone, iron pyrites, iron-lime-silicate slag, arsenic, and other metals-containing compounds associated with lead ores. These particles readily agglomerate, are primarily submicron in size, difficult to wet, cohesive, and will bridge and arch in hoppers. On the average, this dust loading is quite substantial (see Table 245).

Virtually no sulfur dioxide emissions are associated with the various refining operations. However, a small amount of particulates is generated by the dross reverberatory furnace (10 kg/MT of lead). Finally, minor quantities of particulates are generated by ore crushing and material-handling operations.   These emission factors are also presented in Table 245.

Water Pollution

Data on water usage, recirculation practices, and usage per ton of metal reported by lead and zinc producers are given in Table 246. The water-use data for lead and zinc mines permit the following general conclusions and observations.

[a]   Intake and discharge quantities are related to a number of factors. The principal determinants are plant capacity and the availability (and presumably the cost) of water.

[b]   The overall recycle ratio of total water use to raw water intake is generally low (1, no recycle, to 1.5, 50% recycle), although in one case (No. 14, Table 246) a recycle ratio of 2.6 was reported.

[c]   The volume of water reported used in processing and auxiliary operations per ton of metal for both lead and zinc mine-concentrator combinations ranged between 10,660 to 21,600 gallons in a reasonably close pattern. The spread is probably attributable to differences in ore grade and slight variations in plant practice.

[d]   There is wide scattering in the data showing the distribution of intake water to process, cooling, etc. Specific reasons for this are not clearly discernible from the reports supplied to the study. The variability is associated with plant size and a host of other factors such as location, ore grade, climate conditions, water availability, cost, etc.

TABLE 246: WATER USE DATA FOR LEAD AND ZINC MINES AND PLANTS

Number	Total Intake, MGY	Discharge to Waste, MGY	Total Use(a), MGY	Indicated Recycle Ratio(b)	Gallons Used per Ton of Metal	Gallons Consumed per Ton of Metal(c)	Distribution % of Intake(d) Process	Cooling	Boiler Feed	Air Pollution Control	Sanitary	Recirculation Ratios(f) Process	Cooling	Boiler Feed	Air Pollution Control
Lead:															
Mine and Concentrator															
1	330	330	330	1.0	11,200	0	66.6	22.2	0.6	5.6	5	1	1	1	1
2	238	238	306	1.3	12,000	0	63.3	28.3	0.3	5.0	3	1.5	1	1	1
3	357	318	472	1.3	10,660	864	89.2	3.0	UK(f)	7.4	0.5	1.2	2	2	2
4	474(g)	473	473	1	UK	286	88.6	11.1	>0.1	>0.1	0.2	1	1	1	1
5	350	348	69	UK	11,481	41,520	87.5	2.5	UK	UK	10.0	UK	UK	UK	UK
6	99	0	(h)		UK		95.2	UK	UK	UK	2.4	UK	UK	UK	UK
7	2,900	2,470	4,400	1.5	20,600	6,000	98.0	0.4	0.4	0.4	0.8	1.5	2	2.0	2.0
Smelter															
8	894	763	894	1	14,600	2,147	14.2	77.0	>0.1	1.2	—	—	—	—	—
9	264	37.8	UK	UK	UK	1,220	16.2	38.0	7.6	28.6	—	—	—	—	—
Smelter and Pyrometallurgical Refinery															
10(i)	107	50.8	801	9.5	9,210	647	10.4	83.3	—	—	6.3	7	1.8	—	1.7
11(i)	1,230	1,008	5,104	6.6	22,200	986	29.6	40.3	—	29.1	1.1	4	8.8	—	1.9
Pyrometallurgical Refinery															
12	126	126	189	1.5	1,310	0	—	75	—	—	25	—	1.7	—	—
Zinc															
Mine and Concentrator															
13	257	257	319	1.3	21,600	0	49.9	29.2	—	4.2	16.7	1.6	1	—	1
14	303	303	823	2.6	13,500	0	98.6	—	—	—	1.4	2.5	1	—	1
15	1,299(k)	122	691	1.0	17,700	UK	76.0	15.9	>0.1	6.3	1.7	1	1	—	1
Smelter															
16	72.2(l)	19.6(m)	60.2	1.2	1,200	1,050	41.6	19.4	0.1	—	38.9	1.5	1.7	—	1

(continued)

TABLE 246: (continued)

Number	Total Intake, MGY	Discharge to Waste, MGY	Total Use(a), MGY	Indicated Recycle Ratio(b)	Gallons Used per Ton of Metal	Gallons Consumed per Ton of Metal(c)	Distribution % of Intake(d)					Recirculation Ratios(f)			
							Process	Cooling	Boiler Feed	Air Pollution Control	Sanitary	Process	Cooling	Boiler Feed	Air Pollution Control
Smelter, Refinery, and Acid Plant															
17(n)	693	341	18,900	27.3	156,000	2,907	25.1	50.1	20.0	1.7	3.1	20	4.4	–	1
18(o)	4,130	4,078	14,800	3.6	55,200	195	1.1	36.2	1.0	61.0	0.7	36	4.4	–	2.6

(a) Intake plus recycle.
(b) Intake plus recycle divided by intake.
(c) Intake-discharge divided by tons metal produced.
(d) As reported or calculated from data.
(e) Total water used divided by fraction of new water in total, 1 indicates no recycle.
(f) UK = Unknown.
(g) Mine water.
(h) Recirculation practiced but data not reported.
(i) No sulfuric acid plant included.
(j) Sulfuric acid plant included.
(k) Reported intake but does not coincide with reported use.
(l) Reported intake but only 50.4 MGY can be accounted for.
(m) Discharge to ground.
(n) Electrolytic refinery.
(o) Pyrometallurgical refinery.

Source: Report EPA-R2-73-247a.

The data for zinc and lead smelter discharges also reflect the effects of such variables. Recycle ratios tend to be high in most of the lead and zinc smelter operations, reflecting the large percent usage for cooling, the usual operation of cooling towers and cooling-water circuits, and the trend to economize on the use of water treated chemically to protect tubes and piping.

In five of the smelter operations included in the data listed, sulfuric acid plants were operated in association with the smelters, and the operation of these plants apparently exerts no consistent recognizable effects on the external- or internal-water-use patterns. The variations in individual plant systems, design, and practice have greater effects on water usage than the presence or absence of an acid plant.

The lead and zinc smelter-refinery combinations reported relatively lower water requirements. In four out of the five zinc or lead refineries responding, and in the industry in general, smelting and refining are carried out in the same plant. Three out of the five operations for which data are listed include sulfuric acid plants. The water-use characteristics of the zinc refining operations 17 and 18 are distinguishable one from the other on the basis of climate and type of operation. The electrolytic zinc refinery 17, is in an arid climate and shows low intake, low discharge, very high use and consumption, and high recirculation rates. The other zinc refining operation 18, located in an area of plentiful rainfall, which refines zinc by distillation, shows much higher intakes and discharges but lower use and consumption.

The sources of wastewater in the lead-zinc industry are listed in the following tabulation:

> Water Treatment
>   Clarification sludges
>   Sand filter backwash
>   Water-softening backwash or sludges
>   Ion-exchange regeneration wastes
> Sanitary Wastes
> Process Water (smelters and refineries)
>   Slag granulation wastewater
>   Gas scrubbing wastewater
>   Spent electrolyte
> Direct Cooling Water
>   Steam jet ejector condensate
> Cooling Waters
>   Indirect once through cooling water
>   Cooling tower blowdown
> Miscellaneous
>   Boiler blowdown.

The quantity of wastewater discharged from lead and zinc plants has been discussed in the previous section. The compositions of wastewaters and the receiving streams reported in the survey are shown in Table 247.

Although these data represent only a small segment of the lead-zinc operations, they provide an insight into the natures and types of wastes that are generated by the industry. Effluents from the tailings ponds of mine-concentrator facilities apparently do not pose serious problems in general. There is a suggestion however that the effluents from some concentrators might contain concentrations of cadmium and lead at or near the level of concern.

Smelter and refinery effluents have a much greater pollution potential than those from concentrators with respect to both cadmium and lead.

Waste-treatment practice in the smelter-refinery segment of the industry ranged from virtually none, through simple settling, to neutralization-flocculation-settling.

TABLE 247: COMPOSITION OF WASTEWATERS AND RECEIVING STREAMS IN LEAD-ZINC METALLURGICAL PROCESSES
(in mg/l)

Mine Concentrator Effluents(a)

pH	5.0-8.0
As	0.0-0.09
Cd	0.0-0.02
Cu	0.004-0.3
Fe	0.04-1.5
Mn	0.0-0.8
Pb	0.0-0.18
Zn	

Lead Smelters, Refineries

	I(b)			II(c)		III(d)		
	Upstream	Outfall	Downstream	Intake	Receiving Stream	Intake	Neutralized Acid Plant Water	Other Waste Water
pH	7.6-8.1	7.4-8.2	7.6-8.2	7.8	7.8-11.1	7.2	5.0	8.6
As	0.11	0.15-0.46				--	--	--
Cd	0.002	0.02-1.09	0.03-0.113			--	7.7	0.5
Cu	0.13	0.13	0.13	0.34	0.0-0.007	5.0	0.06	4.0
Fe	--	--		0.8	0.0-0.015	--	7.4	1.3
Mn	--	--		0.2	0.0-0.03	--	0.7	0.5
Ni	--	--		--	0.004-0.05	--	0.2	0.06
Pb	0.03-0.03	0.07-0.157	0.03-0.05	0.14	0.0-0.01	--	0.5	11.0
Zn	0.08-0.13	0.11-2.00	0.08-0.43	0.019	0.01-0.048	0.9	8.0	2.0
SO4				6.5	9.2-23.0	126	960	200

(continued)

TABLE 247: (continued)

	I(e) Effluent	II(f) Intake	II(f) Discharge
pH	7.7	5.0-7.0	5.2-6.8
As	--	--	--
Cd	0.39	--	0.9
Cu	--	--	0.02-1.35
Fe	1.0	1.0-1.2	1.3-19.7
Mn	--	0.4-1.0	0.4-1.75
Pb	--	--	--
Zn	8.5	5.7	18-25
SO4	--	136-179	182-291
CN	--	--	1.3-1.5
Dissolved Solids	583	372-391	304-534
Suspended Solids	10	14-18	14-164

(a) From tailings ponds.

(b) Company surveyed effect of waste discharge on stream; outfall was combined process and cooling water. Major contributant to process water impurities was effluent from "gas conditioning" operation in which fumes and dusts were "moisturized" by spraying with water before entering electrostatic precipitator. Spray water collected in sump below conditioner and discharged.

(c) Company studied contribution of waste to stream over and above background impurities by analyzing intake water and the receiving stream below the outfall. Major source of contamination in effluent was water discharged from slag granulation operations.

(d) Company operates lead smelter in conjunction with sulfuric acid plant. Provided data on intake water, neutralized acid plant water prior to discharge, and "other" waste water, which includes slag granulation and cooling water.

(e) Combined waste stream from zinc smelter includes coke plant, cooling water, gas scrubbers, spills, clean-up, etc. No acid plant.

(f) Zinc smelter in conjunction with sulfuric acid plant. Analyses show composition of water supply and combined waste discharge.

Source: Report EPA-R-2-73-247a.

References

(1)   U.S. Environmental Protection Agency, "Compilation of Air Pollutant Emission
        Factors," 2nd Edition, *Publication Number AP-42,* Research Triangle Park, N.C.,
        Office of Air and Water Programs (July 1974).
(2)   Hallowell, J.B., Shea, J.F., Smithson, G.R., Jr.; Tripler, A.B. and Gonser, B.W.
        "Water Pollution Control in the Primary Nonferrous Metal Industry, I—Copper,
        Lead and Zinc Industries," *Report EPA-R2-73-247a,* Washington, D.C., U.S.
        Government Printing Office (September 1973).

LEAD SMELTING, SECONDARY

In the secondary smelting, refining, and alloying of lead, the three types of furnace most
commonly used are reverberatory, blast or cupola, and pot. The grade of metal to be produced, soft, semisoft, or hard, dictates the type of furnace to be used.

Used for the production of semisoft lead, the reverberatory furnace reclaims this metal
from a charge of lead scrap, battery plates, oxides, drosses, and lead residues. The furnace
consists of an outer shell built in the shape of a rectangular box lined with refractory brick.
To provide heat for melting, the charge gas or oil-fired burners are usually placed at one
end of the furnace, and the material to be melted is charged through an opening in the
shell.

The charge is placed in the furnace in such a manner as to keep a small mound of unmelted
material on top of the bath. Continuously, as this mound becomes molten at the operating
temperature (approximately 1250°C), more material is charged. Semisoft lead is tapped
off periodically as the level of the metal rises in the furnace. The amount of metal recovered is about 50 to 60 kilograms per square meter of hearth area per hour.

A similar kind of furnace, the revolving (rotary) reverberatory, is used at several European
installations for the recovery of lead from battery scrap and lead sulfate sludge. Its charge
makeup and operating characteristics are identical to the reverberatories used in the United
States, except that the furnace slowly revolves as the charge is heated.

The blast (cupola) furnace, used to produce "hard" lead, is normally charged with the
following: rerun slag from previous runs (4.5%); cast-iron scrap (4.5%); limestone (3%);
coke (5.5%); and drosses from pot furnace refining, oxides, and reverberatory slag (82.5%).
Similar to an iron cupola, the furnace consists of a steel sheet lined with refractory material. Air, under high pressure, is introduced at the bottom through tuyeres to permit
combustion of the coke, which provides the heat and a reducing atmosphere.

As the charge material melts, limestone and iron form an oxidation-retardant flux that
floats to the top, and the molten lead flows from the furnace into a holding pot at a
nearly continuous rate. The rest (30%) of the tapped molten material is slag, 5% of which
is retained for later rerun. From the holding pot, the lead is usually cast into large ingots
called "buttons" or "sows."

Pot-type furnaces are used for remelting, alloying, and refining processes. These furnaces
are usually gas-fired and range in size from 1 to 45 metric tons capacity. Their operation
consists simply of charging ingots of lead or alloy material and firing the charge until the
desired product quality is obtained. Refining processes most commonly employed are those
for the removal of copper and antimony to produce soft lead, and those for the removal
of arsenic, copper, and nickel to produce hard lead.

Air Pollution

The emissions and controls from secondary lead smelting processes may be conveniently considered according to the type of furnace employed. With the reverberatory furnaces, the temperature maintained is high enough to oxidize the sulfides present in the charge to sulfur dioxide and sulfur trioxide, which, in turn, are emitted in the exit gas. Also emitted are such particulates (at concentrations of 16 to 50 grams per cubic meter) as oxides, sulfides, and sulfates of lead, tin, arsenic, copper, and antimony. The particles are nearly spherical and tend to agglomerate. Emission factors for reverberatory furnaces are presented in Table 248.

Combustion air from the tuyeres passing through the blast furnace charge conveys metal oxides, bits of coke, and other particulates present in the charge. The particulate is roughly 7% by weight of the total charge (up to 44 g/m³). In addition to particulates, the stack gases also contain carbon monoxide. However, the carbon monoxide and any volatile hydrocarbons present are oxidized to carbon dioxide and water in the upper portion of the furnace, which effectively acts as an afterburner.

Fabric filters, preceded by radiant cooling columns, evaporative water coolers, or air dilution jets, are also used to control blast furnace particulates. Overall efficiencies exceeding 95% are common. Representative size distributions of particles in blast and reverberatory furnace streams are presented in Table 249. Compared with the other furnace types, pot furnace emissions are low (see Table 248). However, to maintain a hygienic working environment, pot furnace off gases, usually along with emission streams from other furnaces, are directed to fabric filter systems.

TABLE 248: EMISSION FACTORS FOR SECONDARY LEAD SMELTING FURNACES WITHOUT CONTROLS[a], EMISSION FACTOR RATING: B

Furnace type	Particulates		Sulfur dioxide	
	kg/MT	lb/ ton	kg/MT	lb/ton
Reverberatory	73.5 (28.0 to 156.5)	147 (56 to 313)	40.0 (35.5 to 44.0)	80 (71 to 88)
Blast (cupola)	96.5 (10.5 to 190.5)	193 (21.0 to 381.0)	26.5 (9.0 to 55.0)	53.0 (18 to 110)
Pot	0.4	0.8	Neg	Neg
Rotary reverberatory	35.0	70.0	NA	NA

[a]All emission factors expressed in terms of kg/MT and lb/ton of metal charged to furnace.
NA-no data available to make estimates.

TABLE 249: REPRESENTATIVE PARTICLE SIZE DISTRIBUTION FROM A COMBINED BLAST AND REVERBERATORY FURNACE GAS STREAM[a]

Size range, μm	Fabric filter catch, wt %
0 to 1	13.3
1 to 2	45.2
2 to 3	19.1
3 to 4	14.0
4 to 16	8.4

[a]Reference 1.
[b]These particles are distributed log-normally, according to the following frequency distribution:

$$f(D) = 1.56 \exp \left[\frac{-(\log D - 0.262)^2}{0.131} \right]$$

Source: EPA Publication AP-42, 2nd Edition.

Reference

(1)   U.S. Environmental Protection Agency, "Compilation of Air Pollutant Emission Factors," 2nd Edition, *Publication Number AP-42,* Research Triangle Park, N.C., Office of Air and Water Programs (July 1974).

LEATHER INDUSTRY

Water Pollution

For purposes of establishing effluent limitations guidelines and standards of performance, the leather tanning and finishing industry has been divided into six major subcategories. These subcategories have been derived principally by similarities in process and waste loads. Such factors as age and size of plant, climate, and waste control technologies favor segmentation of the industry into these six subcategories. The following tabulation is a capsule summary of these subcategories:

| | - - - - - - - - - - - - - - - - Primary Processes - - - - - - - - - - - - - - - - | | |
Subcategory	Beamhouse	Tanning	Leather Finishing
1	Pulp hair	Chrome	Yes
2	Save hair	Chrome	Yes
3	Save hair	Vegetable	Yes
4	Hair previously removed	Previously tanned	Yes
5	Hair previously removed or retained	Chrome	Yes
6	Pulp or save hair	Chrome or no tanning	No

Currently, waste from about 60% of the tanneries (and approximately 60% of the production) is discharged to municipal sewer systems, while the remainder is discharged directly to surface waters. Compared with other industries, waste treatment facilities for those plants in the leather tanning and finishing industry discharging directly to rivers or streams are severely lacking. There are no exemplary waste treatment plants handling only tannery waste.

Materials which can appear in tannery wastes include the following:

Hair	Lime	Sugars and starches
Hide scraps	Soluble proteins	Oils, fats, and grease
Pieces of flesh	Sulfides	Surface active agents
Blood	Amines	Mineral acids
Manure	Chromium salts	Dyes
Dirt	Tannin	Solvents
Salt	Soda ash	

The basic parameters used to define waste characteristics are as follows:

BOD_5 (Five-Day Biochemical Oxygen Demand)	Chromium
COD (Chemical Oxygen Demand)	Oil and Grease (Hexane Solubles)
Suspended Solids	Sulfide
Total Nitrogen	pH

Additional data from some sources permitted a check of the various forms of nitrogen and volatile components of total and suspended solids. No significant information was available on color. Information was also collected on total daily waste flow. Data for each of the subcategory types of plants are summarized in Table 250.

TABLE 250: RAW WASTEWATER CHARACTERISTICS* BY CATEGORY

Characteristics	Category 1 No. of Tanneries	Range	Average	Category 2 No. of Tanneries	Range	Average	Category 3 No. of Tanneries	Range	Average
Flow: m³/kg	46	0.007 - 0.156	0.053	14	0.001 - 0.189	0.063	16	0.007 - 0.106	0.050
(gal/lb)		(0.8 - 18.7)	(6.4)		(0.1 - 22.6)	(7.6)		(0.8 - 12.7)	(6.0)
BOD₅	24	4.8 - 270	95	9	22 - 140	69	12	7.4 - 130	67
COD	18	10.5 - 595	260	7	88 - 215	140	9	24 - 695	250
Total Solids	16	36 - 890	525	7	140 - 900	480	9	120 - 800	345
Suspended Solids	23	6.7 - 595	140	9	30 - 350	145	10	20 - 445	135
Total Chromium	18	0.1 - 19	4.3	7	0.3 - 12	4.9	5	0.2 - 0.6	0.2
Sulfides	12	0.1 - 46	8.5	4	0.1 - 2.8	0.8	7	0.1 - 4.2	1.2
Grease	13	0.1 - 70	19	5	0.7 - 105	43	7	0.1 - 160	33
Total Alkinity (as CaCO₃)	12	0.5 - 300	98	1	62 - 85	72	6	4.1 - 135	66
Total Nitrogen (as N)	7	3.1 - 44	17	6	3.6 - 22	13	5	0.9 - 23	9.2
pH	26	1.0 - 13.0	-	8	4.0 - 12.6	-	12	2.0 - 13.0	-
Temperature** °C	15	2.8 - 93.2	21.1	6	1.7 - 27.8	18.3	3	4.4 - 28.9	17.2
(°F)		(37 - 200)	(70)		(35 - 82)	(65)		(40 - 84)	(63)

Characteristics	Category 4 No. of Tanneries	Range	Average	Category 5 No. of Tanneries	Range	Average	Category 6 No. of Tanneries	Range	Average
Flow: m³/kg	10	0.003 - 0.033	0.020	20	0.006 - 0.204	0.063	3	0.014 - 0.056	0.028
(gal/lb)		(0.3 - 3.9)	(2.4)		(0.7 - 24.4)	(7.6)		(1.7 - 6.7)	(3.4)
BOD₅	3	6.7 - 67	37	8	10 - 140	67	2	32 - 160	110
COD	3	5.7 - 63	28	5	11 - 265	170	2	53 - 155	230
Total Solids	2	47 - 285	140	7	52 - 980	490	2	210 - 910	595
Suspended Solids	3	7.0 - 125	47	8	3.1 - 865	88	2	44 - 185	110
Total Chromium	3	0.4 - 4.8	2.6	7	0.1 - 2.1	1.2	1	3.8 - 5.9	4.4
Sulfides	1	2.1	2.1	1	4.5	4.5	1	2.0 - 6.3	3.7
Grease	3	2.2 - 19	7.9	7	0.6 - 46	24	2	1.0 - 19	6.6
Total Alkinity (as CaCO₃)	1	39	39	3	6.6 - 180	69	1	37 - 54	43
Total Nitrogen (as N)	2	0.8 - 6.5	3.7	5	0.6 - 29	6.0	1	14 - 18	16
pH	3	3.4 - 11.2	-	9	1.5 - 12.5	-	2	9.2 - 10.4	-
Temperature** °C	3	10.0 - 26.6	20.7	3	4.4 - 36.6	22.8	-	-	-
(°F)		(50 - 80)	(69)		(40 - 98)	(73)		-	-

*Based on kg/kkg of hide (lb/1,000 lb); flow in m³/kg (gal/lb); temperature in °C(°F).
**Average temperature is average summer and winter values; temperature range is low winter to high summer values.

Source: Report EPA 440/1-73/015.

Reference

(1) U.S. Environmental Protection Agency, "Development Document for Proposed Effluent Limitations Guidelines and New Source Performance Standards for the Leather Tanning and Finishing Point Source Category," *Report EPA 440/1-73/015*, Washington, D.C. (November 1973).

LIME MANUFACTURE

Lime (CaO) is the high-temperature product of the calcination of limestone (CaCO₃). Lime is manufactured in vertical or rotary kilns fired by coal, oil, or natural gas.

Air Pollution

Atmospheric emissions in the lime manufacturing industry include particulate emissions from the mining, handling, crushing, screening, and calcining of the limestone and combustion products from the kilns. The vertical kilns, because of a larger size of charge material, lower air velocities, and less agitation, have considerably fewer particulate emissions. Control of emissions from these vertical kilns is accomplished by sealing the exit of the kiln and exhausting the gases through control equipment.

Particulate emission problems are much greater on the rotary kilns because of the smaller size of the charge material, the higher rate of fuel consumption, and the greater air velocities through the rotary chamber. Methods of control on rotary-kiln plants include simple and multiple cyclones, wet scrubbers, baghouses, and electrostatic precipitators. Emission factors for lime manufacturing are summarized in Table 251 (1).

TABLE 251: PARTICULATE EMISSION FACTORS FOR LIME MANUFACTURING WITHOUT CONTROLS[a], EMISSION FACTOR RATING: B

Operation	Emissions[b]	
	lb/ton	kg/MT
Crushing[c]		
Primary	31	15.5
Secondary	2	1
Calcining[d]		
Vertical kiln	8	4
Rotary kiln	200	100

[a] Emission factors expressed as units per unit weight of lime processed.
[b] Cyclones could reduce these factors by about 70 percent. Venturi scrubbers could reduce these factors by about 95 to 99 percent. Fabric filters could reduce these factors by about 99 percent.
[c] Reference 3.
[d] References 2, 4

Source: EPA Publication AP-42, 2nd Edition

Water Pollution

The raw wastes produced from calcium oxide manufacture consist of fine dusts collected from the plant gas effluent by scrubbing systems. At the exemplary facility this dust removal is achieved by use of bag filters and other dry particulate collection equipment. No wet scrubbing techniques are employed. Wet scrubbing of these dusts is used commonly at nonexemplary plants.

Waste Product	Process Source	kg/kkg of Product (lb/ton)
Dry particulate matter	Kiln gases (Dry collector)	67 (133) (no effects of startup and shutdown)

Exemplary plant water usage is described as follows. All cooling water is recycled and all process water is consumed in the manufacture because with the use of dry waste collection techniques, there is no waterborne effluent from the facility. This plant achieves 95% or better solids collection at the kiln collector. Municipal water intake to the plant amounts to 638 liters of product (153 gal/ton) plus the amount evaporated on the cooling tower. This water is not further treated in the plant prior to use. This amount of water represents the process water, that is the water used in the hydrator. The cooling water flow for the bearings on the tube mill and pistons on the hydrator pump amounts to 1,000 liters per metric ton of product (240 gal/ton) on an average and it is completely recycled with makeup water added to compensate for evaporation. There is no waterborne effluent from an exemplary lime plant.

References

(1) U.S. Environmental Protection Agency, "Compilation of Air Pollutant Emission Factors," 2nd Edition, *Publication Number AP-42,* Research Triangle Park, N.C., Office of Air and Water Programs (April 1973).

(2) Lewis, C. and Crocker, B., "The Lime Industry's Problem of Airborne Dust." *J. Air Pol. Control Assoc. 19:*31-39, (January 1969).

(3) State of Maryland Emission Inventory Data. Maryland State Department of Health, Baltimore, Maryland. (1969).

(4) A Study of the Lime Industry in the State of Missouri for the Air Conservation Commission of the State of Missouri. Reston, Virginia, Resources Research, Incorporated. (December 1967) p. 54.

(5) U.S. Environmental Protection Agency, "Development Document for Proposed Effluent Limitations Guidelines and New Source Performance Standards for the Major Inorganic Products Segment of the Inorganic Chemicals Mfg. Point Source Category," *Report EPA 440/1-73/007* (August 1973).

LIQUEFIED PETROLEUM GAS COMBUSTION

Liquefied petroleum gas, commonly referred to as LPG, consists mainly of butane, propane, or a mixture of the two, and of trace amounts of propylene and butylene. This gas, obtained from oil or gas wells as a by-product of gasoline refining, is sold as a liquid in metal cylinders under pressure, and, therefore, is often called bottled gas. LPG is graded according to maximum vapor pressure with Grade A being predominantly butane, Grade F being predominantly propane, and Grades B through E consisting of varying mixtures of butane and propane. The heating value of LPG ranges from 97,400 Btu/gallon (6,480 kcal/liter) for Grade A to 90,500 Btu/gallon (6,030 kcal/liter) for Grade F. The largest market for LPG is the domestic-commercial market, followed by the chemical industry and the internal combustion engine.

Air Pollution

Emission factors for LPG combustion are presented in Table 252. LPG is considered a "clean" fuel because it does not produce visible emissions (1). Gaseous pollutants such as carbon monoxide, hydrocarbons, and nitrogen oxides do occur, however. The most significant factors affecting these emissions are the burner design, adjustment, and venting (2).

TABLE 252: EMISSION FACTORS FOR LPG COMBUSTION[a]
EMISSION FACTOR RATING: C

| | Industrial process furnaces | | | | Domestic and commercial furnaces | | | |
| | Butane | | Propane | | Butane | | Propane | |
Pollutant	lb/10³ gal	kg/10³ liters	lb/10³ gal	kg/10³ liters	lb/10³ gal	kg/10³ liters	lb/10³ gal	kg/10³ liters
Particulates	1.8	0.22	1.7	0.20	1.9	0.23	1.8	0.22
Sulfur oxides[b]	0.09S	0.09S	0.09S	0.01S	0.09S	0.01S	0.09S	0.01S
Carbon monoxide	1.6	0.19	1.5	0.18	2.0	0.24	1.9	0.23
Hydrocarbons	0.3	0.036	0.3	0.036	0.8	0.096	0.7	0.084
Nitrogen oxides[c]	12.1	1.45	11.2	1.35	(8 to 12)[d]	(1.0 to 1.5)[d]	(7 to 11)[d]	(0.8 to 1.3)[d]

[a]LPG emission factors calculated assuming emissions (excluding sulfur oxides) are the same, on a heat input basis, as for natural gas combustion.
[b]S equals sulfur content expressed in grains per 100 ft³ gas vapor; e.g., if the sulfur content is 0.16 grain per 100 ft³ (0.366 g/100 m³) vapor, the SO₂ emission factor would be 0.09 x 0.16 or 0.014 lb SO₂ per 1000 gallons (0.01 x 0.366 or 0.0018 kg SO₂/10³ liters) butane burned.
[c]Expressed as NO₂.
[d]Use lower value for domestic units and higher value for commercial units.

Source: EPA Publication AP-42, 2nd Edition

Improper design, blocking and clogging of the flue vent, and lack of combustion air result in improper combustion that causes the emission of aldehydes, carbon monoxide, hydrocarbons, and other organics. Nitrogen oxide emissions are a function of a number of variables including temperature, excess air, and residence time in the combustion zone. The amount of sulfur dioxide emitted is directly proportional to the amount of sulfur in the fuel.

References

(1) Air Pollutant Emission Factors. Final Report. Resources Research, Inc. Reston, Va. Prepared for National Air Pollution Control Administration, Durham, N.C., under Contract Number CPA-22-69-119 (April 1970).

(2) Clifford, E.A., "A Practical Guide to Liquified Petroleum Gas Utilization." New York, Moore Publishing Co. (1962).

MAGNESIUM SMELTING, SECONDARY

Magnesium smelting is carried out in crucible or pot-type furnaces that are charged with magnesium scrap and fired by gas, oil, or electric heating. A flux is used to cover the surface of the molten metal because magnesium will burn in air at the pouring temperature (approximately 1500°F or 815°C). The molten magnesium, usually cast by pouring into molds, is annealed in ovens utilizing an atmosphere devoid of oxygen.

Air Pollution

Emissions from magnesium smelting include particulate magnesium (MgO) from the melting, nitrogen oxides from the fixation of atmospheric nitrogen by the furnace temperatures, and sulfur dioxide losses from annealing oven atmospheres. Factors affecting emissions include the capacity of the furnace; the type of flux used on the molten material; the amount of lancing used; the amount of contamination of the scrap, including oil and other hydrocarbons; and the type and extent of control equipment used on the process. The emission factors for a pot furnace are shown in Table 253 (1).

TABLE 253: EMISSION FACTORS FOR MAGNESIUM SMELTING
EMISSION FACTOR RATING: C

Type of furnace	Particulates[a]	
	lb/ton	kg/MT
Pot furnace		
Uncontrolled	4	2
Controlled	0.4	0.2

[a]References 2 and 3. Emission factors expressed as units per unit weight of metal processed.

Source: EPA Publication AP-42, 2nd Edition

References

(1) U.S. Environmental Protection Agency, "Compilation of Air Pollutant Emission Factors," 2nd Edition, *Publication Number AP-42,* Research Triangle Park, N.C., Office of Air and Water Programs (April 1973).

(2) Allen, G.L. et al. Control of Metallurgical and Mineral Dusts and Fumes in Los Angeles County. Department of the Interior, Bureau of Mines, Washington, D.C. Information Circular Number 7627 (April 1952).

(3) Hammond, W.F. Data on Non-Ferrous Metallurgical Operations. Los Angeles County Air Pollution Control District (November 1966).

MEAT INDUSTRY

Air Pollution

Meat smoking is a diffusion process in which food products are exposed to an atmosphere of hardwood smoke, causing various organic compounds to be absorbed by the food. Smoke is produced commercially in the United States by three major methods: (1) by burning dampened sawdust (20 to 40% moisture), (2) by burning dry sawdust (5 to 9% moisture) continuously, and (3) by friction. Burning dampened sawdust and kiln-dried sawdust are the most widely used methods. Most large, modern, production meat smokehouses are the recirculating type, in which smoke is circulated at reasonably high temperatures throughout the smokehouse.

Emissions from smokehouses are generated from the burning hardwood rather than from the cooked product itself. Based on approximately 110 pounds of meat smoked per pound of wood burned (110 kg of meat per kg of wood burned), emission factors have been derived for meat smoking and are presented in Table 254.

Emissions from meat smoking are dependent on several factors, including the type of wood, the type of smoke generator, the moisture content of the wood, the air supply, and the amount of smoke recirculated. Both low-voltage electrostatic precipitators and direct-fired afterburners may be used to reduce particulate and organic emissions. These controlled emission factors have also been shown in Table 254 (1). Air pollution also occurs from meat rendering operations as discussed by Jones (4).

TABLE 254: EMISSION FACTORS FOR MEAT SMOKING[a,b]
EMISSION FACTOR RATING: D

Pollutant	Uncontrolled		Controlled[c]	
	lb/ton of meat	kg/MT of meat	lb/ton of meat	kg/MT of meat
Particulates	0.3	0.15	0.1	0.05
Carbon monoxide	0.6	0.3	Neg[d]	Neg
Hydrocarbons (CH_4)	0.07	0.035	Neg	Neg
Aldehydes (HCHO)	0.08	0.04	0.05	0.025
Organic acids (acetic)	0.2	0.10	0.1	0.05

[a]Based on 110 pounds of meat smoked per pound of wood burned (110 kg meat/kg wood burned).
[b]References 2, 3, and section on charcoal production.
[c]Controls consist of either a wet collector and low-voltage precipitator in series or a direct-fired afterburner.
[d]With afterburner.

Source: EPA Publication AP-42, 2nd Edition

Water Pollution

Water is a raw material in the meat packing industry that is used to cleanse products and to remove and convey unwanted material. The principal operations and processes in meat packing plants where wastewater originates are: animal holding pens, slaughtering, cutting, meat processing, secondary manufacturing (by-product operations) including both edible and inedible rendering and clean-up.

Wastewaters from slaughterhouses and packinghouses contain organic matter (including grease), suspended solids, and inorganic materials such as phosphates, nitrates, nitrites, and salt. These materials enter the waste stream as blood, meat and fatty tissue, meat extracts, paunch contents, bedding, manure, hair, dirt, contaminated cooling water losses from edible and inedible rendering, curing and pickling solutions, preservatives, and caustic or alkaline detergents (4)(5).

The raw wasteload from all four subcategories of the meat packing industry discussed in the following paragraphs includes the effects of in-plant materials recovery which incidentally serves the function of primary waste treatment.

The parameters used to characterize the raw effluent were the flow, BOD_5, suspended solids (SS), grease, chlorides, phosphorus, and Kjeldahl nitrogen. BOD_5 is considered to be, in general, the best available measure of the wasteload. Parameters used to characterize the size of the operations were the kill (live weight) and amount of processed meat products produced. All values of waste parameters are expressed as kg/1000 kg LWK, which has the same numerical value when expressed in lb/1000 lb LWK. In some cases, treated effluents are so dilute that concentration becomes limiting. In these cases, concentration is expressed as milligrams per liter, mg/l. Kill and amount of processed meat products are expressed in thousands of kg.

The following Tables 255 through 258 include a summary of data showing averages, standard deviations, ranges, and number of observations (plants) presented in the following sections for each of the four subcategories of the industry.

TABLE 255: SUMMARY OF PLANT AND RAW WASTE CHARACTERISTICS FOR SIMPLE SLAUGHTERHOUSES

Base	Flow 1/1000 kg LWK	Kill 1000 kg/day	BOD_5 kg/1000 kg LWK	Suspended Solids kg/1000 kg LWK	Grease kg/1000 kg LWK	Kjeldahl Nitrogen as N kg/1000 kg LWK	Chlorides as Cl kg/1000 kg LWK	Total Phosphorus as P kg/1000 kg LWK
(Number of Plants)	(24)	(24)	(24)	(22)	(12)	(5)	(3)	(5)
Average	5,328	220	6.0	5.6	2.1	0.68	2.6	0.05
Standard Deviation	3,644	135	3.0	3.1	2.2	0.46	2.7	0.03
Range, low-high	1,334– 14,641	18.5– 552.	1.5– 14.3	0.6– 12.9	0.24– 7.0	0.23– 1.36	0.01– 5.4	0.014– 0.086

TABLE 256: SUMMARY OF PLANT AND RAW WASTE CHARACTERISTICS FOR COMPLEX SLAUGHTERHOUSES

Base	Flow 1/1000 kg LWK	Kill 1000 kg/day	BOD_5 kg/1000 kg LWK	Suspended Solids kg/1000 kg LWK	Grease kg/1000 kg LWK	Kjeldahl Nitrogen as N kg/1000 kg LWK	Chlorides as Cl kg/1000 kg LWK	Total Phosphorus as P kg/1000 kg LWK
(Number of Plants)	(19)	(19)	(19)	(16)	(11)	(12)	(6)	(5)
Average	7,379	595	10.9	9.6	5.9	0.84	2.8	0.33
Standard Deviation	2,718	356	4.5	4.1	5.7	0.66	2.7	0.49
Range, low-high	3,627– 12,507	154– 1498	5.4– 18.8	2.8– 20.5	0.7– 16.8	0.13– 2.1	0.81– 7.9	0.05– 1.2

Source: Report EPA 440/1-73/012.

TABLE 257: SUMMARY OF PLANT AND RAW WASTE CHARACTERISTICS FOR LOW-PROCESSING PACKINGHOUSES

Base	Flow 1/1000 kg LWK	Kill 1000 kg/day	BOD$_5$ kg/1000 kg LWK	Suspended Solids kg/1000 kg LWK	Grease kg/1000 kg LWK	Kjeldahl Nitrogen as N kg/1000 kg LWK	Chlorides as Cl kg/1000 kg LWK	Total Phosphorus as P kg/1000 kg LWK	Processed Products 1000 kg/day	Ratio of Processed Products to Kill
(Number of Plants)	(23)	(23)	(29)	(22)	(15)	(6)	(5)	(4)	(23)	(23)
Average	7,842	435	8.1	5.9	3.0	0.53	3.6	0.13	54	0.14
Standard Deviation	4,019	309	4.6	4.0	2.1	0.44	2.7	0.16	52	0.09
Range, low-high	2,018-17,000	89-1,394	2.3-18.4	0.6-13.9	0.8-7.7	0.04-1.3	0.5-4.9	0.03-0.43	3.0-244.	0.016-0.362

TABLE 258: SUMMARY OF PLANT AND RAW WASTE CHARACTERISTICS FOR HIGH-PROCESSING PACKINGHOUSES

Base	Flow 1/1000 kg LWK	Kill 1000 kg/day	BOD$_5$ kg/1000 kg LWK	Suspended Solids kg/1000 kg LWK	Grease kg/1000 kg LWK	Kjeldahl Nitrogen as N kg/1000 kg LWK	Chlorides as Cl kg/1000 kg LWK	Total Phosphorus as P kg/1000 kg LWK	Processed Products 1000 kg/day	Ratio of Processed Products to Kill
(Number of Plants)	(19)	(19)	(19)	(14)	(10)	(3)	(7)	(3)	(19)	(19)
Average	12,514	350	16.1	10.5	9.0	1.3	15.6	0.38	191	0.65
Standard Deviation	4,894	356	6.1	6.3	8.3	0.92	11.3	0.22	166	0.39
Range of low-high	5,444-20,261	8.8-1,233.	6.2-30.5	1.7-22.5	2.8-27.0	0.65-2.7	0.8-36.7	0.2-0.63	4.5-631.	0.40-2.14

Source: Report EPA 440/1-73/012.

References

(1) U.S. Environmental Protection Agency, "Compilation of Air Pollutant Emission
       Factors," 2nd Edition, *Publication Number AP-42,* Research Triangle Park, N.C.,
       Office of Air and Water Programs (April 1973).
(2) Carter, E. Communication between Maryland State Department of Health and
       Resources Research, Incorporated. (November 21, 1969).
(3) Polglase, W.L., Dey, H.F., and Walsh, R.T. "Smokehouses." In: *Air Pollution
       Engineering Manual.* Danielson, J. A. (ed.). U.S. DHEW, PHS, National Center
       for Air Pollution Control. Cincinnati, Ohio. *Publication Number 999-AP-40.*
       (1967) p. 750-755.
(4) Jones, H.R., *Pollution Control in Meat, Poultry and Seafood Processing,* Park Ridge,
       N.J., Noyes Data Corp. (1974).
(5) U.S. Environmental Protection Agency, "Development Document for Proposed
       Effluent Limitations Guidelines and New Source Performance Standards for the
       Red Meat Processing Segment of the Meat Product and Rendering Processing Point
       Source Category," *Report EPA 440/1-73/012,* Washington, D.C. (October 1973).

MERCURY MINING AND PRODUCTION

Water Pollution

The water-usage patterns reported contained the elements indicated in Figure 33.

**FIGURE 33: GENERALIZED DIAGRAM OF COMMON FEATURES OF WATER
CIRCUITS OF MERCURY MINES AND PLANTS**

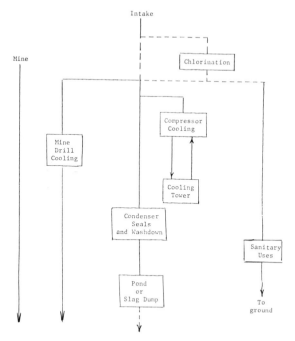

Source: EPA-R2-73-247b

The only process-type water consists of the water pool which forms the seals at the bottoms of the condenser U-tubes, and any wash water used in periodic cleaning of the inside of the condenser tubes for maintenance purposes. One instance was reported of the use of a closed-circuit cooling system for associated air compressors.

Remote locations and an arid climate are characteristics of most of the mercury operations. Thus, one mine operation reported the use of bottled drinking water and chemical toilets, but did have a discharge of mine water. No furnacing operation was associated with this mine. The three operations had capacities to mine and process about 200,000 tons of ore a year and showed the following discharges of water.

Intake, MGY	Discharge MGY	Type of Discharge or Receiver
5	0	Evaporation pond
250	250	Untreated; to surface water
60	46	Mixed mine and process water

In the last example, wastewaters were percolated through the slag or tailings dump and the effluent from the heap was diluted with fresh water. The diluted effluent which constituted the discharge to the receiving water had the following analysis:

	mg/l
Nitrate	2.4
Chloride	63.8
Sulfate	2382.6
Bicarbonate	12.2
Carbonate	0
Sodium	186
Potassium	23.0
Calcium	289.6
Magnesium	294.4
Fluorides	1.7
Silica	18.0
Iron	0.02
Manganese	2.4
Boron	3.7
Cyanide	0.03
Lead	0.000
Arsenic	0.000
Copper	0.08
Mercury	0.02
Total Hardness	1950.0
pH	4.8 - 5.7

Considerable elevation of the contents of dissolved salts and some metals are apparent in this analysis.

The information obtained on the primary mercury industry is so scant as to serve only as an isolated example. However, even among the various aspects of the state of the industry, there is the long-range possibility of one process modification to overcome the problem of mercury emissions to the atmosphere which would affect water usage by the industry. One of the possible alternatives to the processes is hydrometallurgical processing.

Table 259 summarizes some of the aspects of mercury mining and processing emissions and their control.

TABLE 259: POLLUTION FROM MERCURY MINING AND PROCESSING

Nominal or Overall Operation	Unit Operation or Process	Category or Use of Water	Wastewater Characteristics	Assessment of Reported Status of Wastewater Treatment or Water Pollution Control	Recommended Future Action
Mining	Mining	Mine drainage	High dissolved salts; potentially high mercury	Impact uncertain due to remote locations and arid climate conditions; information sample poor due to industry air pollution/economic status.	Suggest watchdog-type exercise to continually assess industry as operations fluctuate; the number of operations is currently decreasing, due to price situation.
Furnacing	Condenser seals and washdown	Process	Mercury and soot contents	One example gave settling and percolation through slag dump.	See above.
Overall	–	–	–	Air pollution problems would possibly result in process changes with water pollution potential.	

Source: EPA-R2-73-247b

Reference

(1) Hallowell, J.B., Shea, J.F., Smithson, G.R., Jr., Tripler, A.B. and Gonser, B.W., "Water Pollution Control in the Primary Nonferrous Metals Industry, II — Aluminum, Mercury, Gold, Silver, Molybdenum and Tungsten," *Report No. EPA-R2-73-247b,* Washington, D.C., U.S. Government Printing Office (September 1973).

MINERAL WOOL MANUFACTURING

The product mineral wool used to be divided into three categories: slag wool, rock wool, and glass wool. Today, however, straight slag wool and rock wool as such are no longer manufactured. A combination of slag and rock constitutes the charge material that now yields a product classified as a mineral wool, used mainly for thermal and acoustical insulation.

Mineral wool is made primarily in cupola furnaces charged with blast-furnace slag, silica rock, and coke. The charge is heated to a molten state at about 3000°F (1650°C) and then fed to a blow chamber, where steam atomizes the molten rock into globules that develop long fibrous tails as they are drawn to the other end of the chamber. The wool blanket formed is next conveyed to an oven to cure the binding agent and then to a cooler.

Air Pollution

The major source of emissions is the cupola or furnace stack. Its discharge consists primarily of condensed fumes that have volatilized from the molten charge and gases such as sulfur oxides and fluorides. Minor sources of particulate emissions include the blowchamber, curing oven, and cooler. Emission factors for various stages of mineral wool processing are shown in Table 260 (1). The effect of control devices on emissions is shown in footnotes to the table.

TABLE 260: EMISSION FACTORS FOR MINERAL WOOL PROCESSING WITHOUT CONTROLS[a], EMISSION FACTOR RATING: C

Type of process	Particulates		Sulfur oxides	
	lb/ton	kg/MT	lb/ton	kg/MT
Cupola	22	11	0.02	0.01
Reverberatory furnace	5	2.5	Neg[b]	Neg
Blow chamber[c]	17	8.5	Neg	Neg
Curing oven[d]	4	2	Neg	Neg
Cooler	2	1	Neg	Neg

[a]Emission factors expressed as units per unit weight of charge.
[b]Negligible.
[c]A centrifugal water scrubber can reduce particulate emissions by 60%.
[d]A direct-flame afterburner can reduce particulate emissions by 50%.

Source: EPA Publication AP-42, 2nd Edition.

Reference

(1)   U.S. Environmental Protection Agency, "Compilation of Air Pollutant Emission Factors," 2nd Edition, *Publication Number AP-42*, Research Triangle Park, N.C., Office of Air and Water Programs (April 1973).

NATURAL GAS COMBUSTION

Natural gas is one of the major fuels used throughout the country. It is used mainly for power generation, industrial process steam and heat production, and domestic and commercial space heating. The primary component of natural gas is methane, although varying amounts of ethane and smaller amounts of nitrogen, helium, and carbon dioxide are also present. The average gross heating value of natural gas is 1,050 Btu/stdft^3 (9,350 kcal/Nm^3). Because natural gas in its original state is a gaseous, homogenous fluid, its combustion is simple and can be precisely controlled. Common excess air rates range from 10 to 15%; however, some large units operate at excess air rates as low as 5% to maximize efficiency and minimize nitrogen oxide (NO_x) emissions.

Air Pollution

Even though natural gas is considered to be a relatively clean fuel, some emissions can occur from the combustion reaction; poor mixing, insufficient air, etc. may cause large amounts of smoke, carbon monoxide, and hydrocarbons to be produced. Moreover, because a sulfur-containing mercaptan is added to natural gas for detection purposes, small amounts of sulfur oxides will also be produced. Nitrogen oxides are the major pollutants of concern when burning natural gas; they are a function of the temperature in the combustion chamber and the rate of cooling of the combustion products. Emission levels generally vary considerably with the type and size of unit and are also a function of loading. Emission factors for natural gas combustion are presented in Table 261. Flue gas cleaning equipment has not been utilized to control emissions from natural gas combustion equipment.

TABLE 261: NATURAL-GAS COMBUSTION—EMISSION FACTOR RATING: A

	Type of unit					
	Power plant		Industrial process boiler		Domestic and commercial heating	
Pollutant	lb/10^6 ft^3	kg/10^6 m^3	lb/10^6 ft^3	kg/10^6 m^3	lb/10^6 ft^3	kg/10^6 m^3
Particulates	5-15	80-240	5-15	80-240	5-15	80-240
Sulfur oxides (SO$_2$)[a]	0.6	9.6	0.6	9.6	0.6	9.6
Carbon monoxide	17	272	17	272	20	320
Hydrocarbons (as CH$_4$)	1	16	3	48	8	128
Nitrogen oxides (NO$_2$)	700[b-d]	11,200[b-d]	(120-230)[e]	(1920-3680)[e]	(80-120)[j]	(1280-1920)[f]

[a]Based on an average sulfur content of natural gas of 2,000 g/10^6 stdft^3 (4,600 g/10^6 Nm^3).
[b]Use 300 lb/10^6 stdft^3 (4,800 kg/10^6 Nm^3) for tangentially fired units.
[c]At reduced loads, multiply this factor by a load reduction coefficient (1).
[d]See text for potential NO_x reductions due to combustion modifications. Note that the NO_x reductions from these modifications will also occur at reduced load conditions.
[e]This represents a typical range for many industrial boilers. For large industrial units (>100 MMBtu/hr) use the NO_x factors presented for power plants.
[f]Use 80 (1,280) for domestic heating units and 120 (1,920) for commercial units.

Source: EPA Publication AP-42, 42nd Edition.

Reference

(1)   U.S. Environmental Protection Agency, "Compilation of Air Pollutant Emission Factors," 2nd Edition, *Publication Number AP-42,* Supplement Number 3, Research Triangle Park, N.C., Office of Air and Water Programs (July 1974).

NITRIC ACID MANUFACTURE

Air Pollution

The emissions derived from nitric acid manufacture consist primarily of nitric oxide, which accounts for visible emissions; nitrogen dioxide; and trace amounts of nitric acid mist. By far, the major source of nitrogen oxides is the tail gas from the acid absorption tower (Table 262). In general, the quantity of NO_x emissions is directly related to the kinetics of the nitric acid formation reaction.

The specific operating variables that increase tail gas NO_x emissions are: (1) insufficient air supply, which results in incomplete oxidation of NO; (2) low pressure in the absorber; (3) high temperature in the cooler-condenser and absorber; (4) production of an excessively high-strength acid; and (5) operation at high throughput rates, which results in decreased residence time in the absorber.

TABLE 262: NITROGEN OXIDE EMISSIONS FROM NITRIC ACID PLANTS[a]
EMISSION FACTOR RATING: B

Type of control	Control efficiency, %	Emissions (NO_2)[b]	
		lb/ton acid	kg/MT acid
Weak acid			
Uncontrolled	0	50 to 55[c]	25.0 to 27.5
Catalytic combustor (natural gas fired)	78 to 97	2 to 7[d]	1.0 to 3.5
Catalytic combustor (hydrogen fired)	97 to 99.8	0.0 to 1.5	0.0 to 0.75
Catalytic combustor (75% hydrogen, 25% natural gas fired)	98 to 98.5	0.8 to 1.1	0.4 to 0.55
High-strength acid	—	0.2 to 5.0	0.1 to 2.5

[a]References 1 and 2.
[b]Based on 100 percent acid production.
[c]Range of values taken from four plants measured at following process conditions: production rate, 120 tons (109 MT) per day (100 percent rated capacity); absorber exit temperature, 90° F (32° C); absorber exit pressure, 7.8 atmospheres; acid strength, 57 percent. Under different conditions, values can vary from 43 to 57 lb/ton (21.5 to 28.5 kg/MT).
[d]To present a more realistic picture, ranges of values were used instead of averages.

Source: EPA Publication AP-42, 2nd Edition.

Aside from the adjustment of these variables, the most commonly used means for controlling emissions is the catalytic combustor. In this device, tail gases are heated to ignition temperature, mixed with fuel (natural gas, hydrogen, or a mixture of both), and passed over a catalyst. The reactions that occur result in the successive reduction of NO_2 to NO and, then, NO to N_2. The extent of reduction of NO_2 to N_2 in the combustor is, in turn, a function of plant design, type of fuel used, combustion temperature and pressure, space velocity through the combustor, type and amount of catalyst used, and reactant concentrations (Table 262). Comparatively small amounts of nitrogen oxides are also lost from acid concentrating plants. These losses (mostly NO_2) occur from the condenser system, but the emissions are small enough to be easily controlled by the installation of inexpensive absorbers.

Acid mist emissions do not occur from a properly operated plant. The small amounts that may be present in the absorber exit gas stream are removed by a separator or collector prior to entering the catalytic combustor or expander. Finally, small amounts of nitrogen dioxide are lost during the filling of storage tanks and tank cars.

Nitrogen oxide emissions (expressed as NO_2) are presented for weak nitric acid plants in Table 262. The emission factors vary considerably with the type of control employed, as well as with process conditions. For comparison purposes, the Environmental Protection Agency (EPA) standard for both new and modified plants is 3.0 pounds per ton of 100% acid produced (1.5 kg per metric ton), maximum 2-hour average, expressed as NO_2 (3). Unless specifically indicated as 100% acid, production rates are generally given in terms of the total weight of product (water and acid). For example, a plant producing 500 tons (454 MT) per day of 55 weight percent nitric acid is really producing only 275 tons (250 MT) per day of 100% acid.

Water Pollution

The raw waste load from nitric acid production at the exemplary plant is listed below. There are no nitrates in the waste. All weak nitric acid lost in manufacture is recycled to the process at this facility. The wastes consist only of water treatment chemicals used for the cooling water.

Waste Products	Process Source	Avg. kg/kkg HNO_3 (lb/ton)*	
Lime	Boiler Feedwater	0.47	(0.94)
Calcium and Magnesium Carbonates	Boiler Feedwater	0.6	(1.2)
Disodium Phosphate	Boiler	0.0016	(0.0032)
Sodium Sulfate	Boiler	0.0008	(0.0016)
Sulfuric Acid	Cooling Tower	0.0016	(0.0032
Chlorine	Cooling Water Treatment	1.0	(2.0)

*Values not affected by startup and shutdown.

References

(1) Control of Air Pollution from Nitric Acid Plants. Unpublished Report. Environmental Protection Agency, Research Triangle Park, N.C.

(2) "Atmospheric Emissions from Nitric Acid Manufacturing Processes." U.S. DHEW, PHS, Division of Air Pollution. Cincinnati, Ohio. *Publication Number 999-AP-27.* (1966).

(3) Standards of Performance for New Stationary Sources. Environmental Protection Agency, Washington, D.C. Federal Register, 36(247): (December 23, 1971).

(4) U.S. Environmental Protection Agency, "Development Document for Proposed Effluent Limitations Guidelines and New Source Performance Standards for the Major Inorganic Products Segments of the Inorganic Chemicals Mfg. Point Source Category," *Report EPA 440-1-73/007* (August 1973).

NITROGEN FERTILIZER MANUFACTURE

Nitrate fertilizers are defined as the product resulting from the reaction of nitric acid and ammonia to form ammonium nitrate solutions or granules. Essentially three steps are involved in producing ammonium nitrate: neutralization, evaporation of the neutralized solution, and control of the particle size and characteristics of the dry product.

Anhydrous ammonia and nitric acid (57 to 65% HNO_3) are brought together in the neutralizer to produce ammonium nitrate. An evaporator or concentrator is then used to increase the ammonium nitrate concentration. The resulting solutions may be formed into granules by the use of prilling towers or by ordinary granulators. Limestone may be added in either process in order to produce calcium ammonium nitrate.

In addition to ammonium nitrate, there are the following nitrogen fertilizers: ammonia, ammonium nitrate, ammonium sulfate and urea. The reader is referred to the section dealing with the manufacture of each of these materials for additional details.

Air Pollution

The main emissions from the manufacture of nitrate fertilizers occur in the neutralization and drying operations. By keeping the neutralization process on the acidic side, losses of ammonia and nitric oxides are kept at a minimum. Nitrate dust or particulate matter is produced in the granulation or prilling operation. Particulate matter is also produced in the drying, cooling, coating, and material handling operations. Additional dust may escape from the bagging and shipping facilities.

Typical operations do not use collection devices on the prilling tower. Wet or dry cyclones, however, are used for various granulating, drying, or cooling operations in order to recover valuable products. Table 263 presents emission factors for the manufacture of nitrate fertilizers.

TABLE 263: EMISSION FACTORS FOR NITRATE FERTILIZER MANUFACTURING WITHOUT CONTROLS, EMISSION FACTOR RATING: B

Type of process[a]	Particulates		Nitrogen oxides (NO_3)		Ammonia	
	lb/ton	kg/MT	lb/ton	kg/MT	lb/ton	kg/MT
With prilling tower						
Neutralizer[b]	—	—	—	—	2	1
Prilling tower	0.9	0.45	—	—	—	—
Dryers and coolers[c]	12	6	—	—	—	—
With granulator						
Neutralizer[b]	—	—	—	—	2	1
Granulator[c]	0.4	0.2	0.9	0.45	0.5	0.25
Dryers and coolers[d]	7	3.5	3	1.5	1.3	0.65

[a]Plants will use either a prilling tower or a granulator but not both.
[b]Controlled factor based on 95% recovery in recycle scrubber.
[c]Use of wet cyclones can reduce emissions by 70%.
[d]Use of wet-screen scrubber following cyclone can reduce emissions by 95 to 97%.

Source: EPA Publication AP-42, 2nd Edition.

Water Pollution

The four process operations, ammonia, urea, ammonium nitrate, nitric acid, in the nitrogen fertilizer category, discharge the following types of wastewater.

> Water treatment plant effluent (includes raw water filtration and clarification, water softening, and water deionization).
> Closed loop cooling tower blowdown.
> Boiler blowdown.
> Compressor blowdown.
> Process condensate.
> Spills and leaks that are collected in pits or trenches.
> Nonpoint source discharges that are "collected" due to rain or snow.

Water Treatment Plant Effluent: The total effluent stream from a combined water treatment system will range from 8 to 20 l/kkg (2 to 5 gal/ton) of product with an ammonia plant having the larger amount due to the large amounts of raw water used. The contaminants in this effluent are mainly due to the initial contaminants in the raw water and therefore would be specific to the area and geographic conditions rather than the process plants involved. If the water treatment plant effluent contains ammonia due to the use of stripped, process condensate as process or boiler water makeup (replacing raw water makeup), then the ammonia-nitrogen discharge allowance is applicable.

Cooling Tower Blowdown: The cooling water requirements and expected blowdown requirements for four process plants in the nitrogen fertilizer industry are listed below.

	Circulation Requirement liter/l/kkg	Circulation Requirement gal/ton	Blowdown Requirement liter/l/kkg	Blowdown Requirement gal/ton
Ammonia	104,000 to 417,000	25,000 to 100,000	1,670 to 2,920	400 to 700
Urea	41,700 to 167,000	10,000 to 40,000	375 to 1,470	90 to 350
Ammonium Nitrate	8,350 to 29,200	2,000 to 7,000	84 to 250	20 to 60
Nitric Acid	104,000 to 146,000	25,000 to 35,000	1,250 to 2,500	300 to 600

In this closed loop cooling tower system, chemicals are added to inhibit scale formation, corrosion and the growth of bacteria. Due to the nature of the makeup water, the inhibitor chemicals and the evaporation water loss from the tower, a quantity of blowdown is required to prevent excessive build up of chemicals and solids in the circulation system. This quantity will vary, as shown in the above table, from plant to plant depending on the total circulation system.

The quality of this cooling system blowdown will vary mostly with makeup water condition and inhibitor chemicals and will not be greatly affected by the process plant associated with it. Any leaks that might develop in process or machinery exchangers should not significantly affect the contaminant concentration of the cooling water. The largest contaminant in the cooling water, that is neither intentionally added as an inhibitor nor comes in with makeup, is ammonia. Due to the proximity of the cooling tower in relation to any of the four nitrogen fertilizer operations, atmospheric ammonia is readily absorbed in the cooling water.

Some possible range of concentration for some of the contaminants that might be contained in the cooling water blowdown are

	mg/l
Chromate	0 - 250
Ammonia	5 - 100
Sulfate	500 - 3,000
Chloride	0 - 40
Phosphate	10 - 50
Zinc	0 - 30
Oil	10 - 1,000
TDS	500 - 10,000
MEA	0 - 10

This blowdown can be either treated by itself if necessary or combined with other effluents for total treatment. However, it is recommended that this stream be treated separately for chromate-zinc reduction since this is the main source of these contaminants (Cr and Zn) to the total plant effluent.

Boiler Blowdown: These four nitrogen fertilizer processes will generate up to 6,000 kg of steam/kkg (12,000 lb of steam/ton) of product depending on what processes are at the plant site. Ammonia will have the highest steam load followed by nitric acid, urea and ammonium nitrate. The pressure of the steam generated by and/or used in these plants will range from atmospheric up to 103 atm (1,500 psig).

Depending on the operating pressure of the steam system, the treatment of the boiler feed water will vary from extensive, including deionization, at 103 atm (1,500 psig) down to not much more than filtration at atmospheric pressure. Inhibitor chemicals are also added to boilers to prevent corrosion and scale formation throughout the system.

The combination of make-up water quality and the addition of inhibitor chemicals necessitate blowdown periodically to remove contaminants from the boiler. Based on the actual steam generated in a nitrogen fertilizer complex, this blowdown quantity will range from 42 to 145 l/kkg (10 to 35 gal/ton) of product.

Some contaminants in boiler blowdown from nitrogen complex boilers are as follows:

	mg/l
Phosphate	5 - 50
Sulfite	0 - 100
TDS	500 - 3,500
Zinc	0 - 10
Suspended solids	50 - 300
Alkalinity	50 - 700
Hardness	50 - 500
SiO$_2$	10 - 50

This effluent stream may be treated separately if necessary or combined with the total effluent for treatment.

Compressor Blowdown: This wastewater effluent stream has been separated out because it should contain the largest proportional amount of oil and grease. Primarily, the blowdown containing oil will come from interstage cooling-separation in the reciprocating compressors operating on ammonia synthesis gas, on ammonia process air and on urea carbon dioxide. If these streams can be contained then oil separation equipment can be kept to a minimum.

Due to the nature and expense of reciprocating compressors they are usually replaced by centrifugal compressors, when the ammonia plant capacity reaches 550 kkg/day (600 ton/day). The use of centrifugal compressors results in much less oil and grease in the blowdown effluent. The quantity of this blowdown will vary and can run up to 208 l/kkg (50 gal/ton) of product.

Process Condensate: Process condensate, although it may have many of the similar contaminants, will be handled separately for each of the four process plants.

Ammonia Process Condensate — Process steam supplied to the primary reformer is in excess of the stoichiometric amount required for the process reactions and, therefore, when the synthesis gas is cooled either by heat recovery or cooling water, a considerable amount of process condensate is generated. The quantity of this condensate will range from 1,500 to 2,500 kg/kkg (3,000 to 5,000 lb/ton) of product. The contaminants in this condensate may be ammonia, methanol, some organics from the CO$_2$ recovery system and possibly some trace metals. The ammonia discharged in this waste stream can range from 1,200 to 1,750 kg/1,000 kkg (2,400 to 3,500 lb/1,000 ton).

Urea Process Condensate — Following the urea forming reactions the pressure is reduced to allow ammonia, carbon dioxide and ammonium carbamate to escape from urea product. Partial condensation of these flashed gases along with the condensation of water vapor from the urea concentration step results in a condensate containing urea, ammonium carbamate, ammonia and carbon dioxide. The quantity of this stream will range from 417 to 935 l/kkg (100 to 225 gal/ton) of product. Ammonia discharge in this stream has been observed at the level of 9,000 kg/1,000 kkg (18,000 lb/1,000 ton) of urea product. Urea discharge at the rate of 33,500 kg/1,000 kkg (67,000 lb/1,000 ton) of urea product has also been cited.

Ammonium Nitrate Process Condensate — The nitric acid-ammonia reaction being highly exothermic causes a large amount of water to be flashed off taking with it ammonia, nitric acid, nitrates and some nitrogen dioxide. If climatic conditions or air pollution regulations require that this stream be condensed then this contaminated condensate will range between 208 and 458 l/kkg (50 and 110 gal/ton) of product. Ammonia discharges in the stream could be at the levels of 150 kg/1,000 kkg (300 lb/1,000 ton) and ammonium nitrate at 7,000 kg/1,000 kkg (14,000 lb/1,000 ton) of ammonium nitrate product.

Nitric Acid Process Condensate — Using the ammonia oxidation process for production of 55 to 65% strength acid there are no process condensate effluent streams.

Collected Spills and Leaks: In all process plants there will be a small quantity of material either spilled, during loading or transferring, or leaking from some pump seal or bad valve. When this material, whether it be cooling water, process condensate, carbon dioxide scrubbing solution, boiler feed water or anything else, gets on a hard surface where it can be collected in a trench, then it will probably have to be treated before being discharged. The quantity of this material is not dependent on plant size, but more on the operating philosophy and housekeeping procedures.

Nonpoint Discharges: Rain or snow can be a collection medium for a sizable quantity of contaminants. These contaminants may be air-borne ammonia that is absorbed as the precipitation falls, or it may be urea or ammonium nitrate prill dust that is lying on the ground around prill towers. Dry fertilizer shipping areas may also have urea and/or ammonium nitrate that can be washed down by rain or snow. Pipe sweat and drip pots are another potential source of contaminants.

The aqueous process waste streams generated by the productions of nitrogen fertilizers are quite similar in composition. In industrial complexes where two or more of the nitrogen fertilizers are produced the aqueous process wastes are usually combined and handled as a single stream for treatment and disposal. Table 264 summarizes the typical range of compositions and volumes of the individual waste streams which could be encountered in a nitrogen fertilizer production complex. Cooling water and boiler water blowdown may also be combined with the process waste streams after chromate removal, but have not been included in the numbers given in Table 264.

TABLE 264: COMPARISON OF AQUEOUS PROCESS WASTE STREAMS FROM NITROGEN FERTILIZER PLANTS

	------------ Aqueous Process Waste Streams* ------------			
	Ammonia Plant	Ammonium Nitrate Plant	Ammonium Sulfate Plant	Urea Plant
Volume (gal/ton product)	100 - 1,000	50 - 1,200	100 - 10,000**	50 - 2,000
Contaminants (mg/l)				
NH$_3$	20 - 100	200 - 2,000	10 - 1,000**	200 - 4,000
CO$_2$	150 - 750			100 - 1,000

(continued)

TABLE 264: (continued)

	- - - - - - - - - - - - - Aqueous Process Waste Streams* - - - - - - - - - - - - - -			
	Ammonia Plant	Ammonium Nitrate Plant	Ammonium Sulfate Plant	Urea Plant
Contaminants (cont'd)				
NO$_3$.		50 - 1,000		
SO$_4$			5 - 500**	
MEA	50 - 100			
Urea				50 - 1,000
Oil	100 - 10,000			10 - 100
BOD	50 - 150	< 20	< 20	30 - 300
COD	60 - 200	< 20	< 20	50 - 500

*Does not include cooling water and boiler water blowdown.
**Wide range due to possible recycle of aqueous scrubber solutions.

Source: EPA Report 12020 FPD 09/71.

References

(1) U.S. Environmental Protection Agency, "Compilation of Air Pollutant Emission Factors," 2nd Edition, *Publication Number AP-42,* Research Triangle Park, N.C., Office of Air and Water Programs (April 1973).
(2) U.S. Environmental Protection Agency, "Development Document for Proposed Effluent Limitations Guidelines and New Source Performance Standards for the Basic Fertilizer Chemicals Segment of the Fertilizer Manufacturing Point Source Category," *Report EPA 440/1-73/011,* Washington, D.C. (November 1973)
(3) Fullam, H.T. and Faulkner, B.P., "Inorganic Fertilizer and Phosphate Mining Industries — Water Pollution and Control," *Report 12020 FPD 09/71,* Washington, D.C., U.S. Government Printing Office (September 1971).

NONFERROUS METALS INDUSTRY

Pollution control in the nonferrous metals industry is primarily a question of air pollution control. Water pollution is of secondary consequence. The industry comprises both the primary or smelting segment in which metals are produced from their ores and the secondary or scrap recovery segment.

The pollution problems center on [1] sulfur dioxide, [2] particulates and [3] minor constituents such as fluorides. Then, as with all other industrial pollution control areas, the question of economics arises. Can pollutants be recovered for profit? Will pollution control render certain plants or processes uneconomic?

The problems, the technological solutions and the economic questions related to pollution control in the nonferrous metals industry with particular attention to air pollution, have been reviewed by H.R. Jones (1). The reader is also referred to other sections of this volume on aluminum, copper, lead and zinc manufacture for additional detail on pollutant sources and emissions.

Air Pollution

Sulfur dioxide is the primary pollutant emitted from the smelting of copper, lead, and zinc ore concentrates. Sources of emission are roasters, reverberatory furnaces, converters, and sinter machines. The emissions are characterized by heavy mass rates and relatively weak flue gas concentrations from reverberatory furnaces and uneven pulsating flows from converters. In addition, the flows are contaminated with small quantities of materials harmful to human health, such as arsenic, bismuth, cadmium and mercury. These factors also make the collection of sulfur pollutants difficult and costly.

Some 2,800,000 short tons of sulfur are contained in the sulfur oxide gases generated annually at U.S. smelters. Of this, 850,000 tons or 31% are recovered, almost all as sulfuric acid, and the remaining 1,920,000 tons are emitted to the atmosphere. Smelters east of the Mississippi generate 350,000 tons and recover 86% of it, while western smelters recover 23% of the 2,420,000 tons that they generate. Western smelters are the source of over 97% of emissions to atmosphere from smelters in this country. Some 76% of smelter emissions to the atmosphere come from copper smelters, 17% from zinc smelters, and 7% from lead smelters. Zinc smelters recover 59% of their emitted sulfur oxides, and lead smelters recover 27%.

Five of the 16 copper smelters studied control sulfur oxide emissions to some degree. Altogether, the copper smelter industry recovers 19% of the sulfur oxides generated, and this is expected to rise several percent. Most of the smelters are poorly located with respect to sulfuric acid markets as noted by Arthur G. McKee & Co. (2).

As pointed out by the U.S. Bureau of Mines (3), the domestic base-metal producing industry has been challenged with the task of reducing the amount of sulfur oxides and other pollutants emitted by smelters treating sulfide ores and concentrates. This challenge has been imposed by legislation aimed at reducing air pollution. Standards for both ambient air quality and allowable emissions have been proposed and already are in effect in some states.

The expanded production of acid which is expected to accompany the sulfur dioxide control efforts is causing considerable concern. In many cases, smelters are not near major markets for sulfuric acid, and it is likely that unmarketable acid will have to be neutralized and disposed of by impounding.

One typical characteristic exists, in almost all the processes in the production of nonferrous metals, the particulates emitted are metallic fumes generally submicron in size. Table 265 (1), shown on the following page summarizes particulate emissions from the primary nonferrous metals industry.

The secondary nonferrous smelting and refining industry comprises establishments primarily engaged in recovering nonferrous metals and alloys from scrap and dross. The secondary nonferrous industry, as used here, will include copper, lead, zinc, and aluminum recovery. Copper, including brass and bronze alloys, and lead production from secondary smelters account for about 50% of the yearly total production of these metals in the United States.

The nature of furnace operations in this industry is such that emissions vary widely during the cycle from charging the scrap to pouring the melt. Peak emission surges occur in nearly all the furnace operations. The principal emission from the secondary copper, lead, zinc and aluminum industries is particulates — smoke, dust and metallic fumes. Small amounts of sulfur oxide may also be evolved. Table 266 summarizes particulate emissions from the secondary nonferrous metals industry.

Table 267 presents typical particle size distribution data for secondary metal processing. Control of these emissions requires highly efficient collection equipment such as baghouses, electrostatic precipitators, and high pressure-drop scrubbers.

Ores and concentrates of copper, zinc and lead have varying amounts of minor constituents that sometimes require the use of special processing methods to recover them or to overcome or avoid the problems they could cause. There is not only considerable variation in the constituents from one smelter to another, but also from time to time in any particular smelter. These constituents include chlorine compounds in lead smelters, fluorine in copper, zinc and lead smelters and arsenic in most ores of the western United States. Cadmium is commonly associated with zinc, and the quantities are large enough so that most plants

TABLE 265: PARTICULATE EMISSIONS – PRIMARY NONFERROUS METALS INDUSTRIES

	Quantity of Material	Emission Factor	Efficiency of Control, C_c	Application of Control, C_f	Net Control, $C_c \cdot C_f$	Emission tons/yr.
I. Aluminum						
A. Preparation of Alumina						
1. Grinding of Bauxite	13,000,000 tons bauxite	6 lb./ton bauxite	-	-	0.80	8,000
2. Calcining of Hydroxide	5,840,000 tons alumina	200 lb./ton alumina	-	-	0.90	58,000
B. Aluminum Mills						
1. Soderberg Cells						
(a) Horizontal stud	800,000 tons aluminum	144 lb./ton aluminum	0.40	1.0	0.40	35,000
(b) Vertical stud	700,000 tons aluminum	84 lb./ton aluminum	0.64	1.0	0.64	10,000
2. Prebake Cells	1,755,000 tons aluminum	63 lb./ton aluminum	0.64	1.0	0.64	20,000
C. Materials Handling	3,300,000 tons aluminum	10 lb./ton aluminum	0.90	0.35	0.32	11,000
Total from primary aluminum industry						142,000
II. Copper						
A. Ore Crushing	170,000,000 tons ore	2 lb./ton ore	0.0	0.0	0.0	170,000
B. Roasting	40% of 1,437,000 tons copper	168 lb./ton copper	0.85	1.0	0.85	7,000
C. Reverberatory Furnace	1,437,000 tons copper	206 lb./ton copper	0.95	0.85	0.81	28,000
D. Converter	1,437,000 tons copper	235 lb./ton copper	0.95	0.85	0.81	33,000
E. Fire Refining						
F. Slag Furnaces						
G. Materials Handling: Ore, Limestone, Slag, etc.	1,437,000 tons copper	10 lb./ton copper	0.90	0.35	0.32	5,050
Total from primary copper industry						243,000
III. Zinc						
A. Ore Crushing	18,000,000 tons ore	2 lb./ton ore	0.0	0.0	0.0	18,000
B. Roasting						
1. Fluidized Bed, Suspension	75% of 1,020,000 tons zinc	2,000 lb./ton zinc	0.98	1.0	0.98	15,000
2. Ropp, Multiple Hearth	15% of 1,020,000 tons zinc	333 lb./ton zinc	0.85	1.0	0.85	4,000
C. Sintering and Sinter Crushing	60% of 1,020,000 tons zinc	180 lb./ton zinc	0.95	1.0	0.95	3,000
D. Distillation	60% of 1,020,000 tons zinc		0.0	0.0	0.0	15,000
E. Materials Handling	1,020,000 tons zinc	7 lb./ton zinc	0.90	0.35	0.32	2,000
Total from primary zinc industry						57,000
IV. Lead						
A. Ore Crushing	4,500,000 tons ore	2 lb./ton ore	0.0	0.0	0.0	4,000
B. Sintering	467,000 tons lead	520 lb./ton lead	0.95	0.90	0.86	17,000
C. Blast Furnace	467,000 tons lead	250 lb./ton lead	0.85	0.98	0.83	10,000
D. Drossing Kettle		-				
E. Softening Furnace		-				
F. Desilvering Kettles		-				
G. Cupeling Furnaces						
H. Refining Kettles		-				
I. Dross Reverberatory Furnace	467,000 tons lead	20 lb./ton lead	0.90	-	0.50	2,000
J. Materials Handling	467,000 tons lead	5 lb./ton lead		0.35	0.32	1,000
Total from primary lead industry						34,000
Total from nonferrous metals industries						476,000

Source: Report PB 203,522 (4).

TABLE 266: PARTICULATE EMISSIONS – SECONDARY NONFERROUS METALS

Source	Quantity of Material	Emission Factor	Efficiency of Control, C_c	Application of Control, C_l	Net Control, $C_c \cdot C_l$	Emission tons/yr.
I Copper						
A. Materials Preparation						
1. Wire burning	300,000 tons insulated wire	275 lb./ton				41,000
2. Sweating furnaces	64,000 tons, auto radiators	15	0.95	0.20	0.19	400
3. Blast furnaces	287,000 tons, scrap and residue	50 lb./ton scrap	0.90	0.75	0.68	2,300
B. Smelting and Refining	1,170,000 tons scrap					
1. In secondary smelters, etc.						
(a) charge		30 lb./ton charge				
(b) refine		39				
(c) pour		0.6				
		70	0.95	0.60	0.57	17,000
				Total from Secondary Copper		60,700
II Aluminum						
A. Sweating Furnaces	500,00 tons of scrap	32 lb./ton scrap	0.95	0.20	0.19	6,000
B. Refining Furnaces	1,015,000 tons of scrap					
1. Reverberatory		4 lb./ton scrap	0.95	0.60	0.57	900
2. Pot (crucible)		2 lb./ton scrap	–	–	–	–
3. Induction		–	–	–	–	–
C. Fluxing	136,000 tons Cl used	1,000 lb./ton Cl used	–	–	0.25	51,000
D. Dross Processing						
1. Hot process						
2. Milling process						
				Total from Secondary Aluminum		57,900
III Lead Furnaces						
A. Pot	53,000 tons scrap	0.8 lb./ton scrap	0.95	0.95	0.90	<100
B. Blast Furnace	119,000 tons scrap	190 lb./ton scrap	0.95	0.95	0.90	1,000
C. Reverberatory		130 lb./ton scrap	–	–	–	–
D. Rotary Reverberatory		70 lb./ton scrap	–	–	–	–
Average, reverberatory	554,000 tons scrap	100 lb./ton scrap	0.95	0.95	0.90	3,000
				Total from Secondary Lead		4,000
IV Zinc						
A. Sweating						
1. Metallic scrap	52,000 tons scrap	12 lb./ton charge	0.95	0.20	0.19	300
2. Residual scrap	210,000 tons scrap	30 lb./ton charge	0.95	0.20	0.19	2,600
B. Distillation Furnace	233,000 tons Zn recovered					
1. Distillation retort		47 lb./ton Zn				
2. Muffle		45 lb./ton Zn				
Average, distillation furnaces		45 lb./ton Zn	0.95	0.60	0.57	2,200
				Total from Secondary Zinc		5,100
				Total from Secondary Nonferrous Metals		127,700

Particle Size Distribution From Secondary Metal Melting Operations

Operation	44 microns or greater, %	20 to 44 microns, %	10 to 20 microns, %	5 to 10 microns, %	Less than 5 microns, %
Aluminum smelting	3	10	23	30	34
Brass smelting	–	–	–	–	100
Bronze smelting	–	–	–	–	100
Lead smelting	–	–	2	3	95
Zinc smelting	–	–	–	–	100

Source: Report PB 203,522 and PHS Publication 999 AP-42.

TABLE 267: ASSESSMENT OF INFORMATION OBTAINED IN TERMS OF UNIT PROCESSES AND FUTURE NEEDS

Metal	Nominal or Overall Operation	Unit Operation or Process	Category or Use of Water	Waste Water Characteristics	Assessment of Reported Status of Waste Water Treatment or Water Pollution Control	Recommended Future Action
Aluminum	Mining	Mining			Two operating sites in Arkansas; no information	Need basic information; maybe best handled at state level.
	Bauxite ore refining; alumina refining	Steam plant, calcining, furnace cooling	Indirect cooling	As previously		As above
		Caustic leaching	Process	Red mud slurry; possible high pH, very high suspended solids.	Mostly lagooned (red mud lake) and recycled; one or more cases involve direct discharge; recycle rate generally high in the industry, no specific needs identified.	This problem of dewatering and solid waste disposal is being studied by U.S. Bureau of Mines.
Aluminum	Smelting	Rectifiers, Melting furnaces	Indirect cooling	As previously		As above
		Metal casting	Direct cooling	As previously		As above
		Anode plant Gas scrubbers	Process	Carbonaceous suspended solids; dissolved volatile hydrocarbons, oils, tar, sulfur oxides.	Currently skimmed, settled and diluted for release; better treatments need development.	Need R&D for water treatment process; need additional analyses to set waste loads, economic, and effluent standards on current state of the art.
		Pot line gas scrubbers	Process	High fluoride, chloride, sodium, aluminum, and sulfate ions.	Recovery and recycle practiced in some cases; but process should be optimized, as should the treatment of necessary small bleed-off streams from recovery operations. The impact of a proprietary dry fume-scrubbing method should be assessed.	Need additional analyses to set waste loads etc. Need R&D for fluoride recovery process efficiency.

(continued)

TABLE 267: (continued)

Metal	Nominal or Overall Operation	Unit Operation or Process	Category or Use of Water	Waste Water Characteristics	Assessment of Reported Status of Waste Water Treatment or Water Pollution Control	Recommended Future Action
Copper	Mining	Open pit	Dust control	No waste water	No problem	None
	Mining	Underground	Mine drainage	Suspended solids, dissolved salts; sometimes sulfates and low pH.	Generally not a problem in the copper industry due to geographical and climatic conditions.	Waste load variable due to geological conditions; treatment has been achieved in Missouri Lead Belt by settling and holding; suggest demonstration on critical case.
	Concentrator	Ore crushing	Dust control	None	Retained in ore	
	Concentrator	Flotation	Process and transport medium	Suspended solids; flotation reagents; traces (< 1 ppm) of metal sulfides, cyanides, arsenic; 2000-4000 ppm total dissolved solids. High (alkaline) pH.	Sent to tailings pond; no problem in arid climates; in wetter areas metals contents may exceed local discharge standards; need technology transfer from desert areas to accomplish closed loop water system.	Suggest research to determine operative mechanisms in best and worst case; need educational effort related to spread of optimum tailings pond design and best practice (holding time, oxidation mechanisms; flocculents); need transfer of knowledge on recycle practice.
				Modification of flotation	Modification of flotation reagent concentrations and compositions have been demonstrated or are being studied.	WQO program on reagent modification should be continued or accelerated.
	Mine-concentrator operations	Sanitary wastes	Domestic supply	Sanitary wastes	Usually sent to ground (i.e., seepage or evaporation); occasionally recycled through tailings pond, thus potential for recycle is demonstrated.	Need R&D on processes to treat sanitary wastes and recycle, both in the context of plant operation.
		Power plants	Boiler blowdown, cooling tower blowdown	High chlorides, suspended solids, chromates	Usually recycled through tailings pond	Current data should allow settling of waste loads, treatment economics, and effluent standards; may need development of economical treatment to allow recycle.
	Leaching	Leaching	Spent liquor	Ferrous sulfates; low (acid) pH	Usually associated with desert operation; evaporated in holding ponds; occasional accidental spills; may be recycled through tailings pond or neutralized by the tailings pond before release.	Possible need for system to treat spills
	Smelting	Roasting	Calcine cooling	None	Evaporated	None

(continued)

TABLE 267: (continued)

Metal	Nominal or Overall Operation	Unit Operation or Process	Category or Use of Water	Waste Water Characteristics	Assessment of Reported Status of Waste Water Treatment or Water Pollution Control	Recommended Future Action
Copper	Smelting	Reverberatory smelting furnace	Indirect cooling	Cooling water blowdown; high chlorides, suspended solids; chromates	A common waste water problem requiring the development of economical treatment methods to allow recycle; application demands high quality water.	See above
	Smelting	Cast metal and mold cooling	Direct cooling	Suspended solids (carbonaceous or oily mold dressing; metal oxides)	Often times wasted; recycle or reuse demands treatment by settling and oil removal.	Need analyses to determine waste loads; treatment economics may be obtained from operating examples; effluent standards could then be set.
	Smelting	Slag granulation	Process water	Suspended solids (silicates; inert oxides; possibly soluble trace element compounds, e.g., arsenates, antimonates).	Usually settled and recycled to same use.	Need analyses to determine waste loads
Copper	Electrolytic refining	Electrolytic refining	Electrolyte solution	Acid sulfate solution	Sometimes discharge with or without neutralization or dilution; strong need for cheap neutralization, alternate disposal or treatment allowing recycle.	Need some additional analyses to determine waste loads; neutralization is already practiced; costs and effluent standards could be developed fairly readily.
		Power supply (DC rectifiers)	Indirect cooling	Same as above; chlorides, chromates, suspended solids	Need for treatment for recycle	See above
		Furnaces Copper melting By-Product refining	Indirect cooling	Same	Same	Same
		Metal casting	Direct cooling	Same as above (suspended solids, oils)	Same as above; needs more widespread application of settling and oil removal.	See prior case
	By-Product metals (Se, Te, Ag, Au)	Gas scrubbing	Process	Dissolved trace element salts: arsenic, selenates, tellurium	Present fate largely unknown.	Need more information on industrial processing and for problem definition.
		Electrolytic refining	Process	Dissolved salts	Usually recycled or carefully treated because of precious metal content.	Same

(continued)

TABLE 267: (continued)

Metal	Nominal or Overall Operation	Unit Operation or Process	Category or Use of Water	Waste Water Characteristics	Assessment of Reported Status of Waste Water Treatment or Water Pollution Control	Recommended Future Action
Lead	Mining	Mining	Mine drainage	Dissolved salts of copper, lead, and zinc at ppm level	Generally discharged through tailings pond; metal contents are marginally near some effluent standards; enforcement of the most stringent standards would require economical treatment of very large volumes of water for low level impurities; tailings pond oxidation mechanisms recommended for study, as are carbonate precipitation mechanisms. Associated climate usually results in plentiful water supply.	See above
Lead	Flotation concentration	Flotation	Process water	Metals, flotation reagents, high pH, suspended solids	Trace metal removal same as above; polyelectrolytes aid settling but are added expense; cyanides readily oxidizable given sufficient holding time; long alcohols must be controlled to avoid eutrophication of some receiving waters; tailings pond mechanisms and design can be used effectively if designer is knowledgeable.	See above
Lead	Smelting	Sintering Blast furnace Drossing Softening Slag fuming	Indirect cooling	Same as above (chlorides, chromates, suspended solids).	Need for recycle treatment	See above
Lead	Electrolytic refining	Electrolytic refining	Spent electrolyte	Acidic	Practice unknown (1 or 2 plants operating in U.S.)	Need basic information on industrial practice and for problem definition.
Lead		Spills, equipment washdown	Process	Acid, high dissolved solids	Could be better controlled by plant drainage, re-design and segregation; some can be recycled.	Once controlled by drainage, segregation, this is generally a neutralization exercise; suggest setting a nil discharge standard for new plants and requiring update of old plants in some reasonable time.

(continued)

TABLE 267: (continued)

Metal	Nominal or Overall Operation	Unit Operation, or Process	Category or Use of Water	Waste Water Characteristics	Assessment of Reported Status of Waste Water Treatment or Water Pollution Control	Recommended Future Action
Zinc	Mining	Mining	Mine drainage	Same as above; trace metal contents and a higher incidence of trace impurities such as arsenic and cadmium.	As above; trace metal levels (ppm) possibly above some local standards.	See above
Zinc	Concentration	Flotation	Process	As above: suspended solids, trace metals, As, Cd, residual flotation reagents, and high pH.	As above; improved tailings pond management needed; or alternative economical treatment of large quantities of water to remove ppm level impurities.	As above
Zinc	Smelting	Roasting Sintering Retorting Refining Distillation Liquation	Indirect cooling	As previously; chlorides, chromates, suspended solids.	Need wider application and/or improved treatment to allow recycle.	As above
Zinc	Smelting	Ingot casting	Direct cooling	As previously: suspended solids, oils.	Treatment needed for recycle.	As above
Zinc	Smelting	By-Product cadmium recovery	Process; acid leach liquors	Acid; trace metal and impurity contents (As, Cd, Zn, Tl, Pb, Zu, Cu, metal sulfides in suspension).	Need for neutralization process to drop all trace impurities.	One plant has reported waste loads on unit process basis during this program; need additional process detail and effluent analyses; need R&D on neutralization or other treatment for complex-impurity streams.
Zinc	Electrolytic zinc	Leaching and electrolysis	Process; spent leach liquors, electrolytes; barren filtrates.	Generally acid; low pH, high sulfate	Information lacking (1 or 2 plants operating in U.S.).	Need basic information on current industrial practice and for problem definition.
Zinc	Electrolytic zinc	Spills, equipment wash water	Process	Acid; high sulfates and metals	Could be better controlled by plant drainage redesign and segregation; some can be recycled.	See above
Copper, lead, zinc	Smelting	Sulfuric acid plant	Indirect cooling	As previously	As previously	As above
Ditto	Smelting	Sulfuric acid plant	Gas cleaning (spray towers, cottrell sump, scrubber liquor, etc.)	Acid; high levels of arsenic, cadmium, copper, lead, zinc, selenium, tellurium, etc.	Practice variable; this stream is so conglomerate as to be only waste; this stream is variously disposed of by mixing, dilution, neutralization, or other means. A constant, recurring, and increasing problem as SO_2 control is advanced. More detailed examination of effectiveness of common neutralization processes is needed relative to many "impurity" type elements.	These waste liquors need R&D for complete impurity removal process; some analyses available; waste loads reported have wide range.

Source: Report EPA-R-2-73-247a and 247b

recover it. Mercury is found in some zinc ores and has been recovered in some instances. Bismuth, selenium and tellurium are often present in ores of most nonferrous metals.

Volatile constituents such as arsenic get into the gases and usually deposit in the flue system. One copper smelter has recovered and purified arsenic for sale as a by-product, according to Arthur G. McKee & Co. (2).

Most of the ores processed in the United States are sulfide mineral concentrates, separated by various roughing and flotation techniques from a wide range of gangue minerals. The separation is of necessity incomplete, and a portion of the gangue minerals accompanies the sulfide minerals through the thermal processing. The gangue minerals frequently contain inorganic fluorides. These fluorides are evolved as gaseous HF in the high temperature zones, the copper reverberatory furnaces, lead refining kettles, and zinc sintering machines and roasting furnaces, where temperatures range from 1400° to 2400°F, and more than sufficient combined hydrogen to satisfy the stoichiometry of the reaction is present. There is no information available in the open literature on the fluoride contents of the various ores and concentrates, and no data have been published on fluoride emissions from United States smelters.

Water Pollution

Table 267 summarizes information gathered by EPA (5) on aluminum, copper, lead and zinc.

References

(1) Jones, H.R., "Pollution Control in the Nonferrous Metals Industry," Park Ridge, N.J. Noyes Data Corp. (1972).
(2) McKee, A.G. and Co. "Systems Study for Control of Emissions: Primary Nonferrous Smelting Industry," *Report PB 184,884,* Springfield, Va., Nat. Tech. Information Service (June 1969).
(3) U.S. Bureau of Mines, "Control of Sulfur Oxides Emissions in Copper, Lead and Zinc Smelting," *Information Circular 8527,* Washington, D.C. (1971).
(4) Midwest Research Institute, "Particulate Pollutant System Study, III, Handbook of Emission Properties," *Report PB 203,522,* Springfield, Va., Nat. Tech. Information Service (May 1971).
(5) Hallowell, J.B., Shea, J.F., Smithson, G.R., Jr., Tripler, A.B. and Gonser, B.W., "Water Pollution Control in the Primary Nonferrous Metals Industry" (2 volumes), *Report Numbers EPA-R2-73-247a and 247b,* Washington, D.C., U.S. Government Printing Office (September 1973).

OPEN BURNING

Open burning can be done in open drums or baskets and in large-scale open dumps or pits. Materials commonly disposed of in this manner are municipal waste, auto body components, landscape refuse, agricultural field refuse, wood refuse, and bulky industrial refuse.

Air Pollution

Ground-level open burning is affected by many variables including wind, ambient temperature, composition and moisture content of the debris burned, size and shape of the debris burned, and compactness of the pile. In general, the relatively low temperatures associated with open burning increase the emission of particulates, carbon monoxide, and hydrocarbons and suppress the emission of nitrogen oxides. Sulfur oxide emissions are a direct function of the sulfur content of the refuse. Emission factors are presented in Table 268 for the open burning of three broad categories of waste: municipal refuse, automobile components, and horticultural refuse.

TABLE 268: EMISSION FACTORS FOR OPEN BURNING
EMISSION FACTOR RATING: B

Pollutant	Municipal refuse	Automobile components[*]	Agricultural field burning	Landscape refuse and pruning	Wood refuse
Particulates					
lb/ton	16	100	17	17	17
kg/MT	8	50	8.5	8.5	8.5
Sulfur oxides					
lb/ton	1	Neg	Neg	Neg	Neg
kg/MT	0.5	Neg	Neg	Neg	Neg
Carbon monoxide					
lb/ton	85	125	100	60	50
kg/MT	42.5	62.5	50	30	25
Hydrocarbons (CH_4)					
lb/ton	30	30	20	20	4
kg/MT	15	15	10	10	2
Nitrogen oxides					
lb/ton	6	4	2	2	2
kg/MT	3	2	1	1	1

*Upholstery, belts, hoses, and tires burned in common.

Source: EPA Publication AP-42, 2nd Edition.

Reference

(1) U.S. Environmental Protection Agency, "Compilation of Air Pollutant Emission Factors," 2nd Edition, *Publication Number AP-42*, Research Triangle Park, N.C., Office of Air and Water Programs (April 1973).

ORCHARD HEATING

Orchard heaters are commonly used in various areas of the United States to prevent frost damage to fruit and fruit trees. The five common types of orchard heaters are pipeline, lazy flame, return stack, cone, and solid fuel. The pipeline heater system is operated from a central control and fuel is distributed by a piping system from a centrally located tank. Lazy flame, return stack, and cone heaters contain integral fuel reservoirs, but can be converted to a pipeline system. Solid fuel heaters usually consist only of solid briquettes, which are placed on the ground and ignited. The ambient temperature at which orchard heaters are required is determined primarily by the type of fruit and stage of maturity, by the daytime temperatures, and by the moisture content of the soil and air.

During a heavy thermal inversion, both convective and radiant heating methods are useful in preventing frost damage; there is little difference in the effectiveness of the various heaters. The temperature response for a given fuel rate is about the same for each type of heater as long as the heater is clean and does not leak. When there is little or no thermal inversion, radiant heat provided by pipeline, return stack, or cone heaters is the most effective method for preventing damage.

Proper location of the heaters is essential to the uniformity of the radiant heat distributed

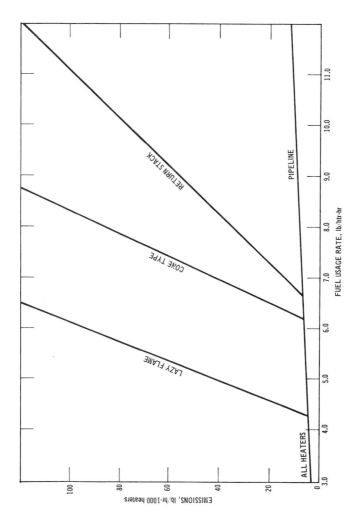

FIGURE 34: PARTICULATE EMISSIONS FROM ORCHARD HEATERS.

Source: EPA Publication AP-42, 2nd Edition.

among the trees. Heaters are usually located in the center space between four trees and are staggered from one row to the next. Extra heaters are used on the borders of the orchard.

Air Pollution

Emissions from orchard heaters are dependent on the fuel usage rate and the type of heater. Pipeline heaters have the lowest particulate emission rates of all orchard heaters. Hydrocarbon emissions are negligible in the pipeline heaters and in lazy flame, return stack, and cone heaters that have been converted to a pipeline system. Nearly all of the hydrocarbon losses are evaporative losses from fuel contained in the heater reservoir. Because of the low burning temperatures used, nitrogen oxide emissions are negligible. Emission factors for the different types of orchard heaters are presented in Table 269 and Figure 34.

TABLE 269: EMISSION FACTORS FOR ORCHARD HEATERS[a]
EMISSION FACTOR RATING: C

Pollutant	Type of heater				
	Pipeline	Lazy flame	Return stack	Cone	Solid fuel
Particulate					
lb/htr-hr	b	b	b	b	0.05
kg/htr-hr	b	b	b	b	0.023
Sulfur oxides[c]					
lb/htr-hr	$0.13S^d$	0.11S	0.14S	0.14S	NA[e]
kg/htr-hr	0.06S	0.05S	0.06S	0.06S	NA
Carbon monoxide					
lb/htr-hr	6.2	NA	NA	NA	NA
kg/htr-hr	2.8	NA	NA	NA	NA
Hydrocarbons[f]					
lb/htr-hr	Neg[g]	16.0	16.0	16.0	Neg
kg/htr-hr	Neg	7.3	7.3	7.3	Neg
Nitrogen oxides[h]					
lb/htr-hr	Neg	Neg	Neg	Neg	Neg
kg/htr-hr	Neg	Neg	Neg	Neg	Neg

[a]Reference (1).
[b]Particulate emissions for pipeline, lazy flame, return stack, and cone heaters are shown in Figure 34.
[c]Based on emission factors for fuel oil combustion.
[d]S = sulfur content.
[e]Not available.
[f]Based on emission factors for fuel oil combustion. Evaporative losses only. Hydrocarbon emissions from combustion are considered negligible. Evaporative hydrocarbon losses for units that are part of a pipeline system are negligible.
[g]Negligible.
[h]Little nitrogen oxide is formed because of the relatively low combustion temperatures.

Source: EPA Publication AP-42, 2nd Edition.

Reference

(1) U.S. Environmental Protection Agency, "Compilation of Air Pollutant Emission Factors," 2nd Edition, *Publication Number AP-42,* Research Triangle Park, N.C., Office of Air and Water Programs (April 1973).

ORGANIC CHEMICALS MANUFACTURE

Air Pollution

While air pollution in the sense of emission of hydrocarbons, H_2S, HCl, HF, and Cl_2 constitute something of a problem for manufacturers of organic chemicals, they do not seem to constitute a problem of the magnitude of the industry's water pollution problem. The organic chemical industry must, however, be concerned with two aspects of the air pollution problem: [1] The effect of discharges on the public ranging from really toxic fumes to soot emissions on washdays. [2] The effect of discharges on the utility of air as a chemical raw material. This is essentially the problem of industry fouling its own nest.

As pointed out by Sittig (1), the requirements of many organic chemicals manufacturing processes for reaction air are considerable indeed. Some examples are as follows:

Process	Tons Air Required/Ton Product
Synthesis gas	4.0
Hydrogen cyanide	10.0
Phthalic anhydride	5.0
Carbon black	22.0

The presence of dust, coke breeze, nitrogen oxides, sulfur oxides or carbon oxides in such air intakes may cause a variety of process problems such as system plugging and catalyst poisoning. A recent discussion of atmospheric emissions from the petroleum industry (2), while not wholly applicable, gives some guidance as to emissions which may be encountered in petrochemical manufacture and their control. Now that the coking of coal may be growing in importance as a source of organic chemicals, the particular air pollution problems presented by coke ovens come into focus as a problem related to organic chemicals manufacture. Some control techniques associated with coke oven operations are detailed by Sittig (3).

Water Pollution

The 1972 amendments to the Federal Water Pollution Control Act impose much more severe and specific structures on the organic chemicals manufacturing industry than did the Clean Air Act of 1970. Thus, substantial attention must be devoted to aqueous wastes from this industry and possible techniques for their treatment. The sources of wastes emanating from refinery and petrochemical operations can be divided into five general categories as noted by Jewell and Ford (4): wastes containing a principal raw material or product resulting from the stripping of the product from solution; by-products produced during reactions; spills, slab washdowns, vessel cleanouts, sample point overflows, etc.; cooling tower and boiler blowdown, steam condensate, water-treatment wastes, and general washing water; and, storm waters, the degree of contamination depending on the nature of the drainage area.

The principal contaminants in the wastewaters include organics from residual products and by-products, oils, suspended solids, acidity, heavy metals and other toxic materials, color, and taste and odor-producing compounds. The concentration of BOD_5 and COD, respectively, in untreated refinery effluents has averaged 108 and 204 lb/1,000 bbl of crude oil refined. The pH of refinery wastewaters is normally alkaline, but may vary considerably depending on disposal of spent acids, acid washes, etc.

The organic chemicals manufacturing industry is a complex one in which interrelated chemicals compete for raw materials and markets via increasingly complex technologies. The water usage and subsequent wastewater discharges are closely related to this mix of products and processes. The diversity of products and manufacturing operations to be covered indicates the need for separate effluent limitations for different segments within the industry. To this end, process-oriented subcategories have been developed by the Environmental Protection Agency (5). They are discussed in some detail in the pages which follow.

Subcategory A: Nonaqueous Processes — Contact between water and reactants or products is minimal. Water is not required as reactant or diluent, and is not formed as a reaction product. The only water usage stems from periodic washes or catalyst hydration. To be considered within Subcategory A, the unit operations and chemical conversions within a process module must be essentially anhydrous. Contact water usage shall be only in the form of periodic washes or steaming used to treat nonaqueous catalysts or working solvents. The other sources of wastewater are from external washing and maintenance operations within the process battery limits. External water sprays utilized to provide cooling on the outside of process pipes are contaminated through contact with chemicals present on the ground within the battery limits; consequently, they should be collected and discharged to a process sewer for subsequent treatment.

Subcategory B: Processes with Process Water Contact as Steam Diluent or Absorbent — Process water is in the form of dilution steam, direct product quench, or absorbent for effluent gases. Reactions are all vapor phase over solid catalysts. Most processes have an absorber coupled with steam stripping of chemicals for purification and recycle. Steam is also used for decoking of catalysts.

Subcategory B processes are characterized by unit operations and chemical conversions where the primary contact between water and chemicals is through vapor-liquid interfaces. Although final separation and discharge of water from the process may be as a liquid from a decant drum, contact within the process is normally: [1] through the mixing of steam with hydrocarbon vapors; [2] gaseous chemicals passing countercurrently through an aqueous absorption or quench medium; or [3] steam used to strip more volatile chemicals from liquid hydrocarbon mixtures. In all of these cases, the ultimate concentration of contaminants in the aqueous stream is governed by the specific vapor-liquid equilibria between the aqueous phases and the chemical phases. Hydrocarbon concentrations as total organic carbon (TOC) are generally less than 1 g/l or 1,000 mg/l in the aqueous streams associated with this type of processing.

Subcategory C: Aqueous Liquid-Phase Reaction Systems — These are liquid-phase reactions where the catalyst is in aqueous medium, e.g., dissolved or emulsified mineral salt, or acid-caustic solution. Continuous regeneration of catalyst system requires extensive water usage. Substantial removal of spent inorganic salt by-products may also be required. Working aqueous catalyst solution is normally corrosive. Additional water may be required in final purification or neutralization of products.

The chemical conversions associated with Subcategory C processes are characterized by intimate contact between water and the reaction mixture or catalyst system. Water is used as a reaction medium in many of these systems because both the chemicals and the catalyst are infinitely soluble. The chemical conversions are generally multistep reactions and generally more complicated than the vapor-phase reactions in Subcategory B. The Subcategory C reactions are also generally less selective in their yield to desired products and subsequently produce more by-products which must be removed from the system.

Typical unit operations involve liquid-solid interfaces where water is used to wash contaminants from solid chemical products. Because of the much larger quantities of chemicals and catalyst present in aqueous solution, most Subcategory C processes utilize many of the same unit operations as in Subcategory B, for the purpose of recovering these materials prior to discharging the water. There is also much more extensive internal recycling of aqueous process streams. The hydrocarbon concentrations (as TOC) in the process wastewaters which are ultimately discharged are in some cases 10-fold those for Subcategory B or approximately 10 g/l or 10,000 mg/l. The amount of contaminants, when expressed on a production basis, is also higher because of the required removal of by-products which are necessarily present in aqueous solutions.

Subcategory D: Batch and Semicontinuous Processes — Processes are carried out in reaction kettles equipped with agitators, scrapers, reflux condensers, etc. depending on the nature of the operation. Many reactions are liquid-phase with aqueous catalyst systems.

Reactants and products are transferred from one piece of equipment to another by gravity flow, pumping, or pressurization with air or inert gas. Much of the material handling is manual with limited use of automatic process control. Filter presses and centrifuges are commonly used to separate solid products from liquid. Where drying is required, air or vacuum ovens are used. Cleaning of noncontinuous production equipment constitutes a major source of wastewater. Waste loads from product separation and purification will be at least ten times those from continuous processes.

Subcategory D refers to batch processes. These operations are characterized by small production volumes and highly variable mixtures of products. A typical batch dye plant manufactures a wide variety of products at any specific point in time. This product mix itself may change completely on a schedule basis as short as one week. The segregation and characterization of process wastewater associated with the production of any one specific dye is not possible, nor is it practical as a basis for establishing effluent limitations. Instead, the total wastewater emanating from the batch plant is considered.

It is an economic necessity that equipment be transferred from the making of one chemical to that of another in multiproduct batch plants. Although certain items may be used for only one type of chemical conversion, product purity requires that process lines and vessels be purged and cleaned between batches. Water is the most common cleaning solvent used in such applications, both because of the relatively low cost associated with its use and because other organic solvents cannot provide the required removal of contaminants. Wastewater from cleaning is, therefore, a major contributor to the raw waste load (RWL) of Subcategory D processes.

Additional considerations include the fact that most of the chemical conversions are carried out in aqueous media and are generally much more complex than those done continuously. The reactions are generally less selective and produce greater quantities of waste by-products. They also frequently require rapid cooling which can be provided only through the direct addition of ice or refrigerated quench water. The raw waste load (RWL) data obtained in field surveys of representative processes in the various categories surveyed by EPA in the study cited above are summarized in Table 270.

TABLE 270: MAJOR RWL'S OF POLLUTANTS BASED ON PROCESS WASTEWATER

Category	Flow RWL gals./1,000 lbs	BOD_5 RWL lbs/1,000 lbs	COD RWL lbs/1,000 lbs	TOC RWL lbs/1,000 lbs
A Conc. Range (mg/l)	0.25 – 2,000	0.1 – 0.13 (400 – 1,000)	0.3 – 3.7 (200 – 10,000)	0.034 – 0.9 (50 – 2,000)
B Conc. Range (mg/l)	50 – 3,000	0.09 – 7.0 (50 – 500)	0.47 – 21.5 (200 – 5,000)	0.2 – 40 (100 – 2,000)
C Conc. Range (mg/l)	30 – 3,000	1.3 – 125 (3,000 – 10,000)	1.9 – 385 (10,000 – 50,000)	1.5 – 150 (3,000 – 15,000)
D Conc. Range (mg/l)	10,000 – 100,000	52 – 220 (100 – 3,000)	180 – 4,800 (1,000 – 10,000)	60 – 1,600 (200 – 2,000)

Source: Report EPA 440/1-73/009

Subcategories B, C, and D were further subcategorized on the basis of raw waste loads. For Subcategories B and C, product processes were classified above and below the median raw waste load (BOD_5 and COD) for the subcategory. Those product processes with raw waste loads above the median subcategory raw waste load were designated B$_2$ and C$_2$, respectively, and those product processes with raw waste loads below the median value were designated B$_1$ and C$_1$, respectively, for each subcategory. Subcategory D includes one subcategory, azoic dyes and components. The raw waste loads (RWL) associated with each of the continuous process Subcategories (A, B, and C) are based on contact process water only. Most continuous processes are able to achieve segregation and do not include noncontact cooling water or steam. Subcategory D includes all water usage associated with the process in that rapid cooling with direct contact is required in the manufacture of dyes.

The subcategorization assigns specific products to specific subcategories according to the manufacturing process by which they are produced. Where more than one process is commercially used to produce a specific chemical, it is possible that the chemical may be listed in more than one subcategory, because the unit operations and chemical conversions associated with different feedstocks may differ drastically in regard to process water usage and associated RWL. The most common chemical conversions used within the industry were examined and are themselves subcategorized in the tabulation below:

Subcategory A	Subcategory B	Subcategory C	Subcategory D
Acylation	Amination	Alcoholysis	Alkylation
Alkylation	Hydration	Ammonolysis	Amination
Aromatization	Dehydration	Dehydration	Condensation
Friedel-Crafts	Hydrogenation	Esterification	Nitration
Reactions	Dehydrogenation	Hydroformulation	
Halogenation	Oxidation	Hydration	
	Pyrolysis	Neutralization	
		Nitration	
		Oxidation	

A breakdown of these subcategories in terms of typical specific organic chemical products is given in Table 271 (3).

TABLE 271: SUBCATEGORIES OF THE ORGANIC CHEMICALS MANUFACTURING
INDUSTRY (PHASE 1 — MAJOR PRODUCT-PROCESSES)

Subcategory A	Process Descriptions
Cyclohexane	Hydrogenation of Benzene
Ethyl Benzene	Alkylation of Benzene with Ethylene
Vinyl Chloride	Addition of HCl to Acetylene
BTX Aromatics	Hydrotreating Pyrolysis Gasoline
BTX Aromatics	Solvent Extraction of Reformate

Subcategory B
B1 Products

Ethylene and Propylene	Pyrolysis of Naptha or liquid Petroleum Gas (LPG)
Butadiene	Co-product of Ethylene
Methanol	Steam Reforming of Natural Gas
Acetone	Dehydrogenation of Isopropanol
Acetaldehyde	Dehydrogenation of Ethanol
Vinyl Acetate	Synthesis of Ethylene and Acetic Acid

B2 Products

Butadiene	Dehydrogenation of n-Butane
Butadiene	Oxidative dehydrogenation on n-Butane
Acetylene	Partial Oxidation of Methane
Ethylene Oxide	Oxidation of Ethylene
Formaldehyde	Oxidation of Methanol
Ethylene Dichloride	Direct Chlorination of Ethylene
Vinyl Chloride	Cracking of Ethylene Dichloride
Styrene	Dehydrogenation of Ethyl Benzene
Methyl Amines	Addition of Ammonia to Methanol

(continued)

TABLE 271: (continued)

Subcategory C
C1 Products

Acetaldehyde	Oxidation of Ethylene
Acetic Acid	Oxidation of Acetaldehyde
Acrylic Acid	Synthesis with CO and Acetylene
Aniline	Nitration and Hydrogenation of Benzene
Bisphenol A	Condensation of Phenol and Acetone
Caprolactam	Oxidation of Cyclohexane
Coal Tar	Pitch forming and Distillation
Ethylene Glycol	Hydrogenation of Ethylene Oxide
Dimethyl Terephthalate	Esterification of Terephthalic acid (TPA)
Oxo Chemicals	Carbonylation and Condensation
Phenol	Oxidation of Cumene
Terephthalic Acid	Oxidation of p-Xylene
Terephthalic Acid (Polymer grade)	Purification of Crude Terephthalic Acid

C2 Products

Acrylates	Esterification of Acrylic Acid
P-Cresol	Sulfonation of Toluene
Methyl methacrylate	Acetone Cyanohydrin Process
Terephthalic Acid	Nitric Acid Process
Tetraethyl Lead	Addition of Ethyl Chloride to Lead Amalgam

Subcategory D

Organic Dyes: Azoic Dyes and Components	Batch Processes

Source: Report EPA 440/1-73/009

References

(1) Sittig, M., "Combining Oxygen and Hydrocarbons for Profit," Houston, Texas, Gulf Publishing Co. (1962).
(2) U.S. Environmental Protection Agency "Atmospheric Emissions from the Petroleum Refining Industry," *Report PB 225,040,* Springfield, Va., Nat. Tech. Information Service (August 1973).
(3) Sittig, M., *Pollution Control in the Organic Chemical Industry,* Park Ridge, N.J., Noyes Data Corp. (1974).
(4) Jewell, W.L. & Ford, D.L., "Preliminary Investigational Requirements–Petrochemical and Refinery Waste Treatment Facilities," *Report 12020 EID 03/71,* Washington, D.C., U.S. Government Printing Office (March 1971).
(5) U.S. Environmental Protection Agency "Development Document for Proposed Effluent Limitation Guidelines and New Source Performance Standards for the Major Organic Products Segment of the Organic Chemicals Manufacturing Point Source Category," *Report EPA 440/1-73/009,* Washington, D.C. (December 1973).

PAINT AND VARNISH MANUFACTURE

Paint Manufacturing

The manufacture of paint involves the dispersion of a colored oil or pigment in a vehicle, usually an oil or resin, followed by the addition of an organic solvent for viscosity adjustment. Only the physical processes of weighing, mixing, grinding, tinting, thinning, and packaging take place; no chemical reactions are involved. These processes take place in large mixing tanks at approximately room temperature..

Varnish Manufacturing

The manufacture of varnish also involves the mixing and blending of various ingredients to produce a wide range of products. However, in this case chemical reactions are initiated by heating. Varnish is cooked in either open or enclosed gas-fired kettles for periods of 4 to 16 hours at temperatures of 200° to 650°F (93° to 340°C).

Varnish cooking emissions, largely in the form of organic compounds, depend on the cooking temperatures and times, the solvent used, the degree of tank enclosure, and the type of air pollution controls used. Emissions from varnish cooking range from 1 to 6 percent of the raw material.

Air Pollution

The primary factors affecting emissions from paint manufacture are care in handling dry pigments, types of solvents used, and mixing temperature. About 1 or 2 percent of the solvents is lost even under well-controlled conditions. Particulate emissions amount to 0.5 to 1.0 percent of the pigment handled.

To reduce hydrocarbons from the manufacture of paint and varnish, control techniques include condensers and/or adsorbers on solvent-handling operations, and scrubbers and afterburners on cooking operations. Emission factors for paint and varnish are in Table 272 (1).

TABLE 272: EMISSION FACTORS FOR PAINT AND VARNISH MANUFACTURING WITHOUT CONTROL EQUIPMENT*

EMISSION FACTOR RATING: C

Type of product	Particulate		Hydrocarbons**	
	lb/ton pigment	kg/MT pigment	lb/ton of product	kg/MT pigment
Paint	2	1	30	15
Varnish				
Bodying oil	—	—	40	20
Oleoresinous	—	—	150	75
Alkyd	—	—	160	80
Acrylic	—	—	20	10

*Afterburners can reduce gaseous hydrocarbon emissions by 99 percent and particulates by about 90 percent. A water spray and oil filter system can reduce particulates by about 90 percent.

**Expressed as undefined organic compounds whose composition depends upon the type of varnish or paint.

Source: EPA Publication AP-42, 2nd Edition

Reference

(1) U.S. Environmental Protection Agency, "Compilation of Air Pollutant Emission Factors," 2nd Edition, *Publication Number AP-42,* Research Triangle Park, N.C., Office of Air and Water Programs (April 1973).

PAPERBOARD MANUFACTURE

Pulpboard manufacturing includes the manufacture of fibrous boards from a pulp slurry. This includes two distinct types of product, paperboard and fiberboard. Paperboard is a general term that describes a sheet 0.012 inch (0.30 mm) or more in thickness made of fibrous material on a paper machine. Fiberboard, also referred to as particle board, is much thicker than paperboard and is made somewhat differently. There are two distinct phases in the conversion of wood to pulpboard: [1] the manufacture of pulp from the raw wood, and [2] the manufacture of pulpboard from the pulp. This section deals only with the latter as the first is covered under the section on pulp and paper manufacture.

In the manufacture of paperboard, the stock is sent through screens into the head box, from which it flows onto a moving screen. Approximately 15 percent of the water is removed by suction boxes located under the screen. Another 50 to 60 percent of the moisture is removed in the drying section. The dried board then enters the calendar stack, which imparts the final surface to the product.

In the manufacture of fiberboard, the slurry that remains after pulping is washed and sent to the stock chests where sizing is added. The refined fiber from the stock chests is fed to the head box of the board machine. The stock is next fed onto the forming screens and sent to dryers, after which the dry product is finally cut and fabricated.

Air Pollution

Emissions from the paperboard machine consist only of water vapor, and little or no particulate matter is emitted from the dryers. Particulates are emitted, however, from the drying operation of fiberboard. Additional particulate emissions occur from the cutting and sanding operations, but no data are available by which to estimate these emissions. Emission factors for pulpboard manufacturing are shown in Table 273.

TABLE 273: PARTICULATE EMISSION FACTORS FOR PULPBOARD MANUFAC-TURING*

EMISSION FACTOR RATING: E

Type of product	Emissions	
	lb/ton	kg/MT
Paperboard	Neg	Neg
Fiberboard*	0.6	0.3

*Emission factors expressed as units per unit weight of finished product.

Source: EPA Publication AP-42, 2nd Edition

Water Pollution

The raw waste load of paperboard from waste paper mills is generated in the stock preparation area and is mainly a function of the type of raw materials and additives used. In general, the higher the percentage of kraft or neutral sulfite waste paper used in the furnish, the higher the BOD_5 value per ton of product. Mills whose wastes have the higher BOD_5 value generally include those that employ an asphalt dispersion system in the stock preparation process in order to melt and disperse the asphalt found in corrugated waste paper. This system subjects the fiber to a heat and pressure environment in a press and digester which contributes to the higher BOD_5 loads. Effluent volume, BOD_5, and total suspended solids data for 42 mills have been collected by the EPA (2) from data from the Michigan Water Resources Commission, the Wisconsin Water Resources Commission, and the NCASI.

The volume of effluent ranged from 13,760 to 100,150 liters per metric ton (3.3 to 24.0 thousand gallons per short ton) of product and it is known that at three mills the effluent has been virtually eliminated through clarification and water reuse. However, these mills manufacture a small number of products of coarse grade which makes this procedure possible.

Total suspended solids losses for the 42 mills listed range from 4.0 to 61.5 kilograms per metric ton (8 to 123 pounds per short ton) of product, 27 containing 20 kilograms per metric ton (40 pounds per short ton) or under. This value depends upon the type of save-all employed for fiber recovery and the application of the more effective types is contingent upon the kinds of waste paper used and the products manufactured. All mills of this type can employ a cylinder-type save-all and, while it is not the most effective type, it serves to separate usable from unusable fiber and ordinarily restricts losses to less than 20 kilograms per metric ton (40 pounds per short ton). It also serves to protect effluent treatment systems from slugs of fiber and clarifiers from flotation problems.

BOD_5 values ranged from 5 to 37.5 kilograms per metric ton (10 to 75 pounds per short ton) of product, 30 of the 42 being less than or equal to 12.5 kilograms per metric ton (25 pounds per short ton). Residual pulping liquor, starch, and other adhesives, such as glutens, account for most of the BOD_5. Reduction of suspended solids is the only in-process control exercised which would reduce BOD_5.

Reference

(1) U.S. Environmental Protection Agency, "Compilation of Air Pollutant Emission Factors," 2nd Edition, *Publication Number AP-42*, Research Trinagle Park, N.C., Office of Air and Water Programs (April 1973).
(2) U.S. Environmental Protection Agency, "Development Document for Proposed Effluent Limitations Guidelines and New Source Performance Standards for the Unbleached Kraft and Semichemical Pulp Segment of the Pulp, Paper and Paperboard Mills Point Source Category," *Report EPA 440/1-74/025*, Washington, D.C. (January 1974).

PERLITE MANUFACTURE

Perlite is a glassy volcanic rock consisting of oxides of silicon and aluminum combined as a natural glass by water of hydration. By a process called exfoliation, the material is rapidly heated to release water of hydration and thus to expand the spherules into low-density particles used primarily as aggregate in plaster and concrete. A plant for the expansion of perlite consists of ore unloading and storage facilities, a furnace-feeding device, an expanding furnace, provisions for gas and product cooling, and product-classifying and product-collecting equipment. Vertical furnaces, horizontal stationary furnaces, and horizontal rotary furnaces are used for the exfoliation of perlite, although the vertical types are the most numerous. Cyclone separators are used to collect the product.

Air Pollution

A fine dust is emitted from the outlet of the last product collector in a perlite expansion plant. The fineness of the dust varies from one plant to another, depending upon the desired product. In order to achieve complete control of these particulate emissions, a baghouse is needed. Simple cyclones and small multiple cyclones are not adequate for collecting the fine dust from perlite furnaces. Table 274 (1) summarizes the emissions from perlite manufacturing.

TABLE 274: PARTICULATE EMISSION FACTORS FOR PERLITE EXPANSION FUR-
NACES WITHOUT CONTROLS*

EMISSION FACTOR RATING: C

Type of furnace	Emissions**	
	lb/ton	kg/MT
Vertical	21	10.5

*Emission factors expressed as units per unit
weight of charge.
**Primary cyclones will collect 80% of the partic-
ulates above 20 micrometers, and baghouses
will collect 96% of the particles above 20 micrometers.

Source: EPA Publication AP-42, 2nd Edition

Reference

(1) U.S. Environmental Protection Agency, "Compilation of Air Pollutant Emission Factors," 2nd Edition, *Publication Number AP-42,* Research Triangle Park, N.C., Office
of Air and Water Programs (April 1973).

PESTICIDE MANUFACTURE

Virtually every pesticide production process produces aqueous or gaseous streams and frequently solid wastes which contain unreacted ingredients, unrecovered products and solvents, and unavoidable or undesired by-products.

Extensive efforts are usually made to minimize by-products and to recover, recycle, or
otherwise prevent these process losses from occurring. For each process, however, a balance point is eventually reached between the expense of recovery and the value of the
recovered product. In the past, the economic considerations were frequently dominant
and process losses were included as unavoidable costs.

Under the recent emphasis on environmental contamination further efforts have been made
to recover many previously lost materials, even when economics indicated that it was more
expensive to do so, and most pesticide manufacturers have invested in or are in the process
of building extensive waste treatment facilities wherein those wastes which cannot be recovered are degraded to acceptable levels or disposed by state approved methods.

Air Pollution

The air pollution aspects of pesticide production are essentially without quantitative data.
The only air pollution data that have been seen on pesticides are for ambient air samples:
i.e., almost no emissions data on specific pesticides from a given plant have been published.
These data are much needed. Gaseous emissions from parathion production contain significant amounts of mercaptans including hydrogen sulfide, for example.

Water Pollution

A summary of the principal aqueous wastes generated and the disposal method employed
by the producers of the key pesticides is shown in Table 275 (2).

TABLE 275: SUMMARY OF MANUFACTURING WASTES AND DISPOSAL

Pesticide	Liquid Wastes		Solid or Other Wastes	
	Source	Disposal	Source	Disposal
DDT	Processing solutions	Evaporative basin	Reactor solutions	County Dump
Aldrin	Floor washings, etc.	Evaporation basin	Lime slurry	Lime pit
Dieldrin	Process solutions	Evaporation basin	Filter solids	Incinerate
Chlordane	Process solutions	Deep well	Filter solids	Clay pit
Toxaphene	Pinene-Camphene plant	Bio-treatment plant		
Disulfonton	Process solutions	Neutralize, hold, discharge	Filter solids	Solid waste
	Process solutions	Secondary treatment plant	Filter solids, etc.	Commercial landfill
			H_2S	Flare
Malathion	Process solutions	Barge to deep sea	Filter solids	Landfill (with lye)
Phorate	Process solutions	Barge to deep sea	Filter solids	Landfill
			Mercaptan losses	Flare
Parathions	Process solutions	Waste treatment plant	H_2S; S	Flare; incinerate
Carbaryl	Process solutions	Secondary waste treatment	H_2, $COCl_2$, amine	Flare
			Heavy residues	Incinerate
Aldicarb	Process solutions	Neutralize, secondary waste treatment	Process vents	Flare
2,4-D (Dow)	Process solutions	Trickling filter; biological waste treatment plant	Filter solids and still bottoms	Incinerate, Scrub
2,4-D (Chipman)	Process solutions	Charcoal absorption/ filtration treatment	-	-
2,4,5-T (Dow)	As per 2,4-D			
2,4,5-T (Thompson-Hayward)	Process solutions	Oxidation pond, discharge	Solids	Landfill
Atrazine	Process solutions	Most to river; some to deep well		
Trifluralin	Process solutions	Biological waste treatment	NO_x	Scrubber
Alachlor	Process solutions	Discharge	Solvent	Fuel
Captan	Process solutions	Hold, discharge	Gas streams	Scrub, vent
Methyl Bromide			Gaseous wastes	Scrub, waste treatment plant
Pyrethrin	Aqueous still bottoms	Sewer	Process solids	Storage
Bacillus t. (Abbott)	Process solutions	Sterilized; biological waste treatment	Filter solids	Landfill
Bacillus t. (Nutrilite)	Process solutions	Evaporation pond	Process air	Incinerate or filter
$HgCl_2$-Hg_2Cl_2	Process solution	Hg-recovery; discharge to sewer	Filter solids NO_x, H_2S	Hg recovery Recovery

Source: Report PB 213,782

The wastes resulting from DDT manufacture include spent acids (hydrochloric and sulfuric), monochlorobenzene, sodium monochlorobenzene sulfonate, chloral, NaOH caustic wash-waters, chlorobenzene, sulfonic acid, and some product (1). The waste streams may contain DDT in the 1 to 5 ppm range with DDE and other related compounds present in amounts up to four times the DDT level. The pH of the waste is low and the salt content is high. A typical waste stream from carbamate manufacture (1) will contain:

COD	10,000 mg/l
BOD_5	Nil
Total solids	40,000
pH	7 to 10
Suspended solids	Nil
Sodium	8,000
Chlorides	100
Phosphates	Nil
Organic nitrogen	500
Sulfates	20,000
Product	Nil
Toxicity	Low
Flow per pound of product	Not available

The waste streams from parathion manufacture may contain sulfur, HCl, sodium chloride, sodium carbamate, trimethyl theophosphate, and other organics including paranitrophenol and small amounts of product (1). A typical waste stream can be characterized by:

COD	3,000 mg/l
BOD_5	700
Total solids	27,000
pH	2.0
Acidity	3,000
Sodium	6,000
Chlorides	7,000
Phosphates	250
Nitrates	20
Sulfates	3,000
Parathion	20
Calcium	High

Concentrated residues and tank bottoms must also be handled since they contain large amounts of intermediates and some product. This portion of the waste may be a slurry with a paste-like consistency which poses severe handling problems, but seldom enters the wastewater stream.

The waste stream from chlorinated hydrocarbon herbicides includes large amounts of sodium chloride, hydrochloric acid, some caustic, and organics including solvents, phenols, chlorophenols and chlorophenoxy acids. These wastes arise from acidification, washing steps, phase separation steps, incomplete yields and chlorination of the phenolic compounds. A typical waste stream can be characterized by:

COD	8,300 mg/l
BOD_5	6,300
Total solids	104,000
Suspended solids	2,500
pH	0.5
Chlorides	52,000
Chlorophenols	112
Chlorophenoxy acids	235
Nitrogen	Low
Phosphorus	Low
Flow	30 lb COD/ lb product

The chlorinated herbicide wastes can vary considerably from plant to plant and even in the same plant as indicated by the data in Table 276 (1).

TABLE 276: WASTE CHARACTERISTICS OF CHLORINATED HERBICIDE WASTES

Analysis	2, 4, 5-T Acid Waste			2, 4-D Acid Waste		
	Sample #1	Sample #2	Sample #3	Sample #1	Sample #2	Sample #3
pH	7.5	7.9		8.5	9.5	
COD (mg/l)	21,700	25,700	19,600	27,500	23,600	22,700
BOD$_5$ (mg/k)*	16,800	*	13,400	16,700	*	13,000
Chlorides (mg/l)	96,300	69,000		144,000	72,000	
Total Solids (mg/l)		172,467			167,221	
Total Volatile Solids (mg/l)		18,150			22,100	
Suspended Solids (mg/l)		700			348	
Volatile Suspended Solids (mg/l)		242			83	
Total Kjeldahl Nitrogen (mg/l)		40			45	
Odors		Not Offensive			Not Offensive	

*16,680 mg/l using 50-50 dilution of 2,4-D and 2,4,5-T and acclimated seed (below toxic levels)

Source: EPA Report 12020 FYE 01/72

A typical waste stream from a diolefin based chlorinated hydrocarbon pesticide manufacture will include:

COD	500 mg/l
BOD$_5$	50
Total solids	1,000
Suspended solids	100
pH	2
Chlorides	High
Nitrates	?
Phosphates	?
Product	100 to 300 ppb as Endrin
Toxicity	High
Flow	0.375 gallons treated wastewater/ lb product

The above list does not include concentrated liquids, tank bottoms, spent catalysts, etc., which are normally landfilled.

Intermedia Aspects

The producers of the persistent chlorinated hydrocarbons use: an evaporative basin, in part for DDT; an evaporative basin for dieldrin; and deep well disposal for chlordane. Therefore, these plants have no discharges subject to the 1899 Act. The evaporative basins require a word of further comment: evaporative and wind blown losses from these facilities require evaluation and the long-term future of the basin should be considered, e.g., what happens if the production site is closed 25 years from now and converted to other (even residential) uses?

Deep well disposal is also used by several pesticide producers in states where that practice is permitted. Sea disposal is also practiced by a number of producers in the eastern seaboard area and one report states that 329,000 tons of pesticide wastes were disposed at sea in 1968, but the amount of active ingredients of any kind were not specified.

References

(1) Atkins, P.R., "The Pesticide Manufacturing Industry—Current Waste Treatment and Disposal Practices," *EPA Report 12020 FYE 01/72,* Washington, D.C., U.S. Government Printing Office (January 1972).
(2) U.S. Environmental Protection Agency, "The Pollution Potential in Pesticide Manufacturing," *Report PB 213,782,* Springfield, Va., National Technical Information Service (June 1972).

PETROLEUM MARKETING AND TRANSPORTATION

The marketing and transportation of petroleum products involves many distinct operations, each of which can represent a source of evaporation loss. For example, after gasoline is refined, it is transported first via pipeline, rail, ship, or barge to intermediate storage and then to regional marketing terminals for temporary storage in large quantities. From here, the product is pumped into tank trucks that deliver it directly to service stations or to larger distributors at bulk plants. From bulk plants, the product is delivered, again in trucks, to commercial accounts (e.g., trucking companies). The final destination for the gasoline is normally a motor vehicle gas tank. A similar distribution path may be developed for fuel oil and other petroleum products.

Air Pollution

Losses from marketing and transportation fall into five categories, depending on the storage equipment or mode of conveyance used: [1] Large storage tanks: breathing, working, and standing storage losses; [2] Railroad tank cars and tank trucks: loading and unloading losses; [3] Marine vessels: loading, unloading, and transit losses; [4] Service stations: loading and unloading losses from tank trucks and underground tanks; and [5] Motor vehicle tanks: refueling losses. Emissions from such sources are given in Table 277 (1).

TABLE 277: ORGANIC COMPOUND EVAPORATIVE EMISSION FACTORS FOR PETROLEUM TRANSPORATION AND MARKETING SOURCES[a]

EMISSION FACTOR RATING: A

Emission source	Product				
	Gasoline	Crude oil	Naphtha jet fuel (JP-4)	Kerosene	Distillate oil
Tank cars/trucks[b]					
Splash loading					
lb/10^3 gal transferred	12.4	10.6	1.8	0.88	0.93
kg/10^3 liter transferred	1.5	1.3	0.22	0.11	0.11
Submerged loading					
lb/10^3 gal transferred	4.1	4.0	0.91	0.45	0.48
kg/10^3 liter transferred	0.49	0.48	0.11	0.054	0.058
Unloading					
lb/10^3 gal transferred	2.1	2.0	0.45	0.23	0.24
kg/10^3 liter transferred	0.25	0.24	0.054	0.028	0.029
Marine vessels[b]					
Loading					
lb/10^3 gal transferred	2.9	2.6	0.60	0.27	0.29
kg/10^3 liter transferred	0.35	0.31	0.072	0.032	0.035
Unloading					
lb/10^3 gal transferred	2.5	2.3	0.52	0.24	0.25
kg/10^3 liter transferred	0.30	0.28	0.062	0.029	0.030
Transit					
lb/wk-10^3 gal load	3.6	3.2	0.74	0.34	0.36
kg/wk-10^3 liter load	0.43	0.38	0.089	0.041	0.043
Underground gasoline storage tanks[c]					
Splash loading					
lb/10^3 gal transferred	11.5	NU[d]	NU	NU	NU
kg/10^3 liter transferred	1.4	NU	NU	NU	NU
Uncontrolled submerged loading					
lb/10^3 gal transferred	7.3	NU	NU	NU	NU
kg/10^3 liter transferred	0.38	NU	NU	NU	NU
Submerged loading with open vapor return system					
lb/10^3 gal transferred	0.80	NU	NU	NU	NU
kg/10^3 liter transferred	0.097	NU	NU	NU	NU
Submerged loading with closed vapor return system					
lb/10^3 gal transferred	Neg	NU	NU	NU	NU
kg/10^3 liter transferred	Neg	NU	NU	NU	NU

(continued)

TABLE 277: (continued)

Emission source	Product				
	Gasoline	Crude oil	Naphtha jet fuel (JP-4)	Kerosene	Distillate Oil
Unloading					
lb/10³ gal transferred	1.0	NU	NU	NU	NU
kg/10³ liter transferred	0.12	NU	NU	NU	NU
Filling motor vehicle gasoline tanks[e]					
Vapor displacement loss					
lb/10³ gal pumped	11.0	NU	NU	NU	NU
kg/10³ liter pumped	1.3	NU	NU	NU	NU
Liquid spillage loss					
lb/10³ gal pumped	0.67	NU	NU	NU	NU
kg/10³ liter pumped	0.081	NU	NU	NU	NU

[a]Reference (1)

[b]Data based on the following conditions:

Storage temperature: 63 °F (17.2 °C)
Saturation of tank existent vapors in loading and unloading tank trucks and cars: 20 percent

	Gasoline	Crude oil	Naphtha jet fuel (JP-4)	Kerosene	Distillate oil
Molecular weight of vapor, lb/lb-mole	56.8	64.5	63.3	72.7	72.7
Reid vapor pressure					
psia	10.5	7.0	2.5	0.5	0.5
Mg/m²	7.4	4.9	1.75	0.35	0.35
True vapor pressure					
psia	5.8	4.6	1.2	0.5	0.5
Mg/m²	4.1	3.2	0.84	0.35	0.35
Liquid density					
lb/gal	6.2	7.0	6.2	6.8	7.2
kg/liter	0.74	0.84	0.74	0.82	0.87

[c]Factors for underground gasoline storage tanks based on an organic compound vapor space concentration of 40 percent by volume, which corresponds to a saturation of nearly 100 percent.
[d]Not used.
[e]Motor vehicle gasoline tank vapor displacement factor based on an average dispensed fuel temperature of 63 °F (17.2 °C), an average displaced vapor temperature of 67 °F (19.4 °C), and a Reid vapor pressure of 10.5 psia (7.4 Mg/m²).

Source: EPA Publication AP-42, 2nd Edition

Reference

(1) U.S. Environmental Protection Agency, "Compilation of Air Pollutant Emission Factors," 2nd Edition, *Publication Number AP-42 Supplement Number 1,* Research Triangle Park, N.C., Office of Air and Water Programs (July 1973).

PETROLEUM REFINING

Although a modern refinery is a complex system of many processes, the entire operation can be divided into four major steps: separating, converting, treating, and blending. The crude oil is first separated into selected fractions (e.g., gasoline, kerosene, fuel, oil, etc.). Because the relative volumes of each fraction produced by merely separating the crude may not conform to the market demands for each fraction, some of the less valuable products, such as heavy naptha, are converted to products with a greater sale value, such as gasoline. This conversion is accomplished by splitting (cracking), uniting (polymerization), or rearranging (reforming) the original molecules. The final step is the blending of the refined base stocks with each other and with various additives to meet final product specifications.

Air Pollution

Estimated emissions from the 262 operating refineries in the United States in 1969 are given in Table 278. Sulfur, one of the impurities in the crude oil, is a principal contributor to pollution problems, and its removal from the crude oil is expensive, although it can be accomplished by a variety of hydrodesulfurization processes.

TABLE 278: ESTIMATED EMISSIONS FROM 262 U.S. REFINERIES (1969)

Pollutant	Emission, 1000 tons	Control, %
Sulfur oxides	2200	0
Nitrogen oxides	61	0
Hydrocarbons	2300	50
Particulates	55	50
Carbon monoxide	2420	75

Source: Report PB 225,040

Other air pollutants from refineries are nitrogen oxides (NO_x), particulates, malodorous compounds, carbon monoxide (CO), smoke, and hydrocarbons. Even though refineries are not the largest industrial emitters of air pollution on a nationwide basis, they do add to local pollution problems because they emit pollutants that can be hazardous if not controlled. These pollutants contribute to the formation of photochemical smog and have harmful effects on public health and property in congested industrial regions, as well as on vegetation as noted by Laster [1]. Table 279 [1] shows major sources of air pollution from a petroleum refinery, with the principal pollutants from these sources.

TABLE 279: MAJOR SOURCES OF POLLUTANT EMISSIONS

Source	Pollutant	Emission factor, lb/1000 bbl
Catalytic cracking regenerator	Sulfur oxides	500 (charge)
	Hydrocarbons	220 (charge)
	Particulates (with ESP)[a]	61 (charge)
	Carbon monoxide	13,700 (charge)
	Nitrogen oxides	63 (charge)

(continued)

TABLE 279: (continued)

Source	Pollutant	Emission factor, lb/1000 bbl
Combustion operations associated with crude oil separation	Sulfur oxides Hydrocarbons Particulates Carbon monoxide Nitrogen oxides	6,400 x %S (oil burned) 140 (oil burned) 800 (oil burned) 2 (oil burned) 2,900 (oil burned)
Storage tanks	Hydrocarbons (cone roof)	400 (throughput)[b]
Miscellaneous factors Loading facilities Sampling Spillage Leaks Barometric condensers Waste oil separator Cooling tower	Hydrocarbons	700 (crude charge)

[a]Electrostatic precipitator

[b]With a floating-roof tank the emission factor would be about 4 pounds per day per 1000 barrels.

Source: Report PB 225,040

The main source of emissions from crude oil preparation processes is the barometric condenser on the vacuum distillation column. This condenser, while maintaining a vacuum on the tower, often allows noncondensable light hydrocarbons and hydrogen sulfide to pass through to the atmosphere. The quantity of these emissions is a function of the unit size, type of feedstock, and the cooling water temperature. Vapor recovery systems reduce these emissions to negligible amounts [see Table 280 (2)].

Emissions from cracking unit regenerators consist of particulates (coke and catalyst fines), hydrocarbons, sulfur oxides, carbon monoxide, aldehydes, ammonia, and nitrogen oxides in the combustion gases. In addition, catalyst fines may be discharged by vents on the catalyst handling systems on both TCC and FCC units. Control measures commonly used on regenerators consist of cyclones and electrostatic precipitators to remove particulates and energy-recovery combustors to reduce carbon monoxide emissions. The latter recovers the heat of combustion of the CO to produce refinery process steam. Waste gas from the hydrocracker contains large amounts of H_2S, which can be processed for removal of sulfur. There are essentially no emissions from reforming operations.

Polymerization, alkylation and isomerization processes, including regeneration of any necessary catalysts, form essentially closed systems and have no unique, major source of atmospheric emissions. However, the highly volatile hydrocarbons handled, coupled with the high process pressures required, make valve stems and pump shafts difficult to seal, and a greater emission rate from these sources can generally be expected in these process areas than would be the average throughout the refinery. The best method for controlling these emissions is the effective maintenance, repair, and replacement of pump seals, valve caulking, and pipe-joint sealer.

Emissions from treating operations consist of SO_2, hydrocarbons, and visible plumes. Emission levels depend on the methods used in handling spent acid and acid sludges, as well as the means employed for recovery or disposal of hydrogen sulfide. Other potential sources of these emissions in treating include catalyst regeneration, air agitation in mixing tanks, and other air blowing operations.

TABLE 280: EMISSION FACTORS FOR PETROLEUM REFINERIES

EMISSION FACTOR RATING: A

Type of process	Particulates	Sulfur oxides (SO_2)	Carbon monoxide	Hydro-carbons	Nitrogen oxides (NO_2)	Alde-hydes	Ammonia
Boilers and process heaters[a]							
lb/10³ bbl oil burned	840	6,720S[b]	Neg[c]	140	2,900	25	Neg
kg/10³ liters oil burned	2.4	19.2S	Neg	0.4	8.3	0.071	Neg
lb/10³ ft³ gas burned	0.02	2s[d]	Neg	0.03	0.23	0.003	Neg
kg/10³ m³ gas burned	0.32	32s	Neg	0.48	3.7	0.048	Neg
Fluid catalytic cracking units							
Uncontrolled							
lb/10³ bbl fresh feed	242	493	13,700	220	71.0	19	54
	(93 to 340)	(313 to 525)			(37.1 to 145.0)		
kg/10³ liters fresh feed	0.695	1.413	39.2	0.630	0.204	0.054	0.155
	(0.267 to 0.976)	(0.898 to 1.505)			(0.107 to 0.416)		
Electrostatic precipitator and CO boiler							
lb/10³ bbl fresh feed	44.7	493	Neg	220	71.0	19	54
	(12.5 to 61.0)	(313 to 525)			(37.1 to 145.0)		
kg/10³ liters fresh fresh feed	0.128	1.413	Neg	0.630	0.204	0.054	0.155
	(0.036 to 0.175)	(0.898 to 1.505)			(0.107 to 0.416)		
Moving-bed catalytic cracking units[a]							
lb/10³ bbl fresh feed	17	60	3,800	87	5	12	6
kg/10³ liters fresh feed	0.049	0.171	10.8	0.250	0.014	0.034	0.017
Fluid coking units							
Uncontrolled							
lb/10³ bbl fresh feed	523	NA[h]	Neg	Neg	Neg	Neg	Neg
kg/10³ liters fresh feed	1.50	NA	Neg	Neg	Neg	Neg	Neg
Electrostatic precipitator							
lb/10³ bbl fresh feed	6.85	NA	Neg	Neg	Neg	Neg	Neg
kg/10³ liters fresh feed	0.0196	NA	Neg	Neg	Neg	Neg	Neg
Compressor internal combustion engines[a]							
lb/10³ ft³ gas burned	Neg	2s	Neg	1.2	0.9	0.1	0.2
kg/10³ m³ gas burned	Neg	32s	Neg	19.3	14.4	1.61	3.2
Blowdown systems[a]							
Uncontrolled							
lb/10³ bbl refinery capacity	Neg	Neg	Neg	300	Neg	Neg	Neg
kg/10³ liters refinery capacity	Neg	Neg	Neg	0.860	Neg	Neg	Neg
Vapor recovery system or flaring							
lb/10³ bbl refinery capacity	Neg	Neg	Neg	5	Neg	Neg	Neg
kg/10³ liters refinery capacity	Neg	Neg	Neg	0.014	Neg	Neg	Neg
Process drains, Uncontrolled							
lb/10³ bbl waste water	Neg	Neg	Neg	210	Neg	Neg	Neg
kg/10³ liters waste water	Neg	Neg	Neg	0.600	Neg	Neg	Neg
Vapor recovery or separator covers							
lb/10³ bbl waste water	Neg	Neg	Neg	8	Neg	Neg	Neg
kg/10³ liters waste water	Neg	Neg	Neg	0.023	Neg	Neg	Neg
Vacuum jets[a]							
Uncontrolled							
lb/10³ bbl vacuum distillate	Neg	Neg	Neg	130	Neg	Neg	Neg
kg/10³ liters vacuum distillate	Neg	Neg	Neg	0.370	Neg	Neg	Neg
Fume burner or waste-heat boiler							
lb/10³ bbl vacuum distillate	Neg	Neg	Neg	Neg	Neg	Neg	Neg

(continued)

TABLE 280: (continued)

Type of process	Particulates	Sulfur oxides (SO$_2$)	Carbon monoxide	Hydro-carbons	Nitrogen oxides (NO$_2$)	Alde-hydes	Ammonia
kg/10^3 liters vacuum distillate	Neg	Neg	Neg	Neg	Neg	Neg	Neg
Cooling towers[a]							
lb/10^6 gal cooling water	Neg	Neg	Neg	6	Neg	Neg	Neg
kg/10^6 liters cooling water	Neg	Neg	Neg	0.72	Neg	Neg	Neg
Pipeline valves and flanges[a]							
lb/10^3 bbl refining capacity	Neg	Neg	Neg	28	Neg	Neg	Neg
kg/10^3 liter refining capacity	Neg	Neg	Neg	0.080	Neg	Neg	Neg
Vessel relief valves[a]							
lb/10^3 bbl refining capacity	Neg	Neg	Neg	11	Neg	Neg	Neg
kg/10^3 liter refining capacity	Neg	Neg	Neg	0.031	Neg	Neg	Neg
Pump seals[a]							
lb/10^3 bbl refining capacity	Neg	Neg	Neg	17	Neg	Neg	Neg
kg/10^3 liter refining capacity	Neg	Neg	Neg	0.049	Neg	Neg	Neg
Compressor seals[a]							
lb/10^3 bbl refining capacity	Neg Neg	Neg	Neg	5	Neg	Neg	Neg
kg/10^3 liter refining capacity	Neg	Neg	Neg	0.014	Neg	Neg	Neg
Miscellaneous (air blowing, sampling, etc.)[a]							
lb/10^3 bbl refining capacity	Neg	Neg	Neg	10	Neg	Neg	Neg
kg/10^3 liter refining capacity	Neg	Neg	Neg	0.029	Neg	Neg	Neg

[a]Reference 1.
[b]S = Fuel oil sulfur content (weight percent): factors based on 100 percent combustion of sulfur to SO$_2$ and assumed density of 336 lb/bbl (0.96 kg/liter).
[c]Negligible emission.
[d]$_S$ = refinery gas sulfur content (lb/100 ft^3): factors based on 100 percent combustion of sulfur to SO$_2$.

Source: EPA Publication AP-42, 2nd Edition

Trace amounts of malodorous substances may escape from numerous sources including settling tank vents, purge tanks, waste treatment units, wastewater drains, valves, and pump seals. Control methods used include: covers for wastewater separators; vapor recovery systems for settling and surge tanks; improved maintenance for pumps, valves, etc; and sulfur recovery plants (3). Emissions associated with blending operations are hydrocarbons that leak from storage vessels, valves, and pumps. Vapor recovery systems and specially built tanks minimize storage emissions; good housekeeping precludes pump and valve leakage.

Water Pollution

After developing an understanding of the fundamental production processes and their inter-relationships in refinery operations, determination of the best method of characterizing of refinery discharges will enhance the interpretation of the industry water pollution profile. If unit raw waste loads could be developed for each production process, then the current effluent wastewater profile could be obtained by simply adding the components, and future profiles by projecting the types and sizes of refineries. However, the information required for such an approach is not available. Essentially all of the available data on refinery waste-waters apply to total API separator effluent, rather than to effluents from specific processes. Another factor detracting from the application of a summation of direct subprocess unit raw waste loads, is the frequent practice of combining specific wastewater streams discharging from several units for treatment and/or reuse. Thus, such streams as sour waters, caustic washes, etc., in actual practice are generally not traceable to a specific unit, but only to a stripping tower or treatment unit handling wastes from several units. The size, sequence, and combination of contributing processes are so involved that a breakdown by units would be extremely difficult to achieve. In view of the limitations imposed by the summation of wastewater data from specific production processes, the evaluation of refinery waste loads was based on total refinery effluents discharged through the API (Oil) separator, which is considered an integral part of refinery process operations for product/raw material recovery prior to final wastewater treatment.

Information on raw waste loading was compiled from the 1972 National Petroleum Refining Wastewater Characterization Studies and plant visits by EPA personnel (4). The data are considered primary source data, i.e., they are derived from field sampling and operating records. The raw waste data for each subcategory of the petroleum refining industry have been analyzed to determine the probability of occurrence of mass loadings for each considered parameter in the subcategory. These frequency distributions are summarized in Tables 281 through 286 for each subcategory. The probability plots are contained in reference (4).

TABLE 281: TOPPING SUBCATEGORY RAW WASTE LOAD

Effluent from Refinery API Separator
net kg/1,000 m³ (pounds/1,000 barrels) of feedstock throughput

Parameter	- - - - - - - - Probability of Occurrence, % less than or equal to - - - - - - - -					
	10%		50% (Median)		90%	
BOD$_5$	0.26	(0.09)	7.1	(2.5)	286	(100)
COD	0.72	(0.25)	24.0	(8.4)	858	(300)
TOC	0.086	(0.03)	4.9	(1.7)	269	(98)
TSS	0.30	(0.105)	6.6	(2.3)	123	(43)
Oil	0.97	(0.34)	5.1	(1.8)	45.8	(16)
Phenols	0.001	(0.00035)	0.029	(0.01)	0.92	(0.32)
Ammonia	0.18	(0.064)	1.43	(0.5)	11.4	(4.0)
Sulfides	0.0037	(0.0013)	1.57	(0.55)	7.2	(2.5)
Flow*	0.024	(1.0)	0.53	(22)	12.0	(500)

*1,000 cubic meters/1,000 m³ feedstock throughput (gallons/bbl)

Source: Report EPA 440/1-73/014

TABLE 282: LOW-CRACKING SUBCATEGORY RAW WASTE LOAD

Effluent from Refinery API Separator
net kg/1,000 m³ (pounds/1,000 barrels) of feedstock throughput

Parameter	- - - - - - - Probability of Occurrence, % less than or equal to - - - - - - - -					
	10%		50% (Median)		90%	
BOD$_5$	15.7	(5.5)	71.5	(25)	343	(120)
COD	54	(19)	200	(70)	744	(260)
TOC	7.4	(2.6)	45.7	(16)	286	(100)
TSS	3.7	(1.3)	27	(9.6)	200	(70)
Oily	5.4	(1.9)	27	(9.6)	137	(48)
Phenols	0.36	(0.125)	2.86	(1.0)	22.9	(8.0)
Ammonia	1.9	(0.67)	10.0	(3.5)	51.5	(18)
Sulfides	0.00049	(0.00017)	1.0	(0.35)	22,308	(7,800)
Flow*	0.10	(4.3)	0.79	(33)	6.48	(270)

*1,000 cubic meters/1,000 m³ feedstock throughput (gallons/bbl)

Source: Report EPA 440/1-73/014

TABLE 283: HIGH CRACKING SUBCATEGORY RAW WASTE LOAD

Effluent from Refinery API Separator
net kg/1,000 m³ (pounds/1,000 barrels) of feedstock throughput

Parameter	- - - - - - -Probability of Occurrence, % less than or equal to- - - - - - -					
	10%		50% (Median)		90%	
BOD₅	28.0	(9.8)	82.9	(29)	235	(82)
COD	45.8	(16)	260	(91)	1,487	(520)
TOC	7.4	(2.6)	52.9	(18.5)	372	(130)
TSS	0.054	(0.019)	32.3	(11.3)	212	(74)
Oil	6.7	(2.35)	31.4	(11.0)	120	(42)
Phenols	0.57	(0.20)	5.1	(1.8)	48.6	(17)
Ammonia	6.6	(2.3)	32.8	(11.5)	166	(58)
Sulfides	0.0049	(0.0017)	1.28	(0.45)	315	(110)
Flow*	0.12	(4.8)	0.62	(26)	48.0	(2,000)

*1,000 cubic meters/1,000 m³ feedstock throughput (gallons/bbl)

Source: EPA 440/1-73/014

TABLE 284: PETROCHEMICAL SUBCATEGORY RAW WASTE LOAD

Effluent from Refinery API Separator
net kg/1,000 m³ (pounds/1,000 barrels) of feedstock throughput

Parameter	- - - - - - -Probability of Occurrence, % less than or equal to- - - - - - -					
	10%		50% (Median)		90%	
BOD₅	34.3	(12)	149	(52)	629	(220)
COD	137	(48)	372	(130)	2,888	(800)
TOC	31.5	(11)	117	(41)	443	(155)
TSS	4.0	(1.4)	44.3	(15.5)	887	(310)
Oily	7.4	(2.6)	45.8	(16)	315	(110)
Phenols	2.2	(0.78)	10.3	(3.6)	48.6	(17)
Ammonia	6.3	(2.2)	34.3	(12)	189	(66)
Sulfides	0.011	(0.004)	1.69	(0.59)	229	(80)
Flow*	0.11	(4.6)	0.96	(40)	12.0	(500)

*1,000 cubic meters/1,000 m³ feedstock throughput (gallons/bbl)

Source: EPA 440/1-73/014

TABLE 285: LUBRICATION SUBCATEGORY RAW WASTE LOAD

Effluent from Refinery API Separator
net kg/1,000 m³ (pounds/1,000 barrels) of feedstock throughput

Parameter	- - - - - - -Probability of Occurrence, % less than or equal to- - - - - - -					
	10%		50% (Median)		90%	
BOD₅	113	(40)	187	(66)	311	(110)
COD	200	(71)	382	(134)	1,750	(618)
TOC	–		–		–	
TSS	3.4	(1.2)	79	(28)	1,769	(625)
Oil	48	(17)	136	(48)	396	(140)
Phenolics	0.2	(0.07)	6.2	(2.2)	20	(7.0)
Ammonia (N)	6.2	(2.2)	22	(7.8)	79	(28)
Sulfides	0.00017	(0.00006)	1.1	(0.4)	35.0	(12.4)
Flow*	0.56	(24.2)	0.91	(39)	13.0	(560)

*1,000 cubic meters/1,000 m³ feedstock throughput (gallons/bbl)

Source: EPA 440/1-73/014

TABLE 286: INTEGRATED SUBCATEGORY RAW WASTE LOAD

Effluent from Refinery API Separator
net kg/1,000 m³ (pounds/1,000 barrels) of feedstock throughput

Parameter	- - - - - - -Probability of Occurrence, % less than or equal to- - - - - - -					
	10%		50% (Median)		90%	
BOD$_5$	45	(16)	238	(84)	413	(146)
COD	120	(42.4)	590	(208)	1,150	(406)
TOC	-		-		-	
TSS	0.6	(0.2)	29	(10.2)	340	(120)
Oil	23.8	(8.4)	133	(47)	750	(26.5)
Phenolics	100	(0.35)	6.5	(2.3)	41	(14.5)
Ammonia (N)	7.4	(2.6)	35.4	(12.5)	170	(60)
Sulfides	0.00028	(0.00010)	1.7	(0.6)	45	(15.9)
Flow*	0.23	(10.0)	1.8	(79)	25.6	(1,100)

*1,000 cubic meters/1,000 m³ feedstock throughput (gallons/bbl)

Source: EPA 440/1-73/014

References

(1) Laster, L.L., "Atmospheric Emissions from the Petroleum Refining Industry," *Report PB 225,040,* Springfield, Va., National Technical Information Service (August 1973).
(2) U.S. Environmental Protection Agency, "Compilation of Air Pollutant Emission Factors," 2nd Edition, *Publication Number AP-42,* Research Triangle Park, N.C., Office of Air and Water Programs (April 1973).
(3) Jones, H.R., *Pollution Control in the Petroleum Industry,* Park Ridge, N.J., Noyes Data Corp. (1973).
(4) U.S. Environmental Protection Agency, "Development Document for Proposed Effluent Limitations Guidelines and New Source Performance Standards for the Petroleum Refining Point Source Category," *Report EPA 440/1-73/014,* Washington, D.C. (December 1973).

PETROLEUM STORAGE

Air Pollution

Petroleum storage evaporation losses are associated with the containment of liquid organics in large vessels at oil fields, refineries, and product distribution terminals. Six basic tank designs, are used for petroleum storage vessels: [1] fixed-roof (cone roof), [2] floating roof (single deck pontoon and double deck), [3] covered floating roof, [4] internal floating cover, [5] variable vapor space, and [6] pressure (low and high). The following are the sources of emissions from petroleum storage:

[1] Evaporation losses: these depend on the true vapor pressure of the liquid stored, the diurnal temperature changes in the tank vapor space, the height of the vapor space (tank outage), the tank diameter, the schedule of tank fillings and emptyings, the mechanical condition of tank; and the type of paint applied to the outer surface.

[2] Breathing losses are associated with fixed roof tanks and consist of vapor expelled from the tank because of thermal expansion, barometric pressure changes, and added vaporization of the liquid.

[3] Working losses consist of hydrocarbon vapor expelled from the vessel as a result of emptying or filling operations. Filling losses represent the amount of vapor (approximately equal to the volume of liquid input) that is vented to the atmosphere through displacement.

After liquid is removed, emptying losses occur, because air drawn in during the operation results in growth of the vapor space. Both filling and emptying (together called working) losses are associated primarily with fixed roof and variable vapor space tanks. Filling losses are also experienced from low pressure tankage, although to a lesser degree than from fixed roof tanks.

[4] Primarily associated with floating roof tanks, standing storage losses result from the improper fit of the seal and shoe to the tank shell.

[5] Wetting losses with floating roof vessels occur when a wetted tank wall is exposed to the atmosphere. These losses are negligible.

[6] Finally, boiling loss is the vapor expelled when the temperature of the liquid in the tank reaches its boiling point and begins to vaporize.

There are two major classifications of tanks used to store petroleum products: fixed-roof tanks and floating-roof tanks. The evaporation losses from both of these types of tanks depend on a number of factors, such as type of product stored (gasoline or crude oil), vapor pressure of the stored product, average temperature of the stored product, tank diameter and construction, color of tank paint, and average wind velocity of the area. In order to estimate emissions from a given tank, an average factor can be obtained by making a few assumptions. These average factors for fixed-roof and floating-roof tanks are presented in Table 287.

TABLE 287: EVAPORATIVE EMISSION FACTORS FOR STORAGE TANKS

Emission Factor Rating: A

Product	Fixed roof						Variable vapor space	
	Breathing loss				Working loss		Working loss	
	"New tank" conditions		"Old tank" conditions					
	lb/day-10^3 gal	kg/day-10^3 liter	lb/day-10^3 gal	kg/day-10^3 liter	lb/10^3 gal throughput	kg/10^3 liter throughput	lb/10^3 gal throughput	kg/10^3 liter throughput
Crude oil	0.15	0.018	0.17	0.020	7.3	0.88	Not used	Not used
Gasoline	0.22	0.026	0.25	0.031	9.0	1.1	10.2	1.2
Naphtha jet fuel (JP-4)	0.069	0.0033	0.079	0.0095	2.4	0.29	2.3	0.28
Kerosene	0.036	0.0043	0.041	0.0048	1.0	0.12	1.0	0.12
Distillate fuel	0.036	0.0043	0.041	0.0048	1.0	0.12	1.0	0.12

Product	Mole wt (M) (lb/mole)	Floating roof			
		Standing storage loss			
		"New tank" conditions		"Old tank" conditions	
		lb/day-10^3 gal	kg/day-10^3 liter	lb/day-10^3 gal	kg/day-10^3 liter
Crude oil	64.5	0.029	0.0034	0.071	0.0086
Gasoline	56.8	0.033	0.0040	0.088	0.011
Naphtha jet fuel (JP-4)	63.3	0.012	0.0014	0.029	0.0034
Kerosene	72.7	0.0052	0.00063	0.012	0.0015
Distillate fuel	72.7	0.0052	0.00063	0.012	0.0015

Source: EPA Publication AP-42, 2nd Edition

Reference

(1) U.S. Environmental Protection Agency, "Compilation of Air Pollutant Emission Factors," 2nd Edition, *Publication Number AP-42 Supplement Number 1,* Research Triangle Park, N.C., Office of Air and Water Programs (July 1973).

PHOSPHATE CHEMICALS MANUFACTURE

Water Pollution

Water is primarily used in the phosphate manufacturing industry for 8 principal purposes: [1] noncontact cooling water, [2] process and product water, [3] transport water, [4] contact cooling or heating water, [5] atmospheric seal water, [6] scrubber water, [7] auxiliary process water, and [8] miscellaneous uses.

Noncontact Cooling Water: Water used without contacting the reactants, such as in a shell and tube heat exchanger, is not contaminated with process effluent. If, however, the water contacts the reactants, then contamination of the water results and the waste load increases. Probably the single most important process waste control technique, particularly for subsequent treatment feasibility and economics, is segregation of noncontact cooling water from process water.

Noncontact cooling water is generally of two types in the industry. The first type is recycled cooling water which is cooled by cooling towers or spray ponds. The second type is once-through cooling water whose source is generally a river, lake or tidal estuary, and the water is returned to the same source from which it was taken.The aqueous process wastes from the phosphate industry arise from the use of wet dust scrubbing equipment for the finely divided solid products, and from processes which use excess process water which may become a waste stream.

Sodium Tripolyphosphate Manufacture: Exemplary plants have no process wastes. The dust collected from the spray dryer gaseous effluent stream is added to the spray dryer solid product stream. The water used for subsequent scrubbing of this gas stream from the spray dryer is then recycled to the mix area and is used as process water in the neutralization step. The cooling air used for the product tempering is vented into the spray dryer vent line upstream of the scrubbing operation.

The neutralization step requires a total of 1,040 l/kkg (250 gal/ton), of which 290 l/kkg (70 gal/ton) are recycled from the scrubber. Make-up water, 750 l/kkg (180 gal/ton), are added since water is evaporated in the product drying step. The make-up water is softened, and regeneration of the softener combined with boiler and cooling tower blowdowns amounts to 210 l/kkg (50 gal/ton); 70% of which is from water treatment regeneration and 30% from blowdowns. These blowdown wastes typically contain 1,500 mg/l of dissolved chlorides.

Calcium Phosphates Manufacture: The raw aqueous wastes from the manufacture of food-grade calcium phosphates are from two primary and approximately equal sources: the centrate or filtrate from dewatering of the dicalcium phosphate slurry, and the effluent from wet scrubbers which collect airborne solids from product drying operations. Both of these sources contain suspended, finely divided calcium phosphate solids. It is considered to be normal practice in an integrated plant to partially utilize the DCP centrate or filtrate as make-up scrubber water. The total raw wastes from this system are typically 4,200 l/kkg (1,000 gal/ton) containing 100 kg/kkg (200 lb/ton) of solids (a concentration of 2.4%. An additional 30 kg/kkg (60 lb/ton) of dissolved solids (0.7% of this waste stream) originates from phosphoric acid mists in the scrubbers and from excess phosphoric acid in the reaction liquid.

For nonfood-grade dicalcium phosphate plants, the water scrubbers which collect airborne solids normally operate at partial recycle. Since there is no waste from a dewatering operation, and since dry dust collection typically precedes wet scrubbing, the raw wastes are considerably smaller than for the food-grade operation. Dry dust collection is typical since only one or two products are made, so that the collected solids may be added directly to the product stream without extensive segregation. Moreover, since purity requirements are considerably less severe, the product stream can tolerate such additions. With the above measures, the wet scrubber wastes are typically 420 l/kkg (100 gal/ton) containing

22.5 kg/kkg (45 lb/ton) of suspended solids (a concentration of 5%) plus 4 kg/kkg (8 lb/ton) of dissolved phosphates from acid mists (0.7%). At one notable plant, this bleed stream from the wet scrubber recirculation system is charged directly to the neutralization reactor; hence, this plant had no discharge whatever. As an added feature, this notable plant used cooling water blowdown as make-up to the airborne-solids scrubbing system, thereby eliminating all aqueous discharges (except for the effluent from regeneration of the water softener).

For the nonfood-grade plants, however, acid defluorination is an additional source of raw wastes (unless already-defluorinated acid is delivered to the plant). Wet-process phosphoric acid (54% P_2O_5) contains approximately 1% fluorine. Upon silica treatment, 13.5 kg per kg of acid (27 lb/ton), or 10.5 kg of silicon tetrafluoride product dicalcium phosphate dihydrate (21 lb/ton), are liberated. When hydrolyzed in the acid scrubber, the raw waste contains 12 kg/kkg product (24 lb/ton) of combined fluosilicic acid (H_2SiF_6), hydrofluoric acid (HF) and silicic acid (H_2SiO_3). These raw wastes are contained in a scrubber water flow of 6,300 l/kkg (1,500 gal/ton), so that the combined concentration of fluosilicic acid, hydrofluoric acid and silicic acid is 1,900 mg/l. For any plant manufacturing calcium phosphates of any grade, noncontact cooling water is used in reactors and/or in dried product coolers.

Other possible sources of aqueous wastes are from regeneration of water softeners and from storm water runoff (all exterior surfaces of calcium phosphate plants become coated with fine lime and/or phosphate dusts). In dry-product plants, a significant housecleaning effort must be continually maintained. In nonfood-grade calcium phosphate plants, the dry-product sweepings (from dust, spills, etc.) are added to the process stream. In food-grade plants, however, the sweepings (consisting of lime, lime grit, and calcium phosphates) are wasted. Typically, this solid waste amounts to 10 kg/kkg (20 lb/ton). Table 288 summarizes some of the characteristics of wastes from phosphate chemicals manufacture.

TABLE 288: SUMMARY OF RAW WASTES FROM PHOSPHATE PLANTS

		Food Grade - - Calcium Phosphates- -		Animal Feed - - - Calcium Phosphates- - -	
	Sodium Tripolyphosphate	Dewatering	Solids Scrubbing	Acid Defluorination	Solids Scrubbing
Process water wasted:					
l/kkg pdt	0	2,100	2,100	6,300	420
gal/ton pdt	0	500	500	1,500	100
Raw waste load, kg/kkg pdt:					
TSS	–	50	50	–	22.5
Dissolved PO_4	–	15	15	–	4
HF, H_2SiF_6, H_2SiO_3	–	–	–	12	–
Raw waste load, lb/ton pdt:					
TSS	–	100	100	–	45
Dissolved P_4	–	30	30	–	8
HF, H_2SiF_6, H_2SiO_3	–	–	–	24	–
Concentrations, mg/l:					
TSS	–	24,000	24,000	–	54,000
Dissolved PO_4	–	7,000	7,000	–	7,000
HF, H_2SiF_6, H_2SiO_3	–	–	–	1,900	–
TDS, mg/l	–	7,000	7,000	1,900	7,000
Solid wastes:					
kg/kkg pdt	0	10	–	–	–
lb/ton pdt	0	20	–	–	–

Source: EPA 440/1-73/006

Reference

(1) U.S. Environmental Protection Agency, "Development Document for Proposed Efflu-
     ent Limitations Guidelines and New Source Performance Standards for the Phos-
     phorus Derived Chemicals Segment of the Phosphate Manufacturing Point Source
     Category," *Report EPA 440/1-73/006,* Washington, D.C. (August 1973).

PHOSPHATE FERTILIZER MANUFACTURE

Nearly all phosphatic fertilizers are made from naturally occurring, phosphorus-containing
minerals such as phosphate rock. Because the phosphorus content of these minerals is not
in a form that is readily available to growing plants, the minerals must be treated to convert
the phosphorus to a plant-available form. This conversion can be done either by the process of acidulation or by a thermal process. The intermediate steps of the mining of phosphate rock and the manufacture of phosphoric acid are not included in this section as they
are discussed in other sections of this volume; it should be kept in mind, however, that
large integrated plants may have all of these operations taking place at one location. In
this section phosphate fertilizers have been divided into three categories: [1] normal superphosphate, [2] triple superphosphate, and [3] ammonium phosphate.

Normal Superphosphate

Normal superphosphate (also called single or ordinary superphosphate) is the product resulting from the acidulation of phosphate rock with sulfuric acid. Normal superphosphate
contains from 16 to 22% phosphoric anhydride (P_2O_5). The physical steps involved in making superphosphate are: [1] mixing rock and acid, [2] allowing the mix to assume a solid
form (denning), and [3] storing (curing) the material to allow the acidulation reaction to
be completed. After the curing period, the product can be ground and bagged for sale,
the cured superphosphate can be sold directly as run-of-pile product, or the material can
be granulated for sale as granulated superphosphate.

Triple Superphosphate

Triple superphosphate (also called double or concentrated superphosphate) is the product
resulting from the reaction between phosphate rock and phosphoric acid. The product
generally contains 44 to 52% P_2O_5, which is about three times the P_2O_5 usually found in
normal superphosphates. Presently, there are three principal methods of manufacturing
triple superphosphate. One of these uses a cone mixer to produce a pulverized product
that is particularly suited to the manufacture of ammoniated fertilizers. This product can
be sold as run-of-pile (ROP), or it can be granulated. The second method produces in a
multistep process a granulated product that is well suited for direct application as a phosphate fertilizer. The third method combines the features of quick drying and granulation
in a single step.

Ammonium Phosphate

The two general classes of ammonium phosphates are monammonium phosphate and diam-
monium phosphate. The production of these types of phosphate fertilizers is starting to
displace the production of other phosphate fertilizers because the ammonium phosphates
have a higher plant food content and a lower shipping cost per unit weight of P_2O_5. There
are various processes and process variations in use for manufacturing ammonium phosphates.
In general, phosphoric acid, sulfuric acid, and anhydrous ammonia are allowed to react to
produce the desired grade of ammonium phosphate. Potash salts are added, if desired, and
the product is granulated, dried, cooled, screened, and stored.

Air Pollution

Emission factors for the various processes involved are shown in Table 289.

TABLE 289: EMISSION FACTORS FOR THE PRODUCTION OF PHOSPHATE FERTILIZERS

EMISSION FACTOR RATING: C

Type of product	Particulates[a]		Fluorides[b]	
	lb/ton	kg/MT	lb/ton	kg/MT
Normal superphosphate				
Grinding, drying	9	4.5	–	–
Main stack	–	–	0.15	0.075
Triple superphosphate				
Run-of-pile (ROP)	–	–	0.03	0.015
Granular	–	–	0.10	0.05
Diammonium phosphate				
Dryer, cooler	80	40	c	c
Ammoniator-granulator	2	1	0.04	0.02

[a]Control efficiencies of 99 percent can be obtained with fabric filters.

[b]Total fluorides, including particulate fluorides. Factors all represent outlet emissions following control devices, and should be used as typical only in the absence of specific plant information.

[c]Included in ammoniator-granulator total.

EPA Publication AP-42, 2nd Edition

Normal Superphosphate Manufacture: The gases released from the acidulation of phosphate rock contain silicon tetrafluoride, carbon dioxide, steam, particulates, and sulfur oxoxides. The sulfur oxide emissions arise from the reaction of phosphate rock and sulfuric acid. If a granulated superphosphate is produced, the vent gases from the granulator-ammoniator may contain particulates, ammonia, silicon tetrafluoride, hydrofluoric acid, ammonium chloride, and fertilizer dust. Emissions from the final drying of the granulated product will include gaseous and particulate fluorides, ammonia, and fertilizer dust.

Triple Superphosphate Manufacture: Most triple superphosphate is the nongranular type. The exit gases from a plant producing the nongranular product will contain considerable quantities of silicon tetrafluoride, some hydrogen fluoride, and a small amount of particulates. Plants of this type also emit fluorides from the curing buildings. In the cases where ROP triple superphosphate is granulated, one of the greatest problems is the emission of dust and fumes from the dryer and cooler. Emissions from ROP granulation plants include silicon tetrafluoride, hydrogen fluoride, ammonia, particulate matter, and ammonium chloride. In direct granulation plants, wet scrubbers are usually used to remove the silicon tetrafluoride and hydrogen fluoride generated from the initial contact between the phosphoric acid and the dried rock. Screening stations and bagging stations are a source of fertilizer dust emissions in this type of process.

Ammonium Phosphate Manufacture: The major pollutants from ammonium phosphate production are fluoride, particulates, and ammonia. The largest sources of particulate emissions are the cage mills, where oversized products from the screens are ground before being recycled to the ammoniator. Vent gases from the ammoniator tanks are the major source of ammonia. This gas is usually scrubbed with acid, however, to recover the residual ammonia.

Water Pollution

The eight process operations, sulfuric acid, phosphate rock grinding, wet process phosphoric acid, phosphoric acid concentration, phosphoric acid clarification, normal superphosphate, triple superphosphate, ammonium phosphates, in the phosphate fertilizer subcategory have the following types of water usage and wastes.

  [a] Water treatment plant effluent — includes raw water filtration and clarification, water softening, and water deionization. All these operations serve only to condition the plant raw water to the degree necessary to allow its use for process water and steam generation.
  [b] Closed loop cooling tower blowdown.
  [c] Boiler blowdown.
  [d] Contaminated water.
  [e] Process water.
  [f] Spills and leaks.
  [g] Nonpoint source discharges. These include surface waters from rain or snow that become contaminated.
  [h] Contaminated water (gypsum pond water) treatment.

Each of the above listed types of water usage and wastes is identified as to flow and contaminant content under separate headings (2).

Water Treatment Plant Effluent: Basically only the sulfuric acid process has a water treatment effluent. This 1,300 to 1,670 l/kkg (310 to 400 gal/ton) effluent stream consists principally of only the impurities removed from the raw water (such as carbonate, bicarbonates, hydroxides, silica, etc.) plus minor quantities of treatment chemicals. The degree of water treatment of raw water required is dependent on the steam pressure generated. Generally medium-pressure 9.5 to 52 atmospheres (125 to 750 psig) systems are used and do require rather extensive make-up water treatment. Hot lime-zeolite water treatment is the most commonly used.

There are phosphate complexes particularly along the Mississippi River which use river water both for boiler make-up and process water. In these plants it is necessary to treat the river water through a settler or clarification system to remove the suspended solids present in the river water before conventional water treatment is undertaken. Effluent limitations for water treatment plant effluent components are not covered in this report.

Closed Loop Cooling Tower Blowdown: The cooling water requirements and normal blowdown quantities are listed below. Effluent limits with respect to thermal components and rust and bacteria inhibiting chemicals in cooling tower blowdown or for once-through cooling water are not covered in this report.

Process	- - - - -Circulation Requirement- - - - -		- - - - Discharge Requirement- - - -	
	l/kkg	gal/ton	l/kkg	gal/ton
Sulfuric acid (per ton 100% H_2SO_4)	75,000 to 83,000	18,000 to 20,000	1,670 to 2,500	400 to 600
Rock grinding (per ton rock)	33 to 625	8 to 150	33 to 625*	8 to 150*
Phosphoric acid (per ton P_2O_5)	0 to 19,000	0 to 4,500	0 to 19,000*	0 to 4,500*
Phosphoric acid concentration (per ton P_2O_5)	None	None	None	None
Phosphoric acid clarification (per ton P_2O_5)	690 to 3,200	165 to 770	690 to 3,200*	165 to 770*
Normal super (per ton product)	None	None	None	None
Triple super (per ton product)	None	None	None	None
Ammonium phosphate (per ton product)	None	None	None	None

*Noncontaminated, only temperature increase in discharge water.

Closed loop cooling systems function with forced air and water circulation to affect water cooling by evaporation. Evaporation acts to concentrate the natural water impurities as well as the treatment chemicals required to inhibit scale growth, corrosion, and bacteria growth. Such cooling systems require routine blowdown to maintain impurities at an acceptable operating level. The blowdown quantity will vary from plant to plant and is dependent upon the overall cooling water circulation system.

The quality of the cooling system blowdown will vary with the make-up water impurities and inhibitor chemicals used. The type of process equipment being cooled normally has no bearing on the effluent quality. Cooling is by an indirect (no process liquid contact) means. The only cooling water contamination from process liquids is through mechanical leaks in heat exchanger equipment. Such contamination does periodically occur and continuous monitoring equipment is used to detect such equipment failures. Listed below are the normal ranges of contaminants that may be found in cooling water blowdown systems.

Contaminant	Concentration, mg/l
Chromate	0 to 250
Sulfate	500 to 3,000
Chloride	35 to 160
Phosphate	10 to 50
Zinc	0 to 30
TDS	500 to 10,000
SS	0 to 50
Biocides	0 to 100

Cooling tower blowdown can be treated separately or combined with other plant effluents for treatment. The method to be employed is dependent upon the chemical treatment method used and cost. Those plants which utilize chromate or zinc treatment compounds generally treat the blowdown stream separately to minimize effluent treatment costs.

Boiler Blowdown: The only steam generation equipment in a phosphate complex other than possibly auxiliary package boilers is in the sulfuric acid plant. Medium pressure, 9.5 to 52 atmospheres (125 to 750 psig), steam systems are the most generally used. Boiler blowdown quantities are normally 1,300 to 1,670 l/kkg (310 to 400 gal/ton). Typical contaminant concentration ranges are listed below. Separate effluent limitations for boiler blowdown with respect to both thermal discharge and specific contaminants are not covered in this report.

Contaminant	Concentration, mg/l
Phosphate	5 to 50
Sulfite	0 to 100
TDS	500 to 3,500
Zinc	0 to 10
Alkalinity	50 to 700
Hardness	50 to 500
Silica (SiO_2)	25 to 80

Contaminated Water (Gypsum Pond Water): Contaminated water is used to supply essentially all the water needs of a phosphate fertilizer complex. The majority of U.S. phosphate fertilizer installations impound and recirculate all water which has direct contact with any of the process gas or liquid streams. This impounded and reused water accumulates sizeable concentrations of many cations and anions, but particularly F and P. Concentrations of 8,500 mg/l F and in excess of 5,000 mg/l P are not unusual. Acidity of the water also reaches extremely high levels (pH 1 to 2). Use of such poor quality water necessitates that the process equipment materials of construction be compatible with the corrosive nature of the water.

Contaminated water is used in practically all process equipment in the phosphate subcategory except sulfuric acid manufacturing and rock grinding. The water requirements of such major water using equipment as barometric condensers, gypsum sluicing, gas scrubbing equipment, and heat exchangers are all supplied by contaminated water. Each time the water is reused, the contaminant level is increased. While this contaminated water is a major process effluent, it is not discharged from the complex. The following list indicates ranges of contaminated water usage for each process.

Process	Usage l/kkg	Usage gal/ton
Sulfuric acid	None	None
Rock grinding	None	None
Wet process phosphoric acid	16,400 to 20,800	3,800 to 5,000
NPK process-nitric acid acidulation	1,000 to 2,300	240 to 540
Phosphoric acid concentration	2,500 to 2,600	550 to 570
Phosphoric acid clarification	690 to 1,040	225 to 250
Normal superphosphate	940 to 1,040	225 to 250
Triple superphosphate	660 to 1,040	158 to 250
Ammonium phosphate	5,000 to 6,500	1,200 to 1,500

Make-up Water: Make-up water in a phosphate complex is defined as fresh water untreated except for suspended solids removal. Normally fresh water use to all process units is held to an absolute minimum. Such restraint is necessary because all make-up water used finds its way into the contaminated water system. Excessive fresh water use will therefore needlessly increase the contaminated water inventory beyond the containment capacity. This in turn means contaminated water must undergo costly treatment before discharge to natural drainage whenever such discharge is permitted. Normal ranges of make-up water use are listed below for each of the process units. There is no discharge except into a process stream or to the contaminated water system.

Process	Make-up Water Usage l/kkg	Make-up Water Usage gal/ton
Sulfuric acid	63 to 83	15 to 20
Rock grinding	None	None
Wet process phosphoric acid	None	None
Phosphoric acid concentration	0.8 to 1.6	0.2 to 0.4
Phosphoric acid clarification	None	None
Normal superphosphate	None	None
Triple superphosphate	None	None
Ammonium phosphates	None	None

Spills and Leaks: Spills and leaks in most phosphate fertilizer process units are collected as part of the housekeeping procedure. The collected material is, where possible, reintroduced directly to the process or into the contaminated water system. Spillage and leaks therefore do not normally represent a direct contamination of plant effluent streams that flow directly to natural drainage.

Nonpoint Source Discharge: The primary origin of such discharges is dry fertilizer material which dusts over the general plant area and then dissolves in rain or melting snow. The magnitude of this contaminant source is a function of dust containment, housekeeping, snow/rainfall quantities, and the design of the general plant drainage facilities. No meaningful data were obtained on this intermittent discharge stream.

Contaminated Water (Gypsum Pond Water) Treatment System: The contaminated water treatment system discharge effluent is the only major discharge stream from a phosphoric acid complex other than the water treatment and blowdown streams associated with the sulfuric acid process. Discharge from this system is kept to an absolute minimum due to the treatment cost involved. In fact, several complexes report that they have not treated and discharged water for several years.

The need to treat and discharge water has been previously mentioned to be dependent upon the contaminated water inventory. As a result, water discharged from the treatment system is not done continuously throughout the year. Once the necessity for treatment occurs, however, the flow is continuous for that period of time required to adjust the contaminated water inventory. Normally, this period is 2 to 4 months per year, but is primarily dependent upon the rainfall/evaporation ratio and occurrence of concentrated rainfall such as an abnormal rainy season or a hurricane. Some phosphate fertilizer installations in the Western U.S. perennially have favorable rainfall/evaporation ratios and never have need to treat or discharge water. The quantity of water discharged from the contaminated water treatment system is strictly dependent upon the design of the treatment system and has no direct connection to production tonnage. Contaminated water treatment systems generally have capacities of 2,085 to 4,170 l/min (500 to 1,000 gpm).

The common treatment system is a two-stage liming process. Three main contaminated water parameters, namely pH, F, and P are addressed. Reported ranges for these parameters after treatment are: pH 6 to 9; F 15 to 40 mg/l; and P 30 to 60 mg/l.

Intermedia Aspects

Normal Superphosphate Production: In the production of normal superphosphate gaseous fluorides (principally silicon tetrafluoride) are evolved from various steps in the process. The principal points of fluoride evolution are the mixer, den, and storage area. Overall fluoride evolution will vary between 10 to 50% of the fluoride in the rock. Assuming a 3.5% fluoride concentration in the rock, the fluoride evolved will be between 4 to 20 pounds per ton of product. The amount of fluoride evolved will depend primarily on the concentration of the sulfuric acid used. The higher the acid concentration the greater the fluoride evolution. The average fluoride evolution is about 35% or approximately 14 pounds per ton of product for a rock containing 3.5% fluoride. Typically, the fluoride initially in the rock would be distributed as follows:

Fluoride Disposition	% of Fluoride in Original Rock
Evolved in mixer	2
Evolved in den and conveyors	30
Evolved in storage (curing)	3
Retained in product	65

The fluoride concentration in the offgas streams from the mixer den and conveyors is relatively high, and they are usually combined and scrubbed in a single scrubber system. The gases evolved in the storage building are usually removed from the building by air circulation. The fluoride concentration in the air stream is normally quite low making fluoride removal by scrubbing impractical. Recent improvements in building design reduce the air flow required and have made scrubbing for fluoride removal practical.

The fluoride is scrubbed from the gas stream with water or dilute fluosilicic acid. Some particulate matter is present in the offgas and it plus the silica formed by the reaction of SiF_4 and water complicate scrubber operation. It is difficult to define the composition of the solution exiting the scrubber. Scrubber design, water source, and ultimate fluoride disposal all affect the type and volume of scrub solution used. When the normal superphosphate plant is operated in conjunction with a wet process phosphoric acid plant gypsum pond water can be used as the scrub solution which is then recycled back to the pond after use.

In some instances a small fresh water scrub, low in fluoride, is used in the final scrub stage to reduce the fluoride in the exit gas to an acceptable level. This stream also recycles to the gypsum pond or is used in first stage scrubbing. When pond water is not available, then a holding system is necessary so the scrub liquor can be recycled (normally the scrub water cannot be used on a once-through basis because of the potential for fluoride contamination of water supplies). The concentration of fluosilicic acid in the recycle water is allowed to build to substantial levels. Make-up fresh water is used in the last stage of scrubbing. The acid solution can be recovered for sale as 20 to 25% fluosilicic acid.

When the acid is not recovered for sale a fraction of the recycle liquor must be bled off to maintain the fluosilicic acid concentration. The make-up fresh water replaces the acid removed as a bleed or for sale. The bleed stream represents the principal aqueous effluent from the normal superphosphate plant which operates independent of an acid plant. This stream will carry from 4 to 20 pounds of fluoride per ton of product. The concentration of fluoride in the stream can vary over a wide range depending on the scrubber design but normally will be at least several percent. The volume can vary from 10 to 200 gal/ton of product. By treating the bleed stream for fluoride removal it can be recycled as make-up water or released to the environment. Ammonium fluoride solution can also be used for fluoride scrubbing. The silicon tetrafluoride reacts with the ammonium fluoride to form ammonium fluosilicate:

$$SiF_4 + 2NH_4F \longrightarrow (NH_4)_2SiF_6$$

The process has the advantage that no silica is formed. However, the process has not been used to any extent commercially. In general, the problems of fluoride scrubbing in a normal superphosphate plant are quite similar to those encountered in a wet process phosphoric acid plant. Scrubbing liquors represent the only aqueous waste stream from most normal superphosphate plants, except for spill cleanup. Solid spills are usually recovered if possible. Liquid spills are washed to a sump for collection and sent to the gypsum pond if a pond is available.

Triple Superphosphate Production: Gaseous fluorides are evolved in the production of triple superphosphate. Fluoride evolution is more constant than in normal superphosphate production and varies between 32 to 40% of the fluoride in the rock. For a 3.5% fluoride rock this corresponds to a fluoride evolution of 10 to 12 pounds per ton of product. In run-of-pile triple superphosphate production fluoride is evolved principally in the mixer and conveyor belt (or den) with a minor amount evolved in the storage area. The fluoride distribution for ROP triple superphosphate production has been reported to be as follows:

Fluoride Disposition	% Total Fluorine Initially in Rock
Evolved at mixer, belt and transfer conveyors	31.4 to 33.3
Evolved in storage	2.0 to 2.8
Retained in product	63.9 to 66.6

In addition, when the triple superphosphate is granulated after curing, scrubbing of the fumes from the dryer may be required as small amounts of fluoride may be evolved. When granular triple superphosphate is made directly from rock and phosphoric acid, fluoride evolution occurs principally in the reactor, mixer, dryer, cooler, and screens; the total off-gas must be scrubbed for fluoride removal.

Almost all triple superphosphate plants are operated in conjunction with a wet process phosphoric acid plant. Therefore, unless the fluoride is to be recovered for sale, the scrubber solution is principally pond water plus some fresh water, both of which are returned to the acid plant gypsum ponds after use. Scrubber operation is similar to that encountered in normal superphosphate production, and the waste stream compositions are similar. Fluoride added to the gypsum pond will amount to about 10 to 12 pounds per ton of product. Fluoride can be recovered for sale by using a dilute fluosilicic acid scrub solution, as in normal superphosphate production, if economics justify the recovery.

Ammonium Phosphate Production: In the preparation of ammonium phosphate substantial quantities of ammonia are volatilized from the neutralizer and granulator. Process economics require that the ammonia be recovered. This is normally done by scrubbing the gas stream from the neutralizer-granulator with phosphoric acid. The phosphoric acid scrub combines with the incoming process acid which feeds the neutralizer. A portion of the mixed acid stream recycles back to the scrubber. Scrubbing removes most of the ammonia from the gas stream. However, the phosphoric acid scrub solution contains 1 to 3% fluoride, and optimum scrubber operation for ammonia recovery results in the stripping of some of the fluoride from the acid. The result is a gas stream containing substantial amounts of fluoride and a small amount of ammonia. The fluoride volatilized ranges from 0.1 to 0.3 pounds per ton of product. Air pollution control regulations usually require that the fluoride be scrubbed from the gas stream. Scrubbing is usually accomplished with water. The scrubber flow required varies from 500 to 3,000 gallons per ton of product; the fluoride concentration in the exit scrub water ranges from 25 to 250 mg/l. The ammonia concentration in the scrub water varies from 5 to 50 mg/l. Scrub solution can be recycled with a resulting increase in contaminant concentrations. When the ammonium phosphate plant is operated in conjunction with a wet process phosphoric acid plant the waste scrub solution can be discharged to the gypsum pond. Pond water can be used as the scrub solution.

Some ammonia is also evolved from the dryer. The gas from the dryer also contains substantial amounts of particulate matter. The solids are removed in a cyclone separator and the gas stream is scrubbed with phosphoric acid to recover the ammonia. The scrubbing can occur in the same unit used to scrub the gas from the neutralizer and granulator or in a separate scrubber. Where a separate scrubber is used, a separate tail gas scrubber for fluoride removal may also be required. The scrub solution from this tail gas scrubber would then be combined with that from the other fluoride scrubber.

Process spills are washed to a sump system with water and returned to the process if possible. Otherwise they are normally transferred to a holding pond and then discharged. If a gypsum pond is available the wash solution may discharge to it.

Ammoniated Superphosphate Production: In the production of granular ammoniated superphosphate ammonia and dust particles are evolved from the ammoniator-granulator and dryer. The dust is removed from the gas stream with cyclones. The ammonia is then removed by scrubbing the gas with recycled water. To control the ammonia concentration in the recycled scrub water at the required level a portion of the scrub is bled off and returned to the process. The bleed is usually mixed with the ammoniating solution or returned directly to the ammoniator. When the scrub is not recycled to the process the scrub solution may be used on a once-through basis. The ammonia concentration is quite low and the solution is discharged to a holding pond or receiving waters. Gypsum pond water can be used as the scrubbing medium when available.

When spills occur they are usually washed to a sump system with a minimum of water. The preferred practice is to recycle the material from the sump to the process. However, in some plants the material is discharged to a pond or receiving water. The composition of the wash solution is difficult to predict but usually contains substantial amounts of ammonia and phosphate.

Granular NPK fertilizers can be produced in equipment similar to the equipment used for ammoniated superphosphate production. Waste stream problems are similar to those encountered in ammoniated superphosphate production. Waste streams generated from process spills are more complex since they may contain potassium, chloride, sulfate, nitrate, and urea in addition to ammonia and phosphate.

References

(1) U.S. Environmental Protection Agency, "Compilation of Air Pollutant Emission Factors," 2nd Edition, *Publication Number AP-42,* Research Triangle Park, N.C., Office of Air and Water Programs (April 1973).
(2) U.S. Environmental Protection Agency, "Development Document for Proposed Effluent Limitations Guidelines and New Source Performance Standards for the Basic Fertilizer Chemicals Segment of the Fertilizer Manufacturing Point Source Category," *Report EPA 440/1-73/011,* Washington, D.C. (November 1973).
(3) Fullam, H.T. and Faulkner, B.P., "Inorganic Fertilizer and Phosphate Mining Industries - Water Pollution and Control," *Report 12020 FPD 09/71,* Washington, D.C., U.S. Government Printing Office (September 1971).

PHOSPHATE ROCK PROCESSING

Phosphate rock preparation involves beneficiation to remove impurities, drying to remove moisture, and grinding to improve reactivity. Usually, direct-fired rotary kilns are used to dry phosphate rock. These dryers burn natural gas or fuel oil and are fired countercurrently. The material from the dryers may be ground before storage in large storage silos. Air-swept ball mills are preferred for grinding phosphate rock.

Air Pollution

Although there are no significant emissions from phosphate rock beneficiation plants, emissions in the form of fine rock dust may be expected from drying and grinding operations. Phosphate rock dryers are usually equipped with dry cyclones followed by wet scrubbers. Particulate emissions are usually higher when drying pebble rock than when drying concentrate because of the small adherent particles of clay and slime on the rock. Phosphate rock grinders can be a considerable source of particulates. Because of the extremely fine particle size, baghouse collectors are normally used to reduce emissions. Emission factors for phosphate rock processing are presented in Table 290.

TABLE 290: PARTICULATE EMISSION FACTORS FOR PHOSPHATE ROCK PROCESSING WITHOUT CONTROLS[a]

EMISSION FACTOR RATING: C

Type of source	Emissions	
	lb/ton	kg/MT
Drying[b,c]	15	7.5
Grinding[b,d]	20	10
Transfer and storage[d,e]	2	1
Open storage piles[e]	40	20

[a]Emission factors expressed as units per unit weight of phosphate rock.

[b]References 2 and 3.

[c]Dry cyclones followed by wet scrubbers can reduce emissions by 95 to 99 percent.

[d]Dry cyclones followed by fabric filters can reduce emissions by 99.5 to 99.9 percent.

[e]Reference 3.

Source: EPA Publication AP-42, 2nd Edition

Water Pollution

The effluents produced in the mining and beneficiation of phosphate rock are contained in the water suspensions leaving the washer plant (4). These suspensions are the phosphate slimes and the sands tailings.

The amounts of these effluents relative to the water balance and water flow rates for a typical Florida plant operation processing 36,000 tons per calendar day of phosphate matrix (approximately 6,000 tons P_2O_5 per day) are given in Table 291.

The flows presented correspond to a water usage of about 13 gallons per minute per daily ton of P_2O_5 production assuming a phosphate matrix containing about 16.5% P_2O_5.

TABLE 291: WATER DISTRIBUTION FOR A TYPICAL FLORIDA PLANT OPERATION

	Gallons per Minute	% of Total Water Used
Source for water used:		
Reclaimed from all ponds	57,200	74
Water from deep wells	20,000	26
Estimated water usage:		
Mining (matrix)	10,000	13
Processing	22,000	28
Slimes disposal	39,000	51
Sand tailings	6,200	8

Source: Report 12020 FPD 09/71

It can be noted that the major effluent is that of the slimes which contain a suspension of clays and very fine solids amounting to about 4 to 6% by weight of the slimes stream. A typical analysis of these slime solids is provided in Table 292.

TABLE 292: APPROXIMATE MINERALOGIC AND CHEMICAL COMPOSITION OF PHOSPHATE SLIME SOLIDS

Mineralogical Weight Compostion of Phosphate Slimes Solids

Mineral	Percent
Carbonate fluorapatite	20 to 25
Quartz	30 to 35
Montmorillonite	20 to 25
Attapulgite	5 to 10
Wavellite	4 to 6
Feldspar	2 to 3
Heavy minerals	2 to 3
Dolomite	1 to 2
Miscellaneous	0 to 1

Chemical Composition of Phosphate Slimes Solids

Chemical	Typical Analyses, percent	Range, percent
P_2O_5	9.06	9 to 17
SiO_2	45.68	31 to 46
Fe_2O_3	3.98	3 to 7
Al_2O_3	8.51	6 to 18
CaO	14.00	14 to 23
MgO	1.13	1 to 2
CO_2	0.80	0 to 1
F	0.87	0 to 1
LOI (1000°C)	10.60	9 to 16
BPL	19.88	19 to 37

Source: Report 12020 FPD 09/71

As previously noted, these slimes are impounded in slimes ponds to allow settling and clarification to occur. Clear water is returned from the ponds to the beneficiation plant.

Intermedia Aspects

When phosphate rock is calcined, fluoride is evolved. The evolved species are silicon tetra-fluoride and hydrogen fluoride. The ratio of HF to SiF_4 increases as the calcining tempera-ture is increased. Even at low calcination temperatures, however, hydrogen fluoride is the principal fluoride species. The amount of fluoride evolved depends on the calcining tem-perature. Fluoride evolution does not begin below about 250°C and the rock must be fused to remove more than 90% of the fluoride. The fluoride evolved is scrubbed with water or dilute hydrofluoric acid. Since very little SiF$_4$ is normally present, the scrubber product is principally hydrofluoric acid:

$$HF(g) \; + \; H_2O \; \longrightarrow \; H_3O^+ \; + \; F^-(aq)$$

To effect the required fluoride removal a small flow of fresh or neutralized scrubber water must be used in the final stage of the scrubber. Recycle scrub solution can normally be used in the first stage of scrubbing. If gypsum pond water is available this can be used for scrubbing. If pond water is not available (or recovery of the fluoride is desired) the scrub water is recycled. The concentration of hydrofluoric acid in the scrub solution builds up, and a bleed stream is required to maintain the acid concentration at an acceptable level. The maximum acceptable acid concentration will depend on a number of factors; primarily gas stream fluoride concentration and scrubber design.

The bleed stream can be used to recover salable fluoride products. If not, then the bleed stream must be treated for fluoride removal prior to discharge. The total fluoride evolved will depend on the calcining conditions. For a 3.5% rock the maximum fluoride evolution would be 65 to 67 pounds per ton of rock treated. Fusion would be required to release this amount of fluoride.

References

(1) Stern, A. (ed.), "Sources of Air Pollution and Their Control," *Air Pollution,* Volume 3, 2nd Edition, Academic Press, N.Y., p. 221-222 (1968).
(2) Unpublished data from phosphate rock preparation plants in Florida, Midwest Research Institute (June 1970).
(3) Control Techniques for Fluoride Emissions, internal document, U.S. Environmental Protection Agency, Office of Air Programs, Durham, N.C. p. 4-46, 4-36, and 4.34.
(4) Fullam, H.T. and Faulkner, B.P., "Inorganic Fertilizer and Phosphate Mining Industries - Water Pollution and Control," *Report Number 12020 FPD 09/71,* Washington, D.C., U.S. Government Printing Office (September 1971).

PHOSPHORIC ACID MANUFACTURE

Phosphoric acid is produced by two principal methods, the wet process and the thermal process. The wet process is usually employed when the acid is to be used for fertilizer production. Thermal-process acid is normally of higher purity and is used in the manufacture of high-grade chemical and food products.

Wet Process

In the wet process, finely ground phosphate rock is fed into a reactor with sulfuric acid to form phosphoric acid and gypsum. There is little market for the gypsum and it is handled as waste material. The phosphoric acid is separated from the insolubles by vacuum filtration. The acid is then normally concentrated to about 50 to 55% P$_2$O$_5$. When superphosphoric acid is made, the acid is concentrated to between 70 and 85% P$_2$O$_5$ (1) (2).

Thermal Process

In the thermal process, phosphate rock, siliceous flux, and coke are heated in an electric furnace to produce elemental phosphorus. The gases containing the phosphorus vapors are passed through an electrical precipitator to remove entrained dust. In the one-step version of the process, the gases are next mixed with air to form P$_2$O$_5$ before passing to a water scrubber to form phosphoric acid. In the two-step version of the process, the phosphorus is condensed and pumped to a tower in which it is burned with air, and the P$_2$O$_5$ formed is hydrated by a water spray in the lower portion of the tower (1).

Air Pollution

Emissions of gaseous fluorides, consisting mostly of silicon tetrafluoride and hydrogen fluoride, are the major problems from wet-process acid. Table 293 summarizes the emission factors from both wet-process acid and thermal-process acid. The principal emission from thermal-process acid is P_2O_5 acid mist from the absorber tail gas. Since all plants are equipped with some type of acid-mist collection system, the emission factors presented in Table 293 are based on the listed types of control.

TABLE 293: EMISSION FACTORS FOR PHOSPHORIC ACID PRODUCTION

EMISSION FACTOR RATING: B

Source	Particulates		Fluorides	
	lb/ton	kg/MT	lb/ton	kg/MT
Wet process (phosphate rock)				
Reactor, uncontrolled	–	–	18[a]	9[a]
Gypsum pond	–	–	1[b]	1.1[b]
Condenser, uncontrolled	–	–	20[a]	10[a]
Thermal process (phosphorus burned[c])				
Packed tower	4.6	2.3	–	–
Venturi scrubber	5.6	2.8	–	–
Glass-fiber mist eliminator	3.0	1.5	–	–
Wire-mesh mist eliminator	2.7	1.35	–	–
High-pressure-drop mist eliminator	0.2	0.1	–	–
Electrostatic precipitator	1.8	0.9	–	–

[a]References 2 and 3.
[b]Pounds per acre per day (kg/hectare-day); approximately 0.5 acre (0.213 hectare) is required to produce 1 ton of P_2O_5 daily.
[c]Reference 4.

Source: EPA Publication AP-42, 2nd Edition

Water Pollution

The production of phosphoric acid by the dry process from elemental phosphorus consumes a total of about 380 liters of water per kkg of product (92 gal/ton) for both the hydration and the acid dilution steps. The cooling water requirements are typically 92,000 liters per kkg of product (22,000 gal/ton); but with recycle of cooling water, the make-up cooling water requirement is approximately 4,600 liters per kkg of product (1,100 gal/ton). There is no aqueous process waste from notable phosphoric acid Plants 003, 006, 042, and 075. However, despite good housekeeping at a notable plant, leaks or spills of phosphoric acid may amount to an average of 1 kg/kkg (2 lb/ton), with a range of 0 to 2.5 kg/kkg (0 to 5 lb/ton).

Where food-grade phosphoric acid is produced, a standard raw waste of 0.1 kg/kkg (0.2 lb/ton) of arsenic sulfide is precipitated by addition of a soluble sulfide (H_2S, Na_2S, NaHS), and filtered out of the acid. An additional 0.75 kg/kkg (1.5 lb/ton) of filter-aid material may accompany the sulfide as a solid waste.

In the production of wet process phosphoric acid by the dihydrate process the gypsum by-product ($CaSO_4 \cdot 2H_2O$) represents a major disposal problem. The quantity of gypsum formed is about 1.5 tons per ton of rock processed which corresponds to about 4.5 to 5 tons per ton of P_2O_5 product. In a few cases the gypsum from the acid plant filters is slurried with water and pumped into the ocean for disposal. Usually, however, the slurried gypsum is pumped to a holding pond (normally called a gypsum pond) where the gypsum settles out to give a clear supernate. The gypsum pond usually is a diked area which may be several hundred acres in area.

Two general rules of thumb are that the pond area should be at least 0.5 acre per daily ton of P_2O_5 production, and the deposited gypsum accumulates at the rate of about one acre-ft per year per daily ton of P_2O_5. Normally all the contaminated waste streams in the acid plant are fed to the gypsum pond. In phosphate fertilizer complexes the waste stream from other processes are usually discharged to the gypsum pond as well. The clear supernate from the gypsum pond is recycled back to supply most of the water requirements for the acid plant and other processes. In well-run acid plants recycled pond water supplies over 80% of the gross water requirements of the plant.

As the pond water is recycled, dissolved impurities build up until an equilibrium composition is reached. The typical range of concentrations for the principal impurities in the pond water (at equilibrium) are given in Table 294. During periods of heavy rainfall the pond water composition may vary somewhat, but not to any great degree.

TABLE 294: TYPICAL EQUILIBRIUM COMPOSITION OF GYPSUM POND WATER

Contaminant	Concentration, mg/l
P_2O_5	6,000 to 12,000
Fluoride	3,000 to 5,000
Sulfate	2,000 to 4,000
Calcium	350 to 1,200
Ammonia	0 to 100
Nitrate	0 to 100
pH	1.0 to 1.5

Source: Report 12020 FPD 09/71

Some fresh make-up water is required for certain process operations such as cooling compressors and cooling the tank where the sulfuric acid is diluted prior to entering the digestor. The trend, however, is to reduce fresh water requirements to a minimum. In areas with little rainfall the evaporation of water from the gypsum pond area makes the discharge of pond water to the environment unnecessary. In most areas some pond water must be discharged. The discharge volume is normally dependent on rainfall and will vary widely during different times of the year for a given plant.

In some localities, such as Florida, the volume of water discharged from the pond is very close to the volume of rainfall collected in the pond during the rainy season. Discharge from the pond of a 600 ton P_2O_5/day acid plant can run as high as 5,000 gpm during periods of high rainfall. Over a year period the discharge should average 500 to 1,000 gal/ton of P_2O_5. In many acid plants the gypsum pond discharge is the only plant waste stream requiring treatment.

The soluble fluoride in the pond water, at equilibrium, will have a significant partial pressure. The following values have been reported: partial pressure of $SiF_4 \cong 0.00007$ mm Hg and partial pressure of $HF \cong 0.0008$ mm Hg. This results in a considerable loss of fluoride to the atmosphere on windy days (up to 200 pounds per day from a typical gypsum pond). However, the fluoride concentration in the air is only about 1 ppb.

Large volumes of cooling water are required in wet process phosphoric acid production. Fresh or salt water is sometimes used for cooling on a once-through basis, and the discharge water contains very few contaminants. The water is usually discharged to the receiving water without treatment, except for cooling and pH monitoring. In most cases the cooling water is recycled, and in most plants the gypsum pond water is used for most cooling purposes. Fresh water for cooling is restricted to certain specific purposes as mentioned earlier. In some plants the cooling water circuit is kept separate from the pond water circuit. Cooling ponds or towers are used, the cooling water is recycled, and only the cooling water blowdown is sent to the gypsum pond. Because of its pH and composition, the gypsum pond water is quite corrosive and special materials of construction are required when the pond water is used for cooling.

The product acid (52 to 55% P_2O_5) from the concentrator normally contains 1 to 3% precipitated solids, principally calcium sulfate and metal fluosilicates. With time, additional precipitates form which consist primarily of complex metal phosphates. They usually amount to about 2 to 5%. On storage these precipitates settle out to form a sludge. When the phosphoric acid is used on site the sludge does not present a problem and is usually not removed from the acid. When the acid is to be shipped, however, the sludge can be a major problem since it can accumulate in shipping containers or interferes with the customer's use of the acid.

Washing of the sludge from the shipping containers can create a significant waste stream. Some plants attempt recovery of the sludge from the wash solutions, but normally it is discharged to the gypsum pond if one is available. The preferred approach when acid is to be shipped is to store the acid in tanks for a substantial period of time. The sludge settles out and is removed as an underflow from the tanks by means of rakes. The recovered sludge, which contains substantial amounts of phosphate, is usually disposed of by blending into a fertilizer, usually triple superphosphate.

Significant quantities of steam are used in wet process phosphoric acid production. In many plants the steam is used on a once-through basis. Uncontaminated steam condensate is discharged to the receiving waters without treatment. Steam condensate which is contaminated, such as that from barometric condensers, and vacuum ejectors, is discharged to the gypsum pond. When steam condensate is recycled only the uncontaminated portion is returned to the steam plant. In the phosphoric acid plant spills occur and they are usually cleaned up by washing to a sump system from which they are moved to the gypsum pond.

Furnace Phosphoric Acid Production: The principal waste stream generated in the production of furnace grade phosphoric acid is a waste water stream contaminated with elemental phosphorus, called phossy water. The phossy water can originate from three principal points in the process: phosphorus condensers, the phosphorus storage area, and phosphorus sludge handling. The elemental phosphorus vaporized from the electric furnace passes through dust collectors and is then condensed with water in one or more condensers. The resulting mixture of liquid phosphorus and water is collected in a sump. The liquid phosphorus settles to the bottom of the sump, while the water forms a cover layer over the phosphorus.

Water from the cover layer is recycled back to the condenser system. Impurities will build up in the water and a bleed is necessary to control the impurity level. The impurities consist principally of colloidal phosphorus particles and soluble fluoride scrubbed out of the gas stream in the condenser. When the phosphorus is condensed, most of the phosphorus collects in a relatively pure form. However, some of the phosphorus collects in a heterogeneous mixture called phosphorus sludge, which is composed of phosphorus, water and inorganic solids. About 5 to 10% of the phosphorus produced ends up in the sludge.

As a general rule, about 5% sludge is produced for every 1% of inorganic solids in the condensed phosphorus. The solids in the sludge are composed of 40 to 60% SiO_2, 5 to 15% CaO, 2 to 4% Fe_2O_3, 1 to 3% Al_2O_3, and 2 to 5% P_2O_5. They will normally amount to between 1 and 5% in the crude phosphorus. The sludge can be separated from the clean phosphorus by gravity difference. When this is done the separated sludge is diluted with water and sent to a centrifuge to separate the contained phosphorus. The water discharged from the centrifuge is a second source of phossy water. The liquid phosphorus is stored under water to prevent oxidation. When the phosphorus is moved, water must be added to or removed from the storage tanks. This water is contaminated with phosphorus and is the third potential source of phossy water. Usually, the various sources of phossy water are combined and handled as a single waste stream. The composition and volume of the phossy water can vary significantly. Typical range of compositions and volume are shown in Table 295.

TABLE 295: TYPICAL PHOSSY WATER COMPOSITION AND VOLUME *

Contaminant	mg/l
Phosphorus	1,000 to 2,000
Phosphate	50 to 500
Fluoride	100 to 1,000
Suspended solids	2,000 to 4,000

*1,000 to 2,000 gal/ton P_2O_5, pH 3 to 6

Source: Report 12020 FPD 09/71

When slag is tapped from the furnace, some fluoride, P_2O_5, and SO_2 may be evolved from the molten slag. If the slag is air cooled the evolution of gas is quite low, amounting to about one pound of total fluoride, SO_2 and P_2O_5 per ton of P_2O_5. When the slag is water quenched fluoride evolution increases and amounts to about 2.5 pounds of fluoride discharge per ton of P_2O_5 product. If the slag is treated to expand it, the fluoride evolution will increase to about eight pounds per ton of P_2O_5 product. When the slag is air cooled scrubbing of the fumes may not be necessary. In other cases, scrubbing will usually be required in order to meet air pollution control regulations. Fluorides are usually scrubbed from the fumes with water; and the water discharging the scrubber can contain up to several thousand ppm fluoride and P_2O_5 if the scrubber water is recycled.

Preparation of the phosphate rock for charging to the furnace may require agglomeration by calcination. This can lead to the generation of a fluoride-containing gas stream which must be scrubbed for fluoride removal. The scrubbing operation and waste stream generated are similar to those discussed in the previous section on phosphate rock calcination. The preparation of phosphoric acid via elemental phosphorus requires large volumes of cooling water (25,000 to 45,000 gal/ton P_2O_5 produced). If water is readily available it can be used on a once-through basis. The once-through cooling water effluent is normally low in impurities and can be discharged to the receiving waters without treatment except for reduction of the temperature to an acceptable level.

When water is in limited supply the cooling water must be recycled. The accumulation of impurities requires a periodic blowdown of the cooling water. The cycles of concentration used normally range from three to seven depending on the purity of the make-up water. The composition of the blowdown water can vary over wide ranges. Typically, the ranges of impurities in the water are as shown in Table 296.

TABLE 296: COMPOSITION OF FURNACE GRADE PHOSPHORIC ACID PLANT COOLING WATER BLOWDOWN*

Impurity	mg/l
Fluoride	10 to 500
Phosphate	10 to 1,000
Sulfate	500 to 5,000
Chromate	0 to 100
Heavy metals	0 to 20
NH_3	0 to 5
Nitrates	0 to 5
SO_2	10 to 100

(continued)

TABLE 296: (continued)

Impurity	mg/l
TDS	500 to 10,000
Oil	0 to 10
Other organics	0 to 100
Suspended solids	50 to 200
pH	5.5 to 6.0

*Cooling water blowdown volume 250 to
1,000 gal/ton P_2O_5

Source: Report 12020 FPD 09/71

Table 297 (5) summarizes some of the characteristics of raw wastes from phosphoric acid manufacture.

TABLE 297: SUMMARY OF RAW WASTE FROM PHOSPHORIC ACID MANUFACTURE

	H_3PO_4, (75%)
Phossy water:	
P_4 concentration, ppm	1,700
l/kkg P_4 consumed	580
kg P_4/kkg P_4 consumed	1
gal/ton P_4 consumed	140
lb/ton P_4 consumed	2
Process water wasted:	
l/kkg pdt	–
gal/ton pdt	–
Raw waste load, kg/kkg pdt:	
HCl	–
H_2SO_3	–
$H_3PO_3 + H_3PO_4$	1
Raw waste load, lb/ton pdt:	
HCl	–
H_2SO_3	–
$H_3PO_3 + H_3PO_4$	2
Concentrations, mg/l:	
HCl	–
H_2SO_3	–
$H_3PO_3 + H_3PO_4$	High
Process water consumed:	
l/kkg pdt	380
gal/ton pdt	92
Cooling water used:	
l/kkg pdt	91,000
gal/ton pdt	22,000
Solid wastes, kg/kkg pdt:	
As compounds	0.1
Total residues	–
Solid wastes, lb/ton pdt:	
As compounds	0.2
Total residues	–

Source: Report EPA 440/1-73/006

Intermedia Aspects

When phosphate rock is reacted with sulfuric acid to make phosphoric acid the fluoride in the rock is also attacked by the acid and freed. The distribution of the fluoride between the various process streams can vary significantly depending on the equipment design used. Table 298 shows the fluoride distribution normally encountered in the dihydrate process.

TABLE 298: FLUORIDE DISTRIBUTION IN WET PROCESS PHOSPHORIC ACID PRODUCTION

Fluoride Disposition	--------- % of Total Fluoride in Rock- ---------			
	Typical Range	Typical Values		ERCO
Evolved from reactor and filters	4 to 7	5	5.5	5
Evolved from concentrator	35 to 45	40	41.9	60 to 65
Remaining in gypsum	25 to 30	30	27.8	20 to 25
Remaining in acid product	20 to 30	25	24.8	10

Source: Report 12020 FPD 09/71

Air pollution control regulations require that the fluoride evolved from the reactor, filters, and concentrator be removed before the gas stream is discharged to the atmosphere. The fluoride is usually scrubbed from the gas streams using water or dilute fluosilicic acid. When the fluoride concentration in the gas stream is high, as in the water vapor-fluoride stream from the acid concentrator, surface condensers are sometimes used. The fluoride is evolved as silicon tetrafluoride and hydrogen fluoride. Silicon tetrafluoride is the predominant species in the vapors from the reactor and filters. The vapor from the concentrator contains both HF and SiF_4. When silicon tetrafluoride is scrubbed it reacts with the water to form fluosilicic acid and silica:

$$3SiF_4 + 2H_2O \longrightarrow 2H_2SiF_6 + SiO_2$$

When a mixture of SiF_4 and HF is scrubbed the HF and SiF_4 combine in the scrub medium to form H_2SiF_6 but no silica is formed if the $HF\text{-}SiF_2$ ratio is 2:1. If the ratio is greater than 2:1 free HF is present in the scrub solution. If the ratio is less than 2:1 some silica is formed. If the fluoride is not to be recovered for sale, gypsum pond water is used to scrub the fluoride from the gas stream. The scrubber solution recycles to the gypsum pond and accounts for a major portion of the fluoride that builds up in the pond water system. Fluoride recovery in the scrubbers usually exceeds 95% and in some localities air pollution control regulations make greater than 98% recovery necessary.

The partial pressure of SiF_4 and HF in the pond water makes it difficult to reach these recovery levels and scrubber design can be quite complex. In some instances a small fresh water stream is used in the final stage of scrubbing to reduce the fluoride in the exit gas to an acceptable level. This stream is then used for earlier scrubbing stages, or is sent directly to the gypsum pond. When economics dictate, fluoride may be recovered as fluosilicic acid for sale or for conversion to other fluoride products. When recovery is practiced the fluoride is scrubbed from the gas stream using a dilute solution of fluosilicic acid. The acid solution is recycled and the concentration can build up as high as 25% H_2SiF_6. A continuous side stream is removed from the recycle system as product to maintain the acid concentration at the desired level.

Normally, a two stage scrubber system is used since the recycle acid solution will not reduce the fluoride in the exit gas to the desired level. The fluosilicic acid solution is used in the first stage scrubbing and pond water or fresh water is used in the second stage. Fluoride recovery for sale is usually practiced on the gas stream from the concentrator. In most cases greater than 80% of the fluoride is recovered for sale from the gas stream, and most of the remaining fluoride enters the gypsum pond water system. The fluoride concentration in the gas stream from the reactor and filters is quite low and fluoride recovery is less practical. When recovery is practiced the fluosilicic acid concentration in the scrubber solution rarely exceeds 10%. Overall fluoride recovery for sale is normally less than 50% of that in the gas stream.

References

(1) Duprey, R.L., "Compilation of Air Pollutant Emission Factors," U.S. DHEW, PHS, National Center for Air Pollution Control, Durham, N.C. *PHS Publication Number 999-AP-42,* p. 16 (1968).
(2) "Atmospheric Emissions from Wet-Process Phosphoric Acid Manufacture", U.S. DHEW, PHS, EHS, National Air Pollution Control Administration, Raleigh, N.C., *Publication Number AP-57* (April 1970).
(3) "Control Techniques for Fluoride Emissions," internal document, U.S. EPA, Office of Air Programs, Research Triangle Park, N.C. (1970).
(4) "Atmospheric Emissions from Thermal-Process Phosphoric Acid Manufacturing," Cooperative Study Project: Manufacturing Chemists' Association, Inc. and Public Health Service, U.S. DHEW, PHS, National Air Pollution Control Administration, Durham, N.C., *Publication Number AP-48,* (October 1968).
(5) Fullam, H.T. and Faulkner, B.P., "Inorganic Fertilizer and Phosphate Mining Industries - Water Pollution and Control," *Report Number 12020 FPD 09/71,* Washington, D.C., U.S. Government Printing Office (September 1971).
(6) U.S. Environmental Protection Agency, "Development Document for Proposed Effluent Limitations Guidelines and New Source Performance Standards for the Phosphorus-Derived Chemicals Segment of the Phosphate Manufacturing Point Source Category," *Report EPA 440/1-73/006,* Washington, D.C. (August 1973).

PHOSPHORUS MANUFACTURE

Phosphorus manufacturing technology involves the following streams emanating from the process (in addition, of course, to the phosphorus product stream): by-products—slag, ferrophosphorus, and carbon monoxide; noncontact cooling water; electrostatic precipitator dust; calciner precipitator dust; calciner and furnace fume scrubber liquor; phosphorus condenser liquor (aqueous phase); phosphorus sludge (or mud); and, slag quench liquor. The following sections discuss each of the above in quantitative detail, and identify which are typically returned to the process and which are classified as raw waste streams from the manufacturing operation. The by-products of the phosphorus manufacturing operation are:

	kg/kkg	lb/ton
Ferrophosphorus	300	600
Slag ($CaSiO_3$)	8,900	17,800
CO gas	2,800	5,600

Both ferrophosphorus and slag are sold, and the carbon monoxide is either used to generate heat in the process or is otherwise burned on site. Hence, none of the above three materials is considered a waste. The quench water used for the by-product slag is separately discussed as a waste stream. The by-product ferrophosphorus is cast as it is tapped from the furnace and air-cooled. The solids are then broken up and shipped. No water is used specifically for ferrophosphorus, and there are no wastes accountable for ferrophosphorus manufacture.

Air Pollution

Electrostatic Precipitator Dust: The high-temperature electrostatic precipitator removes dusts from the furnace gases before these gases are condensed for recovery of phosphorus. These dusts may contain up to 50% P_2O_5, and therefore find value either as a fertilizer for sale or for return to the process. In the latter case, it is transported to the ore blending head end of the plant. One TVA scheme slurries the dust for transport; the slurry is pumped to a settling pond, the settled solids are fed to the ore-blending unit, and the pond overflow is reused in the slurrying operation. The quantity of precipitator dust is approximately 125 kg/kkg of product (250 lb/ton). Regardless of the method of sale or reuse, the precipitator dust is not a waste material to be disposed of from the plant.

Calciner Precipitator Dust: Dry Dust collectors are used in the calcining operation, up-stream of wet scrubbing systems. The dry fine dusts collected are recycled directly to the sizing and calcining operations. The collected and recycled fines may amount to as much as 30% of the net production from the nodulizing process. There is no plant discharge of dry calciner precipitator dusts; therefore this is not a component of the plant's raw waste load.

Water Pollution

Noncontact Cooling Water: Phosphorus production facilities generate huge quantities of heat. The electrical power consumption is approximately 15,500 kwh/kkg (48 million Btu/ton). An additional 8,100 kwh/kkg (25 million Btu/ton) are generated by combustion of the by-product carbon monoxide. Some of this energy, 6,100 kwh/kkg (19 million Btu/ton), is absorbed in the endothermic furnace reaction, and some is absorbed by the endothermic calcining operation. Other portions of this energy are released to the atmosphere by burning of waste carbon monoxide and by convection, radiation and evaporative losses from the equipment and process materials. Still other portions are absorbed by contact waters in the calcining and furnace from scrubbers, in the phosphorus condenser, and in the slag quenching operation.

After accounting for the above energy demands, a significant quantity of heat is absorbed by noncontact cooling water for the furnace shell, the crucible bottom, the fume hood, the tap holes, the electrode fixtures, the electrical transformer, and for any indirect phosphorus condensation. The quantity of this water is highly variable from plant-to-plant, and depends upon the furnace design, the furnace size and the degree of recirculation (through heat exchangers with other water streams or through cooling towers), whether or not cooling water is used in series for different requirements, the inlet temperature of the available cooling water, and the ambient air temperature. One plant uses 325,000 l/kkg of product (78,000 gal/ton); another uses 38,000 l/kkg (9,000 gal/ton); and TVA at Muscle Shoals, Alabama, uses 130,000 l/kkg (31,000 gal/ton).

Calciner and Furnace Fume Scrubber Liquor: Water scrubbers are used for air pollution abatement for the calciner exhaust stream (downstream of dry dust collection), for furnace fumes, for ore sizing dusts, for coke material handling dusts, for raw material feeding operation dusts, and for furnace taphole (slag and ferrophosphorus) fumes. The scrubber liquor contains suspended solids (which are mainly SiO_2 and Fe_2O_3), some phosphates and sulfates as dissolved solids, and a large quantity of fluorides. The average quantity of F in ore is 275 кg/kkg of P_4 (550 lb/ton). Approximately 8% of this quantity of F, or 22 kg (44 lb), is volatilized in the ore calcining operation, and is subsequently a constituent of the scrubber liquor.

This scrubber liquor is highly acidic for three reasons: the sulfur (as SO_3) forms sulfuric acid; the P_2O_5 forms phosphoric acid; and the fluorine, which is released in the form of silicon tetrafluoride, forms fluosilicic acid and silicic acid upon hydrolysis. The quantity of scrubber liquor wasted depends upon the degree of recirculation of this liquor from a sump back to the scrubbers. TVA at Muscle Shoals circulates approximately 21,000 l/kkg of product (5,000 lb/ton) with a portion bled off to control the composition. This scrubber liquor is of the following composition:

Constituent	Concentration, %
F	3.1
SiO_2	1.1
P_2O_5	0.2
Fe_2O_3	0.1
S	1.7

If the fluoride concentration of 3.1% is equated to a standard raw waste load (as previously discussed) of 22 kg/kkg (44 lb/ton), the quantities of other scrubber liquor components may be calculated:

Constituent	Raw Waste Load kg/kkg	Raw Waste Load lb/ton
F	22	44
SiO_2	8	16
P_2O_5	1.5	3
Fe_2O_3	0.5	1
S	12	24

The total $CaCO_3$ acidity of the scrubber liquor, calculated from the above constituent quantities, is 60 kg/kkg (120 lb/ton). Other plants do not recirculate scrubber liquor; the volume wasted is much greater and the constituent concentrations are much smaller, but the raw waste loads (in kg/kkg of product) should be comparable. Plant 181 does not directly recirculate the liquor, and uses 300,000 l/kkg (71,000 gal/ton) for scrubbing.

Phosphorus Condenser Liquor: The furnace gases pass from the electrostatic dust precipitator to the phosphorus condenser, where a recirculating water spray condenses the product. The condenser liquor is maintained at approximately 60°C (140°F), high enough to prevent solidification of the phosphorus [freezing point 44°C (112°F)]. This condenser liquor is phossy water, essentially a colloidal dispersion of phosphorus in water, since the solubility at 20°C (68°F) is only 3.0 mg/l. Depending upon how intimate the water/phosphorus contact was, the phosphorus content of phossy water may be as high as several weight percent.

The condenser liquor also contains constituents other than elemental phosphorus: fluoride, phosphate, and silica. Using the average F content of ore of 275 kg/kkg, plus the estimate that 12% of the F in the ore volatilizes in the furnace and is therefore equivalent to 33 kg/kkg (66 lb/ton), and by accounting for 6 kg of F per kkg (12 lb/ton) which is collected in the precipitator dust and in the phosphorus sludge ash, a raw waste load of F is derived of 27 kg/kkg (54 lb/ton) in the condenser liquor. This condenser liquor is not acidic despite the hydrolysis of P_2O_5 and SiF_4 to H_3PO_4, H_2SiF_6, and H_2SiO_3 because aqueous ammonia or caustic is added to prevent undue corrosion in the condenser.

There are other sources of phossy water within the plant. Storage tanks for phosphorus have a water blanket, which is discharged upon phosphorus transfer. Railroad cars are cleaned by washing with water. Phosphorus may be purified by washing with water. Together, all sources of phossy water wastes amount to about 5,400 l/kkg (1,300 gal/ton), and at a concentration of 1,700 mg/l, the quantity of phosphorus wastes amounts to about 9 kg/kkg produced (18 lb/ton), as reported by TVA.

At TVA, the condenser liquor is recirculated at the rate of 33,000 l/kkg (8,000 gal/ton). Other plants may differ significantly in the quantity of phossy water circulated, but the raw wastes (in kg/kkg of product) should be fairly uniform. For example, Plant 181, which does not directly recirculate its condenser water, uses 84,000 l/kkg (20,000 gal/ton), with an additional 17,000 l/kkg (4,000 gal/ton) for phosphorus handling and storage. To calculate the raw waste loads of phosphate and silica in the condenser liquor, the following TVA recirculated-liquor composition was used:

Constituent	Concentration, %
F	8.3
P_2O_5	5.0
SiO_2	4.2

Equating 8.3% F with the previously-derived 27 kg/kkg of F, the raw waste loads of P_2O_5 and SiO_2 become (respectively) 16.5 kg/kkg (33 lb/ton) and 13.5 kg/kkg (27 lb/ton).

Phosphorus Sludge: In addition to phossy water, the phosphorus condenser sump also collects phosphorus sludge, which is a colloidal suspension typically 10% dust, 30% water and 60% phosphorus. The quantity of sludge formed is directly dependent upon the quantity of dust that escapes electrostatic precipitation; hence the very large investment made for highly efficient precipitators. Using 125 kg of dust (per kkg of product) collected by the electrostatic precipitator, and assuming a 98% collection efficiency, the dust reaching the condenser amounts to 2.5 kg/kkg (5 lb/ton). If all of this dust became part of the sludge, the sludge quantity would be 25 kg/kkg (50 lb/ton) of product, and it would contain 15 kg/kkg (30 lb/ton) of elemental phosphorus.

This sludge is then universally processed for recovery of phosphorus, typically by centrifugation. A 96% recovery has been reported, with the product (subsequently returned to the process) averaging 92 to 96% phosphorus. The remaining 4% of the phosphorus in the sludge is burned in a phosphoric acid unit, so that no wastes emanate from the plant. Other methods for processing the sludge which also result in no plant effluent include heating in a slowly rotating drum in an inert atmosphere to drive off phosphorus vapor, which is then condensed with a water spray into a sump. The solid residue obtained is completely free of elemental phosphorus and can be safely landfilled or recycled to the feed preparation section of the plant.

Slag Quenching Liquor: Slags from phosphorus furnaces are mainly SiO_2 and CaO, and would also contain Al_2O_3, K_2O, Na_2O, and MgO in amounts consistent with the initial ore composition. In addition to these oxides, phosphate rock may contain 0.1 to 0.2 kg/kkg (0.2 to 0.4 lb/ton) of uranium in the ore, and the radiation levels of both the slag and the quench waters must be appropriately noted. Other constituents of the slag presenting problems for quench water pollution control are fluoride and phosphate. Approximately 80% of the original F in the phosphate rock, 220 kg/kkg of P_4 (440 lb/ton) winds up in the slag. About 2.7% of the original P_2O_5 in the phosphate rock, 70 kg/kkg (140 lb/ton), wind up in the slag. At a typical plant, approximately 24,600 l/kkg (5,900 gal/ton) may be used for quenching the slag, with the slag quench liquor having the following composition and raw waste loads:

Constituent	Concentration, mg/l	Raw Waste Load kg/kkg P_4	lb/ton P_4
Total suspended solids	800	20	40
Total dissolved solids	1,700	42.5	85
Phosphates (as P)	12	0.3	0.6
Sulfate (as S)	1,000	25	50
Fe	14	0.35	0.7
F	170	4.5	9
Total Alkalinity	230	5.5	11

Phossy Water Wastes: Because phosphorus is transported and stored under a water blanket, phossy water is a raw waste material at phosphorus using plants as well as at phosphorus producing plants. The standard procedure when liquid phosphorus is transferred from a rail car to the using plant's storage tank is to pump the displaced phossy water from the storage tank back into the emptying rail car. Instead of being wasted at the phosphorus-using plant, the phossy water is shipped back to the phosphorus-producing facility for treatment and/or reuse. Therefore, standard raw phossy water wastes at the phosphorus using plants are due to surges or to anomalies in the storage tank water level control system rather than to the direct wasting of all displaced water.

A more insidious source of phossy water may arise at phosphorus consuming plants. Should reactor contents containing phosphorus ever be dumped into a sewer as a result of operator error, emergency conditions or inadvertent leaks or spills, the phosphorus would remain at the low points in the sewer line generally as a solid [melting point 44°C (111°F)] and would contact all water flowing in that sewer from that time on. Since phosphorus burns when exposed to air (autoignition temperature 93°C), there is general reluctance to clean it out; the common practice is to ensure a continuous water flow to prevent fire. The typical phosphorus loss for phosphorus consuming plants is 1 kg lost to phossy water per kkg consumed (2 lb/ton). Whenever phosphorus is transferred by displacement, 580 liters of water are displaced per kkg of phosphorus (140 gal/ton). These values are equivalent to a phosphorus concentration of 1,700 mg/l. For comparison, a typical phosphorus content in phossy water at a phosphorus-producing plant has also been reported at 1,700 mg/l. Table 299 summarizes some of the characteristics of wastes from phosphorus manufacture.

TABLE 299: SUMMARY OF RAW WASTES FROM PHOSPHORUS MANUFACTURE

	Calciner Scrubber Liquor	Phosphorus Condenser Plus Other Phossy Water	Slag Quenching Water	Composite Waste
Wastewater quantity:				
l/kkg	300,000	100,000	25,000	425,000
gal/ton	72,000	24,000	6,000	102,000
Raw waste load, kg/kkg:				
TSS	8.5	13.5	20.5	42.5
P_4	–	9	–	9
PO_4	2	22	1	25
SO_4	36	–	75	111
F	22	27	4.5	53.5
Total acidity	60	–	–	54.5
Total alkalinity	–	–	5.5	–
Raw waste load, lb/ton:				
TSS	17	27	41	85
P_4	–	18	–	18
PO_4	4	44	2	50
SO_4	72	–	150	222
F	44	54	9	107
Total acidity	120	–	–	109
Total alkalinity	–	–	11	–
Concentrations, mg/l:				
TSS	28	135	820	100
P_4	–	90	–	21
PO_4	7	220	40	59
SO_4	120	–	3,000	260
F	73	270	180	126
Total acidity	200	–	–	128
Total alkalinity	–	–	220	–

Note: Waste water quantities and constituent concentrations are highly variable, depending upon degree of recirculation, but the raw waste loads should be representative.

Source: Report EPA 440/1-73/006

Reference

(1) U.S. Environmental Protection Agency, "Development Document for Proposed Effluent Limitations Guidelines and New Source Performance Standards for the Phosphorus Derived Chemicals Segment of the Phosphate Manufacturing Point Source Category," *Report EPA 440/1-73/006*, Washington, D.C. (August 1973).

PHOSPHORUS OXYCHLORIDE MANUFACTURE

Phosphorus oxychloride, used in the preparation of organic phosphate esters and pharmaceuticals, is made by the reaction of liquid phosphorus trichloride, chlorine and solid phosphorus pentoxide, according to the equation:

$$3PCl_3 + 3Cl_2 + P_2O_5 \longrightarrow 5POCl_3$$

The process is carried out in a batch still. An alternative route involves air oxidation of PCl_3 in a batch reactor with a reflux condenser.

Water Pollution

The water scrubber for the distillation operation in the standard process (using P_2O_5 and Cl_2) typically collects 1.5 kg of HCl (anhydrous basis) and 0.25 kg of H_3PO_4 (100% basis) per kkg of product $POCl_3$ (3 and 0.5 lb/ton), and the scrubber for $POCl_3$ transferring collects about 0.2 kg of HCl and 0.15 kg of H_3PO_4 per kkg of product (0.4 and 0.3 lb/ton). Allowing for small wastes from returnable container cleaning operations, the standard raw waste load is 2 kg of HCl and 0.5 kg of H_3PO_4 per kkg of product (4 and 1 lb/ton). Approximately 2,500 l/kkg (600 gal/ton) of water are used, so that the raw waste concentrations are 800 mg/l HCl and 200 mg/l H_3PO_4.

These waste quantities for $POCl_3$ manufacture are somewhat smaller than for PCl_3 manufacture since $POCl_3$ is less volatile (boiling point 107°C). In the batch process, the refluxing liquid is all PCl_3 at the start, but becomes increasingly richer in $POCl_3$. The air-oxidation process presents a much more difficult task for the reflux condenser, since the vapors are highly diluted with noncondensibles. However, with the use of refrigerated condensers, the measured raw waste load is no different for this process. At Plant 037, data collected over three months from the reactor/still scrubber for $POCl_3$ manufacture, which had an estimated flow rate of 1,800 l/kkg (430 gal/ton), had average net values of: 669 mg/l chloride and 1,213 mg/l $CaCO_3$ acidity.

These data reduce to a raw waste of 1.2 kg/kkg (2.4 lb/ton) of HCl plus 0.35 kg/kkg (0.7 lb/ton) of H_3PO_4; which are extremely close to the corresponding values for Plant 147. Where product $POCl_3$ is filtered, the used filter elements are first washed to hydrolyze the residual $POCl_3$. Disposable elements are then landfilled. The quantity of filtered solids retained on the elements is only a very small fraction of the weight of the used element. The elements are washed in a 55-gallon drum, so that a very small quantity of wastewater (and of acid wastes) is involved compared to the scrubber waste load.

Although there is no continuous withdrawal of residues from $POCl_3$ distillations, very little residue accumulates. Twice a year, this residue (mostly glassy phosphates) is washed out with hot water. The noncontact cooling water requirement for $POCl_3$ manufacture by either the standard or the alternate method is approximately 50,000 l/kkg (12,000 gal/ton). Table 300 summarizes the raw wastes from $POCl_3$ manufacture.

TABLE 300: SUMMARY OF RAW WASTE FROM $POCl_3$ MANUFACTURE

Process water wasted:	
l/kkg pdt	2,500
gal/ton pdt	600
Raw waste load, kg/kkg pdt:	
HCl	2
H_2SO_3	–
$H_3PO_3 + H_3PO_4$	0.5
Raw waste load, lb/ton pdt:	
HCl	4
H_2SO_3	–
$H_3PO_3 + H_3PO_4$	1
Concentrations, mg/l:	
HCl	800
H_2SO_3	–
$H_3PO_3 + H_3PO_4$	200
Process water consumed:	
l/kkg pdt	–
gal/ton pdt	–

(continued)

TABLE 300: (continued)

Cooling water used:	
l/kkg pdt	50,000
gal/ton pdt	12,000
Solid wastes, kg/kkg pdt:	
As compounds	-
Total residues	<0.05
Solid wastes, lb/ton pdt:	
As compounds	-
Total residues	<0.1

Source: Report EPA 440/1-73/006

Reference

(1) U.S. Environmental Protection Agency, "Development Document for Proposed Effluent Limitations Guidelines and New Source Performance Standards for the Phosphorus Derived Chemicals Segment of the Phosphate Manufacturing Point Source Category," *Report EPA 440/1-73/006,* Washington, D.C. (August 1973).

PHOSPHORUS PENTASULFIDE MANUFACTURE

The standard process for the manufacture of phosphorus pentasulfide is by direct union of the elements, according to the equation:

$$P_4 + 10S \longrightarrow 2P_2S_5$$

Water Pollution

The water seals on the batch reactor vent lines accumulate a mixture of phosphorus mud and lower phosphorus sulfides. These seals are cleaned once a week, and the residue amounts to 0.15 kg/kkg (0.3 lb/ton). This residue is hazardous and flammable, and is typically buried. Should any batch be aborted (a rare occurrence) because of agitator failure, cast-iron pot failure or other reason, the material is disposed of by incineration. The dust collected by a cyclone from the P_2S_5 crushing operation amounts to 1 kg/kkg (2 lb/ton).

The still pot for the vacuum distillation step accumulates impurities, which include carbon and iron sulfur compounds and glassy phosphates. Most important, the residues contain arsenic pentasulfide, which is higher-boiling than the corresponding phosphorus pentasulfide. Arsenic occurs naturally with phosphorus (they are both Group V-A elements) at a level of about 0.075 kg/kkg (0.15 lb/ton), of arsenic which is equivalent to 0.05 kg of As_2S_5 per kkg of product P_2S_5 (0.1 lb/ton). The entire still pot residue is about 0.5 kg/kkg (1 lb/ton). Periodically, these residues are removed and the solids are broken up and buried. Approximately 17,000 l/kkg (4,000 gal/ton) of noncontact cooling water is used.

In the casting of liquid P_2S_5, the fumes from burning liquid (molten P_2S_5 autoignited) are scrubbed. Typically, the scrubber water contains 1.25 kg of combined P_2O_5 and SO_2 per kkg of product P_2S_5 (2.5 lb/ton). Because both P_2O_5 and SO_2 are absorbed by a water scrubber only with difficulty, the water flow rate is high, 30,000 l/kkg (7,200 gal/ton). These values reduce the concentrations of PO_2^{-3} and SO_3^{-2} in the scrubber effluent to 17 and 34 mg/l (respectively). Much lower scrubber flow rates could be used should weak caustic or lime be used instead of water. Table 301 summarizes waste solution from P_2S_5 manufacture.

TABLE 301: SUMMARY OF RAW WASTE FROM PHOSPHORUS PENTASULFIDE
MANUFACTURING PLANTS

Phossy water:	
P_4 concentration, ppm	1,700
l/kkg P_4 consumed	580
kg P_4/kkg P_4 consumed	1
gal/ton P_4 consumed	140
lb/ton P_4 consumed	2
Process water wasted:	
l/kkg product	30,000
gal/ton pdt	7,200
Raw waste load, kg/kkg pdt:	
HCl	–
H_2SO_3	1
$H_3PO_3 + H_3PO_4$	0.5
Raw waste load, lb/ton pdt:	
HCl	–
H_2SO_3	2
$H_3PO_3 + H_3PO_4$	1
Concentrations, mg/l:	
HCl	–
H_2SO_3	34
$H_3PO_3 + H_3PO_4$	17
Process water consumed:	
l/kkg product	–
gal/ton product	–
Cooling water used:	
l/kkg product	16,600
gal/ton pdt	4,000
Solid wastes, kg/kkg pdt:	
As compounds	0.05
Total residues	0.7
Solid wastes, lb/ton pdt:	
As compounds	0.1
Total residues	1.4

Source: Report EPA 440/1-73/006

Reference

(1) U.S. Environmental Protection Agency, "Development Document for Proposed Effluent Limitations Guidelines and New Source Performance Standards for the Phosphorus Derived Chemicals Segment of the Phosphate Manufacturing Point Source
Category," *Report EPA 440/1-73/006,* Washington, D.C. (August 1973).

PHOSPHORUS PENTOXIDE MANUFACTURE

To make phosphorus pentoxide, liquid phosphorus is burned in an excess of dry air, according to the equation:

$$P_4 + 5O_2 \longrightarrow 2P_2S_5$$

The product is condensed in a condensation tower or water-cooled barn.

Water Pollution

The wastewater from the tail seals on the condensing towers typically contain 0.25 kg/kkg

(0.5 lb/ton) of H_3PO_4 (100% basis). Approximately 500 l/kkg (120 gal/ton) of water may be used, resulting in a concentration of 470 mg/l for the effluent bleed. The inlet air dryer silica gel is regenerated often, but is renewed very infrequently (perhaps every 10 years). The wasted material is typically landfilled. Approximately 29,000 l/kkg (7,000 gal/ton) of noncontact cooling water is used as noted in Table 302 which summarizes raw waste production from P_2O_5 manufacture (1).

TABLE 302: SUMMARY OF RAW WASTE FROM PHOSPHORUS PENTOXIDE MANUFACTURING

Phossy water:	
P_4 concentration, ppm	1,700
l/kkg P_4 consumed	580
kg P_4/kkg P_4 consumed	1
gal/ton P_4 consumed	140
lb/ton P_4 consumed	2
Process water wasted:	
l/kkg pdt	500
gal/ton pdt	120
Raw waste load, kg/kkg pdt:	
HCl	–
H_2SO_3	–
$H_3PO_3 + H_3PO_4$	0.25
Raw waste load, lb/ton pdt:	
HCl	–
H_2SO_3	–
$H_3PO_3 + H_3PO_4$	0.5
Concentrations, mg/l:	
HCl	–
H_2SO_3	–
$H_3PO_3 + H_3PO_4$	470
Process water consumed:	
l/kkg pdt	–
gal/ton pdt	–
Cooling water used:	
l/kkg pdt	29,000
gal/ton pdt	7,000
Solid wastes, kg/kkg pdt:	
As compounds	–
Total residues	–
Solid wastes, lb/ton pdt:	
As compounds	–
Total residues	–

Source: Report EPA 440/1-73/006

Reference

(1) U.S. Enivironmental Protection Agency, "Development Document for Proposed Effluent Limitations Guidelines and New Source Performance Standards for the Phosphorus Derived Chemicals Segment of the Phosphate Manufacturing Point Source Category," *Report EPA 440/1-73/006,* Washington, D.C. (August 1973).

PHOSPHORUS TRICHLORIDE MANUFACTURE

Phosphorus trichloride is made directly from the elements ($P_4 + 6Cl_2 \longrightarrow 4PCl_3$) in a jacketed batch reactor. Chlorine is bubbled through the charge and PCl_3 is refluxed.

Excess chlorine is avoided to prevent PCl_5 formation.

Water Pollution

The batch or semicontinuous reactor/stills accumulate residues which are periodically but infrequently removed. These residues contain arsenic trichloride, which is higher-boiling than the corresponding phosphorus trichloride. Arsenic occurs naturally with phosphorus (they are both Group V-A elements) at a level of about 0.075 kg/kkg (0.15 lb/ton) of arsenic, which is equivalent to 0.05 kg of $AsCl_3$ per kkg of product PCl_3 (0.1 lb/ton). This is about half the quantity of total residue in the stills (exclusive of residual PCl_3 from the last batch or run before shutdown).

The average noncontact cooling water requirement is 54,000 l/kkg (13,000 gal/ton). Water scrubbers collect PCl_3 vapors from the reaction, the product distillation, the product storage, and the product transfer operations, and hydrolyze these vapors to HCl and to H_3PO_3 (which may subsequently be oxidized to H_3PO_4). The quantity of PCl_3 collected is highly dependent upon the efficiency of the upstream condensers, since PCl_3 is highly volatile:

Temperature, °C	Temperature, °F	PCl_3 Vapor Pressure, mm Hg
20	68	99
40	104	235
60	140	690
76	169	760

At the plant studied (1), sufficient heat transfer area was provided in the condensers to limit the raw waste load to 3 kg of HCl plus 2.5 kg of H_3PO_3 per kkg of product PCl_3 (6 and 5 lb/ton). Approximately 5,000 l/kkg (1,200 gal/ton) of scrubber water were used to collect these wastes. Other smaller waste quantities of HCl and H_3PO_3 generated from tank car and returnable container cleaning operations have been included in these quantities.

These quantities are based upon the most reliable data available; overall material balances of product PCl_3 shipped vs elemental phosphorus received. These data, validated over long periods of time for profitability purposes, show a total loss of phosphorus trichloride of 5 kg/kkg (10 lb/ton). An estimated breakdown of this loss is:

Transfer and storage of phosphorus	1.0 kg/kkg (2 lb/ton)
Reactor/still residues	0.1 kg/kkg (0.2 lb/ton)
Scrubber for distillation tail gases	2.5 kg/kkg (5 lb/ton)
Transfer of PCl_3	1.0 kg/kkg (2 lb/ton)

Other than the estimated loss of elemental phosphorus and the reactor/still residues, the losses which become water-borne raw wastes amount to 3.5 kg/kkg (7 lb/ton). Upon hydrolysis, this stoichiometrically becomes 3 kg/kkg (6 lb/ton) of HCl plus 2.5 kg/kkg (5 lb/ton) of H_3PO_3.

These material-balance data have been used because of their long-term confirmation. Direct measurements of wastewater flow rates and of wastewater constituent analysis were not relied upon in this case since accurate flow rate measurements were not possible in the existing plant configuration and since no statistically meaningful analytical data had been collected.

The acid wastes from washing tank cars and tank trucks, and from washing used $POCl_3$ filter elements, are very small. Water use data, supplemented by independent analyses of the wastewater, yielded the results in Table 303. Total raw waste generated in truck-loading, in tank-car cleaning, and in filter-element washing is 0.014 kg/kkg (0.028 lb/ton) of HCl plus 0.003 kg/kkg (0.007 lb/ton) of total phosphates. Table 304 summarizes the overall raw waste production picture for PCl_3 manufacture.

TABLE 303: MINOR WASTES FROM A PLANT MANUFACTURING PCl₃ AND POCl₃

	Truck-Loading Vent Scrubber	Tank Car Cleanout Water	Filter Element Washout Drum
Water use:			
l/kkg	8.8	10.5	0.46
gal/ton	2.1	2.5	0.11
Constituent analysis, mg/l:			
Chloride	340	715	6,480
Total PO_4	260	26	590
Total acidity	660	–	18,200
Raw waste load, kg/kkg:			
Chloride	0.0030	0.0075	0.0030
Total PO_4	0.0023	0.0003	0.0003
Total acidity	0.0058	–	0.0083
Raw waste load, lb/ton:			
Chloride	0.006	0.015	0.006
Total PO_4	0.005	0.001	0.001
Total acidity	0.012	–	0.017

TABLE 304: SUMMARY OF RAW WASTE FROM PHOSPHORUS TRICHLORIDE MANUFACTURING

Phossy water:	
P_4 concentration, ppm	1,700
l/kkg P_4 consumed	580
kg P_4/kkg P_4 consumed	1
gal/ton P_4 consumed	140
lb/ton P_4 consumed	2
Process water wasted:	
l/kkg pdt	5,000
gal/ton pdt	1,200
Raw waste load, kg/kkg pdt:	
HCl	3
H_2SO_3	–
$H_3PO_3 + H_3PO_4$	2.5
Raw waste load, lb/ton pdt:	
HCl	6
H_2SO_3	–
$H_3PO_3 + H_3PO_4$	5
Concentrations, mg/l:	
HCl	600
H_2SO_3	–
$H_3PO_3 + H_3PO_4$	500
Process water consumed:	
l/kkg pdt	–
gal/ton pdt	–
Cooling water used:	
l/kkg pdt	54,000
gal/ton pdt	13,000
Solid wastes, kg/kkg pdt:	
As compounds	0.05
Total residues	0.05
Solid wastes, lb/ton pdt:	
As compounds	0.1
Total residues	0.1

Source: Report EPA 440/1-73/006

Reference

(1) U.S. Environmental Protection Agency, "Development Document for Proposed Effluent Limitations Guidelines and New Source Performance Standards for the Phosphorus Derived Chemicals Segment of the Phosphate Manufacturing Point Source Category," *Report EPA 440/1-73/006*, Washington, D.C. (August 1973).

PHTHALIC ANHYDRIDE MANUFACTURE

Phthalic anhydride is produced primarily by oxidizing naphthalene vapors with excess air over a catalyst, usually V_2O_5. o-Xylene can be used instead of naphthalene, but it is not used as much. Following the oxidation of the naphthalene vapors, the gas stream is cooled to separate the phthalic vapor from the effluent (1)(2). Phthalic anhydride crystallizes directly from this cooling without going through the liquid phase. The phthalic anhydride is then purified by a chemical soak in sulfuric acid, caustic, or alkali metal salt, followed by a heat soak. To produce 1 ton of phthalic anhydride, 2,500 lb of naphthalene and 830,000 standard cubic feet of air are required (or 1,130 kg of naphthalene and 23,500 standard cubic meters of air to produce 1 MT of phthalic anhydride.

Air Pollution

The excess air from the production of phthalic anhydride contains some uncondensed phthalic anhydride, maleic anhydride, quinones, and other organics. The venting of this stream to the atmosphere is the major source of organic emissions. Table 305 presents emission factor data from phthalic anhydride plants.

TABLE 305: EMISSION FACTORS FOR PHTHALIC ANHYDRIDE PLANTS[a]

EMISSION FACTOR RATING: E

Overall plant	Organics (as hexane)	
	lb/ton	kg/MT
Uncontrolled	32	16
Following catalytic combustion	11	5.5

[a]Reference 3.

Source: EPA Publication AP-42, 2nd edition.

References

(1) Duprey, R.L., "Compilation of Air Pollutant Emission Factors," U.S. DHEW, PHS, National Center for Air Pollution Control, Durham, N.C., *PHS Publication Number 999-AP-42*, p. 17 (1968).
(2) "Phthalic Anhydride," *Kirk-Othmer Encyclopedia of Chemical Technology*, Volume 15, 2nd edition, New York, John Wiley and Sons, Inc. p. 444-485 (1968).
(3) Bolduc, M.J., et al, "Systematic Source Test Procedure for the Evaluation of Industrial Fume Converters," presented at 58th Annual Meeting of the Air Pollution Control Association, Toronto, Canada (June 1965).

PLASTICS INDUSTRY

The manufacture of most resins or plastics begins with the polymerization or linking of the basic compound (monomer), usually a gas or liquid, into high molecular weight noncrystalline solids. The manufacture of the basic monomer is not considered part of the plastics industry and is usually accomplished at a chemical or petroleum plant. The manufacture of most plastics involves an enclosed reaction or polymerization step, a drying step, and a final treating and forming step. These plastics are polymerized or otherwise combined in completely enclosed stainless steel or glass lined vessels. Treatment of the resin after polymerization varies with the proposed use.

Resins for moldings are dried and crushed or ground into molding powder. Resins such as the alkyd resins that are to be used for protective coatings are normally transferred to an agitated thinning tank, where they are thinned with some type of solvent and then stored in large steel tanks equipped with water cooled condensers to prevent loss of solvent to the atmosphere. Still other resins are stored in latex form as they come from the kettle.

Air Pollution

The major sources of air contamination in plastics manufacturing are the emissions of raw materials or monomers, emissions of solvents or other volatile liquids during the reaction, emissions of sublimed solids such as phthalic anhydride in alkyd production, and emissions of solvents during storage and handling of thinned resins (1). Emission factors for the manufacture of plastics are shown in Table 306.

TABLE 306: EMISSION FACTORS FOR PLASTICS MANUFACTURING WITHOUT CONTROLS[a]

EMISSION FACTOR RATING: E

Type of plastic	Particulate		Gases	
	lb/ton	kg/MT	lb/ton	kg/MT
Polyvinyl chloride	35[b]	17.5[b]	17[c]	8.5[c]
Polypropylene	3	1.5	0.7[d]	0.35[d]
General	5 to 10	2.5 to 5	—	—

[a]References 2 and 3.
[b]Usually controlled with a fabric filter efficiency of 98 to 99 percent.
[c]As vinyl chloride.
[d]As propylene.

Source: EPA Publication AP-42, 2nd edition.

Much of the control equipment used in this industry is a basic part of the system and serves to recover a reactant or product. These controls include floating roof tanks or vapor recovery systems on volatile material, storage units, vapor recovery systems (adsorption or condensers), purge lines that vent to a flare system, and recovery systems on vacuum exhaust lines.

Water Pollution

The "Industrial Waste Study of the Plastics and Synthetics Industry" by Celanese Research Company, the Manufacturing Chemists Association survey of the industry, and plant visits by EPA and their representatives provided ranges of pollutants occurring in the different product subcategories of the industry (4). The reported ranges of raw waste loads vary all the way from 0 to 135 units/1,000 units of product for BOD_5, from 0 to 334 for COD, and from 0 to 70 for suspended solids. Data from the above sources are recorded in Tables 307 and 308 for wastewater flows, BOD_5, COD and SS for each of the product subcategories. Other elements, compounds, and parameters which are reported in the wastes from the industry are summarized in Table 309.

Information on raw waste loads for these parameters was not available from the industry with the exception of zinc from rayon manufacture. This range is reported in Table 308. The other elements and compounds listed in Table 309 were based on surveys of the Corps of Engineers permit applications for discharge of wastewaters from a number of plants in the plastics and synthetics industry, reviews with personnel in regional EPA offices, the "Industrial Waste Study of the Plastics Materials and Synthetics Industry" by the Celanese Research Company, the EPA Interim Guideline Document, discussions with industry representatives, literature data on process operations, and internal industrial technical consultants (4).

TABLE 307: WASTEWATER LOADING FOR THE PLASTICS AND SYNTHETICS INDUSTRY

Product	Wastewater Loading (gal/1000 lb.) Observed Flow	Reported Range	Wastewater Loading (cu m/tonne) Observed Flow	Reported Range
Polyvinyl Chloride--Suspension				
Polyvinyl Chloride--Emulsion	1800	(300-5000)	15.0	2.5-41.72
Polyvinyl Chloride--Bulk				
ABS/SAN	2060	(200-3500)		1.67-24.0
Polyvinyl Acetate	1000	(0-3000)	8.3	0-25.03
Polystyrene--Suspension	1100	(0-17,000)	9.2	0-141.8
Polystyrene--Bulk				
Polypropylene	1000	(300-8000)	8.3	2.50-66.7
Lo Density Polyethylene	2130	(0-5,000)	17.8	C-41.72
Hi Density Polyethylene--Solvent	3500	(0-3700)	29.2	0-30.87
Hi Density Polyethylene--Polyform				
Cellophane	29400	(12,000-67,000)	245	100-559
Rayon	16500	((4000-23,000)	138	33.38-191
Polyester Resin	540	(0-20,000)	4.5	0-167
Polyester Resin and Fiber				
Nylon 66 Resin	11250	(0-18,250)	10.4	0-152.3
Nylon 66 Resin and Fiber				
Cellulose Acetate Resin	5000	(2000-50,000)	41.7	16.69-417
Cellulose Acetate Fiber				
Epoxy	430	(300-610)	3.62	2.5-5.1
Phenolics	1480	(60-2400)	12.34	0.5-20
Urea Resins	220		1.8	
Melamine	160		1.3	
Acrylics	3400	(300-6160)	28.4	2.50-50
Nylon 6 Resin and Fiber	6500		54.2	
Nylon 6 Resin	82			

Source: Report EPA 440/1-73/010

TABLE 308: PLASTICS AND SYNTHETICS INDUSTRY RAW WASTE LOADS

All Units Expressed as Kg/kkg (lb/1000 lb of Production)

PRODUCT	BOD$_5$ Reported Range	BOD$_5$ Observed Value	COD Reported Range	COD Observed Value	SS Reported Range	SS Observed Value
Polyvinyl Chloride	0.1 - 48	5.7	0.2-100	25	1 - 30	30
ABS/SAN	2 - 20.7	20.7	5 - 33.5	33.5	0 - 30	8
PVAcetate	0 - 2	1.4	0 - 3	--	0 - 2	--
Polystyrene	0 - 2.2	1.0	0 - 6.0	--	0 - 8.4	--
Polypropylene	0 - 10		0 - 20			
LDPE	0.2 - 4.4	4.4	0.2- 54		0 - 4.1	0.4
HDPE	0 - 1	1.0	0 - 2.4	12	0 - 3.4	
Cellophane	20 -133	22	40 -334	--	6 - 70	48
Rayon (Zinc: 12-50)	20 - 45	22	33 -100	55	--	--
Polyester	3 - 20	20	6 - 45	26	0 - 12	0.7
Nylon 6 & 66 Resins	1 -135	66 <21	1 -300	·66 <14	0 - 8	--
Nylon 6 & 66 Fibers	0.1 - 60	6 <10, 5	0.2- 90	8	0.1- 6	(66)0.2
Cellulose Acetate	6 - 70	55	11 -100	--	2 - 20	--
Expoxy	57 - 82	--	30 -127	--	5 - 24	--
Phenolic Resin	15 - 51	--	90 - 64	--	0.5- 7	--
Urea Resin	--	--	--	--	--	--
Melamine	--	--	--	--	--	--
Acrylics	10 - 40	26	10 - 70	45	0.1 - 1.7	1.7

Source: Report EPA 440/1-73/010

TABLE 309: OTHER ELEMENTS, COMPOUNDS, AND PARAMETERS

Phenolic compounds	Copper
Nitrogen compounds (organic, ammonia, and nitrate nitrogen)	Zinc
	Iron
Phosphates	Titanium
Oil and grease	Cobalt
Dissolves solids	Cadmium
pH	Manganese
Color	Aluminum
Turbidity	Magnesium
Alkalinity	Molybdenum
Temperature	Nickel
Sulfides	Vanadium
Cyanides	Antimony
Mercury	Toxic organic chemicals
Chromium	

Source: Report EPA 440/1-73/010

References

(1) "Air Pollutant Emission Factors," Final Report, Resources Research, Inc., Reston, Va., prepared for National Air Pollution Control Administration, Durham, N.C., (April 1970).
(2) Unpublished data from industrial questionnaire. U.S. DHEW, PHS, National Air Pollution Control Administration, Division of Air Quality and Emissions Data, Durham, N.C. (1969).
(3) Private communication between Resources Research, Incorporated, and Maryland State Department of Health, Baltimore, Maryland (November 1969).
(4) U.S. Environmental Protection Agency, "Development Document for Proposed Effluent Limitations Guidelines and New Source Performance Standards for the Synthetic Resins Segment of the Plastics and Synthetic Materials Manufacturing Point Source Category," Report EPA 440/1-73/010, Washington, D.C. (August 1973).

POTASSIUM CHLORIDE PRODUCTION

Water Pollution

The production of potassium chloride by flotation beneficiation generates three waste streams: a clay slimes waste, a solid sodium chloride waste, and a brine bleed for impurity control. The sylvinite ore may contain up to 1.5% insoluble clay materials. The clay is separated from the NaCl-KCl mixtures as a slime. The slime is washed to remove KCl, dewatered, and discharged to a tailing dump. On a dry weight basis, the clay slimes accumulate at the rate of 30 to 80 lb/ton of KCl product depending on the clay content of the ore.

In some cases the sodium chloride from the process may be sold. Normally, however, the sodium chloride is dewatered to recover the KCl-containing brine and then discharged to the tailings dump. Transfer of the NaCl to the dump is usually as a saturated water slurry. The NaCl discharged can amount to as much as 1.5 tons/ton of KCl product depending on the NaCl-KCl ratio in the sylvinite ore. In some plants the clay slimes and sodium chloride waste are combined and sent to a single tailings dump. In order to control the level of impurities in the recycle brine solution it is necessary to bleed off a portion of the brine from the system on a continuous basis.

The discarded brine is saturated with sodium chloride and potassium chloride. It will also contain lesser amounts of magnesium, calcium, and sulfate, as well as small amounts of organic materials which are added to the brine in the flotation circuit. It is the presence of the magnesium, calcium, and sulfate which makes the bleed necessary. The bleed volume is held to a minimum to reduce the KCl loss. The volume of bleed required depends on the concentration of the critical impurities in the sylvinite ore, and can range as high as 100 gal/ton of product.

In the production of potassium chloride from sylvinite ore by solution-recrystallization the same three waste streams are produced. The only difference is that the brine bleed for the solution-recrystallization process will not contain the organics used in the flotation process. In the production of potassium chloride from brines at Bonneville and Searles Lake any aqueous waste streams generated are not a problem. They are simply returned to the brine source.

Reference

(1) Fullam, H.T. and Faulkner, B.P., "Inorganic Fertilizer and Phosphate Mining Industries—Water Pollution and Control," *Report 12020 FPD 09/71,* Washington, D.C., U.S. Government Printing Office (September 1971).

POTASSIUM DICHROMATE MANUFACTURE

Water Pollution

Potassium dichromate is prepared by reaction of potassium chloride with sodium dichromate. Potassium chloride is added to the dichromate solution, which is then pH-adjusted, saturated, filtered and vacuum cooled to precipitate crystalline potassium dichromate. The product is recovered by centrifugation, dried, sized and packaged. The mother liquor from the product centrifuge is then concentrated to precipitate sodium chloride which is removed as a solid waste from a salt centrifuge. The process liquid is recycled back to the initial reaction tank. The raw wastes from potassium dichromate manufacture are listed below. These are crystalline sodium chloride and filter aids which are solid wastes and are hauled away for landfill disposal by a contractor.

Waste Products	Process Source	Product, kg/kkg (lb/ton)
NaCl	Centrifuge	400 (800)
Filter aid	Filter	0.85 (1.7)

All process waters are recycled. The only wastes currently discharged emanate from contamination of once through cooling water used on the barometric condensers on the product crystallizer. Presently, the only effluent from this plant is cooling water, possibly contaminated with hexavalent chromium in the barometric condenser. Replacement of the condenser with a noncontact heat exchanger will eliminate cooling water contamination, although a larger amount of water will have to be used for the less efficient noncontact heat exchanger. With this change, no process waters will be discharged, however.

Reference

(1) United States Environmental Protection Agency, "Development Document for Proposed Effluent Limitations Guidelines and New Source Performance Standards for the Major Inorganic Products Segment of the Inorganic Chemicals Manufacturing Point Source Category," *Report EPA 440/1-73/007,* Washington, D.C. (August 1973).

POTASSIUM SULFATE MANUFACTURE

Water Pollution

The bulk of the potassium sulfate manufactured in the United States is prepared by reaction of potassium chloride with dissolved langbeinite (potassium sulfate-magnesium sulfate). The langbeinite is mined and crushed and then dissolved in water to which potassium chloride is added. Partial evaporation of the solution produces selective precipitation of potassium sulfate which is recovered by centrifugation or filtration from the brine liquor, dried and sold. The remaining brine liquor is either discharged to an evaporation pond, reused as process water or evaporated to dryness to recover magnesium chloride. The fate of the brine liquor is determined by the saleability of the magnesium chloride by-product (depending on ore quality) and the cost of water to the plant. The table below presents a list of the raw wastes expected for potassium sulfate manufacture.

Waste Product	Process Source	Product, kg/kkg (lb/ton)	
		Average	Range
Muds (silica, alumina, clay and other in- solubles)	Dissolution of langbeinite ore	15 - 30 (30 - 60)	—
Brine liquor (saturated magnesium chloride solution)	Liquor remaining after removal of potassium sulfate	—	0 - 2,000*

*Part of the magnesium chloride is recovered for sale and part of the remaining brine solution is recycled for process water.

The high value corresponds to the case of no recycling or recovery of magnesium chloride. These brines contain about 33% solids. The wastes consist of muds from the ore dissolution and waste magnesium chloride brines and are not affected by startup or shutdown. The latter brine can sometimes be used for magnesium chloride production if high grade langbeinite ore was used. Composition of the brine solutions after potassium sulfate recovery is: potassium, 3.19%; sodium, 1.3%; magnesium, 5.7%; chloride, 18.5%; sulfate, 4.9%; and water, 66.7%.

The amount of brine produced is about 650 kg of solids/kkg of potassium sulfate (1,300 pounds per ton) after evaporation. For higher grade ores, the sodium content is lower. The data presented above were supplied by plant 118 (1). The muds listed above are separated from the brine solutions at an exemplary plant by filtration after dissolution of the langbeinite ore. These are recovered and disposed of as landfill on the plant site.

The brine wastes, containing mostly magnesium chloride, are either disposed of or treated in three different manners: evaporation with recovery of magnesium chloride for sale, this is practiced only when high grade ores are processed; reuse of the brine solution in the process in place of using process water, this is normally done to a considerable extent; and disposal of the brines in evaporation pits. At plant 118, all three of the above options are practiced, depending on the quality of the ore being processed. Water use at plant 118 is described below.

Water Inputs

Type	Quantity		Water Purity
	m³ /day (MGD)	l/kkg (gal/ton)	
Well water	3,790 (1.0)	8,360 (2,000)	40 mg/l total solids

(continued)

Water Flows

- - - - - - - - - - Quantity- - - - - - - - - - -

Type	m³/day (MGD)	l/kkg (gal/ton)	% Recycled
Cooling	13,600 (3.6)	30,000 (7,200)	60 to 70% (remainder evaporated)
Process	2,270 (0.6)	5,010 (1,200)	67% recycled, 33% lost either by evaporation or removal from system with product or by-product.

There are no effluent streams from the plant since much of the water is recycled. Most of the water losses occur during the process evaporation steps.

Reference

(1) U.S. Environmental Protection Agency, "Development Document for Proposed Effluent Limitations Guidelines and New Source Performance Standards for the Major Inorganic Products Segment of the Inorganic Chemicals Manufacturing Point Source Category," *Report EPA 440/1-73/007*, Washington, D.C. (August 1973).

POULTRY PROCESSING

Water Pollution

As described by Gurnham & Associates, pollution control consultants of Chicago, Illinois in an EPA sponsored seminar on upgrading poultry processing facilities to reduce pollution at Little Rock, Arkansas in January 1973, the waterborne wastes from the poultry industry may be summarized as shown in Table 310.

TABLE 310: TYPES OF WATERBORNE WASTES FROM THE POULTRY INDUSTRY

Process Wastes	Ancillary Wastes
Manure	Cleanup wastes
Blood	Sanitary sewage from plant and office personnel
Feathers	Water treatment wastes
Offal	Boiler blowdown wastes
Viscera	Spent cooling waters
Heads, necks, lungs, bones	Kitchen and cafeteria wastes
Fats and greases	Laundry wastes
Decomposed residues of the above	Stormwater and runoff:
	from live bird storage areas
	from parking and trucking areas
	from roofs and other areas

Source: Gurnham & Associates, EPA Technology Transfer Seminar, Little Rock, Arkansas (January 1973)

From the previous table, which is essentially a definition of wastes in industry terms, one progresses to an expression of wastes in the standard parameters of pollution control technology such as are used in waste control regulations. Table 311 gives such a listing in pollution control terminology.

TABLE 311: MAJOR POLLUTANTS FROM THE POULTRY INDUSTRY

Most significant parameters:
  *Total suspended matter
  Biochemical oxygen demand (BOD)
  Grease

Other Significant Parameters:
  Temperature
  Color
  Odor
  pH value
  Acidity and alkalinity
  Turbidity
  *Settleable matter
  *Dissolved matter
  *Total residue on evaporation
  Chemical oxygen demand (COD)
  Total organic carbon (TOC)
  Ammonia nitrogen
  Total Kjeldahl nitrogen
  Phosphate
  Total coliform
  Fecal coliform

*The items starred may be further classified as Volatile or Fixed.

Source: Gurnham & Associates, EPA Technology Transfer Seminar, Little Rock, Arkansas,
         (January 1973)

Table 312 presents annual poultry slaughter statistics and future slaughter projections. The slaughter projections were determined by the use of multiple linear regression techniques based on 15 yr of past slaughter statistics provided by the Department of Agriculture. The BOD wasteload projections were functions of the total slaughter and the per unit wasteload, which in turn was based upon an assumed distribution of technology. Wastewater projections were done in the same manner. From 1970 to 1971 both production and gross wasteloads are shown to increase, while total wastewater volume decreases slightly.

TABLE 312: ANNUAL POULTRY SLAUGHTER—WASTELOADS AND WASTEWATER

Year	Millions of Birds Slaughtered	BOD Millions of lbs./yr.	Total Wastewater MG/yr.
1963	2,419	105	18,206
1966	2,668	125	19,350
1968	2,941	135	23,524
1969	3,051	141	24,411
1970	3,262	147	26,098
1971	3,274	153	24,557
1972	3,385	159	25,388
1977 (est.)	3,976	189	29,817

Source: FWPCA Publication IWP-8

This decrease occurs because it was assumed that by 1971 all processing plants of major significance were using advanced technology. It will be remembered that advanced technology plants have lower wastewater volume per unit of product than typical technology plants have.

Reference

(1) Jones, H.R., *Pollution Control in Meat, Poultry and Seafood Processing,* Park Ridge, New Jersey, Noyes Data Corporation (1974).

POWER PLANTS

Air Pollution

A number of processes have been proposed for removing particulate and SO_2 emissions from stack gases. Some of these processes have been suggested for potential application in fossil fuel power plants. In general the SO_2-removal processes can be categorized as follows.

Alkali scrubbing using calcium carbonate or lime with no recovery of SO_2.
Alkali scrubbing with recovery of SO_2 to produce elemental sulfur or sulfuric acid.
Catalytic oxidation of SO_2 in hot flue gases to sulfur trioxide for sulfuric acid formation.
Dry bed absorption of SO_2 from hot flue gases with regeneration and recovery of elemental sulfur.
Dry injection of limestone into the boiler furnace for removal of SO_2 by gas-solid reaction.

The removal of particulate from stack gases can also be carried out separately, using an electrostatic precipitator or a dry mechanical collector. Wet scrubbing for SO_2 removal can be applied subsequently. Power plants are the largest single source of both particulate and SO_2 emissions. For details of the amount of each of these pollutants from power plants, the reader is referred to the sections in this volume on Particulates (page 117) and Sulfur Oxides (page 159).

Water Pollution

Chemical wastes produced by a steam electric power plant can result from a number of operations at the site. Some wastes are discharged more or less continuously as long as the plant is operating. Some wastes are produced intermittently, but on a fairly regularly scheduled basis such as daily or weekly, but which are still associated with the production of electrical energy.

Other wastes are also produced intermittently, but at less frequent intervals and are generally associated with either the shutdown or startup of a boiler or generating unit. Additional wastes exist that are essentially unrelated to production but depend on meteorological or other factors. Wastewaters are produced relatively continuously from the following sources (where applicable): cooling water systems, ash handling systems, wet scrubber air pollution control systems, and boiler blowdown.

Wastewater is produced intermittently, on a regular basis, by water treatment operations which utilize a cleaning or regenerative step as part of their cycle (ion exchange, filtration, clarification, evaporation). Wastewater produced by the maintenance cleaning of major units of equipment on a scheduled basis either during maintenance shutdown or during startup of a new unit may result from boiler cleaning (water side), boiler cleaning (fire side), air preheater cleaning, stack cleaning, cooling tower basin cleaning and cleaning of miscellaneous small equipment. The efficiency of a power plant depends largely on the cleanliness of its heat transfer surfaces. Internal cleaning of this equipment is usually done by chemical means, and requires strong chemicals to remove deposits formed on these surfaces. Actually the cleaning is not successful unless the surfaces are cleaned to bare metal, and this means in turn that some metal has to be dissolved in the cleaning solution. Cleaning of other facilities is accomplished by use of a water jet only. Rainfall runoff results in drainage from coal piles, floor and yard drains, and from construction activity.

A diagram indicating sources of chemical wastes in a fossil fueled steam electric power plant is shown in Figure 35. A simplified flow diagram for a nuclear plant is shown in Figure 36. Heat input to the boiler comes from the fuel. Recycled condensate water, with some pretreated makeup water, is supplied to the boiler for producing steam. Makeup requirements depend upon boiler operations such as blowdown, steam soot blowing and steam losses. The quality of this makeup water is dependant upon raw water quality and boiler operating pressure. For example, in boilers where operating pressure is below 2,800 kilowatts per square meter (400 psi), good quality municipal water may be used without pretreatment.

On the other hand, modern high pressure, high temperature boilers need a controlled high quality water. The water treatment includes such operations as lime soda softening, clarification, ion exchange, etc. These water treatment operations produce chemical wastes. According to the FPC, the principal chemical additives reported for boiler water treatment are phosphate, caustic soda, lime and alum.

As a result of evaporation, there is a build up of total dissolved solids (TDS) in the boiler water. To maintain TDS below allowable limits for boiler operation, a controlled amount of boiler water is sometimes bled off (boiler blowdown). The steam produced in the boiler is expanded in the turbine generator to produce electricity. The spent steam proceeds to a condenser where the heat of vaporization of the steam is transferred to the condenser cooling system. The condensed steam (condensate) is recycled to the boiler after pretreatment (condensate polishing) if necessary, depending upon water quality requirements for the boiler. As a result of condensate polishing (filtration and ion exchange), wastewater streams are created.

In a nonrecirculating (once-through) condenser cooling system, warm water is discharged without recycle after cooling. The cool water withdrawn from an ocean, lake, river, estuary or ground water source may generate biological growth and accumulation in the condenser thereby reducing its efficiency. Chlorine is usually added to once-through condenser cooling systems to minimize this fouling of heat transfer surfaces. Chlorine is therefore a parameter which must be considered for nonrecirculating cooling water systems.

Cooling devices such as cooling towers are employed in the recirculating cooling systems. Bleed streams (blowdown) must generally be provided to control the build up of certain or all dissolved solids within the recirculating evaporative cooling systems. These streams may also contain chlorine and other additives. According to the FPC the principal chemical additives reported for cooling water treatment are phosphate, lime, alum and chlorine.

As a result of combustion processes in the boiler, residue accumulates on the boiler sections and air preheater. To maintain efficient heat transfer rates, these accumulated residues are removed by washing with water. The resulting wastes represent periodic (intermittent) waste streams.

In spite of the high quality water used in boilers, there is a build up of scale and corrosion products on the heat transfer surfaces over a period of time. This build up is usually due to condenser leaks, oxygen leaks into the water and occasional erosion of metallic parts by boiler water. Periodically, this scale build up is removed by cleaning the boiler tubes with different chemicals, such as acids, alkali, and chelating compounds. These cleaning wastes, though occurring only periodically, contain metallic species such as copper, iron, etc. which may require treatment prior to discharge.

FIGURE 35: TYPICAL FLOW DIAGRAM–STEAM ELECTRIC POWER PLANT (FOSSIL-FUELED) SOURCES OF CHEMICAL WASTES

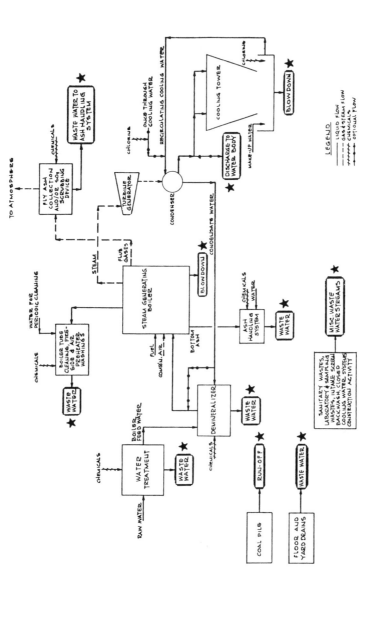

Source: Report EPA 440/1-73/0-29

430 Environmental Sources and Emissions Handbook

FIGURE 36: SIMPLIFIED WATER SYSTEM FLOW DIAGRAM FOR A NUCLEAR UNIT

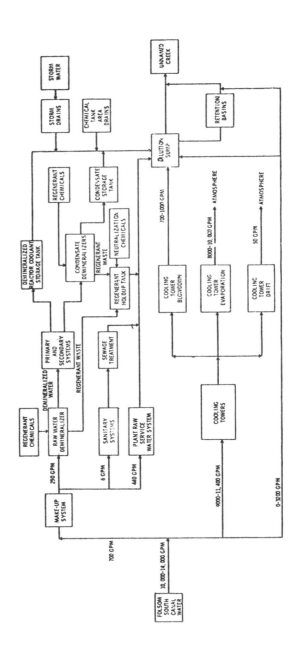

Source: Report EPA 440/1-73/0-29

The build up of scale in cooling tower basins and soot build up in stacks require periodic washings and these operations also give rise to waste streams. For coal fired generating units, outside storage of coal at or near the site is necessary to assure continuous plant operation. Normally, a supply of 90 days is maintained. Coal is stored either in active piles or storage piles. As coal storage piles are normally open, contact of coal with air and moisture results in oxidation of metal sulfides, present in the coal, to sulfuric acid.

The precipitate trickles or seeps through the coal. When rain falls on these piles, the acid is washed out and eventually winds up in coal pile runoff, creating another waste stream. Similarly, contaminated floor and yard drains are another source of pollution within the power plant. Besides these major waste streams, there are other miscellaneous waste streams in a power plant such as sanitary wastes, laboratory and sampling wastes, etc. which are also shown in Figure 35.

In a nuclear fueled power plant, high quality water is used in the steam generating section. Conventional water treatment operations give rise to chemical waste streams similar to those in fossil fueled power plants. Similarly, the cooling tower blowdown is another waste stream common to both fossil fueled and nuclear fueled power plants. Some wastes in a nuclear plant contain radioactive material. The discharge of such wastes is strictly controlled. However, the steam generator in a PWR plant is a secondary system, having a blowdown and periodic cleaning wastes which are not radioactive. The metals listed below are likely to occur in the following waste streams.

Iron
  water treatment: clarification;
  maintenance cleaning: boiler tubes, boiler fireside, and air preheater;
  ash handling: coal fired plants and coal pile drainage.

Copper
  boiler and steam generator (PWR) blowdown;
  chemical cleaning: boiler tubes, air preheater, and boiler fireside.
  condenser cooling water systems: once through, and recirculating.

Mercury
  ash handling: coal fired plants, and coal pile drainage.

Vanadium (oil fired plants only)
  ash handling;
  chemical cleaning: boiler fireside, and air preheater.

Chromium and Zinc
  recirculating condenser cooling system, and closed cooling water system.

Aluminum and Zinc
  coal pile drainage;
  ash handling: coal fired plants;
  water treatment: clarification;
  chemical cleaning: boiler fireside, and air preheater.

In addition to which other materials are found in power plant waste streams as follows.

Phenols: Polychlorinated biphenyls (PCB's) are sometimes used as coolants in large transformers. In case of leaks or spills, these materials could find their way into the yard drainage system. Materials showing up as phenols are also possible in drainage from coal piles, floor and yard drainage, ash handling streams, and cooling tower blowdown.

Sulfates: Sulfates in power plant effluents arise primarily from the regenerant wastes of ion exchange processes. Sulfate may occur in ion exchange and evaporator wastes, boiler fireside and air preheater cleaning, ash handling and coal pile drainage.

Sulfites: Sulfite is used as an oxygen scavenger in the boiler feedwater system in some plants. Plants using sulfite may discharge the sulfite with their boiler blowdown.

Because of its high oxygen demand, sulfite in significant quantities is considered undesirable in a plant discharge. Sulfite may occur in the following waste streams: maintenance cleaning—boiler fireside, air preheater, stack, cooling tower basin; and ash handling—oil fired plants, coal fired plants, coal pile drainage, and air pollution control devices for SO_2 removal.

Boron: Oxidizing agents such as potassium or sodium borate may be contained in cleaning mixtures used for copper removal in the chemical cleaning of boiler and steam generator (PWR) tubes.

Fluoride: Hydrofluoric acid or fluoride salts are added for silica removal in the chemical cleaning of boiler and steam generator (PWR) tubes. Significant thermal discharges from steam electric power plants occur when a power plant utilizes a once through circulating water system to reject the heat not converted into electric energy. The amount of heat energy discharged with the circulating water is equal to the heat value of the fuel less the heat value converted into electric energy and miscellaneous station losses.

The heat energy discharged is therefore directly related to the efficiency of the plant. According to industry practices, the efficiency of a generating unit is expressed as its heat rate, in units of joules per kwh (Btu/kwh). A fossil fired generating unit may be designed for a heat rate of 9.5 million joules per kwh (9,000 Btu/kwh). Since 1 kwh is equivalent to 3.6 million joules per kwh (3,413 Btu/kwh), such a plant would have an efficiency of 38%.

The transfer of heat from the condensing steam to the cooling water results in a temperature rise of the cooling water. For a given amount of heat transfer, the temperature rise of the cooling water is inversely proportional to its flow. That is, a small quantity of water may be heated a great deal or a large quantity of water a small amount. On the average, temperature rises have been centered about $9°C$ ($16°F$) for economic and process considerations. It is clear however, that almost any lower limit on temperature rise can be achieved given a sufficiently large source of cooling water and no economic constraints. It is also clear, however, that a temperature difference reduction does not limit the amount of heat rejection. Table 313 lists heat rates, efficiencies, and waste heat produced for a range of plants typical of the industry. The heat rejection requirements calculated above are satisfied by the heating of the circulating water.

TABLE 313: EFFICIENCIES, HEAT RATES AND HEAT REJECTED BY COOLING WATER

Plant Efficiency,	Plant Heat Rate	Heat Converted to Electricity	Stack and Plant Heat Losses	Heat Rejected to Cooling Water
%	Joules per KWH x 10^{-6}		(Btu/KWH)	
	Fossil-Fueled Units			
38	9.5 (9,000)	3.6 (3,400)	0.95 (900)	4.95 (4,700)
34	10.5 (10,000)	3.6 (3,400)	1.05 (1,000)	5.85 (5,600)
29	12.5 (12,000)	3.6 (3,400)	1.25 (1,200)	7.65 (7,400)
23	15.5 (15,000)	3.6 (3,400)	1.55 (1,500)	10.35 (11,100)
17	21.0 (20,000)	3.6 (3,400)	2.1 (2,000)	15.3 (14,600)
	Nuclear Units			
34	10.5 (10,000)	3.6 (3,400)	0.5 (500)	6.4 (6,100)
29	12.5 (12,000)	3.6 (3,400)	0.6 (600)	8.3 (8,000)

Source: Report EPA 440/1-73/0-29

Intermedia Aspects

As a result of fossil fuel combustion in the boiler, flue gases are produced which are vented to the atmosphere. Depending upon the type of fossil fuel, the flue gases carry certain amounts of entrained particulate matter (fly ash) which are removed in mechanical dust collectors, electrostatic precipitators or wet scrubbing devices. Thus fly ash removal may create another wastewater stream in a power plant.

A portion of the noncombustible matter of the fuel is left in the boiler. This bottom ash is usually transported as a slurry in a water sluicing operation. This ash handling operation presents another possible source of wastewater within a power plant. Depending upon the sulfur content of the fossil fuel, SO_2 scrubbing may be carried out to remove sulfur emissions in the flue gases. Such operations generally create liquid waste streams. Note that SO_2 scrubbing is not required for gas fired plants, or facilities burning oil with a low sulfur content. Nuclear plants, of course, have no ash or flue gas scrubbing waste streams.

The wastewater problems are mainly concerned with wet processes (first three types mentioned above under Air Pollution). The wet processes, alkali scrubbing with and without SO_2 recovery, oxidation of SO_2 for sulfuric acid production, are mainly in pilot plant or prototype stage of development. Of the three processes, sufficient data is available only for the alkali scrubbing process without SO_2 recovery, and consequently only this process is described briefly below.

Flue gas from electrostatic precipitators (optional equipment) is cooled and saturated by water spray. It then passes through a contacting (scrubbing) device where SO_2 is removed by an aqueous stream of lime absorbent. The clean gas is then reheated (optional step) and vented to the atmosphere through an induced draft fan if necessary. The lime absorbent necessary for scrubbing is produced by slaking and diluting quicklime in commercial equipment and passing it to the delay tank for recycle as a slurry through the absorber column(s).

Use of the delay tank provides sufficient residence time for the reaction of dissolved SO_2 and alkali to produce calcium sulfite and sulfate. The waste sulfite/sulfate is then pumped as a slurry to a lined settling pond or mechanical system where sulfite is oxidized to sulfate. The clear supernatant liquid is returned to the process for reuse. The waste sludge containing fly ash (if electrostatic precipitator is not employed) and calcium sulfate is sent for disposal (as a landfill).

The process described above suffers from potential scaling problems. The calcium salts tend to form a deposit, causing equipment shutdown and requiring frequent maintenance. The process is a closed loop type and consequently there is a no net liquid discharge from the process. The disposal of sludge has been covered in the literature. However, depending upon the solids separation efficiency in a pond or mechanical equipment, there may be excess free water associated with the sludge. To dewater this sludge, mechanical filtration equipment may be necessary.

To date eleven utilities have committed themselves to full-scale installation of the alkali scrubbing process without SO_2 recovery. During the course of the recent EPA study (1), visits were made to two plants for observing the scrubbing devices. However, in one plant, the scrubber was not running because of operational problems. The process for the other plant is described in this section.

This plant of 79 Mw capacity burns 0.7% sulfur coal. The boiler gases are split into two streams, approximately 75% going to a scrubber and the remaining 25% going to an electrostatic precipitator. The exhaust gases from the two are then recombined and vented to atmosphere at 210°F. This splitting of the boiler gases is done to reheat the scrubber exhaust gases which are at 124°F (saturated). This stack gas reheating is achieved to minimize scaling problems from moist gases. The scrubber is not specifically used for SO_2 removal. Rather, the primary function is to remove particulates.

On the other hand, some SO_2 pickup is achieved. The flow diagram and the different stream compositions are shown in Figure 37.

FIGURE 37: FLOW DIAGRAM AIR POLLUTION CONTROL SCRUBBING SYSTEM

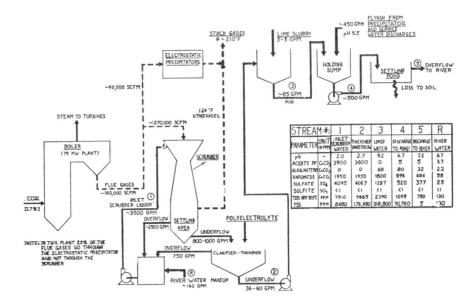

Source: Report EPA 440/1-73/0-29

References

(1) U.S. Environmental Protection Agency, "Air Pollution Aspects of Emission Sources: Electric Power Production—A Bibliography with Abstracts," *Publication No. AP-96,* Washington, D.C., U.S. Government Printing Office (May 1971).

(2) Midwest Research Institute, "Particulate Pollutant System Study, III, Handbook of Emission Properties," *Report PB 203,522,* Springfield, Virginia, Nat. Tech. Information Service (May 1971).

(3) U.S. Department of Health, Education and Welfare, "Control Techniques for Sulfur Dioxide Air Pollutants," *Publication No. AP-52,* Washington, D.C. National Air Pollution Control Administration (January 1969).

(4) U.S. Environmental Protection Agency, "Development Document for Proposed Effluent Limitations Guidelines and New Source Performance Standards for the Steam Electric Power Generating Plant Point Source Category," *Report EPA 440/1-73/0-29,* Washington, D.C. (March 1974).

PRINTING INK MANUFACTURE

There are four major classes of printing ink: letterpress and lithographic inks, commonly called oil or paste inks; and flexographic and rotogravure inks, which are referred to as solvent inks. These inks vary considerably in physical appearance, composition, method of application, and drying mechanism. Flexographic and rotogravure inks have many elements in common with the paste inks but differ in that they are of very low viscosity, and they almost always dry by evaporation of highly volatile solvents (2).

There are three general processes in the manufacture of printing inks: cooking the vehicle and adding dyes, grinding of a pigment into the vehicle using a roller mill, and replacing water in the wet pigment pulp by an ink vehicle (commonly known as the flushing process) (3). The ink varnish or vehicle is generally cooked in large kettles at 200° to 600°F (92° to 315°C) for an average of 8 to 12 hr in much the same way that regular varnish is made. Mixing of the pigment and vehicle is done in dough mixers or in large agitated tanks. Grinding is most often carried out in three-roller or five-roller horizontal or vertical mills.

Air Pollution

Varnish or vehicle preparation by heating is by far the largest source of ink manufacturing emissions. Cooling the varnish components, resins, drying oils, petroleum oils, and solvents, produces odorous emissions. At about 350°F (175°C) the products begin to decompose, resulting in the emission of decomposition products from the cooking vessel. Emissions continue throughout the cooking process with the maximum rate of emissions occurring just after the maximum temperature has been reached. Emissions from the cooking phase can be reduced by more than 90% with the use of scrubbers or condensers followed by afterburners (4)(5).

Compounds emitted from the cooking of oleoresinous varnish (resin plus varnish) include water vapor, fatty acids, glycerine, acrolein, phenols, aldehydes, ketones, terpene oils, terpenes, and carbon dioxide. Emissions of thinning solvents used in flexographic and rotogravure inks may also occur. The quantity, composition, and rate of emissions from ink manufacturing depend upon the cooking temperature and time, the ingredients, the method of introducing additives, the degree of stirring, and the extent of air or inert gas blowing. Particulate emissions resulting from the addition of pigments to the vehicle are affected by the type of pigment and its particle size. Emission factors for the manufacture of printing ink are presented in Table 314.

TABLE 314: EMISSION FACTORS FOR PRINTING INK MANUFACTURING[a]

EMISSION FACTOR RATING: E

	Gaseous organic[b]		Particulates	
Type of process	lb/ton of product	kg/MT of product	lb/ton of pigment	kg/MT of pigment
Vehicle cooking				
General	120	60	—	—
Oils	40	20	—	—
Oleoresinous	150	75	—	—
Alkyds	160	80	—	—
Pigment mixing	—	—	2	1

[a]Based on data from section on paint and varnish.
[b]Emitted as gas, but rapidly condense as the effluent is cooled.

Source: EPA Publication AP-42, 2nd Edition

References

(1)   U.S. Environmental Protection Agency, "Compilation of Air Pollutant Emission Factors," 2nd edition, *Publication Number AP-42,* Research Triangle Park, N.C., Office of Air and Water Programs (April 1973).
(2)   Shreve, R.N., *Chemical Process Industries,* 3rd edition, New York, McGraw Hill Book Co., p. 454-455 (1967).
(3)   Larsen, L.M., *Industrial Printing Inks,* New York, Reinhold Publishing Company (1962).
(4)   Chatfield, H.E., "Varnish Cookers," *Air Pollution Engineering Manual,* Danielson, J.A. (editor), U.S. DHEW, PHS, National Center for Air Pollution Control, Cincinnati, Ohio, Publication Number 999-AP-40, p. 688-695 (1967).
(5)   EPA communication from Interchemical Corporation, Ink Division, Cincinnati, Ohio (November 10, 1969).

PULP AND PAPER INDUSTRY

Chemical wood pulping involves the extraction of cellulose from wood by dissolving the lignin that binds the cellulose fibers together. The principal processes used in chemical pulping are the kraft, sulfite, neutral sulfite semichemical (NSSC), dissolving, and soda; the first three of these display the greatest potential for causing air pollution. The kraft process accounts for about 65% of all pulp produced in the United States; the sulfite and NSSC processes, together, account for less than 20% of the total. The choice of pulping process is determined by the product being made, by the type of wood species available, and by economic considerations.

The kraft process involves the cooking of wood chips under pressure in the presence of a cooking liquor in either a batch or a continuous digester. The cooking liquor, or white liquor, consisting of an aqueous solution of sodium sulfide and sodium hydroxide, dissolves the lignin that binds the cellulose fibers together. When cooking is completed, the contents of the digester are forced into the blow tank. Here the major portion of the spent cooking liquor, which contains the dissolved lignin, is drained, and the pulp enters the initial stage of washing. From the blow tank the pulp passes through the knotter where unreacted chunks of wood are removed. The pulp is then washed and, in some mills, bleached before being pressed and dried into the finished product.

It is economically necessary to recover both the inorganic cooking chemicals and the heat content of the spent black liquor, which is separated from the cooked pulp. Recovery is accomplished by first concentrating the liquor to a level that will support combustion and then feeding it to a furnace where burning and chemical recovery take place.

Initial concentration of the weak black liquor, which contains about 15% solids, occurs in the multiple-effect evaporator. Here process steam is passed countercurrent to the liquor in a series of evaporator tubes that increase the solids content to 40 to 55%. Further concentration is then effected in the direct contact evaporator. This is generally a scrubbing device (a cyclonic or venturi scrubber or a cascade evaporator) in which hot combustion gases from the recovery furnace mix with the incoming black liquor to raise its solids content to 55 to 70%.

The black liquor concentrate is then sprayed into the recovery furnace where the organic content supports combustion. The inorganic compounds fall to the bottom of the furnace and are discharged to the smelt dissolving tank to form a solution called green liquor. The green liquor is then conveyed to a causticizer where slaked lime (calcium hydroxide) is added to convert the solution back to white liquor, which can be reused in subsequent cooks. Residual lime sludge from the causticizer can be recycled after being dewatered and calcined in the hot lime kiln. Many mills need more steam for process heating, for driving equipment, for providing electric power, etc., than can be provided by the recovery furnace alone. Thus, conventional industrial boilers that burn coal, oil, natural gas, and in some cases, bark and wood waste are commonly employed. The production of acid sulfite pulp proceeds similarly to kraft pulping except that different chemicals are used in the cooking liquor. In place of the caustic solution used to dissolve the lignin in the wood, a sulfurous acid-base is employed. To buffer the cooking solution, a bisulfite of sodium, magnesium, calcium, or ammonium is used.

Because of the variety of bases employed in the cooking liquor, numerous schemes for heat and/or chemical recovery have evolved. In calcium-base systems, which are used mostly in older mills, chemical recovery is not practical, and the spent liquor is usually discarded. In ammonium-base operations, heat can be recovered from the spent liquor through combustion, but the ammonium base is consumed in the process. In sodium- or magnesium-base operations (the latter being utilized most frequently in newer sulfite mills) heat, sulfur, and base recovery are all feasible.

When recovery is practiced, the spent liquor proceeds through a multiple-effect evaporator and recovery furnace arrangement similar to that found in the kraft process. The combustion gases from the furnace pass through absorbing (sulfiting) towers where sulfur dioxide is recovered (as bisulfite) for use in subsequent cooks. In magnesium- or sodium-base operations, moreover, the base can also be recovered by feeding the inorganic residue from the furnace (either as smelt or collected ash) into the absorbing tower to react with the sulfur dioxide.

In the neutral sulfite semichemical (NSSC) process, the wood chips are cooked in a neutral solution of sodium sulfite and sodium bicarbonate. The sulfite ion reacts with the lignin in the wood, and the sodium bicarbonate acts as a buffer to maintain a neutral solution. The major difference between this process (as well as all semichemical techniques) and the kraft and acid sulfite processes is that only a portion of the lignin is removed during the cook, after which the pulp is further reduced by mechanical disintegration. Because of this, yields as high as 60 to 80% can be achieved as opposed to 50 to 55% for other chemical processes.

The NSSC process varies from mill to mill. Some mills dispose of their spent liquor, some mills recover the cooking chemicals, and some, which are operated in conjunction with kraft mills, mix their spent liquor with the kraft liquor as a source of makeup chemicals. When recovery is practiced, the steps involved parallel those of the sulfite process.

Air Pollution

Particulate emissions from the kraft process occur primarily from the recovery furnace, the lime kiln, and the smelt dissolving tank. These emissions consist mainly of sodium salts but include some calcium salts from the lime kiln. They are caused primarily by the carry-over of solids plus the sublimation and condensation of the inorganic chemicals.

Particulate control is provided on recovery furnaces in a variety of ways. In mills where either a cyclonic scrubber or cascade evaporator serves as the direct contact evaporator, further control is necessary as these devices are generally only 20 to 50% efficient for particulates. Most often in these cases, an electrostatic precipitator is employed after the direct contact evaporator to provide an overall particulate control efficiency of 85 to ≥99%. In a few mills, however, a venturi scrubber is utilized as the direct contact evaporator and simultaneously provides 80 to 90% particulate control. In either case auxiliary scrubbers may be included after the precipitator or the venturi scrubber to provide additional control of particulates.

Particulate control on lime kilns is generally accomplished by scrubbers. Smelt dissolving tanks are commonly controlled by mesh pads but employ scrubbers when further control is needed. The characteristic odor of the kraft mill is caused in large part by the emission of hydrogen sulfide. The major source is the direct contact evaporator in which the sodium sulfide in the black liquor reacts with the carbon dioxide in the furnace exhaust. The lime kiln can also be a potential source as a similar reaction occurs involving residual sodium sulfide in the lime mud. Lesser amounts of hydrogen sulfide are emitted with the noncondensible off gases from the digesters and multiple-effect evaporators. The kraft process odor also results from an assortment of organic sulfur compounds, all of which have extremely low odor thresholds. Methyl mercaptan and dimethyl sulfide are formed in reactions with the wood component lignin. Dimethyl disulfide is formed through the oxidation of mercaptan groups derived from the lignin. These compounds are emitted from many points within a mill, however, the main sources are the digester/blow tank systems and the direct contact evaporator.

Although odor control devices, per se, are not generally employed in kraft mills, control of reduced sulfur compounds can be accomplished by process modifications and by optimizing operating conditions. For example, black liquor oxidation systems, which oxidize sulfides into less reactive thiosulfates, can considerably reduce odorous sulfur emissions from the direct contact evaporator, although the vent gases from such systems become minor odor sources themselves.

Noncondensible odorous gases vented from the digester/blow tank system and multiple-effect evaporators can be destroyed by thermal oxidation, usually by passing them through the lime kiln. Optimum operation of the recovery furnace, by avoiding overloading and by maintaining sufficient oxygen residual and turbulence, significantly reduces emissions of reduced sulfur compounds from this source. In addition, the use of fresh water instead of contaminated condensates in the scrubbers and pulp washers further reduces odorous emissions. The effect of any of these modifications on a given mill's emissions will vary considerably.

Several new mills have incorporated recovery systems that eliminate the conventional direct contact evaporators. In one system, preheated combustion air rather than flue gas provides direct contact evaporation. In the other, the multiple-effect evaporator system is extended to replace the direct contact evaporator altogether. In both of these systems, reduced sulfur emissions from the recovery furnace/direct contact evaporator reportedly can be reduced by more than 95% from conventional uncontrolled systems.

Sulfur dioxide emissions result mainly from oxidation of reduced sulfur compounds in the recovery furnace. It is reported that the direct contact evaporator absorbs 50 to 80% of these emissions, further scrubbing, if employed, can reduce them another 10 to 20%. Potential sources of carbon monoxide emissions from the kraft process include the recovery furnace and lime kilns. The major cause of carbon monoxide emissions is furnace operation well above rated capacity, making it impossible to maintain oxidizing conditions.

Some nitrogen oxides are also emitted from the recovery furnace and lime kilns although the amounts are relatively small. Indications are that nitrogen oxides emissions from each of these sources are on the order of 1 lb per air-dried ton (0.5 kg/air-dried MT) of pulp produced. A major source of emissions in a kraft mill is the boiler for generating auxiliary steam and power. The fuels used are coal, oil, natural gas, or bark/wood waste. Table 315 presents emission factors for a conventional kraft mill. The most widely used particulate controls devices are shown along with the odor reductions resulting from black liquor oxidation and incineration of noncondensible off-gases.

Significant quantities of particulate emissions will be generated in the sulfite process only if sodium-, magnesium-, or calcium-base liquors are burned. When ammonium-base liquor is burned, few particulates will result because the combustion products are mostly nitrogen, water vapor, and sulfur dioxide.

In magnesium-base recovery systems, high particulate control is necessary because most of the base is swept out of the furnace in the form of magnesium oxide fumes. No particulate emissions will result from these systems, of course, when the spent liquor is not combusted. The major gaseous pollutant is sulfur dioxide. Major potential sources, in probable order of importance, include the digester/blow tank system, absorbing towers, and multiple-effect evaporators.

TABLE 315: EMISSION FACTORS FOR SULFATE PULPING[a]

EMISSION FACTOR RATING: A

Source	Type control	Particulates[b]		Sulfur dioxide (SO$_2$)[c]		Carbon monoxide[d]		Hydrogen sulfide(S=)[e]		RSH, RSR, RSSR(S=)[e,f]	
		lb/ton	kg/MT	lb/ton	kg/MT	lb/ton	kg/MT	lb/ton	kg/MT	lb/ton	kg/MT
Digester relief and blow tank	Untreated[g]	–	–	–	–	–	–	0.1	0.05	1.5	0.75
Brown stock washers	Untreated	–	–	0.01	0.005	–	–	0.02	0.01	0.2	0.1
Multiple effect evaporators	Untreated[g]	–	–	0.01	0.005	–	–	0.1	0.05	0.4	0.2
Recovery boiler and direct contact evaporator	Untreated[h]	150	75	5	2.5	2 - 60	1 - 30	12[i]	6[i]	1[i]	0.5[i]
	Venturi scrubber[j]	47	23.5	5	2.5	2 - 60	1 - 30	12[i]	6[i]	1[i]	0.5[i]
	Electrostatic precipitator	8	4	5	2.5	2 - 60	1 - 30	12[i]	6[i]	1[i]	0.5[i]
	Auxiliary scrubber	3 - 15[k]	1.5 - 7.5[k]	3	1.5	2 - 60	1 - 30	12[i]	6[i]	1[i]	0.5[i]
Smelt dissolving tank	Untreated	5	2.5	0.1	0.05	–	–	0.04	0.02	0.4	0.2
	Mesh pad	1	0.5	0.1	0.05	–	–	0.04	0.02	0.4	0.2
Lime kilns	Untreated	45	22.5	0.3	0.15	10	5	0.5	0.25	0.25	0.125
	Scrubber	3	1.5	0.2	0.1	10	5	0.5	0.25	0.25	0.125
Turpentine condenser	Untreated	–	–	–	–	–	–	0.01	0.005	0.5	0.25
Miscellaneous sources[l]	Untreated	–	–	–	–	–	–	–	–	0.5	0.25

[a]For more detailed data on specific types of mills, consult Reference 1.

[b]Reference 1

[c]Reference 1

[d]Reference 1 Use higher value for overloaded furnaces.

[e]References 1. These reduced sulfur compounds are usually expressed as sulfur.

[f]RSH-methyl mercaptan; RSR-dimethyl sulfide; RSSR-dimethyl disulfide.

[g]If the noncondensible gases from these sources are vented to the lime kiln, recovery furnace, or equivalent, the reduced sulfur compounds are destroyed.

[h]These factors apply when either a cyclonic scrubber or cascade evaporator is used for direct contact evaporation with no further controls.

[i]These reduced sulfur compounds (TRS) are typically reduced by 50 percent when black liquor oxidation is employed but can be cut by 90 to 99 percent when oxidation is complete and the recovery furnace is operated optimally.

[j]These factors apply when a venturi scrubber is used for direct contact evaporation with no further controls.

[k]Use 15(7.5) when the auxiliary scrubber follows a venturi scrubber and 3(1.5) when employed after an electrostatic precipitator.

[l]Includes knotter vents, brownstock seal tanks, etc. When black liquor oxidation is included, a factor of 0.6(0.3) should be used.

Note: Emission factors are for unit weights of air-dried unbleached pulp.

Source: EPA Publication AP-42, 2nd edition.

Because the vented gases from these systems are either scrubbed or fed into the absorbing tower to minimize sulfur losses, actual emissions to the atmosphere will depend on the degree of efficiencies desired. The characteristic kraft odor is not emitted from acid sulfite mills because volatile reduced sulfur compounds are not products of the lignin-bisulfite reaction. No factors for sulfite pulping are presented because of the variety of pulping schemes employed and lack of adequate emissions data.

Particulate emissions are a potential problem in the NSSC process only when recovery systems are employed. Mills that do practice recovery, but are not operated in conjunction with kraft operations often utilize fluidized bed reactors to burn their spent liquor. Because the flue gas contains sodium sulfate and sodium carbonate dust, efficient particulate collection may be included to facilitate chemical recovery.

A potential gaseous pollutant is sulfur dioxide. The absorbing towers, digester/blow tank system, and recovery furnace are the main sources of this pollutant with the amounts emitted dependent upon the capability of the scrubbing devices installed for control and recovery. Hydrogen sulfide can also be emitted from NSSC mills using kraft-type recovery furnaces. The main potential source is the absorbing tower where a significant quantity of hydrogen sulfide is liberated as the cooking liquor is made. Other possible sources include the recovery furnace, depending on the operating conditions maintained, as well as the digester/blow tank system in mills where some green liquor is used in the cooking process. Where green liquor is used, it is also possible that significant quantities of mercaptans will be produced. Hydrogen sulfide emissions can be eliminated if burned to sulfur dioxide prior to entering the absorbing systems. Because the NSSC process differs greatly from mill to mill, and because of the scarcity of adequate data, no emission factors are presented.

Water Pollution

Usage of water and resultant wastewater characteristics for the general operations of wood preparation, pulping processes, and the paper machine are discussed in this section. Since a relatively thorough discussion of wood preparation is presented, it should be noted that raw waste loads resulting from wood preparation are much less than loads resulting from pulping process and paper production and usually utilize wastewater from another unit within the mill.

Wood, the primary fiber raw material for unbleached kraft and NSSC pulps, is received at the mills in various forms and consequently must be handled in a number of different ways. Some mills receive chips from saw mills or barked logs which can be chipped directly. In these instances, little, if any, water is employed in preparation of the wood and no effluent is produced. Most mills receive roundwood in short lengths with the bark remaining on it, and, since the bark interferes with the pulping process and product quality, it must be removed.

Logs are frequently washed before dry or wet barking in order to remove silt. In most installations a water shower is activated by the log itself while on the conveyor so that a minimum of water is used. The actual quantity discharged per unit of wood handled or pulp produced is most difficult to ascertain because of the wide weight variation in stick size and the fact that not all the wood barked at some installations is pulped, a portion going to lumber.

It is established that this effluent is very low in color and BOD_5 and that its suspended solids content is largely silt. Hence, it is generally disposed of on the land together with grits and dregs from the pulp mill and/or ashes from the boiler plants, or combined with the general flowage to the treatment works. Most of the pulpwood used in the United States is small in diameter and is barked dry in drums. However, when large diameter or long wood is used, wet barking is commonly employed. The latter operation is pretty much limited to northern mills and its use is presently declining.

Wet barking of logs is accomplished by one of three methods: by drums, pocket barkers, or hydraulic barkers. Slabs are generally handled by hydraulic units as is the larger diameter and long roundwood. The wet drum barker consists of a slotted drum equipped with internal staves which knock the bark from the wood as the drum rotates in a pool of water.

The bark falls through the slots and is removed with the overflow of water. These units handle from 7 to 45 cords of wood daily. Frequently the water supplied to them is spent process water, and recycling within the barking unit itself is often practiced. Barkers of this type contribute BOD_5 from 7.5 to 10 kg/metric ton (15 to 20 lb/short ton) of wood barked, and from 15 to 20 kg/metric ton (30 to 40 lb/short ton) of suspended solids. Examples of the BOD_5 and suspended solids concentration of this wastewater with the barkers using fresh process water are shown in Table 316. In the table the water source for wet drum barkers is frequently a wastewater which has been recycled from some other source.

Wet pocket barkers are stationary machines which abrade bark from timber by jostling and gradually rotating a confined wood stack against an endless chain belt equipped with projections called dogs which raise the wood pile allowing bark to pass between the chains. Water is sprayed through apertures in the side of the pocket at rates of between 1,254 and 2,280 l/min (330 and 600 gpm) for pockets of 2.8 and 5.7 cords per hours, respectively. The use of this process is rapidly declining in the United States. Hydraulic barkers employ high pressure water jets to blow the bark from the timber which is either conveyed past them or rotated under a moving jet which traverses the log. The volume of water employed is generally from 19,000 to 45,600 liters (5,000 to 12,000 gal) per cord of wood barked depending upon log diameter. Water discharged from all three types of wet barking is generally combined with log washwater, and then coarse screens are used to remove the large pieces of bark and wood slivers which are conveyed away continuously. The flowage then passes to fine screens.

These are of the drum, fixed vertical, or horizontal vibrating type, having wire mesh or perforated plate media with openings in the range of 0.127 to 0.254 cm (0.05" to 0.10"). Screenings are removed and mixed with the coarse materials from the initial screenings, the mixture being dewatered in a press prior to burning in the bark boiler. Press water, which is combined with the fine screen effluent, is very minor in volume. The total waste flow, which amounts to about 19,000 to 26,600 liters (5,000 to 7,000 gal) a cord, generally carries from 0.454 to 4.54 kg (1 to 10 lb) of BOD_5 and 2.72 to 25 kg (6 to 55 lb) of suspended solids per ton of product.

TABLE 316: ANALYSIS OF WET DRUM BARKING EFFLUENTS

MILL	TOTAL SUSPENDED SOLIDS mg/l	%ASH mg/l	BOD5 mg/l	COLOR APHA UNITS
1	2017	--	480	20
2	3171	21	605	50
3	2875	18	987	50

Source: Report EPA 440/1-74/025

The combined discharge contains bark fines and silt, the latter varying greatly in quantity since its presence is due mainly to soil adhering to the logs. In dry weather the percentage of silt in relation to bark fines is low as is the case when logs are stored in or transported by water. However, attachment of mud in wet weather can make this material a major percentage of the total suspended matter passing the fine screens.

Fine screen effluent following hydraulic barkers has been analyzed by several investigators, and examples are shown in Table 317. It can be concluded from the data included in these publications that these effluents have a total suspended solids content ranging from 521 to 2,350 mg/l with the ash content running from 11 to 27%. The latter is generally below 15% for clean logs. BOD_5 values range between 56 and 250 mg/l. These low values are due to the fact that the contact of the water with the bark is short and no grinding action on the wood takes place. Hence, leaching of wood and bark solubles is minimized. The water originally employed is all fresh process water, since the close clearances of the high pressure pumping systems supplying water to the jets will not tolerate the presence of suspended solids in the water.

Such low values are not the case with drum and pocket grinding where attrition in contact with water over an appreciable period of time takes place. Also, spent pulping process waters already high in BOD_5 and color are sometimes used for these barking processes which raise further the ultimate level of organics in the screened effluent.

While wet drum and pocket barker fine screen discharge is not greatly different from that of hydraulic barkers in suspended solids content, the BOD_5 can be considerably higher. BOD_5 values are also greatly affected by the species of wood barked and the season in which the wood was cut since wood juices and water extractables are responsible for it. The BOD_5 contributed by the suspended matter present is a minor fraction of the total BOD_5. The 15 day values are about twice those of the 5 day with little further demand exerted after this period. Table 318 illustrates sewer losses from wet barking operations.

TABLE 317: ANALYSIS OF HYDRAULIC BARKING EFFLUENTS

MILL	TOTAL SUSPENDED SOLIDS _____mg/l_____	% ASH _mg/l_	BOD5	COLOR APHA UNITS
1	2362	27	585	50
2	889	14	101	50
3	1391	17	64	50
4	550	11	99	50
5	521	13	121	50
6	2017	21	56	50
7	2000	19	97	--
8	600	10	250	35

TABLE 318: SEWER LOSSES FROM WET BARKING OPERATIONS

Mill #	Effluent Volume Kiloliter/metric ton (1000 gal/short ton)	BOD5 Kg/metric ton (lbs/short ton)	TSS Kg/metric ton (lbs/short ton)
1	11.3 (2.7)	0.6 (1.2)	3.2 (6.4)
2	10.0 (2.4)	0.9 (1.8)	3.8 (7.6)
3	14.6 (3.5)	6.0 (12.0)	2.75 (5.5)
4	25.0 (6.0)	3.0 ((6.0)	15.0 (30.0)
5	12.5 (3.0)	1.25 (2.5)	11.4 (22.8)
6	4.2 (1.0)	1.0 (2.0)	5.0 (10.0)
7	23.4 (5.6)	9.5 (19.0)	9.0 (18.0)
8	4.2 (1.0)	5.75 (11.5)	15.0 (30.0)
9	31.3 (7.5)	11.05 (20.1)	17.0 (34.0)

Source: EPA 440/1-74/025

The wastewater resulting from unbleached kraft pulping comes primarily from three areas of the process. The effluent from pulp washing, which separates the spent liquor from the pulp, formerly consisted mainly of decker filtrate water containing spent cooking liquor solids and accounts for a high percentage of the total effluent. Today, the use of hot stock washing has considerably reduced the waste load generated in the washing operation. The second area of wastewater sources is condensate streams. Relief condensate from the digesters is condensed and the turpentine is recovered from it by decantation. The residual water from this operation is sewered. Blow and evaporation condensates are contaminated mainly with methanol, ethanol, and acetone, with the extent of their content a function of the wood species pulped. When surface condensers are employed on the evaporators, the volume of this stream is low and its BOD_5 can be reduced by air stripping in a cooling tower or by steam stripping. These condensates are frequently reused for pulp washing. All chemical recovery operations and other minor losses constitute the last BOD_5 source from kraft pulping. Losses per ton of product from kraft pulping itself are difficult to determine because of the common practice of reusing water from integrated papermaking operations into the pulp mill.

Total BOD_5 raw waste load from unbleached kraft mills, including both pulping and papermaking operations, is typically in the 15 to 20 kg/metric ton (30 to 40 lb/short ton) range. The surveyed mills were in the lower region of this range, averaging 15.5 kg/metric ton (31 pounds per short ton). Suspended solids data for 35 mills were within a typical range of 10 to 15 kg/metric ton (20 to 30 lb/short ton).

Surveyed mills, however, averaged 18 kg/metric ton (36 lb/short ton). This difference is most probably explained by the fact that most mills use a filter paper method of determining suspended solids (nonstandard methods, NSM) whereas standard methods (SM) were used in the surveyed mills. Specific relationships between the two methods are difficult to establish because the mills using NSM have many different filter papers presently in use which yield large variations in results. However, SM generally yields higher results than NSM with some reported relationships of up to ten times greater.

Raw waste color APHA color units (CU) are typically in the 500 to 1,500 range, and one of the surveyed mills fell in the low end of this range at 567 units, while a second surveyed mill, on a short term test, measured 286 color units. The impact of in-plant measures is evident in all the surveyed mills as compared with previous typical ranges of data. All of the surveyed mills had reduced their flows to the 40,173 to 54,249 l/metric ton (10,000 to 13,000 gal/short ton) range.

In contrast, an earlier report on 35 mills indicated a range of 83,460 to 125,190 l/metric ton (20,000 to 30,000 gal/short ton). It was reported in 1966 on 19 unbleached kraft mills with a median water usage of 121,017 l/metric ton (29,000 gal/short ton). The details of methods utilized to accomplish this flow reduction, with concomitant reductions in pollution levels in the raw waste, are described in an EPA report (3). Raw waste characteristics of unbleached mill effluent are shown in Table 319.

TABLE 319: RAW WASTE CHARACTERISTICS, UNBLEACHED KRAFT

Mill	Flow kiloliters/kkg (1000 gal/ton)		BOD 5 kg/kkg (lbs/ton)		TSS kg/kkg (lbs/ton)	
	Mill	Survey	Mill	Survey	Mill	Survey
a	43.9 (10.5)	–	13.5 (27)	12 (24)	10.5 (21)	6.5 (13)
b	47.2 (11.3)	–	13.5 (27)	17 (34)	17 (34)	11 (22)
c	39.5 (9.46)	–	14 (28)	9.5 (19)	28 (56)	17 (34)
d	56.3 (13.5)	–	15.5 (31)	22.5 (45)	19.5 (39)	26.5 (53)

Source: Report EPA 440/1-74/025

In most sodium base NSSC mills, liquor is prepared by burning sulfur and absorbing it in soda ash or ammonia, depending on base utilized. This part of the process produces only small quantities of liquid wastes other than floor drainings, equipment washup, and cooling waters which can frequently be used as process water. Digester-relief and blow gases are condensed, and in some mills the condensate is used for pulp washing.

Pulp washwater together with drainings from the blow tank are delivered to the recovery or liquor burning system, or in the case of some sodium base mills to an adjunct kraft recovery system. From the washers the pulp is conveyed to an agitated chest where it is diluted with white water from the paper mill to the desired consistency for feed to the secondary refiners serving the papermaking operation. Other than spent liquor, the pulping and washing operations discharge little wastewater since the small amount of residual liquor solids present in pulp is carried through the machine system passing out with the overflow white water.

The final effluent from sodium base NSSC mills is low in volume because of the high degree of recycle commonly practiced in both the pulping and papermaking operations. For the same reason it is high in BOD_5. Without recovery or incineration of the liquor, effluents would range from 1,500 to 5,000 mg/l with a suspended solids content of from 400 to 600 mg/l. The color and chemical oxygen demand content would be correspondingly high.

Sodium base NSSC mills which practice extensive internal recycle and other in-plant measures, have suceeded in reducing raw waste pollutants to the lower levels shown in Table 320. For example, it has been reported that BOD_5 loadings of 28.5 kg/metric ton (57 lb/short ton) at a flow of 7,094 l/metric ton (1,700 gal/ton). As flow is progressively reduced through more extensive in-plant measures, BOD_5 is reduced to 14.5 kg/metric ton (29 lb/short ton) at 2,921 l/metric ton (700 gal/ton).

TABLE 320: RAW WASTE LOAD

NSSC - Sodium Base
(Liquor not included)

Literature Mill #	Effluent Volume kiloliters/metric ton (1000 gal/short ton)	BOD5 kg/metric ton (lbs/short ton)	TSS kg/metric ton (lb/short ton)
1	36.0 (9.1)	15.0 (30)	7.5 (15)
2	20.0 (4.8)	32.0 (64)	6.0 (12)
3	30.0 (7.2)	21.5 (43)	4.5 (9)
4	25.0 (6.0)	13.5 (27)	8.5 (17)
5	7.1 (1.7)	28.5 (57)	4.0 (8)
6	47.2 (11.3)	30.5 (71)	21.5 (43)
7	41.7 (10.0)	45.0 (90)	14.0 (28)
8	43.4 (10.4)	21.0 (42)	16.5 (33)
9	106.8 (25.6)	23.5 (47)	11.5 (23)
10	83.5 (20.0)	29.5 (69)	23.0 (46)
11	29.2 (7.0)	21.5 (43)	50.0 (100)
12	43.0 (10.3)	11.0 (22)	17.5 (37)
13	100.2 (24.0)	75.0 (150)	20.0 (40)
Exemplary Mill			
f*	44.6 (10.7)	8.5 (17)	8.5 (17)
f**		13 (26)	7.5 (15)

*Mill Records
**Short term survey data (3-7 days)

Source: Report EPA 440/1-74/025.

The lower value cannot be sustained, however, because of operational problems. Also shown in Table 320 is data for the exemplary mill. The data presented is an average of 6 months of daily mill records. It should be noted that mill f may also be included in mills 1 through 13 in Table 320.

Similarly, others have reported a short term average of 5.5 kg/metric ton (11 lb/short ton) for BOD_5. Again, operating difficulties are cited at this low level, and daily variations of BOD_5 range up to 25 kg/metric ton (50 lb/short ton) and higher. For the same mill, one researcher reported a goal of about 50 kg/metric ton (100 lb/short ton) of BOD_5 after installation of liquor recovery. The effluent of a surveyed sodium-base NSSC mill utilizing recycle contained 13 kg/metric ton (26 lb/short ton) of BOD_5 during the survey period lasting several days.

Total dissolved solids is frequently measured in the raw waste from NSSC mills, since it is a relatively rapid indicator of upsets. As dissolved solids exceed 1.5% due to increased recycle, reports of increased operating problems have been reported. A surveyed sodium base mill reported no operating problems due to total dissolved solids at the much lower level of 0.2%. Others reported difficulties in meeting wet strength requirements of the product when total dissolved solids of the recirculated white water reached 3.7%.

The ammonia base NSSC process is similar to the sodium base process described above, except that ammonia is utilized in the preparation of cooking liquor in place of sodium. Wastewater characteristics of the two processes are similar except for the nitrogen concentration in the liquid wastes from the ammonia base mills. The wood preparation step does not generate a significant waste stream since it is essentially a bark removal and chipping operation. This generates a small stream of approximately 10 to 15 gal/min emanating from the chip washer which is directed to a holding pond.

The initial phase of pulp preparation begins with heating the chips in a steaming vessel. The chips are then conveyed by a series of horizontal and vertical screw feeders upward through the cooking liquor and into the digester. The cooking liquor consists of ammonium sulfite, produced on site, and anhydrous ammonia. The pressure and temperature in the digester are controlled by injection of live steam. The digested chips are fed continuously to refiners where they pass between stationary and rotating discs, after which the refined pulp passes into a blow tank to be mixed and diluted to the proper consistency.

The vapor and steam from the blow tank are condensed and used elsewhere. From the blow tank, the pulp goes into a two-stage, countercurrent washer and then into a high density storage chest. From here the pulp is pumped to the secondary refiner and into the blend chest. The weak black liquor from the washers, and any other wasted cooking liquor, goes to the evaporators. The vapor off the evaporators is condensed and goes to the sewer while the remaining black liquor is burned in a liquor disposal unit.

The paper production stage begins after the pulp has been washed, refined and blended. Pulp is removed from the blend chest and processed through refiners, the third and last refining step. Before going to the paper machine, the pulp passes through cleaners and screens. Excess white water from the paper machine flows through a disc-type saveall, where wood fiber is recovered.

There are five sources of wastewater in the ammonia-base NSSC pulp manufacturing process: the evaporators, the powerhouse and maintenance, the pulp mill, the paper machine and the waste paper plant. The latter, however, is an insignificant source. In the surveyed mill which produced 453 metric tons (500 short tons) per day, all chips are washed before entering the digester for removal of sand and dirt. Reuse water from the hot water tank in the pulping area is used as washwater. According to a mill study on July 17, 1972, the chip washer contained the following effluent load: flow, 81.3 l/min (21.5 gpm); SS, 78.8 mg/l; total solids, 98 mg/l; and nonvolatiles, 7.6% (of total solids). The chips washer discharges directly to a drainage ditch leading to the holding pond. The above numbers refer to the raw solids load before discharge into the holding pond.

No raw water is used in the pulping area. The most significant point for water usage is at the washers where reuse water from the paper machine vacuum system is used. Excepting floor drains, the water discharge into the pulp mill sewer comes from the screw feeder and the paper machine saveall. To accomplish sufficient high dry solids content in the chips before the digester, water is pressed out of the chips in the screwfeeder.

The screwfeeder effluent is a low flow high BOD_5 concentrated stream which contributes about 18 to 20% to the total raw BOD_5 load of the mill (January 1972). A study carried out by the mill in November and December 1971 showed the following effluent load from the screw feeder: Flow, 340 l/min (90 gpm); BOD_5, 4,260 mg/l (range 2,180 to 6,080); and BOD_5, 2,090 kg/day (4,600 lb/day).

The saveall overflow is highly variable both in flow and concentrations depending on the amount of clarified water taken for reuse. It is also high in BOD_5 load since it contains the dry solids loss from the washers. This stream discharges to the pulp mill sewer. The weak liquor recovered in the washing plant is evaporated in a quadruple evaporator unit to about 52% dryness. The thick liquor is burned in the recovery boiler, or disposed of on land or sold. The combustion products are gaseous with a negligible residue of inorganic ash. The gaseous products contain significant sulfur dioxide emissions.

Fresh water is used in the evaporation plant vacuum system and in the boiler area as makeup water to the boilers. The cooling water to the surface condenser may be recycled through a cooling tower. The most significant efficient stream is the secondary condensate. The condensate can be separated in three streams: that combined from middle effects, that from the surface condenser, and direct cooling water from the spray condenser and steam ejector. The combined condensate plus the surface condenser condensate can be diverted in one stream and discharged through a boil out tank. This stream contains a high BOD_5 and ammonia load.

The waste loads to the evaporation plant and the effluent from the plant are summarized below for two tests during 1972 and 1973 in Table 321. As can be seen, the condensate BOD_5 and NH_3 concentrations experience wide variations. Effluent is discharged from the following points in the paper machine area: floor drains, gland water, felt conditioners and centricleaners. The effluent discharges to a separate sewer and is metered separately. Table 322 shows raw waste characteristics for the combined condensates sewer, the papermill sewer, and the total mill sewer. Table 323 shows the raw waste characteristics of the exemplary mill for this subcategory.

TABLE 321: EVAPORATION PLANT WASTE LOAD REDUCTION AND SECONDARY CONDENSATE DISCHARGE LOADS

		Weak Black Liquor	Combined Condensate
March 1, 1972			
Flow	liters/min (gpm)	983 (260)	839 (222)
BOD5	mg/l	37,900	7,520
	kg/day (lbs/day)	49,900 (110,000)	8,540 (18,800)
	% of Mill Load		60-75
	% Reduction: Evaporation plant	81	
NH3-N	mg/l	7,000	2,600
	kg/day (lbs/day)	9,260 (20,400)	2,910 (6,400)
	% Reduction Evaporation plant	69	
January 1973			
Flow	liters/min (gpm)	680 (180)	
NH3-N	mg/l	9,600	
	kg/day (lbs/day)	9,400 (20,700)	1,750 (3,860)

Source: Report EPA 440/1-74/025

TABLE 322: RAW WASTE CHARACTERIZATION—NSSC (NH$_3$-N)

	Combined Condensate	Paper Mill Sewer	Total Mill Sewer
Flow kiloliters/day (MGD)	1,020 (0.27)	7,180 (1.9)	12,470 (3.3)
BOD5* mg/l	6,120	620	630
BOD5* kg/day (lbs/day)	6,260 (13,800)	4,470 (9,840)	7,850 (17,300)
Suspended Solids mg/l	5	970	620
Suspended Solids kg/day (lbs/day)	5 (11)	6,950 (15,300)	7,760 (17,100)
Kjeldahl Nitrogen mg/l	2,180	285	210
Kjeldahl Nitrogen kg/day (lbs/day)	2,230 (4,910)	2,050 (4,520)	2,640 (5,810)
Ammonia Nitrogen mg/l	1,700	100	150
Ammonia Nitrogen kg/day (lbs/day)	1,740 (3,830)	750 (1,650)	1,880 (4,130)

* Soluble

TABLE 323: RAW WASTE CHARACTERISTICS—NSSC (NH$_3$-N)

Mill	Flow kiloliters/kkg (1000 gal/ton)	BOD5 kg/kkg (lbs/ton)	TSS kg/kkg (lbs/ton)
e*	34.8 (8.33)	335 (67)	17 (34)
e**		305 (61)	16 (32)

* 13 months of daily mill records
** Short term survey data (3-7 days)

Source: Report EPA 440/1-74/025

References

(1) U.S. Environmental Protection Agency, "Compilation of Air Pollutant Emission Factors," 2nd edition, *Publication No. AP-42,* Supplement No. 3, Research Triangle Park, North Carolina, Office of Air and Water Programs (July 1974).
(2) Jones, H.R., *Pollution Control and Chemical Recovery in the Pulp and Paper Industry,* Park Ridge, New Jersey, Noyes Data Corporation (1973).
(3) U.S. Environmental Protection Agency, "Development Document for Proposed Effluent Limitations Guidelines and New Source Performance Standards for the Unbleached Kraft and Semichemical Pulp Segment of the Pulp, Paper and Paperboard Mills Point Source Category," *Report EPA 440/1-74/025,* Washington, D.C. (January 1974).

RAILWAY OPERATION

Railroad locomotives generally follow one of two use patterns: rail yard switching or road haul service. Locomotives can be classified on the basis of engine configuration and use pattern into five categories: 2-stroke switch locomotive (supercharged), 4-stroke switch locomotive, 2-stroke road service locomotive (supercharged), 2-stroke road service locomotive (turbocharged), and 4-stroke road service locomotive.

The engine duty cycle of locomotives is much simpler than many other applications involving diesel internal combustion engines because locomotives usually have only eight throttle positions in addition to idle and dynamic brake. Emission testing is made easier and the results are probably quite accurate because of the simplicity of the locomotive duty cycle.

Air Pollution

Emissions from railroad locomotives are presented two ways in this section. Table 324 contains average factors based on the nationwide locomotive population breakdown by category. Table 325 gives emission factors by locomotive category on the basis of fuel consumption and on the basis of work output (horsepower hour).

TABLE 324: AVERAGE LOCOMOTIVE EMISSION FACTORS BASED ON NATIONWIDE STATISTICS[a]

Pollutant	Average emissions[b]	
	lb/10^3 gal	kg/10^3 liter
Particulates[c]	25	3.0
Sulfur oxides[d] (SO_x as SO_2)	57	6.8
Carbon monoxide	130	16
Hydrocarbons	94	11
Nitrogen oxides (NO_x as NO_2)	370	44
Aldehydes (as HCHO)	5.5	0.66
Organic acids[c]	7	0.84

[a] Reference 1.
[b] Based on emission data contained in Table 325 and the breakdown of locomotive use by engine category in the United States in Reference 1.
[c] Data based on highway diesel data from Reference 2. No actual locomotive particulate test data are available.
[d] Based on a fuel sulfur content of 0.4 percent from Reference 3.

TABLE 325: EMISSION FACTORS BY LOCOMOTIVE ENGINE CATEGORY[a]

EMISSION FACTOR RATING: B

Pollutant	Engine category				
	2-Stroke supercharged switch	4-Stroke switch	2-Stroke supercharged road	2-Stroke turbocharged road	4-Stroke road
Carbon monoxide					
lb/10^3 gal	84	380	66	160	180
kg/10^3 liter	10	46	7.9	19	22
g/hphr	3.9	13	1.8	4.0	4.1
g/metric hphr	3.9	13	1.8	4.0	4.1
Hydrocarbon					
lb/10^3 gal	190	146	148	28	99
lb/10^3 liter	23	17	18	3.4	12
g/hphr	8.9	5.0	4.0	0.70	2.2
g/metric hphr	8.9	5.0	4.0	0.70	2.2
Nitrogen oxides (NO_x as NO_2)					
lb/10^3 gal	250	490	350	330	470
kg/10^3 liter	30	59	42	40	56
g/hphr	11	17	9.4	8.2	10
g/metric hphr	11	17	9.4	8.2	10

[a] Use average factors (Table 324) for pollutants not listed in this table.

Source: EPA Publication AP-42, 2nd edition.

The calculation of emissions using fuel based emission factors is straightforward. Emissions are simply the product of the fuel usage and the emission factor. In order to apply the work output emission factor, however, an additional calculation is necessary. Horsepower hours can be obtained using the following equation: w = lph where w is work output (horsepower hour); l is load factor (average power produced during operation divided by available power); p is available horsepower; and h is hours of usage at load factor. After the work output has been determined, emissions are simply the product of the work output and the emission factor. An approximate load factor for a line haul locomotive (road service) is 0.4, a typical switch engine load factor is approximately 0.06.

References

(1) Hare, C.T. and Springer, K.J., "Exhaust Emissions from Uncontrolled Vehicles and Related Equipment Using Internal Combustion Engines. Part 1, Locomotive Diesel Engines and Marine Counterparts," Final Report, Southwest Research Institute, San Antonio, Texas. Prepared for the Environmental Protection Agency, Research Triangle Park, North Carolina under *Contract Number EHA 70-108* (October 1972).
(2) Young, T.C., unpublished data from the Engine Manufacturers Association, Chicago, Illinois (May 1970).
(3) Hanley, G.P., *Exhaust Emission Information on Electro-Motive Railroad Locomotives and Diesel Engines,* General Motors Corporation, Warren, Michigan (October 1971).

REFRACTORY METAL (Mo, W) PRODUCTION

Water Pollution

The water usage data reported by these plants is given in Table 326. Total intakes for the plants vary considerably and merely reflect that plant size and operations are considerably diverse. However, the figures show considerable use of water in these complex processing plants. The percentages of water used for the various categories within the plant (Table 327) show that the two categories of process and cooling are the major ones in these plants. Of the four plants submitting data, the quantities of discharge water fell into the following categories.

	Million Gallons/Year
Total discharge	305.7
Sanitary wastewater	18.6
Industrial wastewater	
Treated	105.8
Untreated	181.3

All of the sanitary wastewaters of these plants were discharged to municipal sanitary sewers. The industrial wastewaters were discharged to sanitary sewers, storm sewers, or surface waters. Only one of the plants indicated any problem with discharge wastewater quality and identified heavy metals content as the problem, with the receiver being surface water. The water treated by these plants was all of the type identified as wastewater from acid-leaching operations and was all treated by neutralization and filtration. Lime and sodium carbonate were used as neutralizing reagents.

Within the category of untreated water reported above, the respondents including washing, rinsing, and machining coolants, which were actually treated by simple settling. By the definition used in this report, the water is classified as process water since it contacts the product. However, the respondents stated that none of the processed refractory metals entered this type of water. The settling out of the inert metal particles was classed as a materials recovery measure by these plants, as was the recycling of some of the neutralization sludges.

TABLE 326: WATER DATA FOR REFRACTORY METAL PROCESSING PLANTS

Total Water Intake, MGY(a)	Discharge to Receiving Waters, MGY	Total Water Use, New and Recirculated, MGY	Water Intake per Ton of Metal, gal	Water Used per Ton of Metal, gal	Consumption (Intake minus Discharge) per Ton of Metal, gal
40.0	29.0	40.0	42,200	42,200	3,544
16.7	11.7	27.5	111,000	183,000	33,000
270.	255	NA	NA	NA	NA
22.9	>10	284	196,000	2,430,000	NA

(a) Million gallons per year.

TABLE 327: WATER USES AND RECIRCULATION PRACTICE IN REFRACTORY METAL PROCESSING PLANTS

Percent of New Water Intake						Use Ratio(a) by Process						Overall Use Ratio
Process	Cooling	Boiler Feed	Air Pollution Control	Sanitary	Other	Process	Cooling	Boiler Feed	Air Pollution Control	Sanitary	Other	
90.6	--	6.3	--	3.1	--	0	0	0	0	0	0	1
20.0	65.0	--	5.0	--	10.0	1	2	--	1	--	1	1.6
52.9	17.6	23.5	--	5.9	--	--	--	--	--	--	--	>1(b)
28.4	67.6	--	--	4.0	--	1	18.5	--	--	--	1	12.9

(a) Use Ratio = Total Use ÷ New Water Use; 1 signifies no recycling.

(b) Recirculation reported but not specified.

Source: Report EPA-R2-73-2476

One plant reported better than 99% yield of processed material, which would be understandable in view of the value of refractory metals relative to other metals.

One characteristic of these plants is indicated in the water use diagram in Figure 38. In three of the four plants, water use was segregated and wastewater streams were directed to specific receivers. Examples of this included the splitting of streams of used cooling and process water, partly to sanitary sewers and partly to storm sewers. Table 328 summarizes some of the aspects of water pollution by the refractory metals industry and its control.

FIGURE 38: CHARACTERISTICS OF WATER CIRCUITS OF MOLYBDENUM AND TUNGSTEN PROCESSING PLANTS

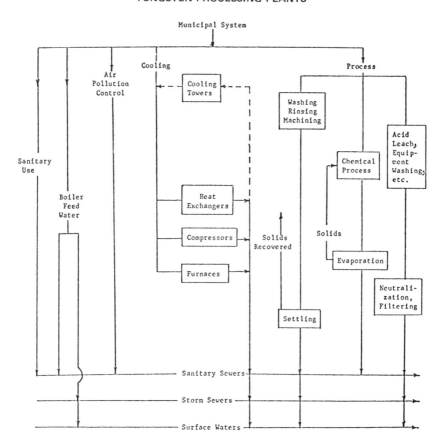

Source: EPA Report R2-73-2476

Reference

(1) Hallowell, J.B., Shea, J.F., Smithson, Jr., G.R., Tripler, A.B. and Gonser, B.W., "Water Pollution Control in the Primary Nonferrous Metals Industry, II—Aluminum, Mercury, Gold, Silver, Molybdenum and Tungsten," *Report Number EPA-R2-73-2476,* Washington, D.C., U.S. Government Printing Office (September 1973).

TABLE 328: WATER POLLUTION IN THE REFRACTORY METALS INDUSTRY

Metal	Operation	Unit Operation	Category or Use of Water	Wastewater Characteristics	Reported Status of Treatment or Control	Future Action
Molybdenum	Mining	Mining	—	—	One of two mines is constructing a total drainage control system.	No action recommended
	Concentration	Froth flotation	Process	High suspended solids, high cyanides	One of two concentrators has developed use of very low cyanides in flotation.	
	Extraction	Roasting	—	—	Potential SO_x pollution problem. May become a water pollution problem by virtue of wet gas scrubbing.	May be studied as portion of acid plant water disposal problem as encountered in copper, lead and zinc industries.
		Hydrogen reduction and powder processing	Powder washing	Spent acid leachants, waste chemical solutions	Well controlled by evaporation or neutralization and filtering due to high value of recovered material; wastes controlled included cyanide components.	No action recommended.
Tungsten	Mining and concentrations	Flotation or gravity concentration	Mine drainage and process	—	Information lacking; operations at high altitudes and remote locations; intermittent due to heavy snows and fluctuating prices.	Basic information needed
	Metal reduction	Pyrometallurgical	—	—	Information lacking; indirect cooling water would be major water use; air pollution control may produce sludges.	Basic information needed
	Scheelite (Tungstic oxide) production	Acid leaching and precipitation	Process	Possible high chlorides and low pH	Information lacking; potential problem.	Basic information needed
	Metal powder production and processing	Hydrogen reduction	—	—	Similar to molybdenum; high metal value results in careful control.	No action recommended
		Metal consolidation			Indirect cooling water major use; machining wastewater treated for materials recovery.	
Vanadium						Too few plant operations

Developments include an inference in a water study report from Arkansas that high chlorides in lake water are attributable to one of the two major vanadium-producing plants. The other major plant is in Colorado, where a chloride air pollution problem may exist. This may become a water problem via the wet scrubber route.

Source: EPA Report R-2-73-2476

RUBBER INDUSTRY, SYNTHETIC

Copolymers of butadiene and styrene, commonly known as SBR, account for more than 70% of all synthetic rubber produced in the United States. In a typical SBR manufacturing process, the monomers of butadiene and styrene are mixed with additives such as soaps and mercaptans. The mixture is polymerized to a conversion point of approximately 60%. After being mixed with various ingredients such as oil and carbon black, the latex product is coagulated and precipitated from the latex emulsion. The rubber particles are then dried and baled.

Air Pollution

Emissions from the synthetic rubber manufacturing process consist of organic compounds (largely the monomers used) emitted from the reactor and blowdown tanks, and particulate matter and odors from the drying operations. Drying operations are frequently controlled with fabric filter systems to recover any particulate emissions, which represent a product loss. Potential gaseous emissions are largely controlled by recycling the gas stream back to the process. Emission factors from synthetic rubber plants are summarized in Table 329 (1).

TABLE 329: EMISSION FACTORS FOR SYNTHETIC RUBBER PLANTS–BUTADIENE-ACRYLONITRILE AND BUTADIENE-STYRENE

EMISSION FACTOR RATING: E

Compound	- - - - -Emissions[a,b] - - - - -	
	lb/ton	kg/Mt
Butadiene	40	20
Methylpropene	15	7.5
Butyne	3	1.5
Pentadiene	1	0.5
Dimethylheptane	1	0.5
Pentane	2	1
Acetonitrile	1	0.5
Acrylonitrile	17	8.5
Acrolein	3	1.5

[a]The butadiene emission is not continuous and is greatest right after a batch of partially polymerized latex enters the blowdown tank.
[b]References (2) and (3).

Source: EPA Publication AP-42, 2nd edition.

Water Pollution

Wastewater characterization data was obtained from literature, EPA documents, and company records (4). Plant visits were made to selected plants to confirm existing data and fill the data gaps. Figures 39, 40 and 41 are generalized flow diagrams of emulsion crumb, solution crumb, and latex production facilities, respectively; they indicate the location of water supply and wastewater generation. Data on total effluent flow and characteristics include utility wastewaters. It is virtually impossible to determine meaningfully total plant effluent flows and characteristics exclusive of utility wastes. It should be noted that utility wastewaters are amenable to treatment by the existing treatment facilities in use and commonly practiced by the industry. Table 330 lists the total effluent flows for plants producing various emulsion crumb rubber products based on a unit of production.

FIGURE 39: GENERAL WATER FLOW DIAGRAM FOR AN EMULSION POLYMERIZED CRUMB RUBBER PRODUCTION FACILITY

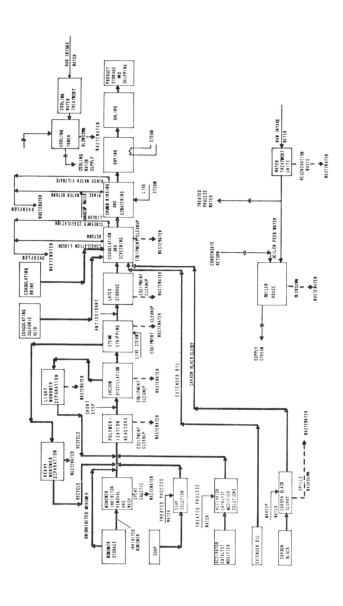

Source: Report EPA 440/1-73/013

FIGURE 40: GENERAL WATER FLOW DIAGRAM FOR A SOLUTION POLYMERIZED CRUMB RUBBER PRODUCTION FACILITY

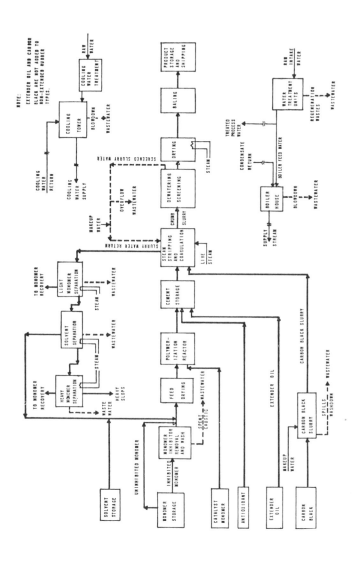

Source: Report EPA 440/1-73/013

FIGURE 41: GENERAL WATER FLOW DIAGRAM FOR AN EMULSION LATEX RUBBER PRODUCTION FACILITY

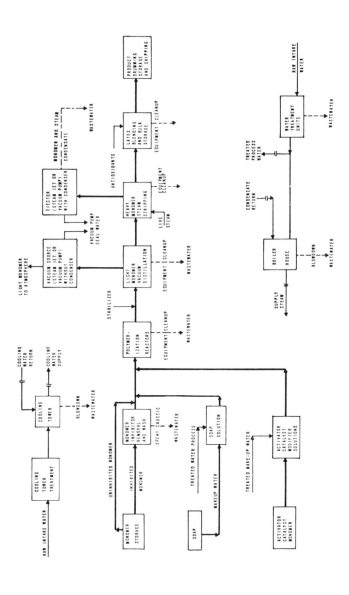

Source: EPA 440/1-73/013

TABLE 330: RAW WASTE LOADS FOR EMULSION CRUMB RUBBER PLANTS[1]

Plant	Product	FLOW L/kkg(gal/ 1,000 lb)	COD kg/kkg(lb/ 1,000 lb)	BOD kg/kkg(lb/ 1,000 lb)	SS[2] kg/kkg,lb/ 1,000 lb)	OIL kg/kkg(lb/ 1,000 lb)
J	SBR and NBR Part Oil and Carbon Black Extended	15,000 (1,860)	11.98	N.A.	3.73	2.09
K	SBR Part Oil Extended	18,500 (2,220)	22.23	2.13	2.30	0.13
K	SBR Oil Extended	18,500 (2,220)	19.76	2.13	11.31	3.54
L	SBR Oil and Carbon Black Extended	16,500 (1,980)	8.72	2.84	3.94	0.48
L	SBR ''Hot'', Non-Extended	15,500 (1,860)	29.24	2.84	N.A.	1.31
L	SBR Non-Extended	15,500 (1,860)	25.87	2.84	11.94	1.45
Median Value			19.63	2.56	6.64	1.5

[1] Includes utility wastewaters.

[2] Raw waste load determined downstream of crumb pits, where the suspended solids and oil levels are reduced.
N.A. Data not available.

Source: Report EPA 440/1-73/013

This data was obtained by plant visits. Although three plants were sampled, six cases of emulsion crumb production were studied. The wastewater contributions of other facilities included solution crumb production and nonrubber commodities. It can be seen from Table 330 that, for similar products, separate plants appear to have different effluent flows. However, different products at the same plant seemingly produce identical wastewater flows.

This is due to the following distinct facts: the water use practices in one plant for different emulsion crumb products are based on one technology, namely that of the company's process design and engineering; and the inability of the sampling team to discern small differences in effluent flows for different products at the same plant. It can also be noted that there is no significant trend in wastewater generation rate between the various types of emulsion crumb rubber product (nonextended, hot, oil extended, and carbon black extended). The average effluent flow rate for emulsion crumb is 16,600 l/kkg (2,000 gal/1,000 lb) of production.

Table 330 also summarizes the raw waste loads for the six cases. It can be seen that the parameter with the highest concentration is COD. The BOD values are generally much lower. The high COD to BOD ratio is indicative of the high resistance of many of the constituents in the wastewaters to biological oxidation. The raw suspended solids concentration in the emulsion crumb wastewaters were determined after separation of the rubber fines in the crumb pits. Since all emulsion crumb plants have separation pits, this raw waste load data is applicable to the industry. Much of the suspended solids contribution is due to uncoagulated latex solids.

The concentration of oil does not appear to be related to the degree to which the crumb rubber is oil extended. The oil analysis cited is really carbon tetrachloride extractables and will also include insoluble monomers. Another significant parameter in emulsion crumb wastewaters is total dissolved solids. This is due to the use of salt in the crumb coagulation process and associated rinse overflows. Surfactants are another characteristic produced by the emulsifying agents. The level of surfactants in the wastewater is considerably lower than the parameters reported in Table 331. Table 331 presents the major constituent loadings of the principal wastewater streams in an emulsion crumb plant. The most significant parameter is total dissolved solids which is produced by the acid and brine coagulation liquors.

TABLE 331: RAW WASTE LOADS OF THE PRINCIPAL INDIVIDUAL WASTEWATER STREAMS IN AN EMULSION CRUMB RUBBER PLANT

Wastewater Stream	COD - kg/kkg (lb/1000 lb.)	BOD - kg/kkg (lb/1000 lb.)	SS[1] - kg/kkg (lb/1000 lb.)	TDS - kg/kkg (lb/1000 lb.)	OIL[1] - kg/kkg (lb/1000 lb.)	SURFACTANTS (lb/1000 lb.)
Monomer Recovery	0.66	0.14	0.08	1.26	0.11	0.0001
Coagulation Liquor Overflow	1.30	N.A.	N.A.	46.25	0.10	N.A.
Crumb Rinse Overflow[2]	6.39	0.46	33.44	42.33	1.46	0.0077
Sub-Total	8.35	0.60	33.52	39.74	1.67	0.0078

[1] Raw waste load determined prior to crumb pit, where the suspended solids and oil levels are reduced.

[2] In one case, the crumb rinse overflow is combined with the coagulation liquor overflow and discharged as one combined stream.

N.A. Data not available.

Source: Report EPA 440/1-73/013

The coagulation liquor and crumb rinse overflows, along with the utility wastes, provide the bulk of the total dissolved solids in the plant effluent. It can be seen that the quantity of surfactants produced are much lower than the other parameters. Surfactants are generated in appreciable quantities only by waste streams included in Table 330. The suspended solids are much higher than in the total effluent since the crumb pits remove much of the suspended solids in the crumb rinse overflow. Removals better than 95% are common. Oil entrained in the rubber is also removed along with rubber crumb solids.

When comparing the subtotal parameter values of Table 331 with the average total effluent loads of Table 330, it can be seen that the three streams listed in Table 331 are the major contributors to the total effluent. The spent caustic scrub solution is an extremely low flow rate wastewater which has very high COD, alkalinity, pH, and color characteristics. It is not, however, a significant waste stream when combined in the total effluent. It is usually bled in at low flow rates into the effluent.

Area washdown and equipment cleanout wastewaters are highly loaded with COD and suspended solids, and, by nature, are intermittent in flow. They cannot be characterized because they are generated on an irregular basis and have greatly variable concentration loadings. Chromium and zinc are present in low concentrations (0.1 mg/l) in the final effluent. They are present due to cooling water treatment, and in some cases can be eliminated or reduced by substitution of chromium-free corrosion inhibitors. Heavy metals from catalysts and other reaction ingredients are not present in measurable concentrations in emulsion plant wastewater effluents.

Table 332 presents the total effluent wastewater flows for facilities producing various solution crumb rubber products. The flow data is given in terms of liters per metric ton (kkg) of production. Five plants were visited and eight types of solution crumb product were sampled. Some plants are multiproduct facilities, and the contributions of the solution crumb facilities were accounted for.

Table 332 shows that there is no discernible difference in the effluent flows between types of product. There appears to be more correlation between products at the same plant site. This is similar to the findings for emulsion crumb rubber production. One plant (Plant M) has a considerably lower effluent flow than all the other facilities. The apparent reason for this difference is the use of a special rubber-finishing process which generates very little or no wastewater. The average effluent flow for solution plants is similar to emulsion plants, and typically approximates 16,600 l/kkg (2,000 gal/1,000 lb) of production.

TABLE 332: RAW WASTE LOADS FOR SOLUTION CRUMB RUBBER PLANTS

Plant	Product	FLOW L/kkg(gal./ 1,000 lb)	COD kg/kkg(lb/ 1,000 lb)	BOD kg/kkg(lb/ 1,000 lb)	SS[1] kg/kkg(lb/ 1,000 lb)	OIL kg/kkg(lb/ 1,000 lb)
K	SBR Oil Extended	10,500 (1,260)	4.04	0.09	0.81	N.A.
K	SBR Carbon Black Extended	17,800 (2.137)	20.80	0.18	2.20	N.A.
L	PBR Oil Extended	28,500 (3,421)	18.40	1.55	5.72	2.43
L	SBR Non-Extended	14,700 (1,765)	13.28	0.82	1.79	1.43
M	PBR Non-Extended	3,400 (408)	0.17.	0.06	0.05	0.07
N	IR Non-Extended	11,900 (1,429)	3.61	1.37	N.A.	0.01
N	PBR Part Oil Extended	11,900 (1,429)	3.01	1.37	5.37	2.32
O	PBR, IR EPDM Part Oil and Carbon Black Extended	29,000 (3,481)	5.33	3.57	3.71	0.23
	Median Value		9.03	1.13	2.81	1.08

[1] Raw waste load determined downstream of crumb pits, where the suspended solids and oil levels are reduced
N.A. Data not available.

Source: Report EPA 440/1-73/013

Table 332 also presents the raw waste loads for the four main parameters. It can be seen that the constituent levels are approximately one-half of those present in emulsion crumb wastewaters. This supports literature and company data which indicate that the solution production processes are cleaner than their emulsion counterparts. The main factor behind this is the absence of coagulation liquor and uncoagulated latex. The COD to BOD ratio is high which indicates that a considerable proportion of the raw wastewater components are not readily biologically oxidizable.

The total dissolved solids content of solution crumb wastewater is considerably lower than for emulsion crumb plants. This again is mainly due to the absence of the coagulation liquor. Surfactant concentrations in the total plant effluent are low. Surfactants are used to deagglomerate the crumb rubber during coagulation and rinsing. The solvent recovery systems do not produce any significant effect on the COD or BOD content of the effluent.

One plant (Plant M) has considerably lower loadings than the others. This is probably due to the fact that rubber for nontire use is produced at this plant. This rubber is used to manufacture impact-resistant resins, and its quality and production controls are extremely critical. In addition, special finishing equipment appears to be used.

The crumb rinse overflow produced at a solution crumb plant is similar to that produced at an emulsion crumb plant, with the exception that uncoagulated latex is not present. The suspended solids, mostly crumb rubber fines, are similar to those in emulsion crumb rinse overflows; the crumb pits produce the same reductions.

The monomer and solvent recovery wastes are comparable to the monomer recovery wastes from emulsion plants. Although heavy slops are produced in some plants, these are usually disposed of by drumming or incineration. Since monomer purities must be high, recovered butadiene, for example, is returned to the monomer supply plant and has no impact on the solution crumb rubber wastewater. Equipment cleanout wastewaters are less of an environmental problem in a solution plant because much of the processing equipment must be kept dry or water-free. Area washdowns are similar in volume, but do not contain latex. These washdowns do pick up rubber solids and oil from pumps and machinery areas.

The spent caustic scrub solution, where used, is identical to that used in emulsion crumb production. In plants where emulsion and solution crumb rubber is produced, the same caustic scrub system is used for both facilities. Catalysts and other reaction ingredients do not produce discernible quantities of heavy metals or toxic constituents. Chromium and zinc in cooling tower blowdown are present in some plant effluents, but in concentrations of 0.1 mg/l or less in the final effluent. These can be eliminated or reduced by using chromium-free corrosion inhibitors.

Table 333 lists the total effluent flows for latex rubber plants. Only two plants are presented, but the similarity between the data values is good. Latex plants are generally part of larger complexes, and flow data for latex operations is difficult to obtain. The flow from latex plants appears to be lower than from either emulsion crumb or solution crumb facilities. The major flow contributions at latex plants originate with equipment cleaning, area washdown operations, and waters from vacuum pump seal systems.

TABLE 333: RAW WASTE LOADS FOR LATEX RUBBER PLANTS

Plant	Product	FLOW L)kkg(gal/1,000lb)	COD kg/kkg(lb/1,000 lb)	BOD kg/kkg(lb/1,000 lb)	SS kg/kkg(lb/1,000 lb)	OIL kg/kkg(lb/1,000 lb)
P	SBR and NBR	14,900 (1,790)	36.37	5.61	6.70	N.A.
Q	SBR	12,000 (1,500)	33.52	5.01	5.63	0.33
Average Value			34.95	5.31	6.17	0.33

N.A. Data not available.

Source: Report EPA 440/1-73/013

The raw waste loads of latex plant wastewaters are considerably higher than either emulsion or solution crumb plants. Equipment cleanout and area washdowns are frequent due to smaller produce runs, and considerable quantities of uncoagulated latex are contained in these wastewaters. The high COD to BOD ratio is typical of all synthetic rubber subcategories and underlines the resistance to biological oxidation of the wastewater constituents. Oil concentration is lower than in emulsion or solution crumb facilities and is contributed by separable monomers, such as styrene, in the wastes. The suspended solids in the effluent are due mainly to uncoagulated latex. Total dissolved solid levels are lower than for emulsion plants because of coagulation liquor stream. Surfactants are present, but in much lower concentrations than the other parameters.

Tank, reactor, and filter cleaning produce considerable quantities of wastewater. These are characterized by high COD, BOD, and suspended solids. In addition, unloading and product loading areas and general plant areas are frequently washed down. The characteristics of these wastes are similar to those produced by usual equipment cleaning in this industry. Vacuum pump seal waters contain small quantities of organics which produce moderate levels of COD from the vacuum stripping operation. The stripping condensates contain condensed monomers. Most of these monomers are decanted from the water and reused. The water layer overflow from the decanter has high COD and BOD concentrations. Spent caustic scrub solution is an extremely low flow waste and has similar characteristics to spent solutions produced in emulsion crumb and solution crumb plants.

References

(1) U.S. Environmental Protection Agency, "Compilation of Air Pollutant Emission

Factors," 2nd edition, *Publication Number AP-42,* Research Triangle Park, N.C., Office of Air and Water Programs (April 1973).

(2) "The Louisville Air Pollution Study," U.S. DHEW, PHS, Division of Air Pollution, Cincinnati, Ohio, p. 26-27 and 124 (1961).

(3) Unpublished data from synthetic rubber plant. U.S. DHEW, PHS, EHS, National Air Pollution Control Administration, Division of Air Quality and Emissions Data, Durham, N.C. (1969).

(4) U.S. Environmental Protection Agency, "Development Document for Proposed Effluent Limitations Guidelines and New Sources Performance Standards for the Tire and Synthetic Segment of the Rubber Processing Point Source Category," *Report EPA 440/1-73/013,* Washington, D.C. (September 1973).

SAND AND GRAVEL PROCESSING

Deposits of sand and gravel, the consolidated granular materials resulting from the natural disintegration of rock or stone, are found in banks and pits and in subterranean and subaqueous beds. Depending upon the location of the deposit, the materials are excavated using power shovels, draglines, cableways, suction dredge pumps, or other apparatus; light charge blasting may be necessary to loosen the deposit. The materials are transported to the processing plant by suction pump, earth mover, barge, truck, or other means. The processing of sand and gravel for a specific market involves the use of different combinations of washers; screens and classifiers, which segregate particle sizes; crushers, which reduce oversize material; and storage and loading facilities (1).

Air Pollution

Dust emissions occur during conveying, screening, crushing and storing operations (2). Because these materials are generally moist when handled, emissions are much lower than in a similar crushed stone operation. Sizeable emissions may also occur as vehicles travel over unpaved roads and paved roads covered by dirt. Although little actual source testing has been done, an estimate has been made for particulate emissions from a plant using crushers: particulate emissions, 0.1 lb/ton (0.05 kg/MT) of product (3).

References

(1) Walker, S., "Production of Sand and Gravel," National Sand and Gravel Association, Washington, D.C. Circular Number 57 (1954).

(2) Schreibeis, W.J. and Schrenk, H.H., "Evaluation of Dust and Noise Conditions at Typical Sand and Gravel Plants," study conducted under the auspices of the Committee on Public Relations, National Sand and Gravel Association, by the Industrial Hygiene Foundation of America, Inc. (1958).

(3) "Particulate Pollutant System Study," volume 1, Mass Emissions, Midwest Research Institute, Kansas City, Mo. Prepared for the Environmental Protection Agency, Research Triangle Park, N.C., under *Contract Number CPA 22-69-104* (May 1971)

SEAFOOD PROCESSING INDUSTRY

Water Pollution

Canned and preserved seafood processing produces a variety of waterborne wastes as described by H.R. Jones (1). Catfish, crab, shrimp and tuna processing have been discussed in detail as regards effluents and their control in an EPA publication (2). Catfish processing will be discussed here as one example. The reader is referred to the complete EPA report for details on crab, shrimp and tuna processing effluents. The farm raised catfish processing industry is relatively new (many plants are less than 5 yr old). There is a large difference in wastewater production depending on whether the fish are delivered in live haul trucks, or on ice, or dry. The samples on which this study is based were taken at five processing plants during April, May and June of 1973. Those months are some of the poorer production months in the industry. Because the peak production season does not come until late summer and fall, mostly small fish were being processed and the additional amount of time required to process smaller fish held the production volume down. The major complication was the severe flooding throughout much of the Mississippi Delta, which hindered or prevented harvesting of the fish, along with other normal industry operations.

There was some difficulty in obtaining samples of the total effluent since the wastewater sources of the processes sampled were quite diverse and often had several exits from the plant. This was usually the case where older buildings designed for other purposes had been converted to catfish processing plants. Depending on the location of the particular plant, a well or city water system supplied the raw water and a city sewer system or local stream were called upon to recieve the final effluent. Figure 42 shows a typical catfish process flow diagram, and Table 334 gives a breakdown of the flow sources.

FIGURE 42: CATFISH PROCESS

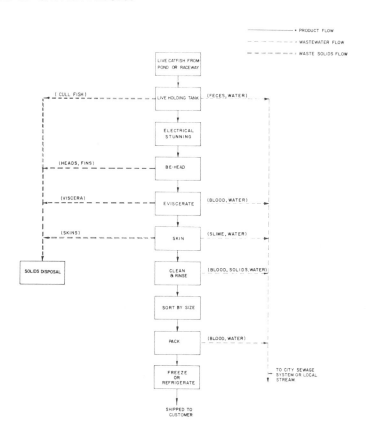

Source: Report EPA 440/1-74/020.

TABLE 334: CATFISH PROCESS MATERIAL BALANCE

Wastewater Material Balance Summary

Average Flow,   116 cu m/day (0.0306 mgd)

	Unit Operation	% of Average Flow	Range, %
a)	live holding tanks	59	55 - 64
b)	butchering (be-heading, eviscerating)	--	-- - --
c)	skinning	4	2 - 7
d)	cleaning	14	9 - 18
e)	packing (incl. sorting)	3	1 - 5
f)	clean-up	7	5 - 9
g)	washdown flows	13	9 - 16

Product Material Balance Summary

Average Raw Product Input Rate, 5.19 kkg/day (5.72 tons/day)

Output	% of Raw Product	Range, %
Food Product	63	-
By-product	27	0 - 32
Waste	10	5 - 37

Source: Report EPA 440/1-74/020

The three main flows formed the effluent and its constituent waste loads. The average wastewater flow from the process plants sampled was 116 m³/day (0.031 mgd) with a moderately large variation of about plus or minus 50% due mainly to holding tank and cleaning differences. The flow from the live holding tank area produced the largest volume of water (59%) and contained the least waste. Conversely, the cleanup flows contributed a relatively small volume of water (7.5%), but contained the highest waste concentrations. The processing flows were the third factor and they contributed a medium volume of water with a medium to heavy waste concentration.

Water reuse was limited to the holding tank and was not a universal practice. Plant 4 retained water in holding tanks for a week or more with an overflow of roughly 0.2 l/sec (3 gpm) from each tank, and as a partial consequence, had one of the lowest total daily flows. Plant 2 had to drain each holding tank completely each time fish were removed from it because of the tank and plant design. Plant 2 had the highest total water usage with over two times the flow of Plant 4. The other plants reused holding tank water in varying degrees. Holding tank flows ran into the tanks from stationary faucets and when the tanks were full the flow drained through standpipe drains. Cleanup flows came almost exclusively from hoses but processing flows were quite diverse in origin.

Processing flows came from skinning machines, washers, chill tanks, the packing area, and eviscerating tables and included water used to flume solids out of the processing area. The by-product solids were removed from the processing area in two ways. They were dry captured in baskets or tubs and removed by that means or flumed to a screening and collection point. All of the plants sampled used the same type of skinning machine, which was designed to operate with a small flow of water. The skins were washed out of the machine.

There is no way to effect dry capture of the skins, short of redesigning the equipment. While the holding tank flow waste was mainly made up of feces, slime, and regurgitated organic matter, the processing and cleanup wastes were made up of blood, fats, small chunks of skin and viscera, and other body fluids or components. A high waste load came from the tanks where the fish were washed, and from the chill tanks. There was no way to dry capture this waste which was composed of blood, fats, and some particulate organic materials.

Table 334 shows the average breakdown of the raw product into food product, by-products and waste. The percent recovered for food depends on the size of the fish and to a slight degree whether manual or mechanical skinning is used. The average is about 63%. Some plants in rural areas dump or bury the waste solids, however, most save the solids and ship them to a rendering plant. The average production rate is about 5.2 kkg/day (5.7 tons/day) with a range from 3 to 7 kkg/day. The average shift length is about 8 hr but is quite variable in some plants due to raw product supply. Table 335 gives the combined average flow and loadings. The average BOD loading was 7.9 kg/kkg with a range from 5.5 to 9.2 kg/kkg. The average BOD concentration was 350 mg/l.

TABLE 335: CATFISH PROCESS SUMMARY

Parameter	Mean	Range		
Flow Rate, cu m/day	116	79	-	170
(mgd)	(0.0306)	(0.021	-	0.045)
Flow Ratio, l/kkg	23,000	15,800	- 31,500	
(gal/ton)	(5510)	(3780	-	7550)
Settleable Solids, ml/l	7.8	--	-	--
Settleable Solids Ratio, l/kkg	180	7.1	-	650
Screened Solids, mg/l	140	--	-	--
Screened Solids Ratio, kg/kkg	3.2	2.5	-	3.9
Suspended Solids, mg/l	400	--	-	--
Suspended Solids Ratio, kg/kkg	9.2	6.8	-	12
5 day BOD, mg/l	340	--	-	--
5 day BOD Ratio, kg/kkg	7.9	5.5	-	9.2
20 day BOD, mg/l	--	--	-	--
20 day BOD Ratio, kg/kkg	--	--	-	--
COD, mg/l	700	--	-	--
COD Ratio, kg/kkg	16	10	-	19
Grease and Oil, mg/l	200	--	-	--
Grease and Oil Ratio, kg/kkg	4.5	3.8	-	5.6
Organic Nitrogen, mg/l	27	--	-	--
Organic Nitrogen Ratio, kg/kkg	0.62	0.51	-	0.75
Ammonia-N, mg/l	0.96	--	-	--
Ammonia-N Ratio, kg/kkg	0.022	0.0045	-	0.045
pH	6.3	5.8	-	7.0

Source: Report EPA 440/1-74/020

In developing the Catfish Process Summary, Table 335, the flow data from Plant 2 was omitted. The excessive water use of 31,500 l/kkg was due to draining the holding tank completely each time the fish were removed. Common practice in the industry includes holding tank water recycle with constant runoff and intermittent drainage.

References

(1)   Jones, H.R., *Pollution Control in Meat, Poultry and Seafood Processing,* Park Ridge, New Jersey, Noyes Data Corporation (1974).
(2)   U.S. Environmental Protection Agency, "Background Document for Proposed Effluent Limitations Guidelines and New Source Performance Standards for the Catfish, Crab, Shrimp and Tuna Segment of the Canned and Preserved Seafood Processing Point Source Category," *Report EPA 440/1-74/020,* Washington, D.C. (January 1974).

SEWAGE SLUDGE INCINERATION

Incineration is becoming an important means of disposal for the increasing amounts of sludge being produced in sewage treatment plants. Incineration has the advantages of both destroying the organic matter present in sludge, leaving only an odorless, sterile ash, as well as reducing the solid mass by about 90%. Disadvantages include the remaining, but reduced, waste disposal problem and the potential for air pollution. Sludge incineration systems usually include a sludge pretreatment stage to thicken and dewater the incoming sludge, an incinerator, and some type of air pollution control equipment (commonly wet scrubbers).

The most prevalent types of incinerators are multiple hearth and fluidized bed units. In multiple hearth units the sludge enters the top of the furnace where it is first dried by contact with the hot, rising, combustion gases, and then burned as it moves slowly down through the lower hearths. At the bottom hearth any residual ash is then removed. In fluidized bed reactors, the combustion takes place in a hot, suspended bed of sand with much of the ash residue being swept out with the flue gas.

Temperatures in a multiple hearth furnace are 600°F (320°C) in the lower, ash cooling hearth; 1400° to 2000°F (760° to 1100°C) in the central combustion hearths, and 1000° to 1200°F (540° to 650°C) in the upper, drying hearths. Temperatures in a fluidized bed reactor are fairly uniform, from 1250° to 1500°F (680° to 820°C). In both types of furnace an auxiliary fuel may be required either during startup or when the moisture content of the sludge is too high to support combustion.

Air Pollution

Because of the violent upward movement of combustion gases with respect to the burning sludge, particulates are the major emissions problem in both multiple hearth and fluidized bed incinerators. Wet scrubbers are commonly employed for particulate control and can achieve efficiencies ranging from 95 to 99+%. Although dry sludge may contain from 1 to 2% sulfur by weight, sulfur oxides are not emitted in significant amounts when sludge burning is compared with many other combustion processes. Similarly, nitrogen oxides, because temperatures during incineration do not exceed 1500°F (820°C) in fluidized bed reactors or 1600° to 2000°F (870° to 1100°C) in multiple hearth units, are not formed in great amounts.

Odors can be a problem in multiple hearth systems as unburned volatiles are given off in the upper, drying hearths, but are readily removed when afterburners are employed. Odors are not generally a problem in fluidized bed units as temperatures are uniformly high enough to provide complete oxidation of the volatile compounds. Odors can also emanate from the pretreatment stages unless the operations are properly enclosed. Emission factors for sludge incinerators are shown in Table 336.

TABLE 336: EMISSION FACTORS FOR SEWAGE SLUDGE INCINERATORS

EMISSION FACTOR RATING: B

| Pollutant | Emissions [a] | | | |
| | Uncontrolled[b] | | After scrubber | |
	lb/ton	kg/MT	lb/ton	kg/MT
Particulate	100	50	3	1.5
Sulfur dioxide	1	0.5	0.8	0.4
Carbon monoxide	Neg	Neg	Neg	Neg
Nitrogen oxides (as NO_2)	6	3	5	2.5
Hydrocarbons	1.5	0.75	1	0.5
Hydrogen chloride gas	1.5	0.75	0.3	0.15

[a]Unit weights in terms of dried sludge.
[b]Estimated from emission factors after scrubbers.

Source: EPA Publication AP-42, 2nd edition.

It should be noted that most sludge incinerators operating today employ some type of scrubber.

Reference

(1)   U.S. Environmental Protection Agency, "Compilation of Air Pollutant Emission Factors," 2nd edition, *Publication Number AP-42,* Supplement Number 3, Research Triangle Park, N.C., Office of Air and Water Programs (July 1974).

SOAP AND DETERGENT MANUFACTURE

Soap Manufacture (1): The manufacture of soap entails the catalytic hydrolysis of various fatty acids with sodium or potassium hydroxide to form a glycerol-soap mixture. This mixture is separated by distillation, then neutralized and blended to produce soap.

Detergent Manufacture (1): The manufacture of detergents generally begins with the sulfuration by sulfuric acid of a fatty alcohol or linear alkylate. The sulfurated compound is then neutralized with caustic solution (NaOH), and various dyes, perfumes, and other compounds are added (2)(3). The resulting paste or slurry is then sprayed under pressure into a vertical drying tower where it is dried with a stream of hot air (400° to 500°F or 204° to 260°C). The dried detergent is then cooled and packaged.

Air Pollution

The main atmospheric pollution problem in the manufacture of soap is odor, and, if a spray dryer is used, a particulate emission problem may also occur. Vent lines, vacuum exhausts, product and raw material storage, and waste streams are all potential odor sources. Control of these odors may be achieved by scrubbing all exhaust fumes and, if necessary, incinerating the remaining compounds. Odors emanating from the spray dryer may be controlled by scrubbing with an acid solution.

The main source of particulate emissions in detergent manufacture is the spray drying tower. Odors may also be emitted from the spray drying operation and from storage and mixing tanks. Particulate emissions from spray drying operations are shown in Table 337 on the following page.

TABLE 337: PARTICULATE EMISSION FACTORS FOR SPRAY-DRYING DETERGENTS
DETERGENTS[a]

EMISSION FACTOR RATING: B

| Control device | Overall efficiency, % | Particulate emissions | |
		lb/ton of product	kg/MT of product
Uncontrolled	–	90	45
Cyclone[b]	85	14	7
Cyclone followed by:			
Spray chamber	92	7	3.5
Packed scrubber	95	5	2.5
Venturi scrubber	97	3	1.5

[a]Based on analysis of data in References 2 through 6.
[b]Some type of primary collector, such as a cyclone, is considered an integral part of the spray-drying system.

Source: EPA Publication AP-42, 2nd edition.

Water Pollution

Numerous organic and inorganic chemical compounds are used in the manufacture of soaps and detergents. As the reactions are carried out some of these materials and their derivatives enter the wastewaters from the processing steps. These materials are then treated as contaminants and processed as wastewater in conventional waste treatment units. In discussion of the individual unit processes that follow, the sources and nature of wastewaters are presented in some detail in an EPA summary document (1). That document discusses the water and wastewater balance and wastewater constituents for a variety of individual processes:

    [1]   Soap manufacture by batch kettle
    [2]   Fatty acids by fat splitting
    [3]   Soap by fatty acid neutralization
    [4]   Glycerine recovery
    [5]   Soap flakes and powders production
    [6]   Bar soap production
    [7]   Liquid soap production
    [8]   Oleum sulfonation and sulfation
    [9]   Air-SO$_3$ sulfonation and sulfation
    [10]   SO$_3$ solvent and vacuum sulfonation
    [11]   Sulfamic acid sulfonation
    [12]   Chlorosulfonic acid sulfonation
    [13]   Neutralization of sulfuric acid esters and sulfonic acid
    [14]   Spray dried detergent manufacture
    [15]   Liquid detergent manufacture
    [16]   Detergent manufacture by dry blending

Wastewater constituents are reviewed here briefly by the process numbers used above.
[1] Leaks, spills and storm runoff or floor washing are invariably collected for recovery of fats and oils by settling and skimming in fat traps. The wastewater will contain some emulsified fats, since the skimmers and settlers do not operate at 100% efficiency.

The fat pretreatment is carried out to remove impurities which would cause color and odor in the finished soap. Acid and/or caustic washing may be used. This results in sodium soaps or sulfuric acid solutions of fatty acids. Other pretreatment steps make use of proprietary chemical formulas which result in water containing the treatment chemicals, fatty impurities and emulsified fats. Clay and carbon treatments give solid wastes and do not directly result in aqueous effluents, but steam is used for heating and the condensate must be removed. Often a barometric condenser is used, and there is carry-over of low molecular weight fatty acids. Wastewaters from the fat skimmer and from the pretreatment steps each contribute about 1.5 kg of BOD_5/1,000 kg of soap (1.5 lb/1,000 lb). Concentrations typically are 3,600 mg/l BOD_5, 4,267 mg/l COD, 250 mg/l of oil and grease with a pH of 5. Saponification of the fats and oils by sodium hydroxide and salting out of the soaps (graining) with salt does not necessarily lead to any effluent. The nigre which comes from washing excess salt and impurities from the neat soap with aqueous caustic is always recycled to some extent.

The organic portion consists of low grade soap. In some plants, the nigre is acidified to convert the soaps to fatty acids which are recovered for sale. Although acidification removes much of the organic contaminant, some is still discharged. Some manufacturers' third alternative for the nigre is the sewer. The stream referred to as sewer lyes arises most often from the reclaiming of scrap soap. The lye and salt water added to separate the soap must be discarded because it contains paper and other dirt. Sewer lyes and nigre are concentrated wastewaters (in mg/l) alkalinity up to 32,000, BOD_5 as high as 45,000, COD up to 64,000, chlorides of 47,000 and a pH of 13.5. Volumes, however, are small, 249 l/kkg of soap (30 gal/1,000 lb).

[2] Process condensate from fat splitting will be contaminated with volatile low molecular weight fatty acids as well as entrained fatty acids and glycerine streams. The barometric condensate will also contain volatile fatty acids. These streams will be settled and skimmed to remove the insoluble fatty acids which are processed for sale. The water will typically circulate through a cooling tower and be reused. To keep emulsified and soluble fatty material at a reasonable level, part of the stream is purged to the sewer. This blowdown contributes about 10 kg of BOD_5 and 18 kg of COD/kkg (10 lb BOD_5 and 18 lb COD/1,000 lb) of fatty acids produced plus some oil and grease.

Treatment of the spent catalyst and water washings stream consists of acidification to break the emulsion and skimming of insoluble fatty acid pitch. The wastewater is neutralized and sent to the sewer. This wastewater will contain salt from the neutralization, zinc and alkaline earth metal salts from the fat splitting catalyst and emulsified fatty acids and fatty acid polymers. One plant had about 0.6 kg of BOD_5 and 0.9 kg of COD/kkg (0.6 lb BOD_5 and 0.9 lb COD/1,000 lb) of fatty acids from this source. Fatty acids and a modest amount of nickel soaps constitute the bulk of contaminants from hydrogenation. Very small amounts of suspended nickel may be present.

[3] The sewer lyes will contain the excess caustic soda and the salt added to grain out the soap. Also, they will contain some dirt and paper not removed by the strainer. Typically, 3 kg of BOD_5 and 5.5 kg of COD/1,000 kg (3 lb of BOD_5 and 5.5 lb of COD/1,000 lb) of soap will be discharged.

[4] The two barometric condensers streams become contaminated with glycerine and salt due to entrainment. Stream 104428 is a water-soluble by-product which is disposed of by washing into the sewer. It contains glycerine, glycerine polymers, and salt. The organics will contribute to BOD_5, COD and dissolved solids. The sodium chloride will also contribute to dissolved solids. Little or no suspended solids, oil and grease or pH effect should be seen.

From the glycerine evaporator barometric condenser about 30 kg of COD and 15 kg of BOD_5/kkg (30 lb of COD and 15 lb of BOD_5/1,000 lb) of product will be discharged. The foots and glycerine still contribute about equal amounts of BOD_5 and COD, 2.5 kg (2.5 pounds) and 5 kg (5 lb) respectively per kkg (1,000 lb) of glycerine produced. Because the barometric condensers use large amounts of water, concentrations are low. The foots are diluted only enough to remove them from the still, and concentrations of organics and salts are about 30,000 to 400,000 mg/l.

[5] In this process the neat soaps from processes [1] or [3] are converted to flakes or powders for packaging and sale. The unit processes produce the dry soap in the physical form desired. There are a number of possible effluents shown on the flow sheet for

process [5]. However, survey of the industry showed that most operating plants either recycled any wastewater to extinction, or used dry cleanup processes. Occasionally water will be used for equipment cleanup. In converting neat soap to flakes, powder or bars, some scrap results. The flow sheets show scrap reclaim in processes [1] and [3]. There is an aqueous effluent, sewer lyes, from this reclaim operation. All existing soap plants both make and process soap. Therefore, the given categorization will handle the effluent sewer lyes under processes [1] or [3]. Should a plant start with neat soap, the sewer lye guidelines should be applied to process [5].

[6] The contaminant of all wastewaters is soap which will contribute primarily to BOD_5 and COD. Concentrations of BOD_5 and COD are typically 1,600 and 2,850 mg/l respectively in the scrubber, which is the major source of contamination. Dryers contribute from 0.3 to 0.7 kg/kkg (0.3 to 0.7 lb/1,000 lb) of soap to wastewaters. Washouts of the filter and of equipment account for the remainder to give an average process total of 2 kg of BOD_5 per kilokilogram (2 lb/1,000 lb) of dry soap.

[7] The liquid soaps will contribute to BOD_5, COD, and dissolved solids. However, amounts are very small (0.1 kg BOD_5 and 0.3 kg COD/kkg of product) (0.1 lb BOD_5 and 0.3 lb COD per 1,000 lb).

[8] Since the source of contaminants is leaks and spills, all the raw materials and products may be present. These are fatty alcohols, dodecylbenzene, sulfuric acid, dodecylbenzene-sulfonic acid, and the esters of sulfuric acid and alcohols. These chemicals will contribute acidity, sulfate ion, MBAS, oil, BOD_5 and COD to the wastewater stream. Concentrations of contaminants will depend on the amount of water used for washdown. Amounts of contaminants average 0.2 kg (0.2 lb) BOD_5, 0.6 kg (0.6 lb) COD and 0.3 kg (0.3 lb) MBAS per 1,000 kilokilogram (1,000 lb) of sulfonated product. The pH will usually be very low (1 to 2) unless the sulfonation wastes are commingled with neutralization wastes.

[9] The stream which receives startup slop will contain unreacted alcohols, alcohol ethoxylates, and alkylbenzenes. These will show up as oil and grease as well as BOD_5 and COD. The other contaminated streams will contain sulfonic and sulfuric acids. These materials affect MBAS, acidity, SO_4, dissolved solids, BOD_5 and COD. Concentrations from the process were observed to range from 380 to 520 mg/l of BOD_5 and 920 to 1,589 mg/l of COD. The pH ranged from 2 to 7. Source of the contamination comes from three streams: leaks and spills 15%, scrubbers 50%, and startup slop 35%.

[10] Leaks and spills will consist of raw materials and products. These will include alkylbenzenes, alcohols and ethoxylated alcohols, sulfonic acids and sulfuric acid esters. These constituents will contribute BOD_5, COD, acidity, dissolved solids, sulfate, sulfite oil, and MBAS to effluent wastewaters.

[11] A small amount of contamination will come from leaks and spills, but a major effluent comes from washing out the reactor between batches. This is necessary since the solvent (water or alcohol) will react with the sulfamic acid. Data submitted by industry indicates that 30 kg (30 lb) of BOD_5 and 60 kg (60 lb) of COD is added to wastewaters/kkg (1,000 pounds) of ammonium ether sulfate produced. This is understandable since the product is viscous and surface to volume ratio of the reactor is high.

[12] The only process water used is for absorbing HCl. Cooling water is noncontact and should remain uncontaminated. Contaminants found in leaks and spills will be the alcohols or alkylphenol and alcohol ethoxylates used for feedstocks. Also, chlorosulfonic acid hydrolysis products (HCl and H_2SO_4) can be expected, plus the sulfated surfactants. The same materials will be in the other wastewaters. Since it is not necessary to clean out the reactor between batches, raw waste loads will be similar to other sulfonation processes, i.e., SO_3-air. An average of 3 kg (3 lb) of BOD_5 and 9 kg (9 lb) of COD/kkg (1,000 lb) of sulfated product was found. Sizeable amounts of acidity and chloride (5 kg/kkg) (5 lb/1,000 pounds) can also be expected if the HCl is sent to the sewer, but many plants recover it and sell it as muriatic acid.

[13] All of the anionic surface active agents used in soaps and detergents will be found in these wastewaters. Also the inorganic salts, such as sodium sulfate, from neutralization of excess sulfuric acids will be found. Alkylbenzenesulfonates, ether sulfates, alcohol sulfates, olefinsulfonates, with the ammonium, potassium, sodium, magnesium and triethanol ammonium cations will be represented. These constituents will contribute to BOD_5 and COD, MBAS dissolved solids, Kjeldahl ammonium nitrogen, sulfate, acidity and alkalinity.

The total amount of contaminants contributed from this process is quite small. The equipment used has stood the tests of time. Of course there will always be occasional leaks and spills. Whether a plant recycles or discharges to a sewer determines the concentration of contaminants and water usage. When recycle is practiced, as little as 10.4 l/kkg (1.25 gal per 1,000 lb) of water is used. Concentrations in this case were high, 6,000 mg/l of BOD_5 and 21,000 mg/l of COD. The other extreme was 4,170 liters (1,100 gal) of water, a BOD_5 of 85 mg/l and a COD of 245 mg/l. Because of the variety of products made quite a range of waste loadings was observed: BOD_5, 0.07 to 0.8 kg/kkg (0.07 to 0.8 lb/1,000 lb) of product; COD, 0.2 to 2.3 kg (0.2 to 2.3 lb); MBAS, 0.1 to 3.1 kg (0.1 to 3.1 lb).

[14] All of the streams will be contaminated with the detergent being produced in the plant at the time. The various surfactants, builders and additives will contribute to BOD_5, COD, MBAS, dissolved and suspended solids, oil and grease and alkalinity. Average raw waste loads for the three types of operations of spray towers are tabulated as follows (kilogram per kilokilogram) (pound per 1,000 pounds).

	BOD5	COD	Suspended Solids	Surfactants	Oil and Grease
Few turnarounds & no air quality problem	0.1	0.3	0.1	0.2	nil
Air quality problems	0.8	2.5	1.0	1.5	0.3
Fast turn-around	0.2	0.4	0.2	0.4	0.03

[15] All of the effluent streams will contain the starting ingredients of the products. These are mainly:

> ammonium, potassium and sodium alkylbenzenesulfonates (LAS)
> ammonium, potassium and sodium alcohol ethoxy sulfates (AES)
> ammonium, potassium and sodium alkylphenol ethoxy sulfates
> ammonium, potassium and sodium olefinsulfonates (OS)
> ammonium, potassium and sodium toluene-and xylenesulfonates (hydrotropes)
> ammonium, potassium and sodium alcohol sulfates
> fatty acid alkanolamine condensates (amides)
> urea (hydrotrope)
> ethanol (hydrotrope)
> polyacrylates and polystyrene (opacifiers)
> dyes and perfume
> phosphate and citrate builders
> silicate builders

These constituents will contribute to alkalinity, BOD_5, COD, nitrogen, MBAS (and undetected surfactants) and dissolved solids. Raw waste loadings vary more in concentration than total amount between plants. The following were observed: BOD_5, 0.5 to 1.8 kg/kkg of detergent, 65 to 3,400 mg/l; COD, 1.0 to 3.1 kg/kkg of detergent, 120 to 7,000 mg/l; and MBAS 0.4 to 1.1 kg/kkg of detergent, 60 to 2,000 mg/l.

As much as 90% of the detergent can come from washout of tanks and lines when changing products. With fewer changes waste loads will decrease, but the lower levels in the above data represent minimum with current practice.

[16] The only appreciable effluent will be similar to, but lower in concentration than the scrubber water from spray drying. Surfactants, builders, and free oils can be expected. These will contribute to BOD_5, COD, alkalinity, MBAS, dissolved solids and oil and grease.

References

(1) "Air Pollutant Emission Factors," Final Report, Resources Research Inc., Reston, Va. Prepared for National Air Pollution Control Administration, Durham, N.C. under *Contract Number CPA 22-69-119* (April 1970).
(2) Phelps, A.H., "Air Pollution Aspects of Soap and Detergent Manufacture," *J. Air Pol. Control Assoc.* 17, 505 to 507 (August 1967).
(3) Shreve, R.N., *Chemical Process Industries,* 3rd edition, New York, McGraw-Hill Book Company, p. 544-563 (1967).
(4) Larsen, G.P., Fischer, G.I. and Hamming, W.J., "Evaluating Sources of Air Pollution," *Ind. Eng. Chem.* 45, 1070-1074 (May 1953).
(5) McCormick, P.Y., Lucas, R.L. and Wells, D.R., "Gas-Solid Systems," *Chemical Engineer's Handbook,* Perry, J.H. editor, New York, McGraw-Hill Book Company p. 59, (1963).
(6) EPA communication with Maryland State Department of Health, Baltimore, Maryland (November 1969).
(7) U.S. Environmental Protection Agency, "Development Document for Proposed Effluent Limitations Guidelines and New Source Performance Standards for the Soap and Detergent Manufacturing Point Source Category," *Report EPA 440/1-74/18,* Washington, D.C. (December 1973).

SODIUM BICARBONATE MANUFACTURE

Sodium bicarbonate is manufactured by the reaction of soda ash and CO_2 in solution, separated by thickening and centrifugation, and then dried, purified and sold.

Water Pollution

A listing of raw wastes produced in bicarbonate manufacture in a specimen plant is shown below. These consist of unreacted soda ash, solid sodium bicarbonate, boiler wastes and ash from power generation equipment. The ash is treated as a solid waste.

Waste Product	Process Source	kg/kkg of Product (lb/ton)	
		Average	Range
Na_2CO_3	Slurry thickener overflow	38.0 (76.0)	0 - 375 (0 - 750)
Ash	Power generation	17.9 (35.8)	
Water purif. sludge	Boiler feedwater purification	0.3 (0.6)	
$NaHCO_3$	Slurry thickener overflow	10.0 (20.0)	

The quantity of slurry thickener overflow depends upon the operation of another plant utilizing this by-product. The overflow is not constant and occurs only when the sister plant cannot absorb the entire flow. Consequently the value shown above is basic on an annual average, with a wide variation in flow over the period. The effluent from this specimen plant contains 20,000 mg/l of dissolved solids (mostly dissolved carbonates), amounting to 5.75 kg/kkg of product (11.5 lb/ton). All of the bicarbonate wastes are treated along with chlor-alkali and soda ash wastes at the 166 facility in a common treatment system prior to discharge. There are no net effluent loads to the cooling water based on average daily operation and no organics in plant effluent. The plant plans to use the weak slurry thickener overflow, constituting its major source of waste, as a source of liquid for the product dryer scrubber and to recycle this liquid (concentrated with respect to sodium carbonate) back to the process. These process changes will eliminate the discharge of process wastewaters.

Reference

(1)   U.S. Environmental Protection Agency, "Development Document for Proposed Efflu-ent Limitations Guidelines and New Source Performance Standards for the Major Inorganic Products Segment of the Inorganic Chemicals Manufacturing Point Source Category, *Report EPA 440/1-73/007* (August 1973).

SODIUM CARBONATE MANUFACTURE

Soda ash is manufactured by three processes: the natural or Lake Brine process, the Solvay process (ammonia-soda), and the electrolytic soda-ash process. Because the Solvay process accounts for over 80% of the total production of soda ash, it will be the only one discussed in this section. In the Solvay process, the basic raw materials are ammonia, coke, limestone (calcium carbonate), and salt (sodium chloride). The salt, usually in the unpurified form of a brine, is first purified in a series of absorbers by precipitation of the heavy metal ions with ammonia and carbon dioxide. In this process sodium bicarbonate is formed. This bicarbonate coke is heated in a rotary kiln, and the resultant soda ash is cooled and conveyed to storage.

Air Pollution

The major source of emissions from the manufacture of soda ash is the release of ammonia (1). Small amounts of ammonia are emitted in the gases vented from the brine purification system. Intermittent losses of ammonia can also occur during the unloading of tank trucks into storage tanks. The major sources of dust emissions include rotary dryers, dry solids handling, and processing of lime. Dust emissions of fine soda ash also occur from conveyor transfer points and air classification systems, as well as during tank car loading and packaging. Emission factors are summarized in Table 338.

TABLE 338: EMISSION FACTORS FOR SODA-ASH PLANTS WITHOUT CONTROLS

EMISSION FACTOR RATING: D

Type of source	Particulates		Ammonia	
	lb/ton	kg/MT	lb/ton	kg/MT
Ammonia recovery[a,b]	—	—	7	3.5
Conveying, transferring, loading, etc.[c]	6	3	—	—

[a]Reference 2.
[b]Represents ammonia loss following the recovery system.
[c]Based on data in References 3 through 5.

Source: EPA Publication AP-42, 2nd edition.

Water Pollution

In the Solvay process, raw sodium chloride brine is purified to remove calcium and magnesium compounds and is reacted with ammonia and carbon dioxide produced from limestone calcination to yield crude sodium bicarbonate which is recovered from the solutions by filtration. The bicarbonate is calcined to yield soda ash and the spent brine-ammonia solution is reacted with slaked lime and distilled to recover ammonia values for process recycle. The calcium chloride formed as a by-product during the distillation is either discharged as a waste or recovered by evaporation. Although all Solvay process plants have high dissolved solids effluents, this plant is unusual in that it recovers a significant amount of an otherwise wasted by-product.

Since the market for calcium chloride will not absorb the by-product generated from such recovery from all Solvay plants, this plant cannot be considered to be exemplary on this basis. The information from it is given here to document such by-product recovery and because it employs generally good processing practices for this industry. The raw waste loads for one facility consist of brine purification muds, unreacted sodium chloride and the calcium chloride by-product as follows.

Waste Products	Process Source	kg/kkg of Soda Ash (lb/ton)
$CaCO_3$	DS, B, P	84.5 (169)
Na_2CO_3	B	0.3 (0.6)
$CaSO_3$	DS	31 (62)
NaCl	DS, B	510.5 (1,021)
$CaCl_2$	DS	1,090 (2,180)
Na_2SO_4	B	0.8 (1.6)
$Fe(OH)_3$	B	0.1 (0.2)
$Mg(OH)_2$	DS, B, P	48.5 (97)
CaO (inactive)	DS, B	109.5 (219)
NaOH	B	0.05 (0.1)
SiO_2	DS, B	58.5 (117)
CaO (active)	DS	24 (48)
NH_3	DS	0.15 (0.3)
H_2S	DS	0.02 (0.04)
Ash and cinders	P	40 (80)

DS is distillation, B is brine, P is power.

The plant effluent after treatment contains 100,000 mg/l DS (NaCl and $CaCl_2$) in the process waste stream and is high in SS. This type of effluent is typical of a Solvay process plant. The only exemplary feature of this facility is its partial recovery and reuse of $CaCl_2$.

References

(1) "Air Pollutant Emission Factors," Final Report, Resources Research, Inc. Reston, Va. Prepared for National Air Pollution Control Administration, Durham, N.C. (April 1970).
(2) Shreve, R.N., *Chemical Process Industries,* 3rd edition, New York, McGraw-Hill Book Company p. 225-230 (1967).
(3) "Facts and Figures for the Chemical Process Industries," *Chemical Engineering News,* 43, 51-118 (September 6, 1965).
(4) Faith, W.L., Keyes, D.B. and Clark, R.L., *Industrial Chemicals,* 3rd edition, New York, John Wiley and Sons, Inc. (1965).
(5) Kaylor, F.B., "Air Pollution Abatement Program of a Chemical Processing Industry," *J. Air Pollution Control Association,* 15, 65-67 (February 1965).
(6) U.S. Environmental Protection Agency, "Development Document for Proposed Effluent Limitations Guidelines and New Source Performance Standards for the Major Inorganic Products Segment of the Inorganic Chemicals Manufacturing Point Source Category," *Report EPA 440/1-73/007* (August 1973).

SODIUM CHLORIDE MANUFACTURE

Sodium chloride is produced by three methods: solar evaporation of seawater, solution mining of natural brines and conventional mining of rock salt.

Water Pollution

In the solar evaporation process, seawater is concentrated by evaporation over a period of 5 years in open ponds to yield a saturated brine solution. After saturation is reached, the brine is then fed to a crystallizer, wherein sodium chloride precipitates, leaving behind a concentrated brine solution (bittern) consisting of sodium, potassium and magnesium salts. The precipitated sodium chloride is recovered for sale and the brine is then further evaporated to recover additional sodium chloride values and is then either stored, discharged back to salt water or further worked to recover potassium and magnesium salts.

In the solar evaporation process, all of the wastes are present in the bittern solution which is stored at all facilities. No bittern is discharged from this facility. The bittern is stored and has been worked for recovery of other materials. At one plant discussed by EPA (1), treatment consists of storage and further use of the bittern materials. There is no waste discharge. The plant water usage is as follows.

Type	Use	Source	cu m/day (MGD)	1/kkg (gal/ton)	Recycle
Process	Refining process	Well	2,270 (0.60)	894 (214)	100%
Process	Raw Material	Bay	327,000 (86.4)	129,000 (30,900)	None

As the bitterns are stored and further worked, there is no discharge. Eventual total evaporation after further bittern use yields only solid wastes. Sufficient land and ponding area are available at the facility to store bitterns for the next 30 to 50 years without difficulty.

In solution mining, saturated brine for the production of evaporated salt is usually obtained by pumping water into an underground salt deposit and removing the saturated salt solution from an adjacent interconnected well, or from the same well by means of an annular pipe. Besides sodium chloride, the brine will contain some calcium sulfate, calcium chloride, magnesium chloride, and lesser amounts of other materials including iron salts and sulfides.

The chemical treatment given to brines varies from plant to plant depending on the impurities present. Typically, the brine is first aerated to remove hydrogen sulfide and, in many cases, small amounts of chlorine are added to complete sulfide removal and oxidize all iron salts present to the ferric state. The brine is then pumped to settling tanks where it is treated with soda ash and caustic soda to remove most of the calcium, magnesium and iron present as insoluble salts. After clarification to remove these insolubles, the brine is sent to multiple-effect evaporators. As water is removed, salt crystals form and are removed as a slurry.

After screening to remove lumps, the slurry is washed with fresh brine to remove fine crystals of calcium sulfate from the mother liquor to the slurry. These solids are returned to the evaporator. The calcium sulfate concentration in the evaporator eventually builds up to the point where it must be removed by boiling out the evaporators. The washed slurry is filtered, the mother liquor is returned to the evaporators, and the salt crystals from the filter are dried and screened. Salt thus produced from a typical brine will be of 99.8% purity or greater.

Some plants do not treat the raw brine, but control the calcium and magnesium impurities by watching the concentrations in the evaporators and bleeding off sufficient brine to maintain predetermined levels. By such methods, salt of better than 99.5% purity can be made. In either case, the final screening of the dried salt yields various grades depending on particle size. A detailed list of the raw wastes and their process sources is shown below. These include wastes from the multiple evaporators and dryers, sludges from basic purification, as well as water treatment chemicals used for the cooling water.

Waste Products	Process Source	Average kg/kkg of Product (lb/ton)
NaOH	Boiler blowdown	0.0055 (0.011)
Na_3PO_4	Boiler blowdown	0.0015 (0.003)
Na_2SiO_3	Boiler blowdown	0.0025 (0.005)
Na_2SO_3	Boiler blowdown	0.0015 (0.003)
NaCl and $CaSO_4$	Purge from multiple evaporator	0.045 (0.090)
NaCl	Evaporator	0.04 (0.08)
NaCl	Barometric condenser	1.1 (2.2)
NaCl	Miscellaneous sources	—
Brine sludges	Brine purification	91 kkg/yr (100 ton/yr)

The brine sludges are returned to the brine wells for settling and disposal.

Reference

(1) U.S. Environmental Protection Agency, "Development Document for Proposed Effluent Limitations Guidelines and New Source Performance Standards for the Major Inorganic Products Segment of the Inorganic Chemicals Manufacturing Point Source Category," *Report EPA 440/1-73/007,* Washington, D.C. (August 1973).

SODIUM DICHROMATE MANUFACTURE

Sodium dichromate is prepared by calcining a mixture of chrome ore ($FeO \cdot Cr_2O_3$), soda ash and lime, followed by water leaching and acidification of the soluble chromates. The insoluble residue from the leaching operation is recycled to leach out additional material. During the first acidification step, the chromate solution pH is adjusted to precipitate calcium salts. Further acidification converts it to the dichromate and a subsequent evaporation step crystallizes sodium sulfate (salt cake) out of the liquor. The sulfate is then dried and sold. The solutions remaining after sulfate removal are further evaporated to recover sodium dichromate. Chromic acid is produced from sodium dichromate by reaction with sulfuric acid. Sodium bisulfate is a by-product.

Water Pollution

The raw waste loads expected from the manufacture of sodium dichromate and its by-product sodium sulfate are given below. The bulk of the waste originates from the undigested portions of the ores used. These materials are mostly solid wastes. The wastes arising from spills and washdowns contain most of the hexavalent chromium. The wastes from water treatment and boiler blowdowns are principally dissolved sulfates and chlorides. The manufacture of chromic acid contributes no additional wastes.

Waste Product	Process Source	Product (lb/ton) Average	Range
1. Chromate wastes (Materials not digested in H2SO4)	Residues	900 (1800)	—
2. Washdowns* spills, etc.	--	0.75 (1.5)	0.5-1 (1-2)
3. Blowdown	Boilers and cooling towers	--	0.5-1 (1-2)

---
*Includes contributions from the chromic acid unit.

Wastewaters are treated with pickle liquor to effect reduction of chromates present and then all effluent waters are lagooned to settle out suspended solids. This treatment removes 99% of the hexavalent chromium and the discharge contains 0.01 mg/l. The lagoon discharges to a nearby river when full.

Chromate waste control in this plant is excellent. All rainwater, washdowns, spills and minor leaks in the part of the plant which handles hexavalent chromium are captured in the area's sumps and used in the process. Storage facilities are provided to contain a heavy rain and return the water either to the process or to treatment. Separate rainwater drainage is provided for areas not handling hexavalent chromium. Sewers are continuously monitored. Even cooling tower and boiler blowdowns go through the process waste treatment, as do all waste sludges. A batch system is used in the treatment process. Each batch is treated and analyzed before release to the lagoon. Data on the effluent from this exemplary chromate treatment facility are presented below.

	Average	Range
Flow, liters/kkg (gal/ton)	8,860 (2,120)	–
Total suspended solids, mg/l	14	1 - 24
Total dissolved solids, mg/l (mostly chlorides)	10,000	5,000 - 13,000
pH	7.2	6.0 - 8.5
Cr^{+3}, mg/l (mostly suspended solids)	0.14	0.01 - 0.31
Cr^{+6}, mg/l	0.01	

The chromium content has been reduced to negligible values. However, the amount of sodium chloride being discharged is significant. Based on the porous nature of the lagoon walls and the high dissolved solids content discharged into the river, this plant is considered exemplary only from the standpoint of chromates control and treatment.

Reference

(1) U.S. Environmental Protection Agency, "Development Document for Proposed Effluent Limitations Guidelines and New Source Performance Standards for the Major Inorganic Products Segment of the Inorganic Chemicals Manufacturing Point Source Category", *Report EPA 440/1-73/007,* Washington, D.C. (August 1973).

SODIUM METAL MANUFACTURE

Sodium is manufactured by electrolysis of molten sodium chloride in a Downs electrolytic cell. After salt purification to remove calcium and magnesium salts and sulfates, the sodium chloride is dried and fed to the cell, where calcium chloride is added to give a low-melting $CaCl_2$-NaCl eutectic, which is then electrolyzed. Sodium is formed at one electrode, collected as a liquid, filtered and sold. The chlorine liberated at the other electrode is first dried with sulfuric acid and then purified, compressed, liquified and sold.

Water Pollution

There is no waste during operation of an individual cell for the molten salt electrolysis step in the Downs cell process. The cells are run in banks, and individual cells are cleaned out and refilled after the electrolyte is depleted. All of the wastes arise from this cleaning and refilling of individual cells. The wastes produced by sodium manufacture are shown below. Several of the expected wastes are not present. This is due to the reuse of materials in other parts of the facility to make other products, for example, the sulfuric acid used in drying the chlorine.

Waste Products	Process Source	kg/kkg of Product (lb/ton)
NaCl	Process	50 - 65 (100 - 130)
Miscellaneous alkaline salts	Process	25 - 35 (50 - 70)
Ca(OCl)$_2$	Chlorine recovery	45 - 75 (90 - 150)
Fe	Cooling tower	0.065 - 0.095 (0.13 - 0.19)

The process does not normally shut down. The discharges result from the replacement of cells. At the exemplary plant, cooling tower blowdowns and residual chlorine from tail gas scrubbers are discharged without treatment. The stream containing calcium hypochlorite wastes is not discharged but is used to treat cyanide wastes. Cooling water is discharged without treatment and tank wash and runoff water are first ponded to settle out suspended materials and then discharged. These stream effluents consist mostly of dissolved sodium chloride and other chlorides.

References

(1) Sittig, M., *Sodium—Its Manufacture, Properties and Uses,* New York, Reinhold Publishing Corporation (1956).
(2) U.S. Environmental Protection Agency, "Development Document for Proposed Effluent Limitations Guidelines and New Source Performance Standards for the Major Inorganic Products Segment of the Inorganic Chemicals Manufacturing Point Source Category," *Report EPA 440/1-73/007* (August 1973).

SODIUM SILICATE MANUFACTURE

Sodium silicate is manufactured by the reaction of soda ash or anhydrous sodium hydroxide with silica in a furnace, followed by dissolution of the product in water under pressure to prepare sodium silicate solutions. In some plants, the liquid silicate solutions are then further reacted with sodium hydroxide to manufacture metasilicates which are then isolated by evaporation and sold.

Water Pollution

The raw waste loads consist mostly of sodium silicate and unreacted silica.

Waste Products	Process Source	Average kg/kkg of Dry Basis Product (lb/ton)
Sodium silicate	Scrubbers	37 (74)
Silica	Scrubbers	2.85 (5.7)
NaOH/Silicates	Washdowns	0.39 (0.78)

Data on in-plant water use could not be obtained from the exemplary plant above. However, the water use data from another plant not exemplary because it has process water discharge is given below on the basis of unit weight of product (dry basis) to indicate the general level. The water intake is 2,900 liters/metric ton (710 gal/ton) which is used as follows.

Water Use	l/kkg (gal/ton)
Process water	1,020 (245)
Boiler blowdown, compressor cooling, washdown, tank cleaning and misc.	610 (147
Steam, evaporation and other losses	1,330 (319)

At the exemplary plant all scrubber and washdown waters are sent to a totally enclosed evaporation pond. There is no plant effluent. Since this exemplary plant is in an area of normal rainfall and humidity for the humid areas of the United States, the evaporation ponding technique appears generally applicable.

Reference

(1)   U.S. Environmental Protection Agency, "Development Document for Proposed Effluent Limitations Guidelines and New Source Performance Standards for the Major Inorganic Products Segment of the Inorganic Chemicals Manufacturing Point Source Category," *Report EPA 440/1-73/007* (August 1973).

SODIUM SULFITE MANUFACTURE

Sodium sulfite is manufactured by reaction of sulfur dioxide with soda ash. The crude sulfite formed in this reaction is then purified, filtered to remove insolubles from the purification step, crystallized, dried and shipped.

Water Pollution

A listing of the raw wastes produced from sodium sulfite production is given below. These consist of sulfides from the purification step and a solution produced by periodic vessel cleanouts containing sulfite and sulfate.

Waste Products	Process Source	kg/kkg of Product (lb/ton) Average	Range
Metal sulfides	Filter wash	0.755 (1.51)	0.19 - 1.44 (0.38 - 2.88)
Na_2SO_3/Na_2SO_4 solution	Dryer ejector	—	—
Na_2SO_3/Na_2SO_4 solution	Process cleanout	—	—

Cleanouts of various process vessels produce shock loads up to 9.1 kkg (10 tons) of sodium sulfite and sulfate (dry basis). Cleanouts are conducted 3 to 6 times per year. For this, separate tanks are used for surge capacity which bleed into the treatment unit over a 5 to 10 day period.

Reference

(1)   U.S. Environmental Protection Agency, "Development Document for Proposed Effluent Limitations Guidelines and New Source Performance Standards for the Major Inorganic Products Segment of the Inorganic Chemicals Manufacturing Point Source Category," *Report EPA 440/1-73/007*, Washington, D.C. (August 1973).

STARCH MANUFACTURE

The basic raw material in the manufacture of starch is dent corn, which contains starch. The starch in the corn is separated from the other components by wet milling. The shelled grain is prepared for milling in cleaners that remove both the light chaff and any heavier foreign material. The cleaned corn is then softened by soaking (steeping) it in warm water acidified with sulfur dioxide. The softened corn goes through attrition mills that tear the kernels apart, freeing the germ and loosening the hull. The remaining mixture of starch, gluten, and hulls is finely ground, and the coarser fiber particles are removed by screening. The mixture of starch and gluten is then separated by centrifuges, after which the starch is filtered and washed. At this point it is dried and packaged for market (1).

Air Pollution

The manufacture of starch from corn can result in significant dust emissions. The various cleaning, grinding, and screening operations are the major sources of dust emissions. Table 339 presents emission factors for starch manufacturing.

TABLE 339: EMISSION FACTORS FOR STARCH MANUFACTURING[a]

EMISSION FACTOR RATING: D

Type of operation	Particulates	
	lb/ton	kg/MT
Uncontrolled	8	4
Controlled[b]	0.02	0.01

[a]Reference 2.
[b]Based on centrifugal gas scrubber.

Source: EPA Publication AP-42, 2nd edition.

Water Pollution

The basic corn wet milling operations are divided into the basic milling operation, starch production, and syrup refining, respectively. The modern wet corn mill, in many respects, is already a bottled up plant, compared to its ancestors of 50 to 75 years ago. Historically, this segment of the industry has succeeded in reducing the fresh water consumption per unit of raw material used in the basic production operations exclusive of cooling waters. The wastewaters from one source are now used as makeup water for other production operations.

Recycled process wastewaters are identified by the symbol PW to distinguish them from wastewaters that are sewered. Fresh water enters the overall corn wet milling production sequence primarily in the starch washing operations. This water then moves countercurrent to the product flow direction back through the mill house to the steepwater evaporators. More specifically, the process wastewaters from starch washing are reused several times in primary starch separation, fiber washing, germ washing, milling, and finally as the input water to the corn steeping operation. The principal sources of wastewaters discharged to the sewer from this sequence of operations are modified starch washing, and condensate from steepwater evaporation.

Additional fresh water is used in the syrup refinery. Although practice varies within the industry, fresh water may be introduced in starch treating, neutralizing, enzyme production, carbon treatment, ion exchange, dextrose production, and syrup shipping. In some plants, evaporator condensate is used to supply many of these fresh water requirements, particularly in carbon treatment and ion exchange regeneration. Other process wastewaters are used in mud separation, syrup evaporation, animal feeds, and corn steeping.

Total water use in this subcategory varies from less than 3,785 up to 190,000 m^3/day (1.0 to 50 mgd) depending, in large measure, on the types of cooling systems employed. Those plants using once-through cooling water have much higher water demands than those using recirculated systems, whether they be surface or barometric condensers. The water use per unit of raw material ranges from about 0.0067 to 0.0745 m^3/kkg of corn grind (45 to 500 gal/MSBu). Those plants that predominantly use once-through cooling water will have total water use values of about 0.045 m^3/kkg of grind (300 gal/MSBu). This number should be contrasted with the several plants that use recirculated cooling water almost exclusively, where the total water use values are about 0.0075 m^3/kkg (50 gal/MSBu).

Information is not available on the water use by individual production processes since these vary from plant to plant. Company preferences, type of equipment, product mix, and other factors all influence the water use in terms of both the individual processes and the total plant.

As indicated in the discussion on water use, many process wastewaters that were discharged to sewers-years ago, are now recycled back into the process. This section is concerned only with those wastes that are generally discharged to the sewer. Major wastes included are those from steepwater evaporators; starch modifying washing and dewatering; syrup refining (cooling, activated carbon treatment, and ion exchange regeneration); syrup evaporation; and syrup shipping.

The condensate from steepwater evaporation constitutes one of the several major wastewater sources in a corn wet mill. Normally, triple effect evaporators are used with either surface or barometric condensers. Vapors from each of the first two effects passes through the subsequent effect before being discharged to the sewer. For those systems using surface condensers, the condensate from the third effect is sewered.

In the case of barometric condensers, the third effect condensate becomes a part of the barometric cooling water discharge and hence, is greatly diluted. Limited data on characteristics of the waste discharges from the first and second effect evaporators were acquired during the sampling program and are presented in Table 340. Selected samples taken from the barometric cooling waters serving the third effect evaporator, indicate much lower waste concentrations, as expected. BOD_5 levels ranged from 10 to 75 mg/l with typical values reported by industry in the range of 25 mg/l.

TABLE 340: FIRST AND SECOND EFFECT STEEPWATER CONDENSATE WASTEWATER CHARACTERISTICS

	Range (mg/l)
BOD_5	723 - 934
COD	1,095 - 1,410
Suspended solids	10 - 28
Dissolved solids	110 - 292
Phosphorus as P	0.5 - 0.7
Total nitrogen as N	2.4 - 2.6
pH	3.0 - 3.5

Source: Report EPA 440/1-73/028

Surface condensers will generate essentially the same total quantity of waste constituents, but in a much smaller volume of water. To reduce wastewater flows, several plants recirculate barometric cooling water and only discharge the blowdown from the cooling tower to the sewers. Measurements of the blowdown from such a system at one plant indicated a BOD_5 of about 440 mg/l and a suspended solids content of 80 mg/l.

Data from a previous study for the Environmental Protection Agency (Table 342) indicate that steepwater evaporation systems using once-through cooling water generate about 4.5 to 13.4 m³/kkg (30 to 90 gal/SBu) of process wastes. Recirculating cooling water systems, on the other hand, generate about 10% of this flow, namely 0.6 to 0.9 m³/kkg (4 to 6 gallons per SBu).

Additional data from the same study related waste characteristics to raw material input and indicated a BOD_5 range of 0.9 to 2.9 kg/kkg (0.5 to 0.16 lb/SBu) and a COD range 1.1 to 3.2 kg/kkg (0.06 to 0.18 lb/SBu). In many, if not most, corn wet mills the waste from the production of modified starches represents the largest single source of contaminants in terms of organic load.

Limited samples taken at two mills indicated very high BOD, COD, and dissolved and suspended solids, as indicated in Table 341.

TABLE 341: FINISHED STARCH PRODUCTION WASTEWATER CHARACTERISTICS

	Range (mg/l)
BOD$_5$	3,549 - 3,590
COD	8,250 - 8,686
Suspended solids	918 - 2,040
Dissolved solids	9,233 - 16,211
Phosphorus as P	25 - 63
Total nitrogen as N	32 - 41
pH	4.2 - 5.7

Source: Report EPA 440/1-73/028

These very high strength wastes are highly variable in both composition, flow and biodegradability. Information from earlier studies on the waste characteristics relative to raw material input is summarized in Table 342. It is important to note that the production of modified starches varies not only from plant to plant, but from day to day and week to week in any given plant. Moreover, the nature of the wastewater generated from starch modification depends on the particular starches being manufactured. For example, mild oxidation with sodium hypochlorite generates a lower dissolved organic load than highly oxidized starch production. No correlation has yet been established between the types and amounts of starches being produced and the waste loads from this operation.

In most mills, wastewaters are discharged from several operations in the syrup refinery. Most of these wastewaters are generated by the series of operations generally referred to as syrup refining, which includes activated carbon and ion exchange treatment. Typically, the so-called sweetening-off procedures require flushing the spent carbon or ion exchange resin with water prior to regeneration. The first flush of such water is usually sent to the syrup evaporator for reclamation. The final rinse water is very dilute in syrup content and is discharged from the plant.

Sampling data indicate that wastewaters from the ion exchange regeneration are high in organic content, with BOD$_5$ levels of 500 to 900 mg/l, and in dissolved solids, 2,100 to 9,400 milligrams/liter. The pH levels of the wastewater were quite low, averaging about 1.8 and the suspended solids averaged 25 mg/l. Other sources of wastewaters in the syrup refinery include: syrup (flash) cooling, evaporation, dextrose production, and shipping. Samples of wastes from the syrup cooling process at one plant gave the results shown in Table 343.

TABLE 343: CORN SYRUP COOLING WASTEWATER CHARACTERISTICS

	Concentration
BOD$_5$	73
COD	177
Suspended solids	44
Dissolved solids	291
Phosphorus as P	0.2
Total nitrogen as N	0.4
pH	6.7

Source: Report EPA 440/1-73/028

The concentration of wastes from syrup evaporation again depends on the type of condensers used, i.e., surface and barometric with recirculation versus barometric with once-

TABLE 342: INDIVIDUAL PROCESS WASTE LOADS — CORN WET MILLING

Manufacturing Process	Flow cu m/kkg	Flow gals/SBu	BOD kg/kkg	BOD lbs/MSBu	COD kg/kkg	COD lbs/MSBu	Dissolved Solids kg/kkg	Dissolved Solids lbs/MSBu	Total Solids kg/kkg	Total Solids lbs/MSBu
Steepwater evaporation	0.6-13.4	4-90	0.9-2.9	50-160	1.1-3.2	60-180	3.4-3.9	190-220	4.3-4.6	240-260
Feed dewatering	0.3-0.6	2-4	0.2	10	-	-	0.2-0.5	10-30	-	-
Oil refining	0.03-3.7	0.2-25	0.04-0.5	2-30	-	-	0.2-0.4	10-20	0.2-0.5	10-30
Starch modifying, washing, etc.	1.5-7.5	10-50	1.8-10.7	100-600	2.3-10.7	130-600	13.0-28.6	730-1600	15.7-44.7	880-2500
Syrup refining (carbon, ion exchange)	3.7	25	2.7	150	9.3	520	10.4	580	10.5	590
Syrup evaporation	0.3-30	2-200	1.3	70	2.5	140	1.1-6.1	60-340	1.1-9.5	60-530
Corn syrup shipping	0.3-0.4	2-3	0.9	50	1.4	80	1.3	70	1.2	70
Dextrose and corn syrup solids	-	-	-	-	-	-	1.8-2.1	100-120	-	-

Source: Report EPA 440/1-73/028

through cooling water. Other data on the wastewaters from the various activities in a syrup refinery are included in Table 341. Wastewater streams of less importance include discharges from feed dewatering, oil extraction and refining, and general plant cleanup. Sampling data taken at one plant indicate that the wastewaters from the feed house contained about 140 milligrams/liter of suspended solids, and negligible amounts of phosphorus and nitrogen, and had a pH of 5.9. Other data are also presented in Table 341.

Most of the data accumulated from various sources during this study relate to the total raw waste characteristics from corn wet mills. Summary data from 12 of the 17 mills are presented in Table 344. Wastewaters from this grain milling subcategory can generally be characterized as high volume, high strength discharges. The BOD varies, 255 to 4,450 mg/l, with a corresponding range in COD. Those plants with very low BOD_5 values typically have barometric condensing systems using once-through cooling water. At the other extreme, the very concentrated wastes are from plants using recirculated cooling water (either surface or barometric condensers). Suspended solids levels in the total waste streams show similar variations, 81 to 2,458 mg/l. Again, the plants with low suspended solids concentrations are those using barometric condensers with once-through cooling water.

Other waste parameters indicate that the pH of the total waste ranges from 6.0 to 8.0. These average pH values, however, are misleading as wide pH fluctuations are common to many plants. Typically, the waste may be somewhat deficient in nitrogen for biological waste treatment. Dissolved solids levels from certain process operations generally do not constitute a problem when combined in the total waste stream. In those plants that have minimized water use, dissolved solids buildup may be a future concern.

TABLE 344: TOTAL PLANT RAW WASTEWATER CHARACTERISTICS

Plant	BOD, mg/l Average	BOD, mg/l Range	COD, mg/l Average	COD, mg/l Range	Suspended Solids, mg/l Average	Suspended Solids, mg/l Range	pH Average	pH Range
1	1400	400–3000	–	–	1200	66–4628	5.9	4.7–9.4
2	1625	464–4320	2100	–	477	100–3490	6.5	1.4–10.4
3	614	–	–	–	700	–	–	–
4	2880	650–7800	–	–	1230	170–5200	6.1	4.1–9.5
5	450	–	–	–	–	–	–	4.5–8.2
6	444	303–526	798	–	288	225–335	–	–
7	2330	288–25,000	4560	347–40,000	895	43–18,000	7.1	5.0–11.3
8	4450	780–11,000	–	–	2458	410–9400	–	–
9	1650	1246–2370	3500	750–10,300	700	–	6.0	–
10	998	146–4618	–	–	259	8–3216	7.2	5.4–9.8
11	225	–	473	–	81	–	–	–
12	2584	67–9592	–	–	862	19–7744	7.9	1.8–12.0

Source: Report EPA 440/1-73/028

References

(1) "Starch Manufacturing," *Kirk-Othmer Encyclopedia of Chemical Technology,* Volume IX, New York, John Wiley and Sons, Inc. (1964).
(2) Storch, H.L., "Product Losses Cut with a Centrifugal Gas Scrubber," *Chem. Eng. Progr.* 62, 51-54 (April 1966)
(3) U.S. Environmental Protection Agency, "Development Document for Proposed Effluent Limitations Guidelines and New Source Performance Standards for the Grain Processing Segment of the Grain Mills Point Source Category," *Report EPA 440/1-73/028,* Washington, D.C. (December 1973).

STATIONARY ENGINE OPERATIONS

In general, engines included in this category are internal combustion engines used in applications similar to those associated with external combustion sources. The major engines within this category are gas turbines and large, heavy-duty, general utility reciprocating engines. Emission data currently available for these engines are limited to gas turbines and natural gas-fired, heavy-duty, general utility engines. Most stationary, off-highway internal combustion engines are used to generate electric power, to pump gas or other fluids, and to compress air for pneumatic machinery.

Stationary Gas Turbines

Stationary gas turbines find application in electric power generators, in gas pipeline pump and compressor drives, and in various process industries. A recent survey revealed that the majority of these engines are used in electrical generation for continuous, peaking, or standby power. The survey also indicated that the primary fuels used are natural gas and No. 2 (distillate) fuel oil, although residual oil is used in a few instances. Stationary gas turbines are adaptations of aircraft turbines that use a power turbine to convert jet thrust into rotational power. Although turbine duty cycles vary with the engine application, many are run continuously at base load, which is customarily about 80% of maximum capability.

Heavy-Duty, General Utility, Gaseous-Fueled Engines

Engines in this category are used in the oil and gas industry for driving compressors in pipeline pressure boosting systems, in gas distribution systems, and in vapor recovery systems (at petroleum refineries). The engines burn either natural gas or refinery gas.

Air Pollution

Tables 345 and 346 contain emission factors for gas turbines burning distillate fuel oil and natural gas, respectively. The emission values reported are for base load (except for sulfur oxides, which are dependent on fuel consumption and sulfur content). Test data reveal little difference between base load emission levels and peak load emissions levels. For this reason the values listed in the tables can be used to estimate peak load emissions as well as base load emissions. The tabulated values do not apply to emissions at low power settings. Most gas turbine applications, however, do not involve the frequent use of low power so that base load factors apply in the majority of cases.

TABLE 345: EMISSIONS FOR GAS TURBINES USING DISTILLATE FUEL OIL

Emission Factor Rating: C

Pollutant[a]	Emissions (base load)			
	lb/10^3 gal[b]	kg/10^3 liter[b]	lb/10^6 Btu	kg/10^6 kcal
Particulate	8.4	1.0	0.06	0.11
Sulfur oxides[c] (SO$_x$ as SO$_2$)	142S	17S	1.0S	1.8S
Nitrogen oxides (NO$_x$ as NO$_2$)	120	14	0.84	1.5

[a] No data are available for carbon monoxide and hydrocarbon mass emissions. These pollutants exist in significant quantities only at lower power settings (Reference 1). Because normal operation of stationary gas turbines is at base load, carbon monoxide and hydrocarbon emissions can be considered negligible.

[b] Particulate and nitrogen oxides emissions were reported as lb/10^6 Btu and have been converted to lb/10^3 gal (kg/10^3 liter) assuming 140,500 Btu/gal.

[c] This factor was calculated assuming that all surfur in the fuel is oxidized to SO$_2$. The SO$_2$ emission factor is calculated using the weight percent sulfur in fuel.

Source: EPA Publication AP-42, 2nd Edition

TABLE 346: EMISSION FACTORS FOR GAS TURBINES USING NATURAL GAS
Emission Factor Rating: C

Pollutant[c]	Emissions (base load)[a,b]			
	lb/10^6 ft^3	kg/10^6 m^3	lb/10^6 Btu	kg/10^6 kcal
Nitrogen oxides (NO$_x$ as NO$_2$)	600	9,600	0.57	1.0
Sulfur oxides[d]	0.6	9.6	0.0006	0.001

[a] Reference (1).
[b] Natural gas heat content of 1050 Btu/ft^3 (9350 kcal/m^3) was assumed.
[c] Carbon monoxide and hydrocarbon emissions can be considered negligible. Emission factors for particulate are not available but are assumed to be negligible.
[d] Values based on an average natural gas sulfur content of 2000 gr/10^6 ft^3 (4600 g/10^6 m^3).

Source: EPA Publication AP-42, 2nd Edition

Emissions from heavy-duty gaseous-fueled internal combustion engines are reported in Table 347. Test data were available for nitrogen oxides and hydrocarbons only; sulfur oxides are calculated from fuel sulfur content. Nitrogen oxides have been found to be extremely dependent on an engine's work output; hence, Figure 43 presents the relationship between nitrogen oxide emissions and horsepower.

FIGURE 43: NITROGEN OXIDES EMISSIONS FROM STATIONARY INTERNAL COMBUSTION ENGINES

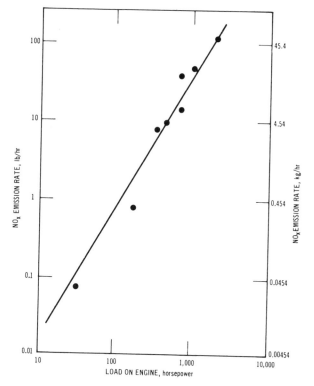

Source: EPA Publication AP-42, 2nd Edition

TABLE 347: EMISSION FACTORS FOR HEAVY-DUTY, GENERAL-UTILITY
STATIONARY ENGINES USING GASEOUS FUELS

Emission Factor Rating: C

Pollutant	Emissions[a]			
	$lb/10^6 ft^3$	$kg/10^6 m^3$	lb/hr	kg/hr
Sulfur oxides[b]	0.6	9.6	—	—
Nitrogen oxides[c]	—	—	—	—
Hydrocarbons[d]	1.2	19	4.2	1.9

[a] Reference 1. Values for $lb/10^6 ft^3$ ($kg/10^6 m^3$) based on 3.37 $10^6 ft^3$/hr heat input.
[b] Based on an average natural gas sulfur content of 2000 $gr/10^6 ft^3$ (4600 $g/10^6 m^3$).
[c] See Figure 43.
[d] Values in Reference 1 were given as tons/day. In converting to lb/hr, 24-hour operation was assumed.

Source: EPA Publication AP-42, 2nd Edition

Reference

(1) U.S. Environmental Protection Agency, "Compilation of Air Pollutant Emission Factors", 2nd Edition *Publication No. AP-42*, Research Triangle Park, N.C., Office of Air and Water Programs (April 1973).

STEEL FOUNDRIES

Steel foundries produce steel castings by melting steel metal and pouring it into molds. The melting of steel for castings is accomplished in one of five types of furnaces: direct electric-arc, electric induction, open-hearth, crucible, and pneumatic converter. The crucible and pneumatic converter are not in widespread use, so this section deals only with the remaining three types of furnaces. Raw materials supplied to the various melting furnaces include steel scrap of all types, pig iron, ferroalloys, and limestone. The basic melting process operations are furnace charging, melting, tapping the furnace into a ladle, and pouring the steel into molds.

An integral part of the steel foundry operation is the preparation of casting molds, and the shake out and cleaning of these castings. Some common materials used in molds and cores for hollow casting include sand, oil, clay, and resin. Shake out is the operation by which the cool casting is separated from the mold. The castings are commonly cleaned by shot-blasting, and surface defects such as fins are removed by burning and grinding.

Air Pollution

Particulate emissions from steel foundry operations include iron oxide fumes, sand fines, graphite, and metal dust. Gaseous emissions from foundry operations include oxides of nitrogen, oxides of sulfur, and hydrocarbons. Factors affecting emissions from the melting process include the quality and cleanliness of the scrap and the amount of oxygen lancing. The concentrations of oxides of nitrogen are dependent upon operating conditions in the melting unit, such as temperature and the rate of cooling of the exhaust gases.

The concentration of carbon monoxide in the exhaust gases is dependent on the amount of draft on the melting furnace. Emissions from the shake out and cleaning operations, mostly particulate matter, vary according to type and efficiency of dust collection. Gaseous emissions from the mold and baking operations are dependent upon the fuel used by the ovens and the temperature reached in these ovens. Table 348 summarizes the emission factors for steel foundries.

TABLE 348: EMISSION FACTORS FOR STEEL FOUNDRIES

Emission Factor Rating: A

Type of process	Particulates[a]		Nitrogen oxides	
	lb/ton	kg/MT	lb/ton	kg/MT
Melting				
Electric arc[b,c]	13 (4 to 40)	6.5 (2 to 20)	0.2	0.1
Open-hearth[d,e]	11 (2 to 20)	5.5 (1 to 10)	0.01	0.005
Open-hearth oxygen lanced[f,g]	10 (8 to 11)	5 (4 to 5.5)	—	—
Electric induction[h]	0.1	0.05	—	—

[a]Emission factors expressed as units per unit weight of metal processed. If the scrap metal is very dirty or oily, or if increased oxygen lancing is employed, the emission factor should be chosen from the high side of the factor range.

[b]Electrostatic precipitator, 92 to 98 percent control efficiency; baghouse (fabric filter), 98 to 99 percent control efficiency; venturi scrubber, 94 to 98 percent control efficiency.

[c]Reference(1).

[d]Electrostatic precipitator, 95 to 98.5 percent control efficiency; baghouse, 99.9 percent control efficiency; venturi scrubber, 96 to 99 percent control efficiency.

[e]Reference(1).

[f]Electrostatic precipitator, 95 to 98 percent control efficiency; baghouse, 99 percent control efficiency; venturi scrubber, 95 to 98 percent control efficiency.

[g]Reference(1).

[h]Usually not controlled.

Source: EPA Publication AP-42, 2nd Edition

Reference

(1)   U.S. Environmental Protection Agency, "Compilation of Air Pollutant Emission Factors, 2nd Edition, *Publication No. AP-42,* Research Triangle Park, N.C., Office of Air and Water Programs (April 1973).

STONE QUARRYING AND PROCESSING

Rock and crushed stone products are loosened by drilling and blasting them from their deposit beds and are removed with the use of heavy earth-moving equipment. This mining of rock is done primarily in open pits. The use of pneumatic drilling and cutting, as well as blasting and transferring, causes considerable dust formation(1). Further processing includes crushing, regrinding, and removal of fines(2). Dust emissions can occur from all of these operations, as well as from quarrying, transferring, loading and storage operations. Drying operations, when used, can also be a source of dust emissions.

Air Pollution

As enumerated above, dust emissions occur from many operations in stone quarrying and processing. Although a big portion of these emissions is heavy particles that settle out within the plant, an attempt has been made to estimate the suspended particulates. These emission factors are shown in Table 349. Factors affecting emissions include the amount of rock processed; the method of transfer of the rock; the moisture content of the raw material; the degree of enclosure of the transferring, processing, and storage areas; and the degree to which control equipment is used on the processes.

TABLE-349: PARTICULATE EMISSION FACTORS FOR ROCK-HANDLING PROCESSES

Emission Factor Rating: C

Type of process	Uncontrolled total[a]		Settled out in plant, %	Suspended emission	
	lb/ton	kg/MT		lb/ton	kg/MT
Dry crushing operations[b,c]					
Primary crushing	0.5	0.25	80	0.1	0.05
Secondary crushing and screening	1.5	0.75	60	0.6	0.3
Tertiary crushing and screening (if used)	6	3	40	3.6	1.8
Recrushing and screening	5	2.5	50	2.5	1.25
Fines mill	6	3	25	4.5	2.25
Miscellaneous operations[d]					
Screening, conveying, and handling[e]	2	1			
Storage pile losses[f]	10	5			

[a] Typical collection efficiencies: cyclone, 70 to 85 percent; fabric filter, 99 percent.

[b] All values are based on raw material entering primary crusher, except those for recrushing and screening, which are based on throughput for that operation.

[c] Reference 3.

[d] Based on units of stored product.

[e] Reference 4.

[f] The significance of storage pile losses is mentioned in Reference 5. The factor assigned here is the author's estimate for uncontrolled total emissions. Use of this factor should be tempered with knowledge about the size of materials stored, the local meteorological factors, the frequency with which the piles are disturbed, etc.

Source: EPA Publication AP-42, 2nd Edition

References

(1)   Air Pollutant Emission Factors, Final Report, Resources Research, Inc. Reston, Va. Prepared for National Air Pollution Control Administration, Durham, N.C., (April 1970).

(2)   Communication between Resources Research, Incorporated, Reston, Virginia, and the National Crushed Stone Association (September 1969).

(3)   Culver, P., Memorandum to files. U.S. DHEW, PHS, National Air Pollution Control Administration, Division of Abatement, Durham, N.C. (January 6, 1968).

(4)   Unpublished data on storage and handling or rock products, U.S. DHEW, PHS, National Air Pollution Control Administration, Division of Abatement, Durham, N.C. (May 1967).

(5)   Stern, A. (ed.) "Sources of Air Pollution and Their Control" *Air Pollution,* Vol. III,. 2nd Ed., New York, Academic Press p. 123-127 (1968).

SUGAR CANE PROCESSING

The processing of sugar cane starts with the harvesting of the crops, either by hand or by mechanical means. If mechanical harvesting is used, much of the unwanted foliage is left, and it thus is standard practice to burn the cane before mechanical harvesting to remove the greater part of the foliage.

After being harvested, the cane goes through a series of processes to be converted to the final sugar product. It is washed to remove larger amounts of dirt and trash; then crushed and shredded to reduce the size of the stalks. The juice is next extracted by one of two methods, milling or diffusion. In milling the cane is pressed between heavy rollers to press out the juice, and in diffusion the sugar is leached out by water and thin juices. The raw sugar then goes through a series of operations including clarification, evaporation, and crystallization in order to produce the final product(1). Most mills operate without supplemental fuel because of the sufficient bagasse (the fibrous residue of the extracted cane) that can be burned as fuel.

Air Pollution

The largest sources of emissions from sugar cane processing are the openfield burning in the harvesting of the crop and the burning of bagasse as fuel. In the various processes of crushing, evaporation, and crystallization, some particulates are emitted but in relatively small quantities. Emission factors for sugar cane processing are shown in Table 350.

TABLE 350: EMISSION FACTORS FOR SUGAR CANE PROCESSING

Emission Factor Rating: D

Type of process	Particulate	Carbon monoxide	Hydrocarbons	Nitrogen oxides
Field burning[a,b]				
lb/acre burned	225	1,500	300	30
kg/hectare burned	250	1,680	335	33.5
Bagasse burning[c]				
lb/ton bagasse	22	—	—	—
kg/MT bagasse	11	—	—	—

[a]Based on emission factors for open burning of agricultural waste.
[b]There are approximately 4 tons/acre (9,000 kg/hectare) of unwanted foliage on the cane and 11 tons/acre (25,000 kg/hectare) of grass and weed, all of which are combustible[2]
[c]Reference 2.

Source: EPA Publication AP-42, 2nd Edition

Water Pollution

The characteristics of the total wastewater effluent from a cane sugar refinery vary widely, depending upon the characteristics of the individual waste stream as described below. However, the following major total raw waste streams can be identified:

The wastewater produced by a crystalline sugar refinery using bone char for decolorization. The majority of waste stream components are char wash water which is a part of the process water stream, condenser cooling water.
The wastewater produced by a crystalline sugar refinery using carbon for decolorization. The major waste streams from this type of refinery are barometric condenser cooling water and process water, including ion-exchange regeneration solutions and carbon slurries.
The wastewater produced by a liquid sugar refinery employing affination and remelt and, therefore, using vacuum pans. The discharge from this refinery is similar to that from the carbon crystalline refinery except that the flow of condenser water is less.
The wastewaters produced by a liquid refinery which does not use affination, does not remelt, and therefore, does not use vacuum pans. The discharge from this refinery is similar to the discharge from the refinery above except that the barometric condenser flow is less.

The wastewaters produced by a refinery which produces both liquid and crystalline sugar by separate processes. The discharge is a combination of the second and third refineries on the previous page.

Figures 44 and 45 are illustrations of the estimated flow and loadings for the process water and barometric condenser cooling water, and total discharge streams for the average crystalline and liquid cane sugar refineries, respectively.

FIGURE 44: RAW WASTE LOADINGS AND WATER USAGE FOR THE AVERAGE CRYSTALLINE CANE SUGAR REFINERY

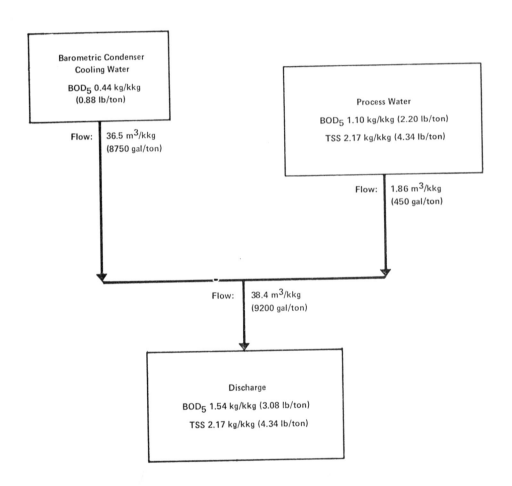

Source: Report EPA 440/1-73/002a

FIGURE 45: RAW WASTE LOADING AND WATER USAGE FOR THE AVERAGE
LIQUID CANE SUGAR REFINERY

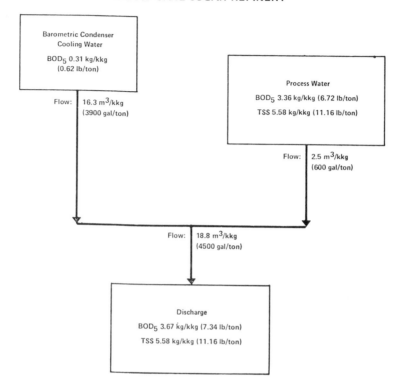

Source: Report EPA 440/1-73/002a

References

(1)   "Sugar Cane," *Kirk-Othmer Encyclopedia of Chemical Technology,* Vol. IX,
      New York, John Wiley and Sons, Inc. (1964).
(2)   Cooper, J., Unpublished data on emissions from the sugar cane industry, Air Pollution
      Control Agency, Palm Beach County, Florida (July 1969).
(3)   U.S. Environmental Protection Agency, "Development Document for Proposed Ef-
      fluent Limitations Guidelines and New Source Performance Standards for the Cane
      Sugar Refining Segment of the Sugar Processing Point Source Category", *Report
      EPA 440/1-73/002a,* Washington, D.C. (December 1973).

SULFUR PRODUCTION

Nearly all of the elemental sulfur produced from hydrogen sulfide is made by the modified
Claus process. The process consists of the multi-stage oxidation of hydrogen sulfide according to the following reaction:

$$\underset{\substack{\text{Hydrogen}\\\text{sulfide}}}{2H_2S} \quad + \quad \underset{\text{Oxygen}}{O_2} \quad \longrightarrow \quad \underset{\text{Sulfur}}{2S} \quad + \quad \underset{\text{Water}}{2H_2O}$$

In the first step, approximately one-third of the hydrogen sulfide is reacted with air in a pressurized boiler (1.0 to 1.5 atmosphere) where most of the heat of reaction and some of the sulfur are removed. After removal of the water vapor and sulfur, the cooled gases are heated to between 400° and 500°F, and passed over a "Claus" catalyst bed composed of bauxite or alumina, where the reaction is completed. The degree of reaction completion is a function of the number of catalytic stages employed. Two stages can recover 92 to 95% of the potential sulfur; three stages, 95 to 96%; and four stages, 96 to 97%. The conversion to sulfur is ultimately limited by the reverse reaction in which water vapor recombines with sulfur to form gaseous hydrogen sulfide and sulfur dioxide. Additional amounts of sulfur are lost as vapor, entrained mist, or droplets and as carbonyl sulfide and carbon disulfide (0.25 to 2.5% of the sulfur fed).

The latter two compounds are formed in the pressurized boiler at high temperature (1500° to 2500°F) in the presence of carbon compounds[1]. The plant tail gas, containing the above impurities in volume quantities of 1 to 3%, usually passes to an incinerator, where all of the sulfur is oxidized to sulfur dioxide at temperatures ranging from 1000° to 1200°F. The tail gas containing the sulfur dioxide then passes to the atmosphere via a stack.

Air Pollution

Virtually all of the emissions from sulfur plants[2] consist of sulfur dioxide, the main incineration product. The quantity of sulfur dioxide emitted is, in turn, a function of the number of conversion stages employed, the process temperature and pressure, and the amounts of carbon compounds present in the pressurized boiler.

The most commonly used control method involves two main steps, conversion of sulfur dioxide to hydrogen sulfide followed by the conversion of hydrogen sulfide to elemental sulfur. Conversion of sulfur dioxide to hydrogen sulfide occurs via catalytic hydrogenation or hydrolysis at temperatures from 600° to 700°F. The products are cooled to remove the water vapor and then reacted with a sodium carbonate solution to yield sodium hydrosulfide. The hydrosulfide is oxidized to sulfur in solution by sodium vanadate. Finely divided sulfur appears as a froth that is skimmed off, washed, dried by centrifugation, and added to the plant product. Overall recovery of sulfur approaches 100% if this process is employed. Table 351 lists emissions from controlled and uncontrolled sulfur plants.

TABLE 351: EMISSION FACTORS FOR MODIFIED-CLAUS SULFUR PLANTS

Emission Factor Rating: D

Number of catalytic stages	Recovery of of sulfur, %	SO_2 emissions[a]	
		lb/ton 100% sulfur	kg/MT 100% sulfur
Two, uncontrolled	92 to 95	211 to 348	106 to 162
Three, uncontrolled	95 to 96	167 to 211	84 to 106
Four, uncontrolled	96 to 97	124 to 167	62 to 84
Sulfur removal process	99.9	4.0	2.0

[a]The range in emission factors corresponds to the range in the percentage recovery of sulfur.

Source: EPA Publication AP-42, 2nd Edition

References

(1) *Kirk-Othmer Encyclopedia of Chemical Technology,* Vol. 19, New York, John Wiley and Sons, Inc. (1969).
(2) Beavon, David K. "Abating Sulfur Plant Tail Gases," *Pollution Engineering,* 4(1):34-35, (January 1972)

(3) U.S. Environmental Protection Agency, "Compilation of Air Pollutant Emission Factors; 2nd Edition, *Publication No. AP-42,* Research Triangle Park, N.C., Office of Air and Water Programs (April 1973).

SULFURIC ACID MANUFACTURE

All sulfuric acid is made by either the lead chamber or the contact process. Because the contact process accounts for more than 97% of the total sulfuric acid production in the United States, it is the only process discussed in this section. Contact plants are generally classified according to the raw materials charged to them: [1] elemental sulfur burning, [2] spent acid and hydrogen sulfide burning, and [3] sulfide ores and smelter gas burning plants. The relative contributions from each type of plant to the total acid production are 68, 18.5, and 13.5%, respectively. All contact processes incorporate three basic operations, each of which corresponds to a distinct chemical reaction. First, the sulfur in the feedstock is burned to sulfur dioxide:

$$[1] \qquad \underset{\text{Sulfur}}{S} \; + \; \underset{\text{Oxygen}}{O_2} \; \longrightarrow \; \underset{\substack{\text{Sulfur} \\ \text{dioxide}}}{SO_2}$$

Then, the sulfur dioxide is catalytically oxidized to sulfur trioxide:

$$[2] \qquad \underset{\substack{\text{Sulfur} \\ \text{dioxide}}}{2SO_2} \; + \; \underset{\text{Oxygen}}{O_2} \; \longrightarrow \; \underset{\substack{\text{Sulfur} \\ \text{trioxide}}}{2SO_3}$$

Finally, the sulfur trioxide is absorbed in a strong, aqueous solution of sulfuric acid:

$$[3] \qquad \underset{\substack{\text{Sulfur} \\ \text{trioxide}}}{SO_3} \; + \; \underset{\text{Water}}{H_2O} \; \longrightarrow \; \underset{\substack{\text{Sulfuric} \\ \text{acid}}}{H_2SO_4}$$

Elemental sulfur, such as Frasch-process sulfur from oil refineries, is melted, settled, or filtered to remove ash and is fed into a combustion chamber. The sulfur is burned in clean air that has been dried by scrubbing with 93 to 99% sulfuric acid. The gases from the combustion chamber are cooled and then enter the solid catalyst (vanadium pentoxide) converter. Usually, 95 to 98% of the sulfur dioxide from the cumbustion chamber is converted to sulfur trioxide, with an accompanying large evolution of heat. After being cooled, the converter exit gas enters an absorption tower where the sulfur trioxide is absorbed with 98 to 99% sulfuric acid. The sulfur trioxide combines with the water in the acid and forms more sulfuric acid.

If oleum, solution of uncombined SO_3 in H_2SO_4, is produced, SO_3 from the converter is first passed to an oleum tower that is fed with 98% acid from the absorption system. The gases from the oleum tower are then pumped to the absorption column where the residual sulfur trioxide is removed.

Two types of plants are used to process this type of sulfuric acid. In one the sulfur dioxide and other combustion products from the combustion of spent acid and/or hydrogen sulfide with undried atmospheric air are passed through gas-cleaning and mist-removal equipment. The gas stream next passes through a drying tower. A blower draws the gas from the drying tower and discharges the sulfur dioxide gas to the sulfur trioxide converter. In a "wet-gas plant," the wet gases from the combustion chamber are charged directly to the converter with no intermediate treatment. The gas from the converter flows to the absorber, through which 93 to 98% sulfuric acid is circulating.

The configuration of a sulfide ore and smelter gas type of plant is essentially the same as that of a spent-acid plant with the primary exception that a roaster is used in place of the

combustion furnace. The feed used in these plants is smelter gas, available from such equipment as copper converters, reverberatory furnaces, roasters, and flash smelters. The sulfur
dioxide in the gas is contaminated with dust, acid mist, and gaseous impurities. To remove
the impurities the gases must be cooled to essentially atmospheric temperature and passed
through purification equipment consisting of cyclone dust collectors, electrostatic dust and
mist precipitators, and scrubbing and gas-cooling towers. After the gases are cleaned and
the excess water vapor is removed, they are scrubbed with 98% acid in a drying tower. Beginning with the drying tower stage, these plants are nearly identical to elemental sulfur
plants.

Air Pollution

Nearly all sulfur dioxide emissions from sulfuric acid plants are found in the exit gases. Extensive testing has shown that the mass of these SO_2 emissions is an inverse function of the
sulfur conversion efficiency (SO_2 oxidized to SO_3). This conversion is, in turn, affected by
the number of stages in the catalytic converter, the amount of catalyst used, the temperature and pressure, and the concentrations of the reactants, sulfur dioxide and oxygen. For
example, if the inlet SO_2 concentration to the converter were 8% by volume (a representative value), and the conversion temperature were 473°C, the conversion efficiency would be
96%. At this conversion, the uncontrolled emission factor for SO_2 would be 55 pounds
per ton (27.5 kg/MT) of 100% sulfuric acid produced, as shown in Table 352.

TABLE 352: EMISSION FACTORS FOR SULFURIC ACID PLANTS[a]

Emission Factor Rating: A

Conversion of SO_2 to SO_3, %	SO_2 emissions	
	lb/ton of 100 % H_2SO_4	kg/MT of 100 % H_2SO_4
93	96	48.0
94	82	41.0
95	70	35.0
96	55	27.5
97	40	20.5
98	27	13.0
99	14	7.0
99.5	7	3.5
99.7	4	2.0
100	0	0.0

[a]Reference 1.
[b]The following linear interpolation formula can be used for
calculating emission factors for conversion efficiencies between 93
and 100 percent: emission factor (lb/ton acid) = -13.65 (percent
conversion efficiency) + 1365.

Source: EPA Publication AP-42, 2nd Edition

For purposes of comparison, note that the Environmental Protection Agency performance
standard for new and modified plants is 4 pounds per ton (2kg/MT) of 100% acid produced,
maximum 2-hour average. As Table 352 indicates, achieving this standard requires a conversion efficiency of 99.7% in an uncontrolled plant or the equivalent SO_2 collection mechanism in a controlled facility. Most single absorption plants have SO_2 conversion efficiencies
ranging from 95 to 98%. In addition to exit gases, small quantities of sulfur oxides are
emitted from storage tank vents and tank car and tank truck vents during loading operations;
from sulfuric acid concentrators; and through leaks in process equipment. Few data are
available on emissions from these sources.

Of the many chemical and physical means for removing SO_2 from gas streams, only the dual absorption and the sodium sulfite-bisulfite scrubbing processes have been found to increase acid production without yielding unwanted by-products. In the dual absorption process, the SO_3 gas formed in the primary converter stages is sent to a primary absorption tower where H_2SO_4 is formed. The remaining unconverted sulfur dioxide is forwarded to the final stages in the converter, from whence it is sent to the secondary absorber for final sulfur trioxide removal. The result is the conversion of a much higher fraction of SO_2 to SO_3 (a conversion of 99.7% or higher, on the average, which meets the performance standard). Furthermore, dual absorption permits higher converter inlet sulfur dioxide concentrations than are used in single absorption plants because the secondary conversion stages effectively remove any residual sulfur dioxide from the primary absorber.

Where dual absorption reduces sulfur dioxide emissions by increasing the overall conversion efficiency, the sodium sulfite-bisulfite scrubbing process removes sulfur dioxide directly from the absorber exit gases. In one version of this process, the sulfur dioxide in the waste gas is absorbed in a sodium sulfite solution, separated, and recycled to the plant. Test results from a 750 ton (680 MT) per day plant equipped with a sulfite scrubbing system indicated an average emission factor of 2.7 pounds per ton (1.35 kg/MT).

Nearly all the acid mist emitted from sulfuric acid manufacturing can be traced to the absorber exit gases. Acid mist is created when sulfur trioxide combines with water vapor at a temperature below the dew point of sulfur trioxide. Once formed within the process system, this mist is so stable that only a small quantity can be removed in the absorber. In general, the quantity and particle size distribution of acid mist are dependent on the type of sulfur feedstock used, the strength of acid produced, and the conditions in the absorber. Because it contains virtually no water vapor, bright elemental sulfur produces little acid mist when burned; however, the hydrocarbon impurities in other feedstocks, dark sulfur, spent acid, and hydrogen sulfide, oxidize to water vapor during combustion. The water vapor, in turn, combines with sulfur trioxide as the gas cools in the system.

The strength of acid produced, whether oleum or 99% sulfuric acid, also affects mist emissions. Oleum plants produce greater quantities of finer, more stable mist. For example, uncontrolled mist emissions from oleum plants burning spent acid range from 0.1 to 10.0 pounds per ton (0.05 to 5.0 kg/MT), while those from 98% acid plants burning elemental sulfur range from 0.4 to 4.0 pounds per ton (0.2 to 2.0 kg/MT). Furthermore, 85 to 95 weight percent of the mist particles from oleum plants are less than 2 microns in diameter, compared with only 30 weight percent that are less than 2 microns in diameter from 98% acid plants.

The operating temperature of the absorption column directly affects sulfur trioxide absorption and, accordingly, the quality of acid mist formed after exit gases leave the stack. The optimum absorber operating temperature is dependent on the strength of the acid produced, throughput rates, inlet sulfur trioxide concentrations, and other variables peculiar to each individual plant. Finally, it should be emphasized that the percentage conversion of sulfur dioxide to sulfur trioxide has no direct effect on acid mist emissions. In Table 353 uncontrolled acid mist emissions are presented for various sulfuric acid plants.

Two basic types of devices, electrostatic precipitators and fiber mist eliminators, effectively reduce the acid mist concentration from contact plants to less than the EPA new-source performance standard, which is 0.15 pound per ton (0.075 kg/MT) of acid. Precipitators, if properly maintained, are effective in collecting the mist particles at efficiencies up to 99% (see Table 354). The three most commonly used fiber mist eliminators are the vertical tube, vertical panel, and horizontal dual-pad types. They differ from one another in the arrangement of the fiber elements, which are composed of either chemically resistant glass or fluorocarbon, and in the means employed to collect the trapped liquid. The operating characteristics of these three types are compared with electrostatic precipitators in Table 354.

TABLE 353: ACID MIST EMISSION FACTORS FOR SULFURIC ACID PLANTS
WITHOUT CONTROLS[a]

Emission Factor Rating: B

Raw material	Oleum produced, % total output	Emissions[b]	
		lb/ton acid	kg/MT acid
Recovered sulfur	0 to 43	0.35 to 0.8	0.175 to 0.4
Bright virgin sulfur	0	1.7	0.85
Dark virgin sulfur	33 to 100	0.32 to 6.3	0.16 to 3.15
Sulfide ores	0 to 25	1.2 to 7.4	0.6 to 3.7
Spent acid	0 to 77	2.2 to 2.7	1.1 to 1.35

[a]Reference 1.

[b]Emissions are proportional to the percentage of oleum in the total product. Use
the low end of ranges for low oleum percentage and high end of ranges for high
oleum percentage.

TABLE 354: EMISSION COMPARISON AND COLLECTION EFFICIENCY OF TYPICAL
ELECTROSTATIC PRECIPITATOR AND
FIBER MIST ELIMINATORS[a]

Control device	Particle size collection efficiency, %		Acid mist emissions			
			98% acid plants[b]		oleum plants	
	>3 μm	≤3 μm	lb/ton	kg/MT	lb/ton	kg/MT
Electrostatic precipitator	99	100	0.10	0.05	0.12	0.06
Fiber mist eliminator						
Tubular	100	95 to 99	0.02	0.01	0.02	0.01
Panel	100	90 to 98	0.10	0.05	0.10	0.05
Dual pad	100	93 to 99	0.11	0.055	0.11	0.055

[a]Reference 1.

[b]Based on manufacturers' generally expected results; calculated for 8 percent sulfur dioxide
concentration in gas converter.

Source: EPA Publication AP-42, 2nd Edition

Water Pollution

At a typical dual absorption plant, only cooling water is discharged. In double absorption
plants, the tail gases are sufficiently depleted to sulfur oxides that there is no need for gas
scrubbers. Also, at this plant, use of extensive maintenance and leak prevention has been
employed to prevent discharge of any product acid. Most water is used for cooling. Process water is consumed to make sulfuric acid and is not discharged. The only plant effluent
is the cooling water used in the heat exchangers and associated water treatment chemicals.

For the single absorption process, the sulfur dioxide is passed through one or more converters and then into one or more absorbers prior to venting to the atmosphere. This arrangement is less effective for both conversion of sulfur dioxide to sulfur trioxide and for
absorption of the sulfur trioxide into the absorber sulfuric acid. As a result, the tail gases
may have to be scrubbed, and this may create a waterborne waste not present for double
absorption plants. For the single absorption sulfur-burning process, there are no wastes
from the sulfuric acid process itself. Wastes arise from the use of water treatment chemicals. The raw wastes are iron, silicon, calcium and magnesium salts from water treatment.

This does not cover spent acid plants based on single absorption.

Most of the cooling water used at this plant is recycled and only 5% emerges from the plant. This is sent to evaporation ponds, from which there is no discharge. The water input is well water in the quantity of 606 cubic meters per day (0.160 MGD) or 1.670 l/kkg of product (400 gallons per ton). This water is used as follows:

Type	Cubic Meters/Day (MGD)	Liters/Kilokilogram (Gallons/Ton)	Recycled
Cooling	560 (0.148)	1,540 (370)	95
Process	45.5 (0.012)	125 (30)	0
Sanitary	Insignificant		0

All waterborne wastes are sent to an evaporation pond. There is no discharge.

References

(1) U.S. Environmental Protection Agency, "Compilation of Air Pollutant Emission Factors," 2nd Edition, *Publication No. AP-42,* Research Triangle Park, N.C., Office of Air and Water Programs (April 1973).
(2) Sittig, M., "Sulfuric Acid Manufacture and Effluent Control," Park Ridge, N.J., Noyes Data Corp. (1971).
(3) U.S. Environmental Protection Agency, "Development Document for Proposed Effluent Limitations Guidelines and New Source Performance Standards for the Major Inorganic Products Segment of the Inorganic Chemical Manufacturing Point Source Category, *Report EPA 440/1-73/007,* Washington, D.C. (August 1973).

SURFACE COATING APPLICATION

Surface-coating operations primarily involve the application of paint, varnish, lacquer, or paint primer for decorative or protective purposes. This is accomplished by brushing, rolling, spraying, flow coating, and dipping. Some of the industries involved in surface-coating operations are automobile assemblies, aircraft companies, container manufacturers, furniture manufacturers, appliance manufacturers, job enamelers, automobile repainters and plastic products manufacturers.

Air Pollution

Emissions of hydrocarbons occur in surface-coating operations because of the evaporation of the paint vehicles, thinners, and solvents used to facilitate the application of the coatings. The major factor affecting these emissions is the amount of volatile matter contained in the coating. The volatile portion of most common surface coatings averages approximately 50%, and most, if not all of this is emitted during the application and drying of the coating.

The compounds released include aliphatic and aromatic hydrocarbons, alcohols, ketones, esters, alkyl and aryl hydrocarbon solvents, and mineral spirits. Table 355 presents emission factors for surface-coating operations.

Control of the gaseous emissions can be accomplished by the use of adsorbers (activated carbon) or afterburners. The collection efficiency of activated carbon has been reported at 90% or greater. Water curtains or filler pads have little or no effect on escaping solvent vapors; they are widely used, however, to stop paint particulate emissions.

TABLE 355: GASEOUS HYDROCARBON EMISSION FACTORS FOR SURFACE-
COATING APPLICATIONS[a]

Emission Factor Rating: B

| Type of coating | Emissions[b] | |
	lb/ton	kg/MT
Paint	1120	560
Varnish and shellac	1000	500
Lacquer	1540	770
Enamel	840	420
Primer (zinc chromate)	1320	660

[a]Reference 1.
[b]Reported as undefined hydrocarbons, usually organic solvents, both
aryl and alkyl. Paints weigh 10 to 15 pounds per gallon (1.2 to 1.9
kilograms per liter); varnishes weigh about 7 pounds per gallon
(0.84 kilogram per liter).

Source: EPA Publication AP-42, 2nd Edition

References

(1) Weiss, S.F., "Surface Coating Operations." *Air Pollution Engineering Manual,*
Danielson, J.A. (ed.), U.S. DHEW, PHS, National Center for Air Pollution Control,
Cincinnati, Ohio, *Publication Number 999-AP-40,* p. 387-390.
(2) Control Techniques for Hydrocarbon and Organic Gases From Stationary Sources,
U.S. DHEW, PHS, EHS, National Air Pollution Control Administration, Washington,
D.C., *Publication Number AP-68* (October 1969).
(3) Air Pollutant Emission Factors, Final Report, Resources Research, Inc. Reston, Va.,
Prepared for National Air Pollution Control Administration, Durham, N.C., (April
1970).

SYNTHETIC FIBER MANUFACTURE

Synthetic fibers are classified into two major categories, semisynthetic and "true" synthetic.
Semisynthetics, such as viscose rayon and acetate fibers, result when natural polymeric materials such as cellulose are brought into a dissolved or dispersed state and then spun into
fine filaments. True synthetic polymers, such as Nylon, Orlon, and Dacron, result from addition and other polymerization reactions that form long chain molecules.

True synthetic fibers begin with the preparation of extremely long, chain-like molecules.
The polymer is spun in one of four ways: [1] melt spinning, in which molten polymer is
pumped through spinneret jets, the polymer solidifying as it strikes the cool air; [2] dry
spinning, in which the polymer is dissolved in a suitable organic solvent, and the resulting
solution is forced through spinnerets; [3] wet spinning, in which the solution is coagulated
in a chemical as it emerges from the spinneret; and [4] core spinning, the newest method,
in which a continuous filament yarn together with short-length "hard" fibers is introduced
onto a spinning frame in such a way as to form a composite yarn.

Air Pollution

In the manufacture of viscose rayon, carbon disulfide and hydrogen sulfide are the major
gaseous emissions. Air pollution controls are not normally used to reduce these emissions,
but adsorption in activated carbon at an efficiency of 80 to 95%, with subsequent recovery
of the CS_2 can be accomplished. Emissions of gaseous hydrocarbons may also occur from

the drying of the finished fiber. Table 356 presents emission factors for semisynthetic and true synthetic fibers.

TABLE 356: EMISSION FACTORS FOR SYNTHETIC FIBERS MANUFACTURING

Emission Factor Rating: E

Type of fiber	Hydrocarbons		Carbon disulfide		Hydrogen sulfide		Oil vapor or mist	
	lb/ton	kg/MT	lb/ton	kg/MT	lb/ton	kg/MT	lb/ton	kg/MT
Semi-synthetic								
Viscose rayon[a,b]	–	–	55	27.5	6	3	–	–
True synthetic[a]								
Nylon	7	3.5	–	–	–	–	15	7.5
Dacron	–	–	–	–	–	–	7	3.5

[a]Reference 1.
[b]May be reduced by 80 to 95 percent adsorption in activated charcoal.

Source: EPA Publication AP-42, 2nd Edition

Water Pollution

Unlike wool and cotton, synthetic fibers of the same type can have different physical and chemical properties. The producers of man-made fibers are continually producing variations of existing fibers and totally new fibers to meet existing or anticipated needs. As a result, the techniques used in finishing synthetic textile fibers are largely dictated by the fiber manufacturer. Therefore, the processing of synthetic fibers can produce a variable wasteload depending on the particular fiber type used, the manufacturer's suggested method of processing, and the process chemicals used. Table 357 lists the significant pollutants found in the wastewater resulting from processing the different types of synthetic fibers. Table 358 lists the pollutional loads and wastewater volume generated.

TABLE 357: SIGNIFICANT POLLUTANTS IN SYNTHETIC FIBER WET PROCESSING

Fiber	Process	Liquid Waste Pollutant
Rayon	Scour and dye	Oil, dye, synthetic detergent, and anti-static lubricants
	Scour and bleach	Synthetic detergent, and hydrogen peroxide
	Salt bath	Synthetic detergent, chloride or sulfate
Acetate	Scour and dye	Antistatic lubricants, dye, sulfonated oils, synthetic detergent, esters, and softeners
	Scour and bleach	Synthetic detergent, hydrogen peroxide or chlorine
Nylon	Scour	Antistatic lubricants, soap, tetrasodium pyrophosphate, soda, and fatty esters.
	Developed disperser dye	Dye, NaNO$_2$, hydrochloric acid, developer and sulfonated oils
	Bleach	Peracetic Acid
Acrylic-Modacrylic	Dye	Dye, formic acid, wetting agent, aromatic amines, retarding agent, and sulfates
	Thermosol dyeing	Acid
	Bleach	Chlorite
	Scour	Synthetic detergent and pin oil

(continued)

TABLE 357: (continued)

Fiber	Process	Liquid Waste Pollutant
Acrylic-Modacrylic	Dyeing with carriers	Chlorobenzenes, hot water, and dye; or phenylmethyl carbinol, dye, and hot water; or ortho-phenylphenol and dye
	Scour	Antistatic lubricants, chlorite or hypochlorite, and nonionic synthetic detergents
	High temperature and pressure dyeing	Dye and hot water
	Bleach	Chlorite, $NaNO_2$, acetic acid, oxalic acid, nitric acid, bisulfite, proprietary bleaches

TABLE 358: POLLUTIONAL LOAD OF SYNTHETIC WET FIBER PROCESSES

Process	Fiber	pH	B.O.D. p.p.m.	B.O.D. lbs./1000 lbs. of cloth	Total Solids p.p.m.	Total Solids lbs./1000 lbs. of cloth	Suspended Solids lbs./1000 lbs. of cloth	Volume in gal per 1000 lbs. of cloth
Scour	Nylon	10.4	1360	30-40	1882	30-50	20-40	6,000-8,000
	Acrylic/Modacrylic	9.7	2190	45-90	1874	12-20	25-50	6,000-8,000
	Polyester	----	500-800	15-25	----	25-35	5-15	3,000-5,000
Scour & Dye	Rayon	8.5	2832	50-70	3334	25-39	0-3	2,000-4,000
	Acetate	9.3	2000	40-60	1778	-----	1-20	4,000-6,000
Dye	Nylon	8.4	368	5-20	641	20-34	2-42	2,000-4,000
	Acrylic/Modacrylic	1.5-3.7	175-2000	2-40	833-1968	6-9	5-20	2,000-4,000
	Polyester	----	480-27,000	15-800	----	30-200	-----	2,000-4,000
Salt Bath	Rayon	6.8	58	0-3	4890	20-200	2-6	500-1,500
Final Scour	Acrylic/Modacrylic	7.1	668	10-25	1191	4-12	3-7	8,000-10,000
	Polyester	----	650	15-25	----	10-50	3-50	2,000-4,000
Special Finishing	Rayon	----	----	20	----	3-100	3-50	500-1,500
	Acetate	----	----	40	----	3-100	3-50	3,000-5,000
	Nylon	----	----	10	----	3-100	3-50	4,000-6,000
	Acrylic/Modacrylic	----	----	60	----	3-100	3-50	5,000-7,000
	Polyester	----	----	2-80	----	3-100	3-50	1,000-3,000

Source: Report 12090 ECS 02/71

The FWPCA has projected the gross wasteload expected to be produced by the synthetic textile finishing industry. These projections are based on the projected growth rate of the industry and on estimates of increased waste treatment and reduction of wastes per unit of operation. These projected figures are given in Table 359 for 1970, 1971, 1972, 1977 and 1982. The synthetic fibers are gaining a large portion of the total textile fiber market. The rather high growth rate anticipated accounts for such high wasteload predictions.

TABLE 359: PROJECTED GROSS WASTELOAD AND WASTEWATER VOLUME FOR SYNTHETIC FIBER FINISHING WASTES

Year	BOD	Suspended Solids	Total Dissolved Solids	Volume (Billion Gals.)
		(Million Pounds)		
1970	288	291	584	28
1971	301	304	608	29
1972	314	316	633	30
1977	375	379	760	36
1982	409	412	827	39

Source: Report 12090 ECS 02/71

References

(1)   U.S. Environmental Protection Agency, "Compilation of Air Pollutant Emission Fac-
      tors", 2nd Edition, *Publication No. AP-42,* Research Triangle Park, N.C., Office of
      Air and Water Programs (April 1973).
(2)   Porter, John J., "State of the Art of Textile Waste Treatment," *Report 12090 ECS
      02/71,* Washington, D.C., U.S. Government Printing Office (Feb. 1971).
(3)   Jones, H.R., "Pollution Control in the Textile Industry," Park Ridge, N.J., Noyes
      Data Corp. (1973).

TEREPHTHALIC ACID MANUFACTURE

The main use of terephthalic acid is to produce dimethyl terephthalate, which is used for
polyester fibers (like Dacron) and films. Terephthalic acid can be produced in various ways,
one of which is the oxidation of p-xylene by nitric acid. In this process an oxygen-contain-
ing gas (usually air), p-xylene, and HNO_3 are all passed into a reactor where oxidation by
the nitric acid takes place in two steps. The first step yields primarily N_2O; the second
step yields mostly NO in the off gas. The terephthalic acid precipitated from the reactor
effluent is recovered by conventional crystallization, separation, and drying operations.

Air Pollution

The NO in the off gas from the reactor is the major air contaminant from the manufacture
of terephthalic acid(1). The amount of nitrogen oxides emitted is roughly estimated in
Table 360.

TABLE 360: NITROGEN OXIDES EMISSION FACTORS FOR TEREPHTHALIC ACID PLANTS[a]

Emission Factor Rating: D

	Nitrogen oxides (NO)	
Type of operation	lb/ton	kg/MT
Reactor	13	6.5

[a]Reference 2.

Source: EPA Publication AP-42, 2nd Edition

References

(1)   Air Pollutant Emission Factors, Final Report, Resources Research, Inc. Reston, Va.
      Prepared for National Air Pollution Control Administration, Durham, N.C. (April
      1970).
(2)   Terephthalic Acid, *Kirk-Othmer Encyclopedia of Chemical Technology,* Vol 9, New
      York, John Wiley and Sons, Inc. (1964).

TEXTILE INDUSTRY

Water Pollution

The reader is also referred to the earlier section of this book on "Synthetic Fiber Manufacture." This section will deal primarily with cotton and wool processing but will conclude with a few words on textile finishing as applied to both synthetic and natural fibers.

Table 361 lists the range of pollutional loads of the various cotton textile wet-processing operations. The ranges have been compiled from the values given by several authors in different studies. Wide variance in the values of each process is caused by the following: the problems of representative sampling in a highly variable waste stream; the several methods of technology available, for example, batch versus continuous processing; the interchangeability of process chemicals, for example, hypochlorite versus peroxide in bleaching; and the wide variety of cotton fabrics manufactured. Any particular cotton processing mill should be producing effluents with values falling within the ranges listed in Table 361. Each process produces a waste with different characteristics. Table 362 characterizes the wastes from cotton wet processing.

TABLE 361: POLLUTION EFFECT OF COTTON PROCESSING WASTES

Process	pH	Wastes (p.p.m.) B.O.D.	Total Solids	Gallons Waste per 1,000 lbs. goods	Pounds B.O.D. per 1,000 lbs. goods	Pounds Total Solids per 1,000 lbs. goods
Slashing, sizing yarn*--------------	7.0-9.5	620-2,500	8,500-22,600	60-940	0.5-5.0	47-67
Desizing-------------	-------	1,700-5,200	16,000-32,000	300-1,100	14.8-16.1	66-70
Kiering-------------	10-13	680-2,900	7,600-17,400	310-1,700	1.5-17.5	19-47
Scouring------------	------	50-110	--------------	2,300-5,100	1.36-3.02	-------
Bleaching (range)----	8.5-9.6	90-1,700	2,300-14,400	300-14,900	5.0-14.8	38-290
Mercerizing----------	5.5-9.5	45-65	600-1,900	27,900-36,950	10.5-13.5	185-450
Dyeing:						
Aniline Black------	-------	40-55	600-1,200	15,000-23,000	5-10	100-200
Basic--------------	6.0-7.5	100-200	500-800	18,000-36,000	15-50	150-250
Developed Colors---	5-10	75-200	2,900-8,200	8,900-25,000	15-20	325-650
Direct-------------	6.5-7.6	220-600	2,200-14,000	1,700-6,400	1.3-11.7	25-250
Naphthol-----------	5-10	15-675	4,500-10,700	2,300-16,800	2-5	200-650
Sulfur-------------	8-10	11-1,800	4,200-14,100	2,900-25,600	2-250	300-1,200
Vats--------------	5-10	125-1,500	1,700-7,400	1,000-20,000	12-30	150-250

*Cloth Weaving Mill Waste. (Composite of all waste connected with each process.)

TABLE 362: CHARACTERISTICS OF COTTON PROCESSING WET WASTES

Process	Significant Pollutants
Desizing	High BOD, neutral pH, high total solids
Scouring	High BOD, high alkalinity, high total solids, high temperature
Bleaching	High BOD, alkaline pH, high solids
Mercerizing	Low BOD, alkaline pH, low solids
Dyeing and Printing	High BOD, high solids, neutral to alkaline pH

Source: Report 12090 ECS 02/71

In treating 1,000 pounds of cotton fabric, a composite waste stream has the following characteristics: pH of 8 to 11; the color either a gray or that of the predominant dye used; the BOD, total solids and suspended solids with concentrations of 200 to 600 ppm, 1,000 to 1,600 ppm and 30 to 50 ppm, respectively. These wastes will be contained in a volume of 30,000 to 93,000 gallons.

Based on estimates of product mix; that is, the amount of 100% cotton fabrics and cotton-synthetic blends; and the degree of technology mix; that is, the number of modern plants and out-of-date plants, and the projected growth rate of the cotton finishing industry, the Federal Water Pollution Control Administration has predicted a gross wasteload and wastewater volume for selected years. Table 363 list these estimates for BOD, suspended solids, total dissolved solids and volume of wastewater for 1970, 1971, 1972, 1977 and 1982.

TABLE 363: PROJECTED GROSS WASTELOAD AND WASTEWATER VOLUME FOR COTTON FINISHING WASTES

Year	BOD	Suspended Solids	Total Dis- solved Solids	Volume (Billion Gallons)
		(Million Pounds)		
1970	529	165	497	99
1971	528	164	496	100.1
1972	526	163	495	100.1
1977	516	160	494	102.3
1982	502	154	483	105.6

Source: Report 12090 ECS 02/71

The reason for the gradual decrease of the gross pollutional load in the coming years is based on these assumptions: new machinery, which tends to produce less pollution per unit of cloth, due to water reuse and countercurrent flow designs will continue to be purchased; trends in process modification, new chemical manufacture and better housekeeping will continue and a larger percentage of the wastes will be treated due to increased efficiency of treatment facilities and increased pressure by State, local, and Federal agencies. In the processing of woolen fibers, five sources of pollutional load exist—scouring, dyeing, washing after fulling, neutralizing after carbonizing, and bleaching with optical brighteners. Table 364 contains average values of the pollution load of each of these processes, and Table 365 lists the significant pollutants from each process.

In Table 364, the pollutional load values vary greatly due to different methods of accomplishing the same process; for example, dyeing can use either acetic acid or ammonium sulfate as a dye assistant, the latter having a lower BOD loading. Variance in volume is caused by differences between batch and continuous processes. In the pH values, the large variance in neutralization is due to the fact that the first soapings' function is to remove residual acid from the fabric; whereas, the second soaping adds alkali to the bath.   For scouring, washing and neutralizing, the values are for the entire process, not separate bowls.

In treating 1,000 pounds of scoured wool, the total wasteload will contain 1,000 pounds of grease, suet, and dirt, plus up to 500 pounds of process chemicals (detergents, alkali, softeners, etc.). This wasteload is broken down into 250 pounds of BOD, 188 pounds of alkalinity, and 40.5 pounds of acidity. The volume of water used will range from

55,725 to 133,060 gallons. A composite waste stream produced by the finishing of wool fibers would have the following pollutional loadings: pH-9.0 to 10.5; BOD and total solids concentrations of 432 to 1,200 ppm and 6,470 ppm respectively. The effluent will have a brown colloidal color, tinged with the most used dye color.

TABLE 364: POLLUTIONAL LOADS OF WOOL WET PROCESSES

Process	pH	p.p.m.	lbs/1,000 lbs. cloth	Total Solids p.p.m.	Volume (gallons)
			B.O.D.		
Scouring	9.0-10.4	30,000-40,000	104.5-221.4	1,129-64,448	5,500-12,000
Dyeing	4.8-8.0	380-2,200	9.0-34.3	3,855-8,315	1,900-2,680
Washing	7.3-10.3	4,000-11,455	31-94	4,830-19,267	40,000-100,000
Neutral- ization	1.9-9.0	28	1.7-2.1	1,241-4,830	12,500-15,700
Bleaching	6.0	390	1.4	908	300-2,680

TABLE 365: CHARACTERISTICS OF WOOL PROCESSING WASTES

Process	Significant Pollutants
Scouring	High BOD, high grease content, high alkalinity, turbid brown color and a temperature of 115° to 125° F.
Dyeing	Acid pH, highly colored, relatively high in BOD and possibly toxic
Washing	High BOD, high oil content and temperatures of 110° to 150° F.

Source: Report 12090 ECS 02/71

As in cotton finishing, the Federal Water Pollution Control Administration has estimated the gross wasteload, in pounds of BOD, and the wastewater volume for the selected years in the period 1970-1982. These estimates are based on the following factors:

Anticipated growth rate of the wool finishing industry.

The assumptions that the industry will continue to purchase processing machinery and process chemicals that tend to produce a lower pollutional load per unit of fabric processed.

Treatment facilities will achieve more efficient pollution load removal.

Increased pressure by Federal, State and local regulatory agencies concerning pollution.

The Administration's estimates are contained in Table 366. Unlike cotton, the total waste discharges of wool are expected to increase slightly in the twelve year period. Although no reason is given for this increase, a reasonable conclusion would be that it is impossible to foresee any major process changes in wool finishing. It can be concluded that each of the three textile finishing groups produces a highly variable wasteload. The constant introduction of new fibers, finishes, chemicals, processes, machinery and techniques and an ever-changing consumer demand for different styles and colors of fabric are the principal causes of this variation.

TABLE 366: PROJECTED GROSS WASTELOAD AND WASTEWATER VOLUME FOR
WOOLEN FINISHING WASTES

Year	BOD (Million lbs.)	Volume (Billion Gals.)
1970	133.0	28.35
1971.	133.2	28.40
1972	133.2	28.40
1977	134.2	28.6
1982	136.0	29.0

Source: Report 12090 ECS 02/71

The wastes from a textile finishing mill come from two sources: the natural impurities present in the fibers and the process chemicals used. In synthetic fibers, only the process chemicals are the source of pollution as they are relatively free of chemical impurities. In the processing of cotton and wool wastes, the natural impurities are the largest source of pollution, and are contained in a highly concentrated effluent.

Wool finishing wastes have high BOD loads, high solid concentrations, and a high grease content that arises from the scouring process. The wool wastes are usually considered the strongest textile wet wastes per gallon of process water used. The Federal Water Pollution Control Administration's predictions of future wool wet waste discharges call for a slight increase in total BOD discharge; however, due to wool's low demand, these wastes will be the lowest for the industry. Although its waste is not as strong as that from the wool industry, due mainly to the absence of grease, the cotton finishing industry produces the largest volume of wet wastes. Cotton textile wastes can have high color content. The Federal Water Pollution Control Administration's gross wasteload projections are for a slight decrease in cotton waste discharges.

The synthetic fiber finishing wastes are lower in volume and concentration than the natural fiber wastes per gallon of water used. The synthetic fiber market's rapid growth has indicated that on a total volume basis, these wastes will be the largest for the industry in a short period of time. One significant difference between synthetic fiber wastes and natural fiber wastes is the presence of toxic materials in the waste stream when dye carriers are used for synthetic dyeing.

References

(1) Porter, J.J., "State of the Art of Textile Waste Treatment," *Report No. 12090 ECS 02/71*, Washington, D.C., U.S. Government Printing office (Feb. 1971).
(2) Jones, H.R. "Pollution Control in the Textile Industry," Park Ridge, N.J., Noyes Data Corp. (1973).

TIMBER INDUSTRY

Water Pollution

For a discussion of the water pollution problems related to the debarking of logs, the reader is referred to the earlier section of this volume on "Pulp and Paper Industry." With regard to wood preserving, wastewater characteristics vary with the particular preservative used, the volume of stock that is conditioned prior to treatment, the conditioning method used, and the extent to which effluents from retorts are diluted with water from other sources. Typically, wastewaters from creosote and pentachlorophenol treatments have high phenolic, COD, and oil contents and may have a turbid appearance that results from emulsified oils. They are always acid in reaction, the pH values usually falling within the range of 4.1 to 6.0. The COD for such wastes frequently exceeds 30,000 mg/l, most of which is attributable to entrained oils and to wood extractives, principally simple sugars, that are removed from wood during conditioning.

Reference

(1) U.S. Environmental Protection Agency "Development Document for Proposed Effluent Limitations Guidelines and New Source Performance Standards for the Plywood, Hardboard and Wood Preserving Segment of the Timber Products Processing Point Source Category," *Report EPA 440/1-73/023,* Washington, D.C. (Dec. 1973).

TIRE AND INNER TUBE MANUFACTURE

Water Pollution

A general process flow diagram for a typical tire production facility is presented in Figure 46. Figure 47 presents a typical inner tube production process diagram. The primary water usage in a tire and inner tube facility is for noncontact cooling and heating. Discharges from service utilities supplying cooling water and steam are the major source of contaminants in the final effluent. Characteristics of these wastewaters are COD, BOD, suspended solids, and dissolved solids.

Table 367 presents the raw waste loading for the combined process and recirculating cooling water, nonprocess wastewaters of the plants visited. Flow variations are due mainly to the use of once-through cooling water in certain plants as opposed to recirculated cooling water in others. In order to adequately estimate the wastewater discharge flow rates, the plant effluent was divided into process wastewaters and nonprocess wastewaters. The process wastewaters consist of mill area oily waters, soapstone slurry and latex dip wastes, area washdown waters, emission scrubber waters, contaminated storm waters from raw material storage areas, etc. The nonprocess wastewaters are sanitary and clean storm waters, utility wastewaters such as once-through cooling water, boiler blowdown, cooling tower blowdown, water treatment wastes and uncontaminated contact cooling water like thread cooling waters.

Plants A and B are new plants using totally recirculated cooling water. Plants E and G are old facilities also using recirculated water. A comparison of these four plants indicates that no significant variations in flow exist due to age of the plant. Plant F typifies a plant using once-through water as its primary source of process cooling. COD and BOD loadings vary to a great degree by the type and amount of chemicals used in the treatment of boiler and cooling tower makeup waters. Larger loadings for older plants indicate an increased amount of process wastewater pollutants in the effluent. Loadings measured in Plant A are high due to the practice of discharging washdowns of soapstone and latex dip areas noticed during the sampling period. This plant uses holding lagoons.

Because all wastes are contained within the plant's boundaries, Plant A discharges contaminants which other exemplary plants (using different technologies) cannot accomplish. These contaminants lead to a correspondingly higher loading. Suspended solid loadings evolve primarily due to water treatment blowdowns, wastes, and boiler blowdowns. In addition, the suspended solid loadings in process wastewater can increase due to spills, leakage, and soapstone discharge. Loadings for old plants tend to be higher than those for new plants.

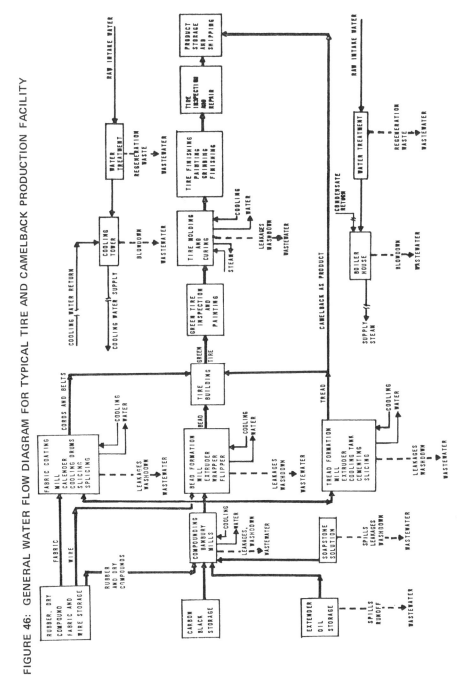

FIGURE 46: GENERAL WATER FLOW DIAGRAM FOR TYPICAL TIRE AND CAMELBACK PRODUCTION FACILITY

Source: Report EPA 440/1-73/013

FIGURE 47: GENERAL WATER FLOW DIAGRAM FOR A TYPICAL INNER TUBE PRODUCTION FACILITY

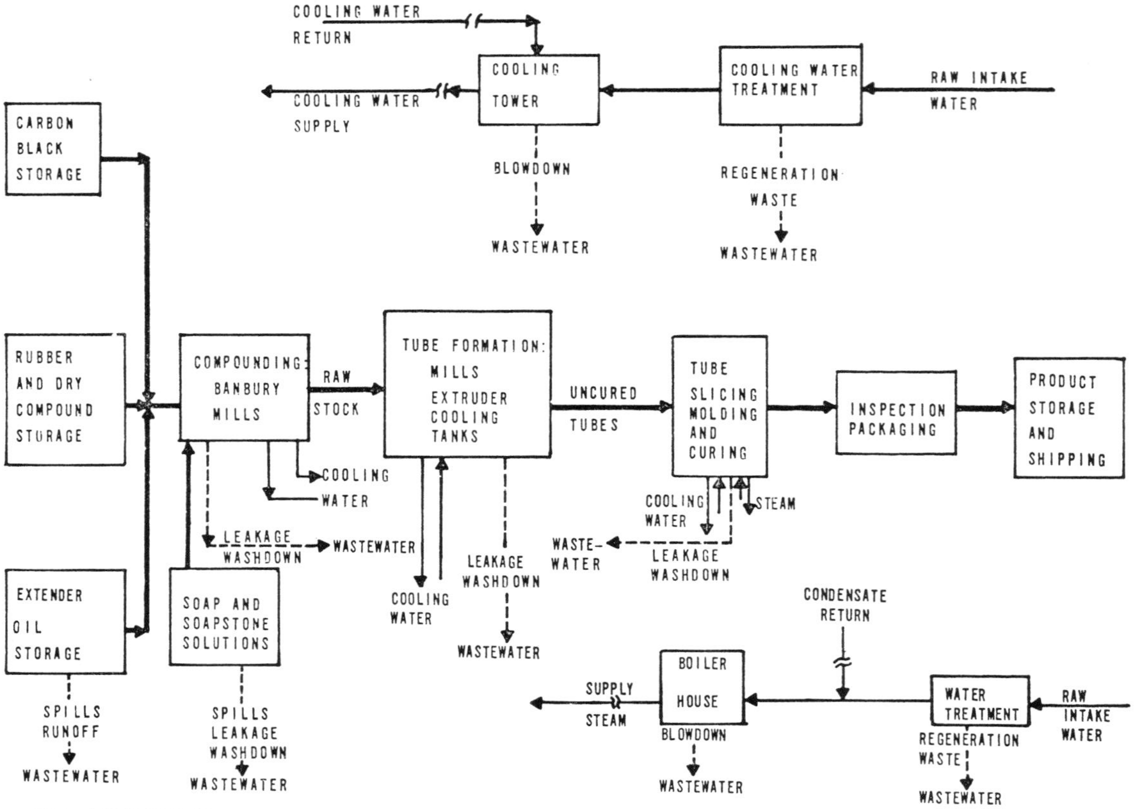

Source: Report EPA 440/1-73/013

This is due in part to the use of older water treatment techniques and the larger volumes of process wastewater containing solids discharged by older facilities. The quantity of dissolved solids discharged is related to the amount of recirculated nonprocess water and the water supply source. Plants using well water typically have higher dissolved solid loadings, than those using municipal or river water sources.

Table 367 also shows that the plant's final end product has no significant effect upon the raw waste loading in the final effluent. Data from Plant H, which produces primarily truck and industrial tires, is not substantially very different from Plants E, F, or G, which, while producing a combination of products, produce mainly passenger tires. Loadings from Plant I are similar to the others, even though this plant's primary product is the manufacture of inner tubes. To substantiate the data and conclusions on total final effluent, Corps of Engineers water discharge permit applications were obtained for a large segment of the tire and inner tube industry.

Comparison of Corps permits for plants considered old and new revealed that the above findings and conclusions are substantially correct. Table 368 lists the main characteristics and the loadings corresponding to a typical old and typical newer tire production facility. Raw waste loads in the process wastewaters leaving the production facility are presented in Table 369. Flow rates are estimates only, mainly due to the intermittent nature of the waste discharges. Although there appears to be no significant difference in the measured flow rates as shown by data, the composition of the flows originating from old and new plants differs greatly. New Plant A uses large amounts of washdown water which comprises the bulk of their process wastewaters. New Plant B process wastewaters consist largely of discharges from an extensive wet air pollution train.

The discharges from this equipment are the primary constituent of the process wastewaters. The process wastewater flow rates leaving older plants are due to other factors such as spills, leakage, runoff from storage areas and inherent plant practices of older facilities. Therefore the data indicate that, given the same housekeeping policies and the same degree of wet air pollution equipment and controls, the process wastewater flow rates from older plants will be higher than from newer plants.

TABLE 367: RAW WASTE LOADS OF TOTAL EFFLUENT FROM TIRE AND INNER TUBE FACILITIES[i]

Plant	Cate-gory	FLOW L/kkg (gal/1000lb) of raw material	COD kg/kkg(lb/1000lb) of raw material	BOD kg/kkg(lb/1000lb) of raw material	SS kg/kkg(lb/1000lb) of raw material	TDS kg/kkg(lb/1000lb) of raw material	OIL kg/kkg(lb/1000lb) of raw material
A	New	6344 (762)	1.890	0.067	0.960	4.900	0.249
B	New	3430 (412)	0.184	0.002	0.047	0.159	0.075
C[2]	New	8251 (991)	NA	NA	1.155	NA	0.794
D	New	10883 (1,306)	0.142	0.012	0.092	0.879	0.009
E	Old	5453 (655)	0.100	0.001	0.440	0.001	0.027[3]
F	Old	123480 (14,824)	3.398	0.296	1.358	0.001	0.650
G	Old	3220 (387)	0.645	0.093	3.429	0.000	0.267
H	Old	72427 (8,695)	0.001	0.036	0.676	0.000	0.167
I	Old	10610 (1,274)	0.615	0.148	2.912	1.810	0.172

[i]Includes utility wastewaters.

[2]Estimated, raw material consumption not known.

[3]Includes treatment by in-plant sumps.

NA - Data not available

Source: Report EPA 440/1-73/013

TABLE 368: INDUSTRY-WIDE AVERAGE VALUES OF FINAL EFFLUENT LOADS FOR TIRE AND INNER TUBE PLANTS[1]

		FLOW – L/kkg. (gal./1000 lb.) of raw material	COD – kg/kkg (lb./1000 lb.) of raw material	SS – kg/kkg (lb./1000 lb.) of raw material	TDS – kg/kkg (lb./1000 lb.) of raw material	OIL – kg/kkg (lb./1000 lb.) of raw material
Old Tire Facilities (to 1958)						
	Minimum	6,267 (752)	0.088	0.001	0.349	0.003
	Maximum	386,478 (46,396)	7.294	5.854	105.660	3.363
	Average	37,000 (4,442)	0.790	1.000	12.000	0.120
New Tire Facilities (1959 to Present)						
	Minimum	2,303 (276)	0.083	0.032	0.387	0.011
	Maximum	48,070 (5,771)	2.020	1.397	12.980	0.187
	Average	10,500 (1,261)	0.580	0.310	5.535	0.042

[1] Includes utility wastewaters.

TABLE 369: RAW WASTE LOADS OF PROCESS WASTEWATERS FROM TIRE AND INNER TUBE FACILITIES

Plant	Cate- gory	FLOW L/kkg(gal/1000lb) of raw material	COD kg/kkg(lb/1000lb) of raw material	BOD kg/kkg(lb/1000lb) of raw material	SS kg/kkg(lb/1000lb) of raw material	TDS kg/kkg(lb/1000lb) of raw material	OIL kg/kkg(lb/1000lb) of raw material
A	New	290 (35)	1.57	0.067	0.982	5.12	0.248
B	New	165 (198)	NA	0.001	0.013	0.3739	0.075
C[2]	New	NA	NA	NA	NA	NA	NA
D	New	NA	0.193	0.010	0.064	3.210	0.009
E	Old	110 (13)	0.046	0.000	0.017	0.019	0.027
F	Old	700 (84)	NA	NA	0.57	NA	0.650
G	Old	590 (71)	NA	NA	0.001	NA	0.260
H	Old	4300 (516)	2.199	0.356	0.520	19.765	0.163
I	Old	769 (92)	NA	NA	2.610	NA	0.172

NA – Data not available

Source: Report EPA 440/1-73/013

Two important characteristics of the process wastewaters are suspended solids and oil. The suspended solids are generally higher from older plants due to greater maintenance and poorer housekeeping and control practices. The same can be said for the oil. Suspended solids evolve from the powdered substances used in the compounding area and from the collection of particulates by wet air pollution control equipment.

The oil is primarily lubrication and hydraulic oils from in-plant sources, and extender and fuel oil from run-off in storage areas. Both parameters can be treated successfully. Plant B is using a sedimentation lagoon to settle solids collected in the compounding area from wet air pollution equipment. It has been demonstrated by Plants A and E that solids collected in other areas can be separated easily by conventional equipment. American Petroleum Institute (API) type separators are being used to treat oily waste effluents of Plants B, D, and E.

Reference

(1)   U.S. Environmental Protection Agency, "Development Document for Proposed Ef-fluent Limitations Guidelines and New Source Performance Standards for the Tire and Synthetic Segment of the Rubber Processing Point Source Category," *Report EPA 440/1-73/013,* Washington, D.C. (Sept. 1973).

TITANIUM DIOXIDE MANUFACTURE

In the sulfate process, ground ilmenite ore is digested with concentrated sulfuric acid at relatively high temperature. The acid used is normally about 150% of the weight of the ore. In some cases, small amounts of antimony trioxide are also added. The resulting sulfates of titanium and iron are then leached from the reaction mass with water, and any ferric salts present are then reduced to ferrous by treatment with iron scrap to prevent coloration of the final titanium dioxide product.

After these operations, the resulting solutions are clarified, cooled and sent to a vacuum crystallizer. There, ferrous sulfate crystallizes out and is then separated from the mother liquor by centrifugation. This material is either sold or disposed of as a solid waste. The mother liquor is then clarified by filtration after addition of filter aid and is further concentrated by vacuum evaporation. Seed crystals or other nucleating agents are added, and the concentrated liquor is then treated with steam to hydrolyze the titanyl sulfate present. The resulting precipitate is collected by filtration, washed several times and then calcined to yield titanium dioxide. The calcined product is ground, quenched and dispersed in water. The coarse products are separated in a thickener to which caustic soda is added to maintain a constant pH. These coarse particles are reground and further processed to yield a purer product.

Water Pollution

Table 370 gives a generalized listing of the raw wastes from titanium dioxide manufacture by the sulfate process. Data in this table are in a form applicable to the effluent from any of the five existing sulfate process plants. Each of these five facilities have slightly different raw wastes due to differences in compositions of the raw ores. Effluents from four titanium dioxide sulfate process facilities are listed in Table 371. The raw wastes from a plant for producing TiO_2 by the chloride process consist of heavy metal salts, waste coke and hydrochloric acid. In the raw waste stream, these are actually metal chlorides before waste treatment.

TABLE 370: SULFATE PROCESS WASTE STREAMS—TiO_2 MANUFACTURE

(1) Dissolving and filtration	Ore and scrap iron plus flocculants	0.07 x total ore and scrap iron discharged
	H_2SO_4	0.0016 x ore
	Organic carbon	0.0004 x ore + 0.1 x C in ore
(2) Copperas (if produced)	$FeSO_4 \cdot 7H_2O$ (as Fe)	$(Fe^{+2} + 1.50\ Fe^{+3})$ in ore −0.33 x TiO_2 in ore
	Total sulfate	1.76 x iron in copperas

(continued)

TABLE 370: (continued)

(3) Strong acid	FeSO$_4$ (as Fe)	0.67 x (iron in ore minus iron in copperas)
	H$_2$SO$_4$	1.07 x ore
	Other ore impurities	0.67 x impurities in ore
	TiO$_2$	0.03 x TiO$_2$ in ore
	Organic carbon	0.0022 x ore + 0.81 x C in ore
(4) Weak acid	FeSO$_4$ (as Fe)	0.33 x (iron in ore minus iron in copperas)
	H$_2$SO$_4$ (total)	0.53 x ore + 0.25 x TiO$_2$ in ore
	Other ore impurities	0.33 x impurities in ore
	TiO$_2$	0.02 x TiO$_2$ in ore
	Organic carbon	0.00025 x ore plus 0.09 x C in ore
(5) Vent and kiln scrubbing	H$_2$SO$_4$	0.01 x ore
(6) TiO$_2$ losses	TiO$_2$	0.016 x TiO$_2$ in ore
	Na$_2$SO$_4$	0.03 x TiO$_2$ in ore

Note: Effluents also contain traces of Pb and Cu from process equipment. Silica and zircon do not react and are discharged with the sludge.

TABLE 371: PARTIAL DISCHARGE DATA FROM TiO$_2$ SULFATE PLANTS

Parameter*	Plant A No. 1	Plant A No. 2	Plant B No. 1	Plant B No. 2	Plant B No. 3	Plant C No. 1	Plant C No. 2	Plant C No. 3	Plant D No. 1
BOD5	10	3	6	3	--	--	0.3	0.5	--
COD	71	145	--	--	--	287	42	27	--
pH	8.0	1.2	6.5	5.6	--	1.0	2.6	5.0	5 min
Alkalinity	220	--	--	--	--	--	--	--	--
Total Dissolved Solids	1660	22,371	15,316	21,300	14,000	15,400	3,000	2,700	5,000
Iron	0.02	823	0.5	1.7	31,000	1,000	45	15	100
Sulfate	1,170	12,377	1,617	1,378	131,000	6,800	187	125	--
Chloride	51.5	105	6,394	7,900	--	625	2,480	2,830	--
Acidity	--	11,435	36	--	--	20,000	160	1000	--
Flow, cu m/day	10,200	Combined	20,000	123,400	6,100	20,000	40,900	30,300	
(MGD)	(2.7)	--	(5.5)	(32.6)	(1.6)	(5.5)	(10.8)	(8.0)	

[1] One plant of one manufacturer is not listed here. Data on titanium dioxide and chromate concentrations were provided.

[2] The corporation owning one facility is currently developing a process for recovery and recycle of the sulfuric acid used. This process is still under testing on the pilot plant scale.

[3] One plant barges its strong acid wastes out to sea for disposal. This method of disposal of highly acid wastes containing large amounts of dissolved heavy metals is not considered satisfactory.

*mg/l unless otherwise specified

Source: Report EPA 440/1-73/007

Reference

(1) U.S. Environmental Protection Agency, "Development Document for Proposed Effluent Limitations Guidelines and New Source Performance Standards for the Major Inorganic Products Segment of the Inorganic Chemical Manufacturing Point Source Category," *Report EPA-440/1-73/007* (August 1973).

UREA MANUFACTURE

Water Pollution

There are several possible aqueous waste streams which may be generated in a urea plant. They include cooling water or cooling water blowdown, stream condensate or blowdown, and wastewater used to scrub ammonia from off gas streams.

The production of urea can require large volumes of water for cooling. Values from 0 to 85,000 gallons per ton of urea have been reported. Most plants, however, use 10,000 to 40,000 gallons per ton of product. The once-through urea processes require the least amount of cooling, while total recycle processes require the largest volumes. Normally the volumes of cooling water required make once-through cooling water use impractical. When once-through cooling is used, however, the impurities in the effluent water are low. Treatment prior to discharge is normally not required except for ponding to reduce the temperature. The cooling water is normally kept separated from other waste streams. When the cooling water is recycled impurities build up and blowdown is required.

As in any cooling water recycle system the number of cycles of concentration is normally determined by the purity and availability of the feed water. The number of cycles of concentration used in urea production usually varies from 3 to 7 as in most recycle cooling water systems. The concentrations of impurities in the blowdown stream can vary over wide ranges but will be quite similar to those given from the blowdown from ammonia plant cooling water systems. The only significant difference is that urea plant blowdown streams will contain small amounts of urea (up to 50 mg/l), and little or no MEA. The volume of the blowdown stream can vary from 50 to 2,000 gallons per ton of product with the usual range being 200 to 800 gallons per ton.

Urea production requires significant volumes of steam (up to 8,000 pounds per ton of product). Steam condensate used on a once-through basis presents no disposal problem since the condensate is very low in contaminants. Blowdown from recycle boiler water systems can be a problem. Normally 10 to 20 cycles of concentration can be used because treated makeup water is usually used. Boiler water blowdown will normally amount to 5 to 10% of the boiler water. The blowdown volume will range from 20 to 100 gallons per ton of urea. The typical range of impurity concentrations are shown in Table 372.

TABLE 372: COMPOSITION OF UREA PLANT BOILER WATER BLOWDOWN

Contaminant	mg/ℓ	Contaminant	mg/ℓ
Phosphate	5-50	EDTA	0-50
Zinc	0-10	Misc. organics	0-200
Heavy metals	0-10	TDS	500-3500
Sulfite	0-100	Alkalinity	50-700
Suspended solids	50-300	Hardness	50-500

Source: Report 12020 FPD 09/71

Several gas streams are generated in urea production which can contain ammonia. Air pollution regulations may require that these gas streams be scrubbed for ammonia removal prior to discharge. In the total recycle process inert gases build up in the recycle system and must be purged. The purge stream will contain small amounts of ammonia and carbon dioxide. The urea solution from the last decomposer stage will contain small amounts of ammonia and possible carbon dioxide.

In the concentrator this ammonia will be evolved with the water vapor. In addition, any biuret formed will result in additional ammonia release. Ammonia can also be released due to hydrolysis of urea in the concentrator. As a result, the vapor stream from the concentrator can contain significant quantities of ammonia. Ammonia releases as large as 3 to 4 pounds per ton of product are possible. Additional ammonia release is possible in the prilling towers and the air stream exiting the prilling tower may contain small amounts of ammonia. When the ammonia loss is high, as from the concentrator, ammonia recovery and recycle to the process is possible. However, even when ammonia recovery is practiced, considerable ammonia containing waste volumes may still be generated due to gas scrubbing requirements.

Process spills can generate significant volumes of wastewater. Wash solutions used to clean up spills are usually combined with other contaminated waste streams and not recycled to the process. Little information is available on the individual contaminated waste streams from urea production since most plants do not analyze the individual streams. Compositions of the combined waste streams (exclusive of cooling water and boiler water blowdown) are available and typical compositions and volumes are presented in Table 373.

TABLE 373: TYPICAL UREA PLANT WASTEWATER VOLUME AND COMPOSITION*

Contaminant	mg/ℓ	Contaminant	mg/ℓ
Ammonia	200-4000	Oil	10-100
Urea	50-1000	BOD	30-300
CO_2	100-1000	COD	50-500

Volume   50-2000 gal/ton urea

*Does not include cooling water and boiler water blowdown

Source: Report 12020 FPD 09/71

Reference

(1)   Fullam, H.T. and Faulkner, B.P., "Inorganic Fertilizer and Phosphate Mining Industries, Water Pollution and Control," *Report 12020 FPD 09/71,* Washington, D.C., U.S. Government Printing Office (September 1971).

WOOD VENEER AND PLYWOOD PRODUCTS INDUSTRY

Plywood is a material made of several thin wood veneers bonded together with an adhesive. Its uses are many and include wall sidings, sheathing, roof-decking, concrete-formboards, floors, and containers. During the manufacture of plywood, incoming logs are sawed to desired length, debarked, and then peeled into thin, continuous veneers of uniform thickness. (Veneer thicknesses of ¼₅ to ⅕ inch are common.)

These veneers are then transported to special dryers where they are subjected to high temperatures until dried to a desired moisture content. After drying, the veneers are sorted, patched, and assembled in layers with some type of thermosetting resin used as the adhesive. The veneer assembly is then transferred to a hot press where, under pressure and steam heat, the plywood product is formed. Subsequently, all that remains is trimming, sanding, and possibly some sort of finishing treatment to enhance the usefulness of the plywood.

Air Pollution

The main sources of emissions from plywood manufacturing are the veneer drying and sanding operations. A third source is the pressing operation although these emissions are considered minor(1). The major pollutants emitted from veneer dryers are organics. These consist of two discernable fractions: condensibles, consisting of wood resin acids, and wood sugars, which form a blue haze upon cooling in the atmosphere, and volatiles, which are comprised of terpenes and unburned methane, the latter occurring when gas-fired dryers are employed. The amounts of these compounds produced depends on the wood species dried, the drying time, and the nature and operation of the dryer itself. In addition, negligible amounts of fine wood fibers are also emitted during the drying process.

Sanding operations are potential sources of particulate emissions. It is estimated that about 1,000 lb of sanderdust may result for every 10,000 square feet of plywood produced. Hence, even if only a fraction of this is discharged to the atmosphere, an air pollution problem may exist. Few data exist to determine the actual magnitude of these emissions although efficient cyclonic collectors reportedly remove large portions of this dust. Emission factors for plywood veneer dryers without controls are given in Table 374.

TABLE 374: EMISSION FACTORS FOR PLYWOOD MANUFACTURING

Emission Factor Rating: B

| Source | Organic compound[a,b] | | | |
| | Condensible | | Volatile | |
	lb/10⁴ ft²	kg/10³ m²	lb/10⁴ ft²	kg/10³ m²
Veneer dryers	3.6	1.9	2.1	1.1

[a]Emission factors expressed in pounds of pollutant per 10,000 square feet of 3/8-in. plywood produced (kilograms per 1,000 square meters on a 1-cm basis).
[b]References 2 and 3.

Source: EPA Publication AP-42, 2nd Edition

Water Pollution

Water usage varies widely in the veneer and plywood segment of the industry, depending on types of unit operations employed and the degree of recycle and reuse of water practiced. In general, total water usage is less than 3.15 l/sec (50 gal/min) for a mill producing 9.3 million m²/yr (100 million sq ft/yr). There are plants presently being designed to recycle all wastewater, however, none are now in operation. Considerable effort can, in any case, be made to reduce the amount of wastewater to be discharged or contained. The amount of information available on volumes and characteristics of wastewaters from the industry is minimal. Data cited in wastewater characterization is based mostly on data from the literature, information supplied by individual mills, and sampling and analyses conducted for the purposes of this study.

Since the volumes that are involved are small, attention has been directed to finding methods for reducing the volumes and ways of handling process water in such a way as to eliminate discharges. In veneer and plywood mills, water is used in the following operations: Log conditioning; cleaning of veneer dryers; washing of the glue lines and glue tanks; and cooling. Figure 48 presents a detailed process flow diagram. The water used and waste characteristics for each operation are discussed below.

Veneer and plywood manufacturing operations use two distinct types of log conditioning systems. These systems are referred to as steam vats and hot water vats. In the South about 50% of the plants use steam vats and 50% use hot water vats. In the West, however, only about 30% of the plants use any kind of conditioning and these plants use steam vats almost exclusively.

The only wastewater from a steam vat is from condensed steam. This water carries leach-ates from the logs as well as wood particles. Table 375 presents the results of analyses of wastewaters from steam vats. The magnitudes of these flows vary according to the size and number of vats. A plant producing 9.31 million m²/yr (100 million sq ft/yr) of plywood on a 9.53 mm (⅜") basis has an effluent of about 1.58 to 3.15 l/sec (25 to 50 gal/min). A southern plywood mill produces a BOD load of 2,500 kilograms per million square meters (515 pounds per million square feet) of board on a 9.53 mm (⅜") basis, and a total solids load of 29,200 kilograms per million square meters of board on a 9.53 mm (6,000 pounds per million square feet) (⅜") basis.

A hot water vat conditions the log with hot water heated either directly with steam or by means of heating coils with steam, oil, or other heat sources. When the vat is heated in-directly, there is no reason for a constant discharge. Hot water vats are usually emptied periodically, regardless of heating method, and the water is discharged and replaced with clean water. Some plants settle spent wastewater and pump it back into the vats. Chemi-cal characteristics for hot water vats for a series of veneer and plywood plants are given in Table 376.

FIGURE 48: WATER BALANCE FOR A PLYWOOD MILL

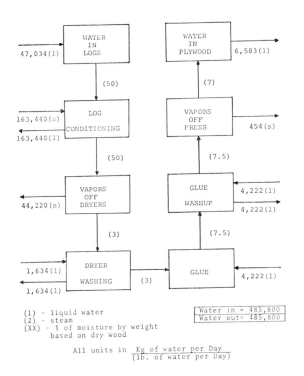

(1) - liquid water
(2) - steam
(XX) - % of moisture by weight
        based on dry wood

All units in $\frac{\text{Kg of water per Day}}{\text{(lb. of water per Day)}}$

Water in = 485,800
Water out= 485,800

Source: Report EPA 440/1-73/023

TABLE 375: CHARACTERISTICS OF STEAM VAT DISCHARGES

Plant	BOD	COD	DS	SS	TS	Turb.	Phenols	Kjld-N	T-PO$_4$-P	pH
A	470	8,310	2,430	2,940	5,370	450	0.69	56.8	5.70	4.12
B	3,117	4,005	--	86	--	--	--	16.5	14	4.1-6.1
C	2,940	8,670	5,080	370	5,450	245	0.57	39.3	--	5.38
D	1,499	3,435	2,202	389	2,591	249	--	--	--	5.3
E	1,298	3,312	2,429	107	2,536	30	0.30	1.87	.173	
F	476	1,668	917	74	991	28	0.20	4.73	1.93	

Note: All units are in mg/l except Turbidity, which is in JTU's and pH.

TABLE 376: CHARACTERISTICS OF HOT WATER STEAM VAT DISCHARGES

Plant	BOD	COD	DS	SS	TS	Turb.	Phenols	Kjld-N	T-PO$_4$-P	pH
A	4,740	14,600	3,950	2,520	6,470	--	0.40	26.4	--	5.4
B	3,100	9,080	1,570	460	2,030	--	--	23.4	--	3.8
C	326	1,492	1,948	72	2,020	800	<1.0	16.2	< 1.0	6.9
D*	1,000			160	1,000					4.5
E*	1,900	4,000	319	1,462	1,781					4.4

Note: All units are in mg/l except Turbidity, which is in JTU, and pH

*Analyses for plants 'D' and 'E' were provided by the respective plants, and figures for plant 'E' represent an average for several mills owned by one company.

Source: Report EPA 440/1-73/023

Veneer dryers accumulate wood particles. Volatile hydrocarbons will also condense on the surface of dryers to form an organic deposit called "pitch." In order to avoid excessive buildup of these substances, dryers must be cleaned periodically. Wood particles can be removed either by flushing with water or by blowing with air. While some of the pitch can be scraped off, generally a high pH detergent must be applied to dissolve most of the pitch and then it must be rinsed off with water. The nature of the dryer washwater varies according to the amount of water used, the amount of scraping prior to application of water, condition of the dryer, operation of the dryer, and, to some extent, the species of wood that is being dried.

The amount of water used varies from plant to plant and from operator to operator. One drying operation was observed to use about 23,000 liters (6,000 gal) of water per dryer over a period of 80 hr. At this plant there were six dryers which were washed every three weeks. The washing operation consisted of removing the bulk of the wood residue by blowing it out with air and hauling it away, and then washing the dryer with water for about ¾ of an hour to remove more wood particles. After this water cleaning, caustic detergent was applied. Finally, the detergent was rinsed off with water for another 45 minutes. Samples of spent water were taken during both applications of water, and the analyses of these samples are shown in Table 377. The effluent from this washing operation was averaged over a 7-day period and expressed on a unit of production basis as shown in Table 378.

TABLE 377: ANALYSIS OF DRYER WASHWATER

Plant	BOD	COD	DS	SS	TS	Turb.	Phenols	Color	Kjld-N	T-PO4-P
A										
Part I	210	1,131	643	113	756	19	1.31	32	17.7	1.93
Part B	840	6,703	1,095	5,372	6,467	50	0.20	43	211	11.0
B	60	1,586	1,346	80	1,426	6	4.68	51	2.91	0.495

Note: All units are concentrations in mg/l except for Turbidity in JTU's and Color in Pt-Cobalt Units.

TABLE 378: WASTE LOADS FROM VENEER DRYERS

Plant	BOD	COD	DS	SS	TS	Phenols	Kjld-N	T-PO4-P
A	60.94	412	99.7	319	418	0.018	13.2	0.18
B	2.33	60.6	52.3	3.09	55.2	0.014	0.112	0.019

Note: All units are in kilograms per million square meters.

Source: Report EPA 440/1-73/023

Various industry contacts emphasized that pitch buildup can be minimized by proper maintenance of the dryers. In addition, the volume of water necessary to wash the dryers can be greatly reduced. One Oregon plant of about ½ the size of the one described previously was observed to use $1/12$ as much water per week to clean its dryers. Wastewater characteristics from this plant are also given in Tables 377 and 378. It must be noted, however, that this plant provides settling and screening for the spent washwater before discharge, and samples were taken at the point of discharge.

Most dryers are equipped with deluge systems to extinguish fires that might be generated inside the dryer. Fires in dryers are quite common, especially in those that are poorly maintained. This water is usually handled in a manner similar to the handling of dryer washwater, and many plants actually take advantage of fires to clean the dryers. Fire deluge water can add significantly to the wastewater problems in some cases. In addition to the two wastewater sources from veneer dryers that have been mentioned, water is occasionally used for flooding the bottom of the dryers.

Many operators question the logic behind this practice, while some claim that it prevents fires and reduces air pollution problems. In any event, this water does not have to add to the wastewater problems of a mill. Several plants recycle all flood, wash, and fire water, and because the flooding results in substantial evaporation of water, these plants have found that fresh water can be used to clean the dryers and still keep the system closed.

Presently there are three types of glues in use in the veneer and plywood industry: phenolic formaldehyde resin; urea formaldehyde; and protein glue. Protein glues are slowly being phased out of the industry, while phenolic glues are becoming more widely used. The main source of wastewater from a glue system results from the washing of the glue spreaders and mixing tanks. The wastewaters from the washing operations are diluted at a ratio of about 20:1 with water to yield the concentrations shown in Table 379.

TABLE 379: AVERAGE CHEMICAL ANALYSIS OF PLYWOOD GLUE WASHWATER (ASSUMING A 20:1 DILUTION WITH WATER)

Analysis And Units	Phenolic Glue	Protein Glue	Urea Glue
COD, mg/kg	32,650	8,850	21,050
BOD, mg/kg	--	440	9,750
TOC, mg/kg	8,800	2,600	4,500
Total Phosphate, mg/kg, as P	6.00	13	37.8
Total Kjeldahl Nitrogen, mg/kg as N	60	600	1,065
Phenols, mg/kg	25.7	90.5	
Suspended Solids, mg/kg	15,250	5,900	10,200
Dissolved Solids, mg/kg	15,250	5,900	10,200
Total Solids, mg/kg	19,850	8,850	27,500
Total Volatile Suspended Solids, mg/kg	4,200	1,700	17,300
Total Volatile Solids, mg/kg	8,600	6,850	27,500

Source: Report EPA 440/1-73/023

References

(1) Hemming, C.B. "Encyclopedia of Chemical Technology," 2nd Edition Volume 15, New York, John Wiley and Sons, p.896-907 (1968).
(2) Monroe, F.L. et al, Investigation of Emissions from Plywood Veneer Dryers, Final Report, Washington State University, Pullman, Washington. Prepared for the Plywood Research Foundation and the U.S. Environmental Protection Agency, Research Triangle Park, N.C. *Publication No. APTD-1144* (February 1972).
(3) Mick, Allen and Dean McCargar. "Air Pollution Problems in Plywood, Particleboard, and Hardboard Mills in the Mid-Willamette Valley", Mid-Willamette Valley Air Pollution Authority, Salem, Oregon (March 24, 1969).
(4) U.S. Environmental Protection Agency, "Development Document for Proposed Effluent Limitations Guidelines and New Sources Performance Standards for the Plywood, Hardboard and Wood Preserving Segment of the Timber Products Processing Point Source Category," *Report EPA 440/1-73/023*, Washington, D.C. (December 1973).

WOOD WASTE COMBUSTION

The burning of wood/bark waste in boilers is largely confined to those industries where it is available as a by-product. It is burned both to recover heat energy and to alleviate a potential solid waste disposal problem. Wood/bark waste may include large pieces such as slabs, logs, and bark strips as well as smaller pieces such as ends, shavings, and sawdust. Heating values for this waste range from 8,000 to 9,000 Btu/lb, on a dry basis; however, because of typical moisture contents of 40 to 75%, the as-fired heating values for many wood/bark waste materials range as low as 4,000 to 6,000 Btu/lb. Generally, bark is the major type of waste burned in pulp mills; whereas, a variable mixture of wood and bark waste, or wood waste alone, is most frequently burned in the lumber, furniture and plywood industries.

A variety of boiler firing configurations are utilized for burning wood/bark waste. One common type in smaller operations is the Dutch Oven, or extension type of furnace with a flat grate. In this unit the fuel is fed through the furnace roof and burned in a cone-shaped pile on the grate. In many others, generally larger, operations, more conventional boilers have been modified to burn wood/bark waste. These units may include spreader stokers with traveling grates, vibrating grate stokers, etc., as well as tangentially fired or cyclone fired boilers. Generally, an auxiliary fuel is burned in these units to maintain constant steam when the waste fuel supply fluctuates and/or to provide more steam than is possible from the waste supply alone.

Air Pollution

The major pollutant of concern from wood/bark boilers is particulate matter although other pollutants, particularly carbon monoxide, may be emitted in significant amounts under poor operating conditions. These emissions depend on a number of variables including: the composition of the waste fuel burned, the degreee of fly-ash reinjection employed, and furnace design and operating conditions.

The composition of wood/bark waste depends largely on the industry from whence it originates. Pulping operations, for instance, produce great quantities of bark that may contain more than 70% moisture (by weight) as well as high levels of sand and other noncombustibles. Because of this, bark boilers in pulp mills may emit considerable amounts of particulate matter to the atmosphere unless they are well controlled. On the other hand, some operations such as furniture manufacture, produce a clean, dry (5 to 50% moisture) wood waste that results in relatively few particulate emissions when properly burned. Still other operations, such as sawmills, burn a variable mixture of bark and wood waste that results in particulate emissions somewhere in between these two extremes.

Fly-ash reinjection, which is commonly employed in many larger boilers to improve fuel-use efficiency, has a considerable effect on particulate emissions. Because a fraction of the collected fly-ash is reinjected into the boiler, the dust loading from the furnace, and consequently from the collection device, increases significantly per ton of wood waste burned. It is reported that full reinjection can cause a 10-fold increase in the dust loadings of some systems although increases of 1.2 to 2 times are more typical for boilers employing 50 to 100% reinjection. A major factor affecting this dust loading increase is the extent to which the sand and other noncombustibles can be successfully separated from the fly-ash before reinjection to the furnace.

Furnace design and operating conditions are particularly important when burning wood and bark waste. For example, because of the high moisture content in this waste, a larger area of refractory surface should be provided to dry the fuel prior to combustion. In addition, sufficient secondary air must be supplied over the fuel bed to burn the volatiles that account for most of the combustible material in the waste. When proper drying conditions do not exist, or when sufficient secondary air is not available, the combustion temperature is lowered, incomplete combustion occurs, and increased particulate, carbon monoxide, and hydrocarbon emissions will result.

Emission factors for wood waste boilers are presented in Table 380. For boilers where fly-ash reinjection is employed, two factors are shown: the first represents the dust loading reaching the control equipment; the value in parenthesis represents the dust loading after controls assuming about 80% control efficiency. All other factors represent uncontrolled emissions.

TABLE 380: EMISSION FACTORS FOR WOOD AND BARK WASTE COMBUSTION
Emission Factor Rating: B

Pollutant	Emissions	
	lb/ton	kg/MT
Particulates[a]		
Bark[b]		
With fly-ash reinjection[d]	75 (15)	37.5 (7.5)
Without fly-ash reinjection	50	25
Wood/bark mixture[b]		
With fly-ash reinjection[d]	45 (9)	22.5 (4.5)
Without fly-ash reinjection	30	15
Wood[f]	5-15	2.5-7.5
Sulfur oxides (SO$_2$)[h]	1.5	0.75
Carbon monoxide[i]	2-60	1-30
Hydrocarbons[k]	2-70	1-35
Nitrogen oxides (NO$_2$)	10	5

[a] These emission factors were determined for boilers burning gas or oil as an auxiliary fuel, and it was assumed all particulates resulted from the waste fuel alone. When coal is burned as an auxiliary fuel, the appropriate emission factor should be used in addition to the above factor.
[b] These factors based on an as-fired moisture content of 50%.
[d] This factor represents a typical dust loading reaching the control equipment for boilers employing fly-ash reinjection. The value in parenthesis represents emissions after the control equipment assuming an average efficiency of 80%.
[f] This waste includes clean, dry (5 to 50% moisture) sawdust, shavings, ends, etc., and no bark. For well designed and operated boilers use lower value and higher values for others. This factor is expressed on an as-fired moisture content basis assuming no fly-ash reinjection.
[h] This factor is calculated by material balance assuming a maximum sulfur content of 0.1% in the waste. When auxiliary fuels are burned, the appropriate factors should be used in addition to determine sulfur oxide emissions.
[i] This factor is based on engineering judgment and limited data. Use lower values for well designed and operated boilers.
[k] This factor is based on limited data. Use lower values for well designed and operated boilers.

Source: EPA Publication AP-42, 2nd Edition

Reference

(1) U.S. Environmental Protection Agency, "Compilation of Air Pollutant Emission Factors", 2nd Edition *Publication No. AP-42,* Supplement No. 3, Research Triangle Park, N.C., Office of Air and Water Programs (July 1974).

ZINC PROCESSING, SECONDARY

Zinc processing includes zinc reclaiming, zinc oxide manufacturing, and zinc galvanizing. Zinc is separated from scrap containing lead, copper, aluminum, and iron by careful control of temperature in the furnace, allowing each metal to be removed at its melting range. The furnaces typically employed are the pot, muffle, reverberatory, or electric induction. Further refining of the zinc can be done in retort distilling or vaporization furnaces where the vaporized zinc is condensed to the pure metallic form. Zinc oxide is produced by distilling metallic zinc into a dry air stream and capturing the subsequently formed oxide in a baghouse. Zinc galvanizing is carried out in a vat or in bath-type dip tanks utilizing a flux cover. Iron and steel pieces to be coated are cleaned and dipped into the vat through the covering flux.

Air Pollution

A potential for particulate emissions, mainly zinc oxide, occurs if the temperature of the furnace exceeds 1100°F (595°C). Zinc oxide (ZnO) may escape from condensers or distilling furnaces, and because of its extremely small particle size (0.03 to 0.5 micron), it may pass through even the most efficient collection systems. Some loss of zinc oxides occurs during the galvanizing processes, but these losses are small because of the flux cover on the bath and the relatively low temperature maintained in the bath.

Some emissions of particulate ammonium chloride occur when galvanized parts are dusted after coating to improve their finish. Another potential source of emissions of particulates and gaseous zinc is the tapping of zinc-vaporizing muffle furnaces to remove accumulated slag residue. Emissions of carbon monoxide occur when zinc oxide is reduced by carbon(1). Nitrogen oxide emissions are also possible because of the high temperature associated with the smelting and the resulting fixation of atmospheric nitrogen. Table 381 summarizes the emission factors from zinc processing.

TABLE 381: PARTICULATE EMISSION FACTORS FOR SECONDARY ZINC SMELTING[a]

Emission Factor Rating: C

Type of furnace	Emissions	
	lb/ton	kg/MT
Retort reduction	47	23.5
Horizontal muffle	45	22.5
Pot furnace	0.1	0.05
Kettle sweat furnace processing[b]		
Clean metallic scrap	Neg	Neg
General metallic scrap	11	5.5
Residual scrap	25	12.5
Reverberatory sweat furnace processing[b]		
Clean metallic scrap	Neg	Neg
General metallic scrap	13	6.5
Residual scrap	32	16
Galvanizing kettles	5	2.5
Calcining kiln	89	44.5

[a]References 2 through 4. Emission factors expressed as units per unit weight of metal produced.
[b]Reference 5.

Source: EPA Publication AP-42, 2nd Edition

References

(1) Air Pollutant Emission Factors, Final Report, Resources Research, Inc. Reston, Va. Prepared for National Air Pollution Control Administration, Durham, N.C., (April 1970).
(2) Allen, G.L. et al, Control of Metallurgical and Mineral Dusts and Fumes in Los Angeles County, U.S. Department of the Interior, Bureau of Mines, Washington, D.C., *Information Circular Number 7627* (April 1952).
(3) Restricting Dust and Sulfur Dioxide Emissions from Lead Smelters (translated from German), Kommission Reinhaltung der Luft, Reproduced by U.S. DHEW, PHS, Washington, D.C., VDI Number 2285 (September 1961).
(4) Hammon, W.F. Data on Non-Ferrous Metallurgical Operations, Los Angeles County Air Pollution Control District (November 1966).
(5) Herring, W. Secondary Zinc Industry Emission Control Problem Definition Study (Part I), Environmental Protection Agency, Office of Air Programs, Research Triangle Park, N.C. *Publication Number APTD-0706* (May 1971).

ZINC SMELTING, PRIMARY

As stated previously, most domestic zinc comes from zinc and lead ores. Another important source of raw material for zinc metal has been zinc oxide from fuming furnaces. For
efficient recovery of zinc, sulfur must be removed from concentrates to a level of less than
2%. This is done by fluidized beds or multiple-hearth roasting occasionally followed by
sintering. Metallic zinc can be produced from the roasted ore by the horizontal or vertical
retort process or by the electrolytic process if a high-purity zinc is needed.

Air Pollution

Dust, fumes, and sulfur dioxide are emitted from zinc concentrate roasting or sintering
operations(1). Particulates may be removed by electrostatic precipitators or baghouses.
Sulfur dioxide may be converted directly into sulfuric acid or vented. Emission factors
for zinc smelting are presented in Table 382.

TABLE 382:   EMISSION FACTORS FOR PRIMARY ZINC SMELTING WITHOUT
CONTROLS[a]

Emission Factor Rating: B

Type of operation	Particulates		Sulfur oxides	
	lb/ton	kg/MT	lb/ton	kg/MT
Roasting (multiple-hearth)[b]	120	60	1100	550
Sintering[c]	90	45	d	d
Horizontal retorts[e]	8	4	–	–
Vertical retorts[e]	100	50	–	–
Electrolytic process	3	1.5	–	–

[a]Approximately 2 unit weights of concentrated ore are required to
produce 1 unit weight of zinc metal. Emission factors expressed as units
per unit weight of concentrated ore produced.
[b]References 3 and 4.
[c]References 2 and 3.
[d]Included in SO_2 losses from roasting.
[e]Reference 3.

Source: EPA Publication AP-42, 2nd Edition

Water Pollution

The reader is referred to the earlier section of the volume dealing with "Lead Smelting,
Primary" for a discussion of the water pollution problems associated with the lead-zinc
industries.

References

(1)   Duprey, R.L. "Compilation of Air Pollutant Emission Factors," U.S. DHEW, PHS, National Center for Air Pollution Control, Durham, N.C. PHS *Publication Number
999-AP-42*, p. 26-28 (1968).
(2)   Stern, A. (ed), "Sources of Air Pollution and Their Control", *Air Pollution*, Volume
III, 2nd Edition, New York, Academic Press p. 182-186 (1958).
(3)   Sallee, G. Private communication on Particulate Pollutant Study, Midwest Research
Institute, Kansas City, Mo., Prepared for National Air Pollution Control Administration, Durham, N.C. (June 1970).
(4)   "Systems Study for Control of Emissions in the Primary Nonferrous Smelting Industry," 3 Volumes, San Francisco, Arthur G. McKee and Company (June 1969).

POLLUTION DETECTION
AND MONITORING HANDBOOK 1974

by Marshall Sittig

Environmental Technology Handbook No. 1

This handbook contains methods for the detection and monitoring of pollutants in industrial effluents, notably air and water.

Such methods are prerequisites for any kind of management of environmental quality. Of equal importance are the data handling systems that must be used to collect and interpret the measurements, regardless of whether manual or automated identifying and monitoring techniques are chosen.

Generally the most important single factor in determining whether manual or automated methods of analysis and level monitoring should be used, is the required frequency interval. In the manual approach costs vary almost directly with the measurement frequency. At monitoring intervals oftener than once a day, or when a continuous watch must be maintained, installation of automatic sensing is more economical, provided reliable automatic sensors are available.

For meaningful control and prevention of pollution the accepted standards and parameters for air and water quality must be known, whether they have been legislated or not. Acceptable levels for each type of impurity from methylmercury to dissolved chlorides or carbon particles must be known by the industrialist or public health official, and this book, with its 1633 references makes a massive attempt to communicate such levels.

As a guideline for the industrial user, data on the toxicity of the various pollutants are included. Such data on the deleterious effects were taken from various government publications. This book is not a treatise of pharmacology or industrial toxicology, but a guide to the accurate evaluation of pollutant levels in a given effluent or ambient substrate.

Whenever applicable, the name of the pollutant is followed by chemical and other information. After this come directions for sampling in air or water or both, as the case may be. Thereafter are given qualitative and quantitative analytical methods suitable for identification and measurement of the degree or level of pollution. This comprehensive information is followed by measurement techniques suitable for repeated or continuous monitoring. Preference is given to methods giving reproducible results in environmental quality control, as recommended by the EPA and other U.S. government agencies. This is followed in each case by many up-to-date references to government publications, patents, and journal articles.

Arrangement is encyclopedic. The book is thus a worthy companion to the author's POLLUTANT REMOVAL HANDBOOK available from the same publisher.

Pollution control is a serious business, and in the design of a monitoring program the specific objectives to be served must be clearly in mind from the very beginning. Otherwise the effort most probably will not result in an efficient and meaningful program. The length of the initial survey, selection of monitoring points, parameter coverage, and sampling frequency all depend on the specific objectives and the timeframe in which the objectives must be achieved. These variables also have a very significant impact on monitoring costs.

Monitoring for the purpose of long-term trend identification, evaluation of standards compliance, and total management of environmental quality must, of course, be a continuing program without end.

It is hoped sincerely that this book, together with its companion volume POLLUTANT REMOVAL HANDBOOK, will do much to establish and sustain such antipollution programs.

A partial and condensed table of contents follows here in which subheadings are indicated below for only the first few entries. The book contains a total of 88 subject entries arranged in an alphabetical and encyclopedic fashion. The subject name refers to the polluting substance, and the text underneath each entry tells how to measure and monitor pollution by said substance:

This book should prove to be very useful as an overall reference volume. The organization is clear, and the text is especially well arranged. The tables and figures are extensive and will be of value to students, engineers, and pollution control authorities.

ISBN 0-8155-0529-9

POLLUTANT REMOVAL HANDBOOK 1973

by Marshall Sittig

The purpose of this handbook is to provide a one volume practical reference book showing specifically how to remove pollutants, particularly those emanating from industrial processes. This book contains substantial technical information.

This volume is designed to save the concerned reader time and money in the search for pertinent information relating to the control of specific pollutants. Through citations from numerous reports and other sources, hundreds of references to books and periodicals are given.

In this manner this book constitutes a ready reference manual to the entire spectrum of pollutant removal technology. While much of this material is presumably available and in the public domain, the locating thereof is a tedious, time-consuming, and expensive process.

The book is addressed to the industrialist, to local air and water pollution control officers, to legislators who are contemplating new and more stringent control measures, to naturalists and conservationists who are interested in exactly what can be done about the effluents of local factories, to concerned citizens, and also to those eager students who can foresee new and brilliant careers in the fields of antipollution engineering and pollution abatement.

During the past few years, the words "pollution", "environment" and "ecology" have come into more and more frequent usage and the cleanliness of the world we live in has become the concern of all people. Pollution, for example, is no longer just a local problem involving litter in the streets or the condition of a nearby beach. Areas of the oceans, far-reaching rivers and the largest lakes are now classified as polluted or subject to polluting conditions. In addition, very surprisingly, lakes and streams remote from industry and population centers have been found to be contaminated.

This handbook therefore gives pertinent and concise information on such widely divergent topics as the removal of oil slicks in oceans to the containing of odors and particulates from paper mills.

Aside from the practical considerations, including teaching you where to look further and what books and journals to consult for additional information, this book is also helpful in explaining the new lingo of pollution abatement, which is developing new concepts and a new terminology all of its own, for instance: "particulates, microns, polyelectrolytes, flocculation, recycling, activated sludge, gas incineration, catalytic conversion, industrial ecology, etc."

In order to have a safe and healthful environment we must all continue to learn and discover more about the new technology of pollution abatement. Every effort has been made in this manual to give specific instructions and to provide helpful information pointing in the right direction on the arduous and costly antipollution road that industry is now forced to take under ecologic and sociologic pressures. The world over, technological and manpower resources are being directed on an increasing scale toward the control and solution of contamination and pollution problems.

In the United States of America we are fortunate in receiving direct help from the numerous surveys together with active research and development programs that are being supported by the Federal Government to help industry and municipalities control their wastes and harmful emissions.

A partial and condensed table of contents is given here. The book contains a total of 128 subject entries arranged in an alphabetical and encyclopedic fashion. The subject name refers to the polluting substance and the text underneath each entry tells how to combat pollution by said substance:

CARBON MONOXIDE
CARBONYL SULFIDE
CEMENT KILN DUSTS
CHLORIDES
CHLORINATED HYDROCARBONS
CHLORINE
CHROMIUM
CLAY
COKE OVEN EFFLUENTS
COLOR PHOTOGRAPHY EFFLUENTS
COPPER
CRACKING CATALYSTS
CYANIDES
CYCLOHEXANE OXIDATION WASTES
DETERGENTS
DYESTUFFS
FATS
FERTILIZER PLANT EFFLUENTS
FLOUR
FLUORINE COMPOUNDS
FLY ASH
FORMALDEHYDE
FOUNDRY EFFLUENTS
FRUIT PROCESSING INDUSTRY
  EFFLUENTS
GLYCOLS
GREASE
HYDRAZINE
HYDROCARBONS
HYDROGEN CHLORIDE
HYDROGEN CYANIDE
HYDROGEN FLUORIDE
HYDROGEN SULFIDE
IODINE
IRON
IRON OXIDES
LAUNDRY WASTES
LEAD
LEAD TETRAALKYLS
MAGNESIUM CHLORIDE
MANGANESE
MEAT PACKING FUMES
MERCAPTANS
MERCURY
METAL CARBONYLS
MINE DRAINAGE WATERS
NAPHTHOQUINONES
NICKEL
NITRATES
NITRITES
NITROANILINES
NITROGEN OXIDES
OIL
OIL (INDUSTRIAL WASTE)

OIL (PETROCHEMICAL WASTE)
OIL (PRODUCTION WASTE)
OIL (REFINERY WASTE)
OIL (TRANSPORT SPILLS)
OIL (VEGETABLE)
ORGANIC VAPORS
OXYDEHYDROGENATION PROCESS
  EFFLUENTS
PAINT AND PAINTING EFFLUENTS
PAPER MILL EFFLUENTS
PARTICULATES
PESTICIDES
PHENOLS
PHOSGENE
PHOSPHATES
PHOSPHORIC ACID
PHOSPHORUS
PICKLING CHEMICALS
PLASTIC WASTES
PLATING CHEMICALS
PLATINUM
PROTEINS
RADIOACTIVE MATERIAL
RARE EARTH
ROLLING MILL DUST & FUMES
ROOFING FACTORY WASTES
RUBBER
SELENIUM
SILVER
SODA ASH
SODIUM MONOXIDE
SOLVENTS
STARCH
STEEL MILL CONVERTER EMISSIONS
STRONTIUM
SULFIDES
SULFUR
SULFUR DIOXIDE
SULFURIC ACID
TANTALUM
TELLURIUM HEXAFLUORIDE
TETRABROMOETHANE
TEXTILE INDUSTRY EFFLUENTS
THIOSULFATES
TIN
TITANIUM
TRIARYLPHOSPHATES
URANIUM
VANADIUM
VEGETABLE PROCESSING INDUSTRY
  EFFLUENTS
VIRUSES
ZINC

ISBN 0-8155-0489-6   **528 pages**

POLLUTION CONTROL
IN THE ORGANIC CHEMICAL INDUSTRY
1974

by Marshall Sittig

Pollution Technology Review No. 9

Wastes from plants manufacturing identical compounds may be quite dissimilar, because of the difference in processes and raw materials. Since the organic chemical industry is now largely petrochemical, preference has been given to waste treatment from such operations.

Detailed treatability evaluation of each waste stream has become a prerequisite for any profitable product.

Physical methods for waste and by-product removal include gravity separation, air flotation, filtration, evaporation and adsorption techniques.

Chemical methods include precipitation, polyelectrolyte treatment, oxidation and other chemical conditioning such as neutralization.

Biological treatment methods comprise activated sludge and its modifications, trickling filters, aerated lagoons, and waste stabilization ponds.

This book intends to assist the chemical engineer in treating chemical waste products and effluents in conjunction with prudent raw material choices and thus bring about process economics. It presents condensed vital data from government and other sources of information that are scattered and difficult to pull together.

A partial and condensed table of contents follows here.

Dinitrotoluene from Nitrotoluene

Disulfoton from Ethanol

Dyes and Pigments from Aromatics

Epichlorohydrin from Allyl Chloride

Ethylbenzene from Benzene

Ethyl Chloride by
  Hydrochlorination of Ethylene
  Chlorination of Ethane
  Hydrochlorination of Ethanol

Ethylene or Propylene from Paraffins

Ethylene Dichloride from Ethylene

Ethylene Glycol from Ethylene Oxide

Ethylene Oxide from Ethylene

Formaldehyde from Methanol

Long Chain Alcohols from
  Ethylene Oligomers

Malathion® from Methanol

Methanol from Natural Gas

Methylamines from Methanol + Ammonia

Methyl Bromide from Methanol

Methyl Methacrylate from
  Acetone Cyanohydrin

Nitrobenzene from Benzene

Nitrochlorobenzene from Chlorobenzene

Nitroparaffins from Paraffins

Oxo Products from Olefins

Parathion from p-Nitrophenol

Perchloroethylene from Propane

Phenol from Chlorobenzene

Phenol from Cumene

Phorate from Ethanol

Phosgene from $CO + Cl_2$

Phthalic Anhydride from
  Naphthalene or Xylene

Propylene Glycol from Propylene Oxide

Propylene Oxide from Propylene

Styrene from Ethylbenzene

SNG from Crude Oil

2,4,5-T from Trichlorophenol

Terephthalic Acid from Xylene

Tetraethyl Lead from Ethyl Chloride

Trichloroethane from Vinyl Chloride

Trichloroethylene by
  Chlorination, then by
  Dehydrochlorination of Acetylene
  or by Oxyhydrochlorination
  of Dichloroethane

Trifluralin from Perfluoromethylchloro-
  benzene

Vinyl Acetate from
  Ethylene and Acetic Acid

Vinyl Chloride from Acetylene

Vinyl Chloride from Ethylene Dichloride

AIR POLLUTION CONTROL
  Halogen Acids
  Halogenated Hydrocarbons
  Halogens
  Hydrocarbons
  Particulates
  Sulfur Compounds
  Coking of Coal

WATER POLLUTION CONTROL
Primary
  Oil Separation
  Equalization
  Neutralization
  Sedimentation
  Flotation
  Flocculation
  Nutrient Addition
Secondary
  Activated Sludge
  Extended Aeration
  Trickling Filters
  Aerated Lagoons
  Waste Stabilization Ponds
  Chemical Oxidation
  Nitrification—Denitrification

Tertiary
  Chemical Precipitation
  Gas Stripping
  Microstraining
  Carbon Adsorption
  Electrodialysis
  Ion Exchange
  Evaporation
  Reverse Osmosis
  Chlorination
  Rapid Sand Filtration
Sludge Handling
  Aerobic Digestion
  Anaerobic Digestion
  Wet Oxidation
  Thickening
  Lagooning
  Sand Drying Beds
  Vacuum Filtration
  Filtration
  Land Disposal
  Incineration
  Sea Disposal
Ultimate Disposal
  Thermal Oxidation
  Deep Well Disposal

OVERALL WATER POLLUTION
CONTROL MODELS
  BPCTCA: Best Practicable Control
    Technology Currently Available
  BATEA: Best Available Technology
    Economically Available
  BADCT: Best Available Demonstrated
    Control Technology

THE ECONOMICS OF POLLUTION
CONTROL

FUTURE TRENDS

ISBN 0-8155-0536-1

INDUSTRIAL WATER PURIFICATION 1974

by Louis F. Martin

Pollution Technology Review No. 16

The Federal Water Pollution Control Act (now made into law) will have far-reaching effects on all industry. The pretreatment standards, about to be published as a result of this act, will exert increasing constraints on industrial process design and expansion plans in the years ahead.

The *Guidelines for the Pretreatment of Discharges of Publicly-Owned Works* prohibit the introduction of industrial pollutants which would pass through a municipal facility "inadequately treated," therefore, all existing industrial facilities and new plant designs must include specific downstream water purification processes, before the effluent is sewered or discharged into public systems.

This book describes over 160 recently devised processes for the treatment of contaminated industrial waters.

A partial and condensed table of contents follows here. Numbers in parentheses indicate the number of processes per topic, chapter headings are given, followed by examples of important subtitles.

While conventional separation processes for solid-liquid and water-oil effluents are applicable to many industrial streams, the process literature clearly highlights the areas of greatest immediate concern as shown by this table of contents:

ISBN 0-8155-0554-X

300 pages

RECYCLING AND RECLAIMING
OF MUNICIPAL SOLID WASTES 1975

by F. R. Jackson

Pollution Technology Review No. 17

More and more cities, large and small, are exploring new technological systems for recycling solid municipal wastes.

Naturally the greatest interest centers upon those recycling systems that permit profitable reclamation of saleable items and materials from the town's refuse.

Articles made of metal can be cleaned, washed, compressed, and sold to the scrap metals industry. Glass can be comminuted and even separated as to color.

Paper and cardboard is a promising target for recycling, because paper fiber is the largest single component of municipal wastes and potentially very valuable.

Water-insoluble organic matter, such as plastics and wood, can be reclaimed and processed for conversion into energy in a variety of ways.

As these new technologies are developed, they will add to the alternatives available to a community, when it decides to recycle some of its wastes as part of its total solid waste management program. Numerous alternate ways and means are needed, because local conditions vary widely, and no single line of approach is capable of meeting every community's needs.

This Pollution Technology Review surveys information available up to the middle of 1974, and is based primarily upon studies conducted by industrial and engineering firms under the auspices of the EPA and other government agencies. Forward-thinking municipal administrators will find this book most useful, because it collates comprehensively the results of governmental and industrial research.

A partial and condensed table of contents follows here.

ISBN 0-8155-0560-4

342 pages